International Joint Conference on Materials Science and Mechanical Engineering (CMSME 2018)

AA006455

IOP Conference Series: Materials Science and Engineering
Volume 383

Bangkok, Thailand
24 – 26 February 2018

ISBN: 978-1-5108-6752-9
ISSN: 1757-8981

Printed from e-media with permission by:

Curran Associates, Inc.
57 Morehouse Lane
Red Hook, NY 12571

CURRAN ASSOCIATES INC. proceedings.com

Some format issues inherent in the e-media version may also appear in this print version.

Copyright© (2018) by the Institute of Physics
All rights reserved. The material featured in this book is subject to
IOP copyright protection, unless otherwise indicated.

Printed by Curran Associates, Inc. (2018)

For permission requests, please contact the Institute of Physics
at the address below.

Institute of Physics
Dirac House, Temple Back
Bristol BS1 6BE UK

Phone: 44 1 17 929 7481
Fax: 44 1 17 920 0979

techtracking@iop.org

Additional copies of this publication are available from:

Curran Associates, Inc.
57 Morehouse Lane
Red Hook, NY 12571 USA
Phone: 845-758-0400
Fax: 845-758-2633
Email: curran@proceedings.com
Web: www.proceedings.com

TABLE OF CONTENTS

PREFACE

We are very pleased to introduce the proceedings of the CMSME2018. A major conference theme was related to the field of Materials Science and Mechanical Engineering. The conference was held in Asian Institute of Technology, Bangkok, Thailand, from February 24 to February 26, 2018.

This conference provides opportunities for the delegates to exchange new ideas face-to-face, to establish business or research relations as well as to find global partners for future collaborations. Their contributions helped to make the Conference as outstanding as it should be. The conference results have lead to significant contributions to the knowledge in these up-to-date scientific fields, encouraged the interaction of researchers and developing academics with the more established academic community in an informal setting to present and to discuss new and current work.

Five respected and international professors who were recognized as outstanding and leading experts delivered keynote lectures to an erudite and enthusiastic audience, sharing their deep insights on future challenges and trends, leading to a remarkable conference. Our participants have the privilege of hearing from these key researchers in their field, as they present the latest findings of their works.

The conference has been a great experience for participants from all over the world, with an excellent technical program as well as social activities. In the event, the conference was a highly successful since more than 60 presented papers maintained the high promise from presenters on a global scale suggested by the written abstracts. The number of delegates of the conference was also highly gratifying, showing the high level of international interest in the subject.

These Proceedings will furnish the scientists of the world with an excellent reference book. It will provide the permanent record of what was presented and an impetus to stimulate further study and research in all these areas. We thank all the contributors for their support and participation in make this conference a great success.

Professor Dr. Juntao Fei
On behalf of the Conference Committee

Content from this work may be used under the terms of the Creative Commons Attribution 3.0 licence. Any further distribution of this work must maintain attribution to the author(s) and the title of the work, journal citation and DOI.
Published under licence by IOP Publishing Ltd

IOP Conf. Series: Materials Science and Engineering **383** (2018) 011002 doi:10.1088/1757-899X/383/1/011002

Peer review statement

All papers published in this volume of *IOP Conference Series: Materials Science and Engineering* have been peer reviewed through processes administered by the proceedings Editors. Reviews were conducted by expert referees to the professional and scientific standards expected of a proceedings journal published by IOP Publishing.

Content from this work may be used under the terms of the Creative Commons Attribution 3.0 licence. Any further distribution of this work must maintain attribution to the author(s) and the title of the work, journal citation and DOI.
Published under licence by IOP Publishing Ltd

IOP Conf. Series: Materials Science and Engineering **383** (2018) 012001 doi:10.1088/1757-899X/383/1/012001

Mechanical properties of poly(propylene fumarate)/nanotube composites: experimental results and theoretical predictions

Ana M. Diez-Pascual*

Analytical Chemistry, Physical Chemistry Chemical Engineering Department, Alcalá University, E-28871 Alcala de Henares, Madrid, Spain

*E-mail: ana.dpascual@gmail.com

Abstract. Biopolyesters such as synthetic poly(propylene fumarate) (PPF) are widely used in the biomedical field. In order to be used as orthopaedic scaffolds, their stiffness and strength have to be improved. Inorganic boron nitride nanotube (BNNT) 1D nanomaterials have recently attracted a lot of interest as reinforcing fillers of polymers due to their outstanding physical properties. This article deals with the preparation and mechanical characterization of PPF/BNNT composites with nanotube concentrations in the range of 0.1-5 wt%. Results demonstrate that the composites exhibit rigidity enough to be used for tissue engineering applications.

1. Introduction

The metastable allotropic form of hexagonal boron nitride nanotubes was first proposed by Rubio *et al.* [1] in 1994 based on the analogies between the lattice structure of graphite and hexagonal boron nitride (Figure 1, left). They are structural analogs of carbon nanotubes (CNTs) but with boron and nitrogen atoms instead of carbon atoms. BNNTs can be visualized as a rolled up hexagonal boron nitride layer, and have different chiralities [2]. The chiral vector that can be expressed as: $C_h = na_1+ma_2 \approx (n,m)$, where integers *n* and *m* represent the chiral indices. According to this, three types of single-walled BNNTs have been described (Figure 1, right): (a) zizag (m = 0); (b) armchair (m = n); (c) chiral (m $\neq$ 0). The B–N bonds have somewhat ionic character since the electron density of B is attracted to the N atoms owed to their higher electronegativity. This provokes a large band gap between the valence and conduction bands of ~5.5 eV, hence they are electrically insulating.

BNNTs can be synthesized by a variety of techniques. The first works on BNNTs employed techniques similar to those used for the synthesis of CNTs such as arc-discharge, laser heating and vaporization, chemical vapor deposition (CVD) and high temperature ball milling. Complete reviews on the synthesis approaches of BNNTs can be found elsewhere [3,4]. Among different paths employing the CVD approach, it should be highlighted those reported by Tang *et al.* [5], who employed metal oxides like MgO, FeO and Li_2O as precursors to synthesize reactive boron oxide (BxOy) vapour that reacted under an NH_3 atmosphere in an induction heating chamber at 1300 °C to produce the BNNTs. This approach has been further improved by Zhi *et al.* [6] to produce high purity BNNTs. This procedure was successful for synthesizing BNNTs in a lab, although commercialization and practical use of BNNTs in industry is hampered by their low production rate and the necessity for tailored chamber design.

Content from this work may be used under the terms of the Creative Commons Attribution 3.0 licence. Any further distribution of this work must maintain attribution to the author(s) and the title of the work, journal citation and DOI.
Published under licence by IOP Publishing Ltd

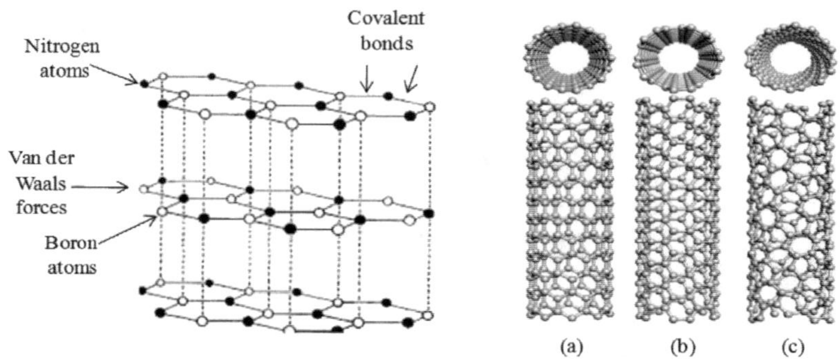

Figure 1. Left: Lattice structure of hexagonal boron nitride; Right: Atomic models of BNNTs: (a) zizag; (b) armchair; (c) chiral.

Using an analogous method, Lee *et al.* [7] reported the preparation high-purity and high-quality BNNTs at 1200 °C in a traditional tube furnace. The fundamental characteristic of this technique is the use of a quartz tube closed at the end to entrap the vapours for the growth of the BNNTs. This method is denominated the growth vapour trapping (GVT) approach, and can be tailored via using catalytic nanoparticles coated on Si substrates. The raw BNNTs produced by this CVD/GVT method are vertically-aligned and display super-hydrophobicity due to their high surface roughness and reduced surface energy. BNNTs can be synthesized by a variety of techniques. The first works on BNNTs employed techniques similar to those used for the synthesis of CNTs such as arc-discharge, laser heating and vaporization, chemical vapor deposition (CVD) and high temperature ball milling. Complete reviews on the synthesis approaches of BNNTs can be found elsewhere [3,4]. Among different paths employing the CVD approach, it should be highlighted those reported by Tang *et al.* [5], who employed metal oxides like MgO, FeO and Li_2O as precursors to synthesize reactive boron oxide (BxOy) vapour that reacted under an NH_3 atmosphere in an induction heating chamber at 1300 °C to produce the BNNTs. This approach has been further improved by Zhi *et al.* [6] to produce high purity BNNTs. This procedure was successful for synthesizing BNNTs in a lab, although commercialization and practical use of BNNTs in industry is hampered by their low production rate and the necessity for tailored chamber design. Using an analogous method, Lee *et al.* [7] reported the preparation high-purity and high-quality BNNTs at 1200 °C in a traditional tube furnace. The fundamental characteristic of this technique is the use of a quartz tube closed at the end to entrap the vapours for the growth of the BNNTs. This method is denominated the growth vapour trapping (GVT) approach, and can be tailored via using catalytic nanoparticles coated on Si substrates. The raw BNNTs produced by this CVD/GVT method are vertically-aligned and display super-hydrophobicity due to their high surface roughness and reduced surface energy.

On the other hand, polyesters based on fumaric acid, a constituent of the Krebs cycle, have involved a lot of attention for biomedical uses due to their superior biocompatibility and biodegradability [8]. Amongst them, the most extensively investigated is poly(propylene fumarate) (PPF), an linear polyester incorporating two ester bonds and one unsaturated carbon-carbon double bond (Figure 2) that allows cross-linking either by free radical polymerization with monomers of methyl methacrylate (MMA) or *N*-vinyl pyrrolidinone (NVP) or by means of photoinitation in the presence of photoinitiators like bisacylphosphine oxide (BAPO) [9]. Cross-linked PPF can suit a number of medical requirements such as biocompatibility, osteoconductivity, sterilizability, and handling characteristics. PPF degrades by simple hydrolysis of the ester bonds and the degradation time depends on polymer characteristics such as molecular weight, type of cross-linkers, and cross-linking density. The degradation products are non-toxic fumaric acid and propylene glycol.

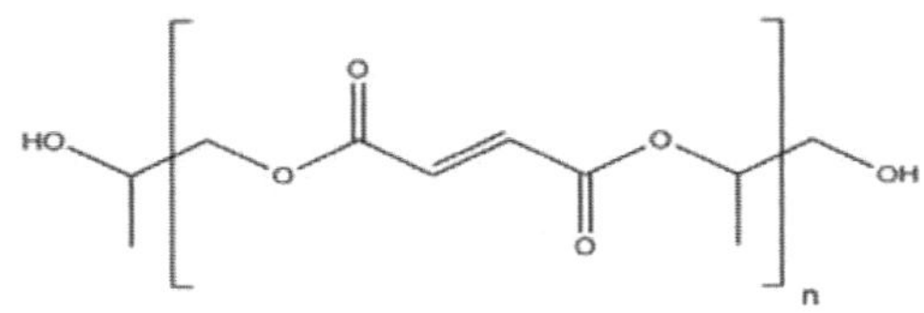

Figure 2. Chemical structure of poly(propylene fumarate) (PPF).

PPF is an amorphous polymer with a glass transition temperature (T_g) raging between -30 and 32 °C depending on the molecular weight, thermal decomposition temperature of 345 °C, density of 0.998 g/cm^3 and melt viscosity in the range of 10^2–10^4 Pa•s. It is an injectable polymer that can be crosslinked *in situ* through chemical reaction or UV laser. Its mechanical properties are dependent on the molecular weight and degree of crosslinking [10]. In previous studies, polyethylene-glycol (PEG)-grafted boron nitride nanotubes (PEG-g-BNNTs) were incorporated into a PPF matrix [11,12]. The present article deals with the reinforcement of a PPF matrix with raw BNNTs, and focuses on the mechanical characterization of the resulting composites.

2. Preparation of PPF/BNNT composites

The inorganic nanotubes were synthesized by CVD following an alternative of the process reported previously [13]. Amorphous boron, NH_4NO_3 and Fe_2O_3 powders (15:15:1 mass ratio) were mixed in an alumina boat and heated to 650°C. The temperature was increased to 1200 °C under a nitrogen flow and kept constant for 3 h. The nanotubes were purified by washing with concentrated HCl solution and finally washed with deionized water and dried overnight. The nanotubes were uniform, showing a characteristic bamboo-like structure and outer diameters of 40–100 nm, with a mean value of 82 nm, and length ≥5 μm, as revealed by scanning electron microscopy SEM (Figure 3).

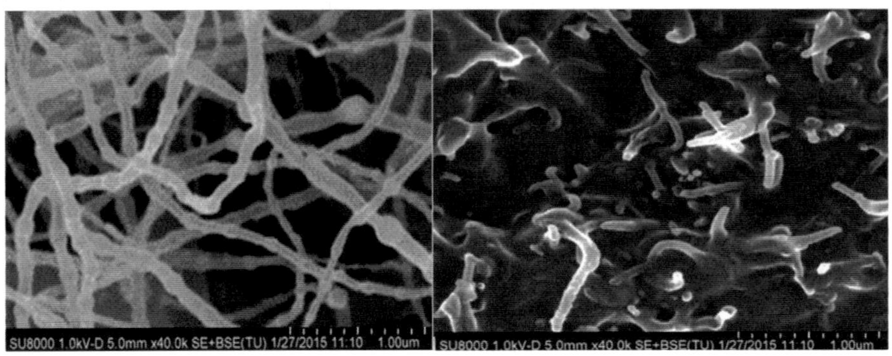

Figure 3. Left: SEM image of the as-synthesized boron nitride nanotubes. Right: SEM image of PPF/BNNT nanocomposite with 5.0wt% inorganic nanotube content.

PPF composites incorporating different amounts (0.1–5.0 wt%) of the inorganic nanotubes were prepared via ultrasonication and curing. Firstly, the nanomaterial was dispersed in chloroform by ultrasonication for 15 min. Separately, PPF and the cross-linker, N-vinyl-pyrrolidone (NVP), were mixed (1:1 w/w) and afterward the corresponding amount of the nanotube dispersion was added. The mixture was then ultrasonicated for another 15 min, and the free radical initiator, benzoyl peroxide (BP), was added to the mixture to begin the polymerization. The mixture was finally casted onto Teflon molds and cured under reduced pressure at 80 °C for one day. The dispersion of the inorganic nanotubes within the matrix was visualized by SEM. Homogenous distribution was found for all the nanotube concentrations investigated, with randomly oriented and curled tubes, as shown in Figure 3.

3. Mechanical properties of PPF/BNNT composites

The study of the mechanical properties of composite biomaterials is crucial for their applications; an adequate balance between flexibility and strength is required. The tensile properties of PPF/BNNT composites were investigated on a 4204 mechanical tester at 23°C and 50% RH, with a speed of 1 mm/min, following the UNE-EN ISO 527-1 standard, and the results are shown in Figure 4.

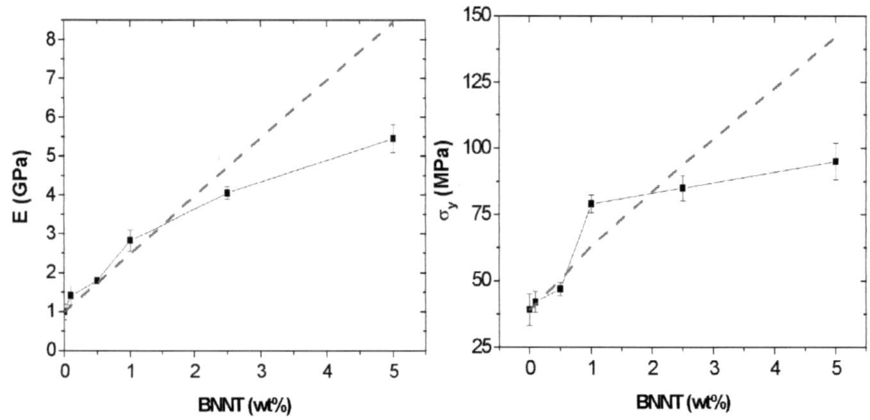

Figure 4. Young's modulus (E) and tensile strength (σ_y) of PPF/BNNT nanocomposites as a function of the nanotube content. The dashed red lines represent the predictions by the Krenchel's rule of mixtures for discontinuous reinforcement.

PPF has a Young´s modulus value close to 1 GPa, which increases gradually with increasing nanofiller content, leading to a 4.4 fold-increase for the composite with 5.0 wt% BNNTs. This exceptional rise should be ascribed to the homogenous filler dispersion and strong PPF-BNNT interfacial adhesion via hydrogen bonding and Van der Waals interactions along with the high elastic modulus of BNNTs [14]. Remarkably, the reinforcement effect attained herein is even higher than reported for the addition of similar amounts of other nanofillers such as single-walled carbon nanotubes (SWCNTs) to PPF [15], despite the carbon based nanotubes present higher modulus, demonstrating the importance of the strong interfacial adhesion to attain an effective stress transfer from the matrix to the filler. Similar trend is found for the tensile strength, resulting in 1.4-fold rise at 5.0 wt% BNNT. Regarding the ductility of the composites, a significant decrease was found on increasing nanotube concentration, particularly at loadings ≥ 2.5 wt%, and the nanocomposite with 5.0 wt% exhibits around 50% drop in ductility. This behaviour is expected to be associated with the intense PPF-BNNT H-bonding interactions. With regard to the toughness of the composites measured as the area under the tensile curve, a slight decrease was at the highest loadings tested.

The modulus (and strength) of the composites can be evaluated by the Krenchel's rule of mixtures (Equation 1) for discontinuous reinforcement [16]:

$$Y_c = (n_o n_l Y_f - Y_m)V_f + Y_m \qquad (1)$$

where n_o is the orientation factor, n_l the length efficiency factor, Y_f the filler modulus, Y_m the matrix modulus and V_f the filler volume fraction. n_l approaches to 1 for $l/D > 10$, and $n_o = 1/5$ for randomly oriented fillers. Taking reported data for the modulus (750 GPa) and strength (10 GPa) of BNNTs [14], it was possible to estimate the theoretical values for the concentration range studied (see red dashed lines in Figure 4). Experimental Young's modulus of composites containing nanotube loadings ≤2.5 wt% are in good agreement with the predictions (differences <10%), whereas the value of the composite with higher loading is well below the predictions (about 55% lower). Similar trend is found for the tensile strength. Notice that the model assumes perfect nanotube dispersion and tube-matrix interfacial adhesion. Dispersion is probably the most crucial issue. To attain a uniform stress distribution and minimize the presence of stress concentration centers, BNNTs have to be uniformly dispersed to the level of isolated tubes individually coated with the biopolymer. However, the nanotubes still remain gathered in small bundles, as revealed by SEM micrograph (Figure 3), and it may occur shear slippage of individual nanotubes within the bundle, which limits the stress transfer; this could explain the discrepancies between the experimental and the theoretical values for the composite with 5.0 wt% BNNTs.

4. Conclusions

Inorganic boron nitride nanotubes were synthesized by CVD and incorporated via ultrasonication and curing into a biocompatible PPF matrix at loadings of 0.1-5.0 wt%. Scanning electron microscopy images revealed a random and homogenous distribution of the inorganic nanotubes within PPF. The nanotubes led to an extraordinary increase in the stiffness and strength of the polymer, ascribed to the strong PPF-BNNT interfacial adhesion via hydrogen bonding and Van der Waals interactions. Experimental Young's modulus and tensile strength of composites containing nanotube loadings $\leq$2.5 wt% were consistent with the predictions, while the values of the composite with 5.0 wt% were considerably below the predictions. These composites are suitable to be used as orthopaedic scaffolds in particular for the replacement of soft tissues like muscles, tendons, ligaments, etc.

Acknowledgments

Dr. A.M. Diez-Pascual wishes to acknowledge the Spanish Ministerio de Economia y Competitividad (MINECO) for a "Ramón y Cajal" Research Fellowship cofinanced by the EU.

References

[1] A. Rubio, J. L. Corkill, M. L. Cohen, Phys. Rev. B 49, 5081 (1994)
[2] J. Wang, C. H. Lee, Y. K. Yap, Nanoscale 2, 2028 (2010)
[3] C. Zhi, Y. Bando, C. Tan, D. Golberg, Mater. Sci. Eng. R 70, 92 (2010)
[4] S. Kalay, Z. Yilmaz, O. Sen, M. Emanet, E. Kazanc, M. E. Culha, Beilstein J. Nanotechnol. 6, 84 (2015)
[5] C. Tang, Y. Bando, T. Sato, K. A. Kurashima, Chem. Commun. 12, 1290 (2002)
[6] C. Zhi, Y. Bando, C. Tan, D. Golberg, Solid State Commun. 135, 67 (2005)
[7] C. H. Lee, J. Wang, V. K. Kayatsha, J. Y. Huang, J. K. Yap, Nanotechnology 19, 455605 (2008)
[8] F. K. Kasper, K. Tanahashi, J. P. Fisher, A. G. Mikos, Nat. Protoc. 4, 518 (2009)
[9] J. P. Fisher, T. A. Holland, D. Dean, P. S. Engel, A. G. Mikos, J. Biomater. Sci. Polym. Ed. 12, 673 (2001)
[10] D. D. Frazier, V. K. Lathi, T. N. Gerhart, D. E. Altobelli, W. C. Hayes, *Proc. Material Research Society Symp.* 394, 15 (1995)
[11] A. M. Diez-Pascual, A. L. Diez-Vicente, RSC Advances 6, 79507 (2016)
[12] A. M. Diez-Pascual, Polymers 9, 260 (2017)
[13] T. H. Ferreira, P. R. Ornelas da Silva, R. Gouvêa dos Santos, E. M. Barros de Sousa, J. Biomater. Nanobiotechnol. 2, 426 (2011)
[14] M. Santosh, P. K. Maiti, A. K. Sood, J. Nanosci. Nanotechnol. 9, 5425 (2009)
[15] X. Shi, J. L. Hudson, P. P. Spicer, J. M. Tour, R. Krishnamoorti, A. G. Mikos, Nanotechnology 16, S531 (2005)
[16] H. Krenchel, Fibre reinforcement (Akademisk Forlag, Copenhagen, 1964)

IOP Conf. Series: Materials Science and Engineering **383** (2018) 012002 doi:10.1088/1757-899X/383/1/012002

The influence of double pitting corrosion on casing collapsing strength based on different ellipsoid shape

Yifei Yan[1,*], Xiang Zhou[2], Shengli Song[3], Bing Shao[1], Xiujuan Yang[4] and Xiangzhen Yan[2]

[1]College of Mechanical and Electronic Engineering, China University of Petroleum, Qingdao, China
[2]Oil and Gas CAE Technology Research Center, China University of Petroleum, Qingdao, China
[3]Petroleum Engineering Technology Research Institute of Zhongyuan Oilfield. Sinopec, China
[4]College of Pipeline and Civil Engineering, China University of Petroleum, Qingdao, China

*E-mail: yanyf163@163.com

Abstract. Acidic environment always contributes on casing corrosion phenomenon, the local corrosion and pitting lead to stress concentration, which decreases the strength and is an important factor to cause damage to the casing. The influences of distance of two pitting, corrosion depth, section parameter on collapsing strength of casing are quantitatively analyzed under the condition of double pitting axial distribution and ring distribution. The results showed that the maximum Von Mises stress of casing first increases, and then goes down to a steady state as the distance of double pitting grows under the condition of double pitting axial distribution, while the stress goes down to a steady state as the distance of double pitting grows under the condition of double pitting ring distribution. It can provide the theoretical foundation for the casing design through analyzing and comparing the influence of double pitting on the collapsing strength of casing under different condition.

1. Introduction

Acidic environment usually leads to the casing corrosion phenomenon, and the corrosion damage form has two types: the uniform corrosion and local corrosion. The uniform corrosion which occurs along the whole casing wall generally causes less damage, while the local corrosion occurs on the certain areas of casing and the rest of the parts are in good condition. The common local corrosion is pitting corrosion(holes)[1-4]. Pitting corrosion reduces the thickness of the casing wall, and causes stress concentration which seriously reduces the strength of casing, bring serious potential danger to the production of oil and gas[5-8]. Therefore the analysis and study the influence of pitting corrosion defect on casing strength has a great significance on selecting materials and extending the service life of the gas well.

Aiming at the influence of pitting corrosion defect on the casing strength, the present study focused the casing collapsing strength of a single pitting defect[9-12]. Using the finite element software ANSYS with the actual casing size and the material parameters, we analyzed the influence of pitting

Content from this work may be used under the terms of the Creative Commons Attribution 3.0 licence. Any further distribution of this work must maintain attribution to the author(s) and the title of the work, journal citation and DOI.
Published under licence by IOP Publishing Ltd

hole spacing, hole section parameter and pitting depth on casing collapsing strength on two working conditions—— Double pitting holes distributed along the axial of casing wall and along the ring of casing wall, providing the theoretical basis for the safety assessment of pitting corrosion during the oil and gas production.

2. Failure criterion determination

Failure criterion is the basis of judging the failure, it depends on failure modes, and there are two criterion which are widely used:

1) Based on elastic failure criterion[13], casing failed when the equivalent stress of corrosion area (using Von Mises equivalent stress) reaches the yield strength of pipe material.

2) Based on plastic failure criterion[14], minimum stress of corrosion area reaches the tensile strength, casing failed.

Using the elastic failure criterion based on the security considerations, in the three dimensional principal stress space, the Von Mises condition[15] is expressed as in equation (1):

$$\sigma_v = \sqrt{\left\{\frac{1}{2}[(\sigma_1-\sigma_2)^2+(\sigma_2-\sigma_3)^2+(\sigma_3-\sigma_1)^2]\right\}} < [\sigma] \quad (1)$$

Where: σ_v is Von Mises equivalent stress , MPa; $[\sigma]$ is allowable stress, MPa.

3. The simplification of pitting corrosion defect

The common form of casing pitting holes can be simply divided into cylindrical spherical or ellipsoidal holes through reviewing literatures and investigation. But in the actual situation, the pitting holes are closer to the ellipsoid through a large number of pitting corrosion pit morphology observation. So we mainly consider the effects of ellipsoidal pitting on the mechanical behavior of casing. Depending on the pitting distribution location, the double pitting defects can be divided into two classes:

1) Pitting holes distribute along the axial, the adjacent pitting located on the same axis direction of casing wall, and circular projection overlap, as is shown in Figure 1. (a), the distance of two hole section center is h, written as the double hole axial spacing.

2) Pitting holes distribute along the ring, the adjacent pitting located on the same height of casing wall, the distribution of axial projection overlap, as is shown in Figure 1.(b), the distance of two hole section center isθ, written as the double hole ring spacing.

3) Pitting hole geometric description: Figure 2 is the three-dimensional topography of a ellipsoid pitting ,the pitting cross section is oval ,the half width is b, the half long is c, the maximum depth is a. Ellipsoid shape parameters is determined by the a,b,c three values. a depicts the depth of pitting, the d=c/b which is section coefficient controls the shape of the pitting section, t is wall thickness, t0 is the maximum pitting depth, t′=t0/t is relative corrosion depth.

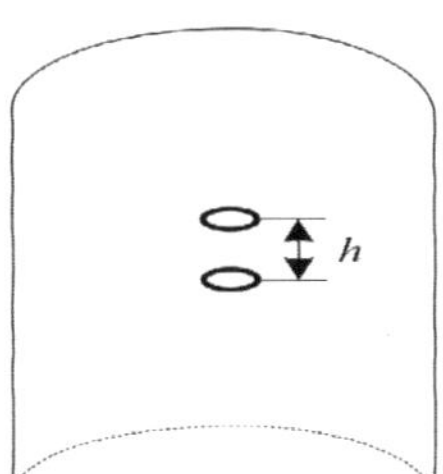

(a)

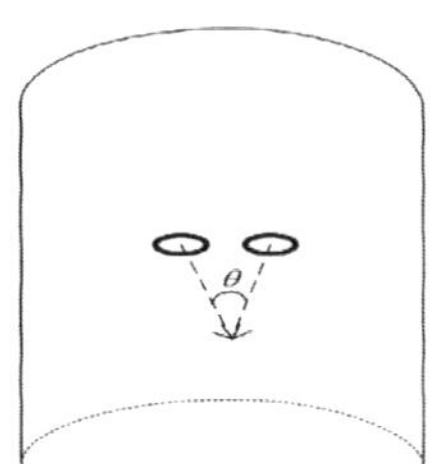

(b)

Figure 1. Casing pitting corrosion model:(a) Double pitting axial distribution, (b) Double pitting ring distribution.

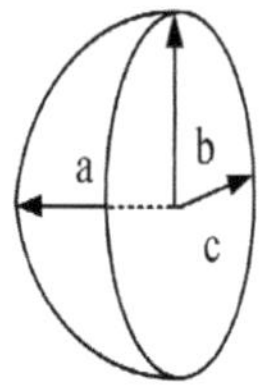

Figure 2. Ellipsoid pitting three-dimensional topography.

4. The establishment of the finite element model

In this paper, two pitting corrosion along the axial and ring distribution are chosen as examples respectively. Three dimensional finite element model of casing pitting corrosion is established using commercial available software ANSYS.

The casing is chosen in a super deep well section which is located in the Tarim Basin formation depth of 5000 m. The equivalent external extrusion pressure is about 70 MPa. Casing type is P110 with the pipe parameters shown in Table 1.

Table 1. Parameters of casing.

Pipe diameter(m)	Pipe thickness(m)	elasticity modulus (GPa)	Poisson's ratio	yield strength
0.1778	0.00919	2.06×1011	0.3	837MPa

5. The calculation results analysis

In order to study the differences between corrosion section, pitting depth and pitting space the influence law of distribution of casing collapsing strength, section coefficient d is selected as 0.5, 1, 2, 3, and relative pitting depth d is chosen as 10%, 20%, 30%, 40%, 50%, the specific calculation condition as shown in Table 2.

Table 2. Working condition of calculation table.

Working condition	Section coefficient	Relative pitting depth	Pitting space
Axial distribution	0.5、1、2、3	20%	2mm~50mm
	3	10%、20%、30%、40%、50%	2mm~50mm
Ring distribution	0.5、1、2、3	20%	2°~50°
	3	10%、20%、30%、40%、50%	2°~50°

A working condition with relative pitting depth t'=20%, pitting section coefficient $d=2$ is taken as an example. When the outer pressure is 50 MPa, the local stress cloud picture of casing pitting area is calculated. When two corrosion holes are along the axial distribution and the corrosion holes are intersected, the intersecting area between two holes is in high stress area, the maximum Von Mises stress zone is located in the intersection area of two holes, as pitting spacing increases, the high stress area gradually changed from the pitting hole edge to the pitting hole center. The maximum Von Mises stress is located in the pitting hole edge; When two corrosion holes are along the annular distribution, the intersecting area between two holes is in the low stress area and the center of two holes and a fraction of hole edges are in the high stress area. The maximum Von Mises stress lies in the edge of pitting hole. As the angle of pitting hole increases, the position of high stress area is unchanged.

5.1. The influence of pitting section on casing strength

When the relative pitting depth is fixed on 20%, the relationships between the maximum Von Mises stress of casing and pitting spacing under the different corrosion section coefficient are shown in Figure 3 an Figure 4. As shown in Figure 3, when double pitting holes is along the axial distribution, the relationship between the maximum Von Mises stress of casing and pitting space obey the Gaussian distribution. When the two hole is intersecting, the maximum stress increases with the increase of the axial spacing and reaches the maximum at the smallest intersection area. When the section coefficient is small, the Von Mises stress will exceed the casing yield strength and the casing is easy to be damaged by pitting corrosion. When the two holes are separated, the interaction of double pitting corrosion defect is weakened. When the spacing reaches the limit axial spacing, the maximum Von Mises stress changes very little. When the pitting section coefficient is 0.5, 1, 2, 3, the limit of the axial spacing is: 16 mm, 20 mm, 22 mm and 22 mm respectively. As is shown in Figure 4, when the double pitting holes are along the ring distribution, the maximum Von Mises stress of casing exponential decreases as the double pitting ring intersection angle increases. When the pitting section coefficients are 0.5, 1, 2 and 3, the limit toroidal angle are 18 °, 20 °, 22 ° and 22 ° respectively. When the pitting spacing is small and the double holes spacing is the same, the maximum Von Mises stress of double holes along the axial distribution is significantly larger than that of the double holes along the ring distribution. The double pitting holes along the axial distribution is more likely to damage the casing.

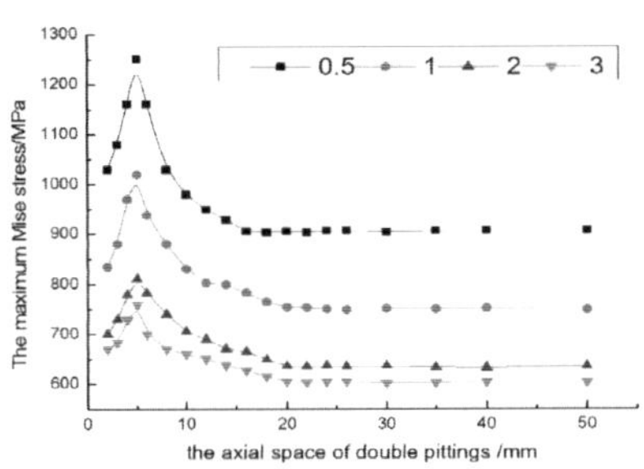

Figure 3. Variation curves of the maximum Von Mises stress with pitting space.

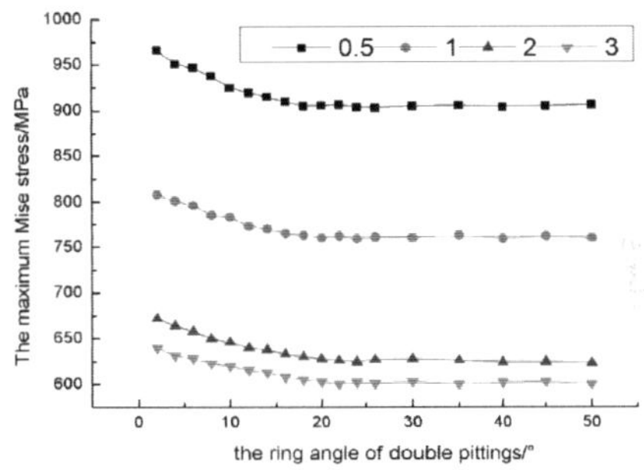

Figure 4. Variation curves of the maximum Von Mises stress with pitting angle.

The relationship between the maximum Von Mises stress of the defects and section coefficient is shown in Figure 5 when the double pitting axial spacing is 2 mm, 6 mm, 12 mm and 20 mm respectively and double pitting ring angle is 2°, 6°, 12°and 20° respectively. It shows that the maximum Von Mises stress decreases exponentially as the cross section coefficient increases. A case example is carried out when the axial spacing is 2mm and ring spacing angle is 2°.

When the cross section coefficient is 0.5, the Von Mises stress of casing exceeds the yield strength under the different pitting spacing, which leads to casing failure. Therefore in actual working condition, the small pitting defect causes a lot of damage to the casing. When the cross section coefficient *d* increases from 2.0 to 3.0, the maximum Von Mises stress of the pitting in axial distribution reduces to 5.6%, the maximum Von Mises stress of pitting in ring distribution reduced to 4.7%. There is less influence of pitting section on casing stress and pitting section coefficient is no longer the main factor which determines casing strength.

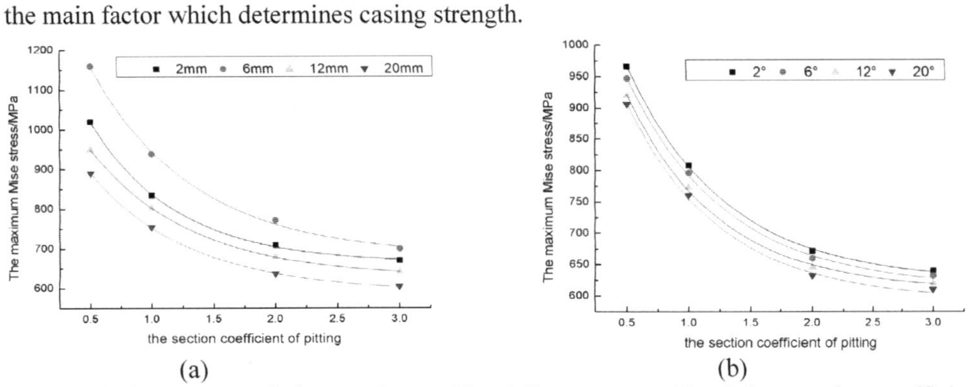

Figure 5. Variation curves of the maximum Von Mises stress with pitting section coefficient, (a)when double pitting axial distribution is different, (b) when double pitting ring distribution is different.

IOP Conf. Series: Materials Science and Engineering **383** (2018) 012002 doi:10.1088/1757-899X/383/1/012002

6. Conclusion

(1) The stress distribution of P110 casing when the double pitting is along axial distribution and ring distribution is studied by the finite element software ANSYS. The limit spacing of double hole axial pitting and double hole ring pitting are obtained. The limit spacing of double hole increases as the increase of the pitting section coefficient.

(2) Through the regression curve it can be obtained that the maximum Von Mises stress exponential decreases as the pitting section coefficient enlarges in the condition of double pitting along axial distribution and ring distribution. When the pitting section coefficient exceeds a certain value, the maximum Von Mises stress of casing does not decrease with the increase of the coefficient of cross section.

(3) Under the condition of the same pitting, double pitting holes along the axial distribution causes more damage than double pitting holes along the ring distribution while the pitting spacing is small.

Acknowledgments

The authors wish to acknowledge the financial support of China Postdoctoral Science Foundation (Grant No. 2017M612375), National Natural Science Foundation of China (GrantNo. 51374228), National key technology project(Grant No. 2016ZX05017-003-01) , and CNPC key laboratory Project (Contract No. 2016A-3905).

References

[1] Lu W Y. 2014. P110 casing corrosion cracking mechanism research. *Journal of Yangtze university (natural)* **11** pp. 78-80

[2] DNV. RP-F101 2010 Corroded Pipelines (Oslo:Det Norske Veritas)

[3] ASME B31G-2009 2009 Manual for determining the remaining strength of Corroded pipeline-Supplement to ASME B31 Code for Pressure Piping(New York: ASME B31 Committee)

[4] Dong S R, He D S and Zhang P 2005 Pairs of pitting corrosion of pipelines of elastic-plastic finite element analysis *Journal of design and research* **9** pp 20-22

[5] Wanees S A, Radwan A B, Alsharif M A and Haleem A E 2017 *Materials chemistry and physics* **190** pp. 79-95

[6] Xu Z Q, Yan X Z, Yang X J 2014 *Journal of mechanical design* **8** pp 75-78

[7] Ossai C I, Boswell B and Davies I 2016 *Engineering failure analysis* **60** pp 209-28

[8] Papavinasam S 2017 *Trends in oil and gas research and technologies* pp 663-88

[9] Deng K H, Lin Y H, Qiang H, Zeng D Z and Lin X X 2015 New high collapse model to calculate collapse strength for casing *Engineering failure analysis* **58** pp 295-306

[10] Liu S H, Zhang H L, Zhu X H and Tong H 2014 *Engineering failure analysis* **42** pp 240-51

[11] Huang X, Mihsein M, Kibble K and Hall R 2000 Collapse strength analysis of casing design using finite element method *Engineering failure analysis* **77** pp 359-67

[12] Kuriyama K,Tsukano Y, and Mimaki T 1992 Effect of wear and bending on casing collapse strength *SPE* **24597** pp 1-10

[13] Tian X and Zhang H 2017 Failure criterion of buried pipelines with dent and scratch defects *Engineering failure analysis* **80** pp 278-89

[14] Dolinski M and Rittel D 2015 Journal of mechanics and physics of solids 83 pp. 1-18

[15] Jaeger J C, Cook N G W, Zimmerman R W 2007 *Fundamental of rock mechanics* (Blackwell Publishing) Chapter 9 p 254

IOP Conf. Series: Materials Science and Engineering **383** (2018) 012003 doi:10.1088/1757-899X/383/1/012003

Safety evaluation of casing string based on BP artificial neural network

Yifei Yan[1,*], Xiujuan Yang[2], Chao Kong[1], Xiaoqi Zhou[3] and Bing Shao[1]

[1]College of Mechanical and Electronic Engineering, China University of Petroleum, Qingdao 266580, China
[2]College of Pipeline and Civil Engineering, China University of Petroleum, Qingdao 266580, China
[3]Petroleum Engineering Technology Research Institute of Zhongyuan Oilfield. Sinopec, 457001, China

*E-mail: yanyf163@163.com

Abstract. In the traditional safety assessment of casing string, some influencing factors are ignored for modelling convenience, which makes the casing string safety assessment effect of oil and gas well not very ideal. For complexity and randomness of casing load and its properties parameters in complex well conditions, BP artificial neural network is created in MATLAB based on the analysis of the influencing factors of casing string security. Casing string section whose safety assessment is more mature is taken as a sample to train the BP neural network. The trained network is applied to make case assessment. At the same time, GUI interface is applied to implement the visualization. The results show that safety evaluation of the casing string can be achieved by using BP neural network. The accuracy of the casing string network safety evaluation is high. It will realize the visualization of safety evaluation and provide more accurate and effective reference for the design of casing string.

1. Introduction

In recent years, the growing global demand for oil makes the trend that drilling of oil become deeper and more complex. High temperature, high pressure and high stress in complex deep wells have becoming the main factors which influence the strength of the casing[1-3]. Complex petroleum exploitation environment make the downhole casing more prone to damage. With the increase of the number of deep wells, casing damage has become a problem that can not be ignored. How to effectively analyze casing damage in a variety of complex factors and how to provide scientific design basis and to provide effective protection for the stable development of oil and gas field have become an arduous task of workers on petroleum science and technology[4-6].

For complexity and randomness of casing load and its properties parameters in complex well conditions, BP artificial neural network method is created for safety assessment of casing string. MATLAB GUI interface is applied to implement the visualization based on the analysis of the influencing factors of casing string security, which lay the foundation of in-depth development of the security and reliability of casing under complex well conditions.

2. Basic principle of BP neural network and its algorithm

Content from this work may be used under the terms of the Creative Commons Attribution 3.0 licence. Any further distribution of this work must maintain attribution to the author(s) and the title of the work, journal citation and DOI.
Published under licence by IOP Publishing Ltd

2.1. Basic principle

Artificial neural networks[7,8] is a kind of information processing system which imitates biological brain structure and function. Artificial neural network has better approximation ability for nonlinear function, which has now become a kind of technology for nonlinear system modeling, identification and control, and is increasingly widely used around the world.

The BP Neural Network[9,10] is a one-way transmission of multi-layer forward Neural Network, it refers to the algorithm of error back propagation. BP network is a typical multi-layer network, composed of input layer, hidden layer and output layer. Layer to layer is adopted full connection and there is no connection between each neuron in the same layer. Figure 1 is a BP network topology structure containing a hidden layer. In the Figure 1, "O" represents the neurons, and a straight line segment represents connecting the two layers of neurons represents the connection weights between them.

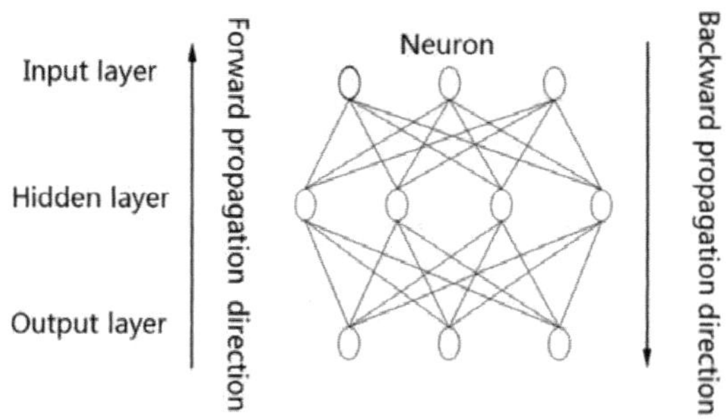

Figure 1. BP neural network model.

2.2. The general theory of BP algorithm

The general theory of BP algorithm can be summarized as forward propagating mode and backward propagating mode[11-13], which are shown below:

(1) Forward propagating mode

The input pattern vector is set as $A_k = (a_1^k, a_2^k, ...a_n^k)$ and the desired output vector is $Y_k = (y_1^k, y_2^k, ...y_n^k)$, k = 1,2,···,m, in which n is the number of input layer units; q is the number of output layer units; m is the number of learning mode.

① Calculating the input of each hidden layer unit:

$$s_j^k = \sum_{i=1}^{n} w_{ij} a_i^k - \theta_j; j = 1,2,...p \tag{1}$$

In the equation (1), w_{ij} is the connection weight from input layer to hidden layer; θ_j is threshold value of the hidden layer unit; p is number of the hidden layer unit

② Calculating the output of the each hidden layer unit b_j^k:

$$b_j^k = f(x) = f(s_j^k); j = 1,2,..., p \tag{2}$$

In the equation (2): $f(x) = \frac{1}{1+e^{-x}}$; b_j^k is the activation value of the hidden layer.

③ Calculating the input and output of output layer of each unit:

$$L_t^k = \sum_{j=1}^{p} v_{jt} b_j^k - \gamma_t; t = 1,2, . . . , q, \tag{3}$$

$$c_t^k = f(L_t^k); t = 1,2,...,q, \tag{4}$$

In the equation: v_{jt} is connection weight from hidden layer to output layer, γ_t is threshold value of the output layer unit.

IOP Conf. Series: Materials Science and Engineering **383** (2018) 012003 doi:10.1088/1757-899X/383/1/012003

(2) Backward Propagation

① Calculating the correction error of output layer d_t^k:

$$d_t^k = (y_t^k - c_t^k) c_t^k (1 - c_t^k) \tag{5}$$

② Calculating the correction error of each element of the hidden layer e_j^k:

$$e_j^k = \left(\sum_{t=1}^{q} d_t^k v_{jt} \right) b_j^k (1 - b_j^k), j = 1,2,\ldots,p \tag{6}$$

③ Fixing the weights between hidden layer and output layer v_{jt} and various units of the output layer threshold γ_t:

$$v_{jt}(N+1) = v_{jt}(N) + \alpha d_t^k b_j^k \tag{7}$$

$$\gamma_t(N+1) = \gamma_t(N) = \alpha d_t^k; \tag{8}$$

In the equation : $j = 1,2,\cdots,p; t = 1,2,\cdots,q; 0 < \alpha < 1$

④Fixing the weights between input layer and hidden layer w_{ij} and threshold of hidden layer units θ_j;

$$w_{ij}(N+1) = w_{ij}(N) + \beta e_j^k \partial_i^k \tag{9}$$

$$\theta_j(N+1) = \theta_j(N) + \beta e_j^k \tag{10}$$

In the equation: $i = 1,2,\cdots,n; j = 1,2,\cdots,p; 0 < \beta < 1$

In the equations (7)-(10): α, β are the learning rate.

3. Application of BP neural network on casing string safety evaluation

A safety evaluation model of casing string was established by using BP artificial neural network in 50 wells of an oil production plant in North China oil field.

3.1. Establishing the network topology structure

(1) Determining the number of nodes in the input layer

There are many factors which affect the safe operation of the casing string, such as geometry, mechanical properties and load of casing and cement ring. Among them, casing yield strength and equivalent stress will eventually affect casing reliability. They are associated with the safe operation of the casing string at different levels. In order to simplify the problem, only factors which have high correlation with the safe operation of the casing string are considered, while ignoring secondary factors. Therefore, correlation analysis is carried out on input variables. The optimized subset which is constituted by the significant variables is taken as the input variables of the network. In order to accurately extract the main factors, SAS system is used to analyze the significance of each factors. Research shows that the main factors which affect the safety operation of the casing string are as follows: in-situ stress σ_x, σ_y, pipe yield strength S_0, external extrusion pressure P_0, casing wall thickness T, casing diameter D, casing elastic modulus E. Thus the number of nodes which are determined are seven.

(2)Determining the number of nodes in the output layer

By setting different degrees of casing damage number of output layer neurons as the target amount, the operational safety of the casing string can be objectively characterized. While carrying out the casing string safety evaluation, weighted average of the objective data is rerunning after the sample data is normalized. The output layer nodes number is 4, which respectively are especially serious damage, serious damage, damage and minor damage.

(3) Determining the number of layers of the network

The main reason that BP neural network can identify the nonlinear model is the hidden layer between the input layer and output layer, which is determined by the practical problems. The number of layers can be selected as 1, 2 or multiple. After many experiments the BP neural network which has a hidden layer is chosen as the best, and the node number is 15.

3.2. Establishing BP network learning samples

The data of 50 wells in an oil production plant in North China oil field is selected as the study sample of BP artificial neural network. Part of the data is in Table 1.

Table 1. Part of the learning samples.

Well number	σ_x	σ_y	S_0	P_0	t	D	E
1	31.25	37.29	758	21.5	10.2	69.85	$2.11*10^5$
2	32.31	42.03	758	22.3	10.4	69.94	$2.11*10^5$
3	35.74	47.90	758	23.1	11.5	69.95	$2.11*10^5$
4	36.84	52.31	758	23.5	11.3	70.12	$2.11*10^5$
5	39.45	49.31	758	23.8	11.6	70.63	$2.11*10^5$
6	40.21	53.42	758	24.2	11.7	70.25	$2.11*10^5$

3.3. Learning and testing of BP network

After the learning process is finished, the structure of the network is determined. A group of data is taken to test the trained network and the output of test vector is A=[0.3661,0.1284,12344,0.1070]. The error between the actual output and the target output is shown in Table 2. It is seen clearly that the error of output value of the test data is smaller, with an average error of 0.004. Under such a sample size, the precision is perfectly acceptable. Therefore BP neural network model can be used to evaluate casing string accurately.

Table 2. The BP neural network output error.

Target Output	Actual Output	Error
0.3656	0.3661	0.0013
0.1274	0.1281	0.0055
0.2334	0.2344	0.0043
0.1065	0.1070	0.0047

3.4. The predicted results evaluation of BP neural network

The input vector dimension of BP artificial neural network is 7 and the output vector is different degrees of casing damage, which are especially serious damage, respectively, serious damage, general damage and minor damage. Using a block of 20 Wells in North China oil field as the test data of BP neural network, the evaluation result which is carried out by commercial software Matlab is shown in Figure 2[14 -16]. The detailed method is illustrated by Sun[15]. As is shown in Figure 2, the maximum error between the predicting result and the actual value is 7.5%. Calculation speed of safety assessment is enhanced on complex deep well casing string by using the BP neural network.

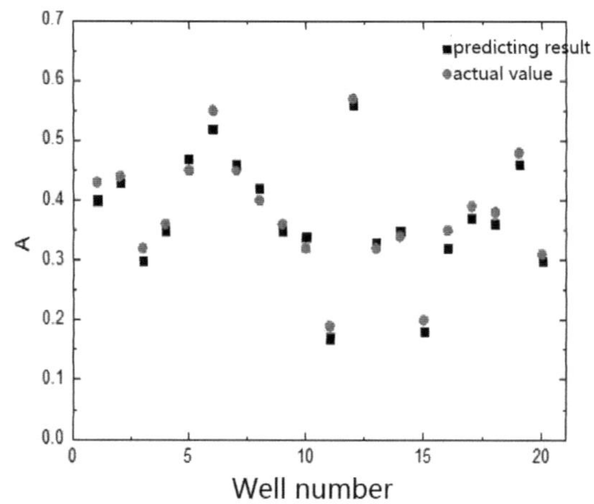

Figure 2. BP network predicting result of 20 wells compared with the actual value.

4. Conclusions

(1)Through the analysis of casing loads, geometry and mechanical properties of casing, cement ring and stratum rocks, seven main factors are determined which are the in-situ stress σ_x, σ_y, pipe yield strength S_0, external extrusion pressure P_0, casing wall thickness t, outer diameter of the sleeve D and casing elastic modulus E. They are treated as evaluation index to carry on safety evaluation of the casing string.

(2)Through the analysis, I came to the following conclusions. Safety evaluation based on BP artificial neural network provides a convenient, effective way for casing safety evaluation. Compared with the traditional method, the evaluation results of calculation are more comprehensive and scientific. Calculation of the casing reliability has higher credibility, which can provide some calculation basis for the scientific and comprehensive analysis of casing.

Acknowledgments
The authors wish to acknowledge the financial support of China Postdoctoral Science Foundation (Grant No. 2017M612375), National key technology project(Grant No. 2016ZX05017-003-01), National Natural Science Foundation of China (GrantNo. 51374228) and CNPC key laboratory Project (Contract No. 2016A-3905).

References

[1] Kaldal G K, Jonsson M T, Palsson H and Karlsdottir S N 2015 *Geothermics* **55** pp 126-37
[2] Cirimello P G, Otegui J L, Carfi G and Morris W 2017 *Engineering failure analysis* **75** pp 1-14
[3] Bachu S 2017 *International journal of greenhouse gas control* **61** pp 146-54
[4] Shen C 2015 *Natural gas industry B* **2** pp 347-53
[5] Yu Z H, Wang L Y and Liu D W 2009 Journal of hydrodynamics 21 pp. 77-83
[6] Chiotis E, Vrellis G 1995 *Geothermics* **24** pp 695-705
[7] Himmelblau D M, Barker R W and Suewatanakul W 1992 Fault classification with the aid of artificial neural networks IFAC/IMACS Symposium, Baden-Baden, Germany pp 541-45
[8] Jr G S S and Yoon Y 1992 *Financial analysts Journal* **48** pp 78-80
[9] Wei H K 2005 Theory and method of neural network structure Design(Beijing: National defence industry press)
[10] Gao J 2003 Artificial neural network principle and simulation examples (Beijing:Mechanical industry press)
[11] Li M S and Chen W C 2012 *Physics Procedia* **25** pp 1212-17
[12] Jing G L, Du W T and Guo Y Y 2012 *Desalination* **291** pp 78-93
[13] B, Dan H C and Li L 2017 *Applied thermal engineering* **120** pp 568-80
[14] Lan X M, Mi J and Huang C M 2003 Achiving BP networks by MATLAB *Research and design of microcomputer application* **12** pp 6-8
[15] Sun L 2002 Index prediction system based on artificial neural network design and implementation(Xi'an: Xi'an jiaotong university)
[16] Su J M 2004 MATLAB7.0 practical guide (Beijing:Electronic industry press)

Synthesis and characterization of rare earth activated barium zirconate ceramic

Sadhana Agrawal*

Department of Physics, National Institute of Technology Raipur, Raipur-492010, India

*E-mail: sagrawal.phy@nitrr.ac.in

Abstract. This paper reports synthesis of rare earth doped BZO perovskite ceramic by facile solid state reaction synthesis route. The phosphor confirms its nanocrystalline structure with high order purity and homogeneity. Characterized by XRD, SEM, EDX, FTIR and TL, a complete investigation of its structural, morphological and thermoluminescent properties is done. The kinetic parameters estimated using peak shape method confirms the second order kinetics of synthesized ceramic powder. With the emerging global concern regarding UVR exposure protection, the BZO perovskite ceramic are emerging as dynamic alternatives for TL mapping.

1. Introduction

With distinct refractory properties, good chemical resistance, high dielectric constant and good structural integrity, ceramic materials have emerged as potential candidate for automotive, sensing, optoelectronics and refractory applications. Perovskite structured ceramic materials with a general chemical formula ABO_3 form a wide range of alkaline earth metal zirconates widely used in manufacturing of MLCC, frequency filters, batteries, fuel cells, thermal barrier coatings and sensors [1]. Barium zirconate, $Ba_{II}Zr_{IV}O_3$, is a perovskite electroceramic material with a high MP (> 2600°C), wide band gap (> 4 eV) and high dielectric constant (~ 50). On doping with rare earth (RE) impurities, the perovskite material show drastically amended physical, chemical, optical, thermal, electrical and magnetic characteristics [2]. Prepared via facile solid state reaction sythesis route, the environmental friendly barium zirconate ceramic with crystalline nature and irregular morphology is obtained. The synthesized ceramic material is also subjected to thermoluminescent analysis in order to confirm its suitability as TL phosphor.The TL phosphors have emerged as the most suitable, sensitive and precise means for radiation detection [3]. The characteristics of a phosphor that makes it worth efficient are- Linear dose response, Spatial resolution, Sensitivity, Fading, Directional dependence, etc. TL phosphors make use of the phenomena of Thermoluminescence which is nothing but Thermally Stimulated Luminescence. This phenomena involves thermal stimulation of the material which is pre-exposed to some type of radiation, so as to get visible light emitted as the output. The name of this process is a misnomer because thermal treatment is given as stimulation and not as excitation. In proper words, Thermoluminescence can be defined as Thermally Stimulated luminescence (TSL) [4].

2. Synthesis

The Synthesis of desired perovskite is done here using solid state reaction synthesis route which is the typical, easy, environment friendly and less sophisticated method [5]. $BaCO_3$, ZrO_2, Eu_2O_3 and Tb_2O_3

Content from this work may be used under the terms of the Creative Commons Attribution 3.0 licence. Any further distribution of this work must maintain attribution to the author(s) and the title of the work, journal citation and DOI.
Published under licence by IOP Publishing Ltd

are taken here as the starting materials in stoichiometric ratio. All chemicals consumed here were of Sigma Aldrich with homogeneity and high purity (>99.99%).The Solid State reaction synthesis route involves two sub processes- Calcination and Sintering [6]. In the process of Calcination, $BaCO_3$ get thermally decomposed to BaO and CO_2, when heated at 1000 ^{0}C for 2 hrs. This BaO obtained here combines with $ZrO_2$to form $BaZrO_3$ in the process of Sintering. During Sintering the firing of rare earth impurities, Eu_2O_3 and Tb_2O_3, in the host lattice also takes place when heated at 1350^{0}C [7]. The RE doped BZO phosphors withEu^{3+} varying from 0.1-2 mole % and keeping Tb^{3+}fixed as 1 mole % are considered here for characterization [8].

3. Characterization results and interpretation

3.1. XRD analysis

XRD analysis is carried out using X-ray diffractometer system PANalytical - XPERT-Powder, which employs X-ray beam of wavelength 0.154nm (Cu-K-alpha) with copper as anode source. The observed X-ray pattern matches with the JCPDS card no. 06-0399 with the reflection peaks recorded at 30.59°, 44.04°, 54.88° and 61.99° corresponding to (110), (200), (211) and (220) planes respectively (in Figure 1) [9]. To estimate the crystallite size Debye Scherrer Formula is used:

$$D = \frac{K\lambda}{\beta Cos\theta}$$

Where, D=Average crystallite size, Λ=Wavelength of X-rays, β =Full width at half maximum FWHM, Θ=Angle of incidence, K=Scherer constant (0.9 - spherical particles) [10].

So, for given sample, K=0.9, λ = 0.154 nm, β = 0.0061394 and 2θ = 30.59°. The synthesized phosphor is orthorhombic with average crystallite size as 23.404 nm [11].

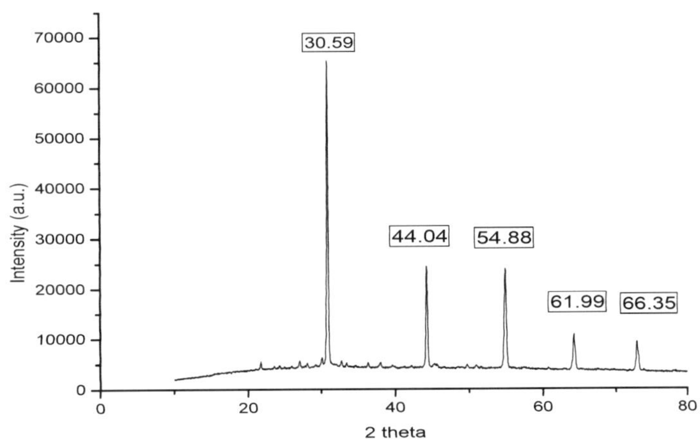

Figure 1. PXRD pattern of Eu^{3+}, Tb^{3+} doped $BaZrO_3$ phosphor.

3.2. SEM and EDXS analysis

In order to extract the topographical and compositional information of synthesized phosphor, SEM analysis is done using Oxford's SEM Analyser. The SEM micrographs show highly agglomerated clusters with spherical morphology of grain size nearly 68 nm (in Figure 2(a)). This confirms crystalline nature of prepared perovskite phosphor. The EDXS analysis also confirms its desired composition (in Figure 2(b))[12].

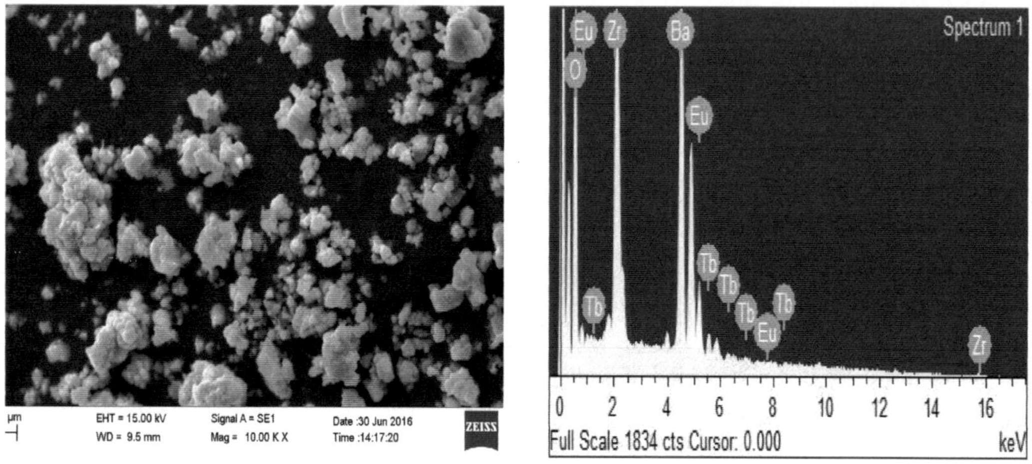

(a) (b)

Figure 2. (a) SEM micrograph and (b) EDXS spectrum of Eu^{3+}, Tb^{3+} doped $BaZrO_3$ phosphor.

3.3. FTIR analysis

The FTIR is done here for functional group analysis of the synthesized phosphor. In FTIR synthesized phosphor irradiated with infrared radiation and the respective vibrational modes are studied. The sharp peaks obtained at 505–597cm^{-1}, which correspond to vibrational stretching of Zr–O. The peak at 1466.73 cm^{-1}corresponds to existence of carbonate CO_3^{2-}group.The strong peak at 1126.67 cm^{-1} corresponds to Ba–O bonding (in Figure 3). The IR spectra confirm the formation of the $BaZrO_3$: Eu^{3+}, Tb^{3+}phosphor [13].

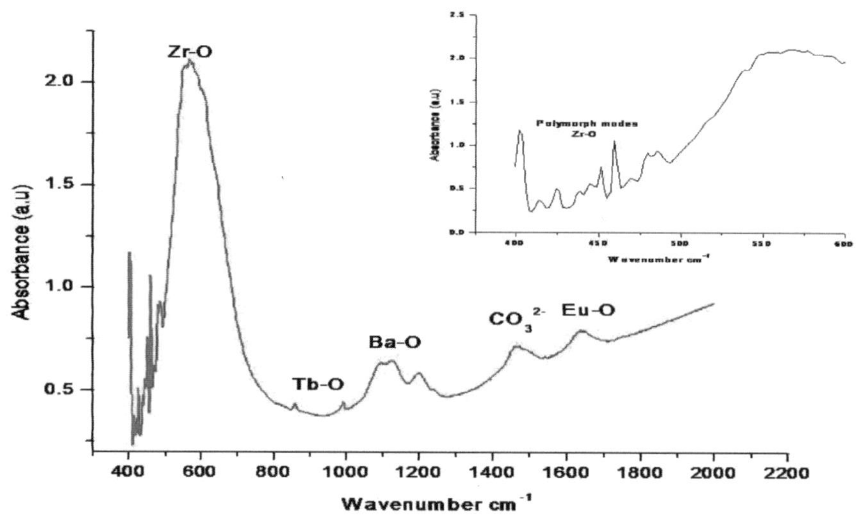

Figure 3. FTIR spectra of Eu^{3+}, Tb^{3+} doped $BaZrO_3$ phosphor.

3.4. Thermoluminescence characteristics analysis

The TL analysis is carried out here by recording TL thermograms of the prepared phosphors when subjected to 10 minutes UVC exposure of 254 nm. Thermoluminescence Reader TL10091 is used here for TL mapping. The heating strips are programmed to heat the samples with a constant heating rate of

5^0C/ sec up to a maximum temperature range of 500^0C. This analysis is done so as to estimate the various trapping parameters of the phosphors synthesized that contribute towards the observed TL emission. The kinetic parameters like activation energy E (eV), order of kinetics and trap depth are computed from the recorded glow curves using Peak shape method [14]. The most intense peak is obtained for Eu^{3+}(1 mol%) and Tb^{3+}(0.5 mol%) doped $BaZrO_3$ phosphor which is shown in Figure 4.

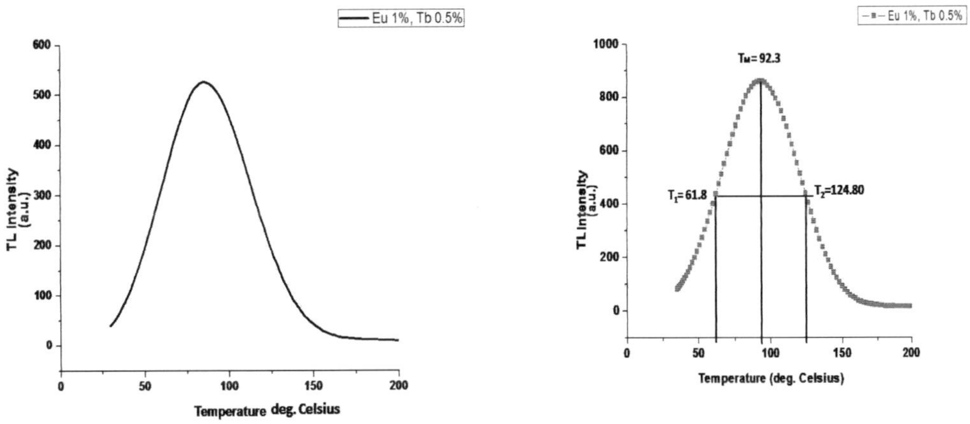

(a) (b)

Figure 4. (a) Recorded TL glow curve and (b) Deconvoluted TL glow curve of Eu^{3+}, Tb^{3+} doped $BaZrO_3$ phosphor.

The average kinetic parameters estimated using peak shape method are- τ =30.50, δ =32.50, ω = 63.0 and μ =0.5158 [15]. The synthesized phosphorsshow second order of kinetics with the average activation energy as 0.419 eV clearly indicating that the irradiated electrons get trapped in shallow traps with less trap depth and hence, require less activation energy to stimulate them resulting into TL emission.[16]

4. Conclusion

This work investigated the TLcharacteristic of Eu^{3+} and Tb^{3+} $BaZrO_3$ nanophosphor synthesized by Solid State Synthesis route. The synthesized phosphor reports nanocrystalline size of 23.404 nm with orthorhombic structure. The SEM analysis confirms spherical morphology of the particles forming agglomerated nanoclusters with grain size of 68 nm. The FTIR spectra confirm the compositional characteristics of the nanophor. The TL glow curves recorded here proves the nanophosphor appropriate for UVR exposure assessment with second order of kinetics and low value activation energy ***E*eff**.

References

[1] J. L. Zhang, J.E. Evetts, *J. Mater. Sci.* **29** (1994) pp 778-85
[2] S. Katyayan, S. Agrawal, Materials today: proceedings 4(8):8016-8024, doi:10.1016/j.matpr.2017.07.139
[3] S. Gopalan, A.V. Virkar, *J. Am. Ceram. Soc.* **82** (1999) pp 2887-89
[4] R. Vassen, X. Cao, F. Tietz, D. Basu, D. Stover, *J. Am. Ceram. Soc.* **3** (2000) pp 2023-28
[5] S. Katyayan, S. Agrawal, Journal of Materials Science Materials in Electronics, doi: 10.1007/s10854-017-7791-8
[6] H.G. Bohn, T. Schober, *J. Am. Ceram. Soc.* **83** (2000) pp 768-72
[7] M.Viviani, M.T. Buscaglia,V. Buscaglia, M. Leoni, P. Nanni, *J. Eur. Ceram. Soc.* **21** (2001) pp 1981-84

IOP Conf. Series: Materials Science and Engineering **383** (2018) 012004 doi:10.1088/1757-899X/383/1/012004

[8] A.M. Azad, S. Subramaniam, *Mater. Res. Bull.* **37** (2002) pp 11-21
[9] A. Erb, E. Walker, R. Fl¨ukiger, *Physica C* **245** (1995) pp 245-51
[10] D. Ravichandran, R. Roy, P. Ravindranathan, W.B. White, *J. Am. Ceram. Soc.* **82** (1999) pp 1082-84
[11] A.A. Athawale, A.J. Chandwadkar, P. Karandikar, R. Pasricha, M.S. Bapat, *Radiat. Phys. Chem.* **75** (2006) pp 755-59
[12] J.J. Kingsley, K.C. Patil, *Mater. Lett.* **6** (1988) pp 427-32
[13] P. Pramanik, *Bull. Meter. Sci.* **22** (1999) pp 335-39
[14] H. Stetson, B. Schwartz, *J. Am. Ceram. Soc.* **44** (1961) pp 420-21
[15] S. Katyayan, S. Agrawal, Journal of Materials Science Materials in Electronics, doi: 10.1007/s10854-017-8156-z
[16] I. Levin, T.G. Amos, S.M. Bell, L. Farber, T.A. Vanderah, R.S. Roth, B.H. Toby, *J. Solid State Chem.* **175** (2003) pp 170-81

Phase evolution criteria for AlCoCrCuFeMnSix (x = 0, 0.3, 0.6 and 0.9) high entropy alloys based on experiment and thermodynamic calculation

Anil Kumar, Anant Rajimwale and Manoj Chopkar*
Department of Metallurgical Engineering, National Institute of Technology, Raipur, Chhattisgarh INDIA

*Email: mchopkar.met@nitrr.ac.in

Abstract. Present study investigates the phase evolution behaviour of AlCoCrCuFeMnSix (x = 0, 0.3, 0.6 and 0.9) high entropy alloys, synthesized by mechanical alloying and arc melting. The phase evolution, microstructure, and mechanical properties were analysed to understand the effect of Si addition in AlCoCrCuFeMn high entropy alloy. The X-ray diffraction, after 15h mechanical alloying, reveals the mixture of body-centered and face-centered cubic structure. The addition of Si favours body-centered cubic structure formation during mechanical alloying. The phase evolution observed after XRD are in accord with the thermodynamically predicted phases.

1. Introduction

High Entropy Alloys or HEAs for short, are alloying frameworks made out of no less than 5 components. The idea of HEAs was presented by Achard in the late of eighteen century, further, in the mid of 1900s, Yeh and his partners investigated the universe of multi-component alloy [1]. The extraordinary capacity of HEAs to form often single solid solution with their incredible properties, for example, wear resistance, high hardness, softening resistance at higher temperature, corrosion and oxidation resistance have made tremendous enthusiasm among scientists.

Previous literatures suggests that researchers mainly focused on the improvement of mechanical properties of HEAs by varying the content of any one transition element of the composition. Though, limited studies have been reported the effect of non-metallic element such as Si addition and their influence on the phase evolution and mechanical properties. As Si is always favourable for structural and mechanical properties in conventional materials like steel. Liu et al., [2] observed that increasing the Si content on $Al_{0.5}CoCrCuFeNiSi_x$ (x = 0, 0.4, 0.8) endorses the BCC phase evolution compared to FCC phase. A substantial improvement in wear resistance was reported with increase in Si contents because of high negative enthalpy of mixing between Si and other elements of composition. Kumar et al.,[3] also observed that, with increase in amount of Si from 0 to 0.9, facilitates the BCC phase formation over FCC phase. Therefore, in the present study, the effect of Si on the phase evolution characteristic of AlCoCrCuFeMnSix (x = 0, 0.3, 0.6, 0.9) HEAs, synthesized by mechanical alloying and arc melting were investigated.

Since the principal reports of high entropy alloys (HEAs), various endeavours have been made to foresee phase evolution in these alloys, specifically to predict compositions, which would support

Content from this work may be used under the terms of the Creative Commons Attribution 3.0 licence. Any further distribution of this work must maintain attribution to the author(s) and the title of the work, journal citation and DOI.
Published under licence by IOP Publishing Ltd

IOP Conf. Series: Materials Science and Engineering **383** (2018) 012005 doi:10.1088/1757-899X/383/1/012005

evolution of solid solution phases. Despite the fact that advancement of alloys based on the multiple elements with equiatomic or near to equiatomic ratio without a doubt broadened the perspective of materials researchers on the plan of new materials, the role of entropy for the formation of solid solution not be over-underlined. Initially, it is suggested that the higher entropy of mixing results into the formation of solid solution phase [4] as expressed in equation (1).

$$(\Delta S_{mix}) = -R\sum_{i=1}^{n} C_i lnC_i \quad (1)$$

Where R is the gas constant (8.314 J/ mol. K), C_i is the atomic fraction of element i and n is the number of components. From the above equation, the high ΔS_{mix} can significantly lower the free energy, in this way bringing down the affinity to order and segregate, which accordingly makes solid solution more effectively form and more stable than intermetallic. However, disregarding having high ΔS_{mix} values, many HEAs are multi-phase and can likewise contain IM phases [5].

A few extra-empirical criteria for anticipating solid solution phase formation in HEAs were presented as of late [6]. Like Hume-Rothery rules for substitutional solid solutions, these criteria concentrate on the atomic size difference (δ), mixing enthalpy (ΔHmix), electronegativity ($\Delta\chi$), valence electron concentration (VEC) and a parameter Ω were also used to develop criteria for the stability of SS, IM and/or amorphous phases in as-cast HEAs. Equations reported by Dong et al., [7] and Zhang et al., [8] has been used to calculate the design parameters.

Another important parameter formulated by Mansoori et al., [9] based on the mismatch entropy for bulk metallic glasses was also used in the present study for phase formation prediction and is given as (equation (2)).

$$\frac{\Delta S_\sigma}{k} = \left[\frac{3}{2}(\zeta-1)y_1 + \frac{3}{2}(\zeta-1)^2 y_2 - \left\{\frac{1}{2}(\zeta-1)(\zeta-3) + ln\zeta\right\}(1-y_3)\right] \quad (2)$$

To verify all these design parameters for phase prediction criterion, a series of AlCoCuCrFeMnSix (x = 0, 0.3, 0.6 and 0.9) HEAs were prepared through MA. The calculated values of these design parameters are listed in table 2. The aim of the present study is to analyze the effect of Si on AlCoCrCuFeMn HEAs and to confirm the phase evolution based on the previously developed thermodynamic calculation for the HEA.

2. Materials and methods

The elemental powders of Al, Co, Cu, Cr, Fe, Ni, and Si were mechanically alloyed for 15 h to prepare HEAs. High energy ball mill (HEMB) (Retsch: PM-400) was used for milling the powder. The ball to powder ratio was 10:1 with 250 ml capacity tungsten carbide vials and 10 mm diameter balls. The rotation speed of 300rpm was set throughout the milling process. Phase evolution are characterized through X-ray diffractometer (XRD, PANAlytical, X' pert powder) with Cu Kα radiation. Finally, the experimental results were compared with the thermodynamic parameters.

Pseudo-Voigt function was used for fitting XRD peak profile. The crystallite sizes and lattice strain of the milled powders were calculated based on Scherrer's formula [10] from XRD peaks and Si was used as the standard sample to subtract the instrumental broadening.

3. Results and discussions

Figure 1 shows the XRD peaks obtained during mechanical alloying of AlCoCuCrFeMnSix powders, as a function of milling time. For as milled samples, the crystalline peaks for principal elements are observed before 5 h. As the milling time increases, the intensities of XRD peaks of pure constitutional elements decreases, which indicates dissolution of pure elements. In addition, further milling of the powder leads to overlapping of peaks to each other compared with the beginning of the milling. From the diffraction pattern it is evident that, after 15 h of milling, mixture of FCC and BCC phase were obtained.

The decrease in intensities and disappearance of most of the XRD peaks by increasing the milling time might be due to the crystal size refinement, higher lattice strain and the formation of solid solution[11].

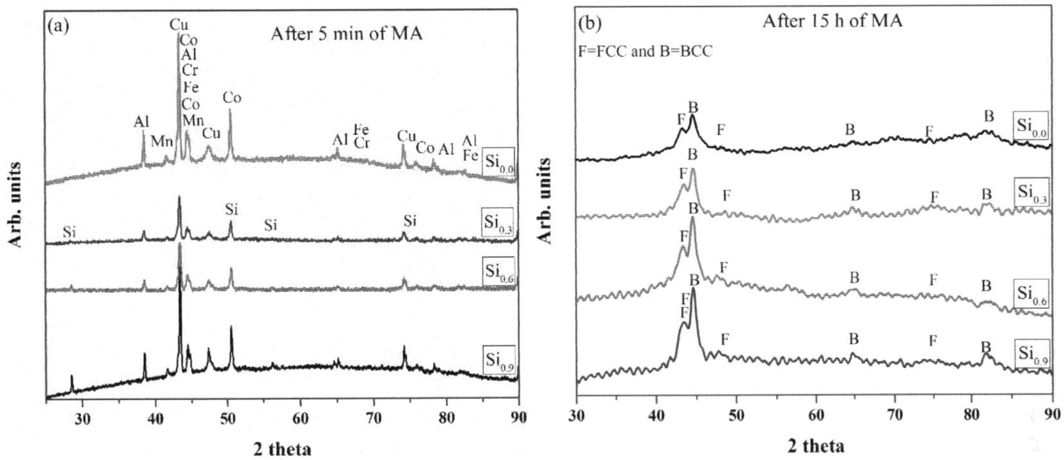

Figure 1. XRD patterns of (a) AlCoCrCuFeMn and (b) $AlCoCrCuFeMnSi_{0.3}$ (c) $AlCoCrCuFeMnSi_{0.6}$ (d) $AlCoCrCuFeMnSi_{0.9}$ mechanical alloyed powders at different time intervals.

The crystallite size and lattice strain are illustrated in Table 1. The increase in milling time during mechanical alloying, contributed to decrease in crystallite size of BCC phase, reaching 24, 21, 20 and 16 nm for the value of x = 0, 0.3, 0.6 and 0.9 respectively. No significant change observed in the crystalline size with increase in milling time from 10 h to 15 h. The regular increase in the lattice strain observed with increase in the milling time, which have constructive impact on the solid solution strengthening. The increase in lattice strain is attributed to (1) mismatch effect of size between constituent elements, as Si has small atomic radius compared to other elements in the compositions, (2) increase in grain boundary fraction and (3) mechanical deformation [12,13]. Growth in dislocation density due to severe plastic deformation also contributes to higher lattice strain [14].

Table 1. Crystallite size (CS) and lattice strain (LS) of BCC phase in AlCoCrFeNiSix (x = 0, 0.3, 0.6, 0.9) high entropy powder with milling time.

Milling	AlCoCrCuFeMn		$AlCoCrCuFeMnSi_{0.3}$		$AlCoCrCuFeMnSi_{0.6}$		$AlCoCrCuFeMnSi_{0.9}$	
time (h)	CS (nm)	LS (%)	CS (nm)	LS (%)	CS (nm)	LS (%)	CS (nm)	LS (%)
5	82	0.89	79	1.12	77	1.18	72	1.51
10	24	2.70	21	2.84	20	2.91	16	2.97
15	25	2.81	21	2.90	21	2.94	16	2.99

To verify the phase evolution during mechanical alloying, mixing entropy (ΔS_{Mix}), mixing enthalpy (ΔH_{Mix}), valence electron concentration (VEC), atomic size mismatch (δ), electronegative difference ($\Delta\chi$), Ω and mismatch entropy $\Delta S_{\sigma}/k$ in HEA according to Hume- Ruthery rule have been calculated and shown in Table 2.

Guo et al. investigated the effect of valence electron concentration (VEC) for the solid solution formation on HEA. They observed that BCC phase will be stable at VEC$\leq$6.8 whereas, for 6.8$\leq$VEC$\leq$8 mixed phases of BCC and FCC occurs [15]. The above-discussed phase prediction criteria based on the VEC is in accord with the alloys studied in this work. The criteria given by Yang et al. based on the atomic size mismatch (δ) for the prediction of stabilized solid solution ($\delta \leq$6.6%) also satisfied in the present studied alloys. Furthermore, the range of $\Delta\chi$ values for the prediction of solid solution phase formation calculated by Guo and Liu also satisfied by alloys prepared in this study [16].

IOP Conf. Series: Materials Science and Engineering **383** (2018) 012005 doi:10.1088/1757-899X/383/1/012005

Liu and Yang et al defined a new parameter Ω to classify solid solution phases and intermetallic, they observed that, if $\Omega > 1$, the solid solution formation will be more and if $\Omega \leq 1$ intermetallic compound formation occurs. This criteria is also satisfied in the present alloys.

Raghvan et al.[17] suggested that the BCC phase is favoured than the FCC phase for the high value of $\Delta S_\sigma/k$ and correspondingly less value of ΔH_{mix}. Intermediate values of $\Delta S_\sigma/k$ and ΔH_{mix} were accepted for mixture of BCC and FCC phases.

Table 2. Calculated parameters ΔHmix, ΔSmix, VEC, δ, Δχ, Ω, $\Delta S_\sigma/k$, $\sqrt{\langle\varepsilon^2\rangle}$ and $\sqrt{u_e}$ for prepared HEAs.

Alloys	ΔS_{Mix}	ΔH_{Mix}	VEC	δ %	Δχ	Ω	$\Delta S_\sigma/k$	$\sqrt{\langle\varepsilon^2\rangle}$	$\sqrt{u_e}$
AlCoCrCuFeMn	14.89	-3.77	7.3	5.33	0.1372	6.28	0.047	0.053	0.048
$AlCoCrCuFeMnSi_{0.3}$	15.77	-9.26	7.1	5.77	0.1381	2.72	0.054	0.057	0.052
$AlCoCrCuFeMnSi_{0.6}$	16.07	-13.75	7.0	6.10	0.1386	1.87	0.060	0.061	0.054
$AlCoCrCuFeMnSi_{0.9}$	16.17	-17.44	6.8	6.36	0.1389	1.48	0.070	0.063	0.055

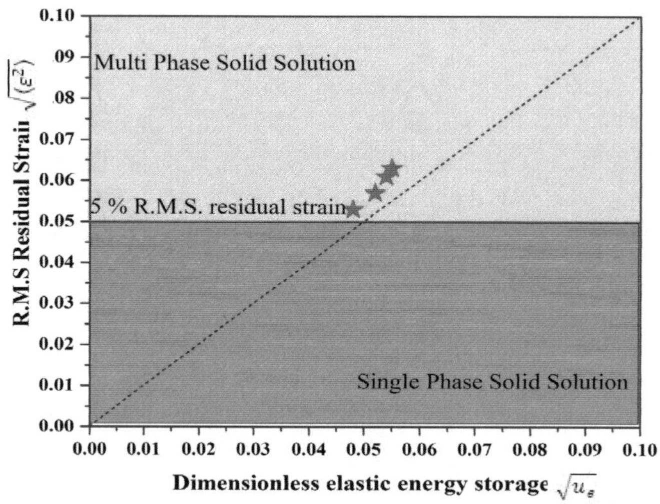

Figure 2. The R.M.S. residual strain and dimensionless elastic energy storage.

Additionally, to verify the phase formation in HEAs after mechanical alloying we have also implemented newly proposed geometric model based on the atomic stress theory proposed by Ye et al., [18]. Ye et al. considered the root mean square residual strain ($\sqrt{\varepsilon^2}$) and square root of dimensionless elastic energy storage ($\sqrt{u_e}$) to classify single phase, multi-phase for as-cast HEAs and compared results with experimental observations. First time, an effort has been made in the present study to extend the model proposed by Ye et al. on the high entropy alloys prepared by mechanical alloying process (Table 2).

Ye et al. extended the Lindemann benchmarks to single and multi-phase alloys. It is claimed that at 5% residual strain, single phase solid solution converts into multi-phase solid solution. During this conversion, all the strain is not released. The extension of Ye et al. theory, which was proposed for HEA alloys prepared by casting route also holds good for HEAs prepared by mechanical alloying route (Figure 2).

4. Conclusion

The $AlCoCrCuFeMnSi_x$ HEA alloys with different Si content (x = 0, 0.3, 0.6 and 0.9) were successfully synthesized via mechanical alloying. During MA, solid solution mixture of BCC and FCC structure formed after 15h of milling. The valence electron concentration, mixing enthalpy, atomic size difference, electronegativity and mismatch entropy play significant role to control FCC and BCC phase formation. It is important to point out that calculation of design parameter for composition under consideration are in good accord with the reported data in literature for solid solution formation. The relationship between experimental data and calculated data is in agreement with model based on the R.M.S. residual strain and dimensionless elastic energy storage. This correlation suggests that conversion from single phase to multi-phase structure occurs at around ~ 5% R.M.S. residual strain.

References

[1] Murty BS, Yeh JW, Ranganathan S (2014) A Brief History of Alloys and the Birth of High-Entropy Alloys. *High Entropy Alloy* pp 1-12

[2] Liu X, Lei W, Ma L, et al (2015) On the microstructures, phase assemblages and properties of Al0.5CoCrCuFeNiSix high-entropy alloys. *J Alloys Compd* **630**:pp 151-57

[3] Kumar A, Dhekne P, Kumar A, Kumar M (2017) Analysis of Si addition on phase formation in AlCoCrCuFeNiSix high entropy alloys. *Mater Lett* **188**: pp 73-76

[4] Yeh J-W, Lin S-J, Chin T-S, et al (2004) Formation of simple crystal structures in Cu-Co-Ni-Cr-Al-Fe-Ti-V alloys with multi-principal metallic elements. *Metall Mater Trans A* **35**: pp 2533-36

[5] Porter DA, Easterling KE (1992) Phase Transformations in Metals and Alloys. Structure. Doi: 10.1007/978-1-4899-3051-4

[6] Guo S (2015) Phase selection rules for cast high entropy alloys: an overview. *Mater Sci Technol* **31**: pp 1223-30

[7] Dong Y, Lu Y, Kong J, et al (2013) Microstructure and mechanical properties of multi-component AlCrFeNiMo x high-entropy alloys. *J Alloys Compd* **573**:pp 96-101.

[8] Yang X, Zhang Y (2012) Prediction of high-entropy stabilized solid-solution in multi-component alloys. *Mater Chem Phys* **132L** pp 233-38

[9] Mansoori G a.m Carnahan NF, Starling KE, Leland TWJ (1971) Equilibrium Thermodynamic Properties of the mixture of Hard Spheres. *J Chem Phys* **54**: p 1523

[10] Varalakshmi S, Kamaraj M, Murty BS (2008) Synthesis and characterization of nanocrystalline AlFeTiCrZnCu high entropy solid solution by mechanical alloying. *J Alloys Compd* **460**: pp 253-57

[11] Zhang KB, Fu ZY, Zhang JY, et al (2010) Characterization of nanocrystalline CoCrFeNiTiAl high-entropy solid solution processed by mechanical alloying. *J Alloys Compd* **495**: pp 33-38

[12] Sui HX, Zhu M, Qi M, et al (1992) the enhancement of solid solubility limits of AlCo intermetallic compound by high-energy ball milling. *J Appl Phys* **71**: pp 2945-49

[13] Fu Z, Chen W, Xiao H, et al (2013) Fabrication and properties of nanocrystalline Co0.5FeNiCrTi0.5 high entropy alloy by MA-SPS technique. *Mater Des* **44**: pp 535-39

[14] Graner F, Glazier J a (1992) Mechanically Driven Alloying of Immiscible Elements. *Phys Rev Lett* **69**: pp 2013-16

[15] Guo S, Ng C, Lu J, Liu CT (2011) Effect of valence electron concentration on stability of fcc or bcc phase in high entropy alloys. *J Appl Phys*

[16] Guo S, Lin CT (2011) Phase stability in high entropy alloys: Formation of solid-solution phase or amorphous phase. *Prog. Nat Sci Mater Int* **21**: pp 433-46

[17] Raghavan R, Hari Kumar KC, Murty BS (2012) Analysis of phase formation in multi-component alloys. *J Alloys Compd* **544**: pp 152-58

[18] Ye YF, Liu CT, Yang Y (2015) A geometric model for intrinsic residual strain and phase stability in high entropy alloys. *Acta Mater* **94**: pp 152-61

Temperature-dependent electrical transport characteristics of p-SnS/n-WO_3:Sb heterojunction diode

T Gaewdang[*] and Ng Wongcharoen

Physics Department, King Mongkut's Institute of Technology Ladkrabang, Bangkok 10520, Thailand

*Email: thitinai.ga@kmitl.ac.th

Abstract. The p-SnS/n-WO_3:Sb heterojunction diode was successfully obtained by thermal evaporating SnS thin films on WO_3 doped with 2.0 mol% Sb_2O_3 of 1 mm thick ceramic pellet substrate. The electrical properties of p-SnS/n-WO_3:Sb heterojunction were investigated by forward current–voltage–temperature (I–V–T) characteristics in the temperature range of 20-300 K. It was found that at low forward bias voltage (<0.25V), the conduction mechanism of the diode exhibits Ohmic conduction. At intermediate voltage (0.25<V<1.0V), the conduction mechanism is dominated by thermionic emission (TE). At bias voltage above 1.0 V, the current transport is due to space charge limited current (SCLC) controlled by an exponential trap distribution in the band gap of WO_3:Sb. The temperature dependence of the saturation current and ideality factor are well described by tunneling enhanced recombination model occurring at junction interface with E_a and E_{00} values about 1.565 eV and 99.5 meV, respectively.

1. Introduction

Tungsten oxide (WO_3) is of intense interest and has been investigated extensively for its distinctive properties. With outstanding electrochromic, photochromic, gas sensor, photo-catalyst and photoluminescence properties, tungsten oxide has been used to construct smart-window, anti-glare rear view mirrors for automobiles, non-emissive displays, optical recording devices, solid-state gas sensors, humidity and temperature sensors, biosensors, photonic crystals and so forth [1]. SnS is an important II-VI group chalcogenide semiconductor because of their potential applications in the fabrications of photovoltaic, solar cells and optoelectronic devices [2]. Heterostructures implicated two different member of II-VI compounds have received much attention. Many of the heterostructures have now become important part of electronic and optoelectronic devices. Recently, various types of heterostructure based on SnS films such as CdS/SnS, SnS_2/SnS, SnS/SnO_2, ZnO/SnS and SnS/graphene were investigated. In this paper, the formation of p-SnS/n-WO_3:Sb heterojunction elaborated by thermal evaporating SnS thin films on WO_3:Sb ceramic substrate was obtained. We investigated I-V characteristics of p-SnS/n-WO_3:Sb heterojunction at various temperatures for evaluating some important parameters such as junction barrier height, ideality factor and series resistance values by using thermionic emission theory and Cheung's method. The carrier concentration, at room temperature, could be obtained from the slope of $1/C^2$-V plot.

2. Materials and methods

Content from this work may be used under the terms of the Creative Commons Attribution 3.0 licence. Any further distribution of this work must maintain attribution to the author(s) and the title of the work, journal citation and DOI.
Published under licence by IOP Publishing Ltd

Polycrystalline powder of WO_3 doped with 2 mol% Sb_2O_3 was ground in an agate mortar thoroughly for 2 h and then pressed into pellet form by hydrostatic pressure about 100 MPa. Then, the WO_3:Sb pellet was sintered in air at 1000^oC for 48 h. The as-sintered WO_3:Sb pellet with 10 mm in diameter and 1 mm in thickness was used as substrate for SnS thin film deposition by thermal evaporation technique in a vacuum of $3x10^{-5}$ mbarr using a tungsten boat as source holder. The thickness of the SnS film using a digital quartz crystal monitoring system (FTM6) and was found to be about 1,000 nm. Silver electrodes were used as electrical contacts on SnS and WO_3:Sb substrate. The I-V characteristics were measured at various temperatures by using a computer interfaced Keithley 236 current/voltage source. The C-V measurements were performed by using Agilent E4980A precision LCR meter at frequency 100 kHz.

3. Results and discussion

The measured I-V plot of p-SnS/n-WO_3:Sb heterojunction at room temperature is shown in Figure 1. This I-V characteristic confirm that a p-n heterojunction diode is formed at the interface of the p-SnS thin films and n-WO_3:Sb pellet. According to the thermionic emission (TE) theory, the current in Schottky diode can be expressed as [3]

$$I = I_s\left\{\exp\left[\frac{q(V-IR_s)}{nkT}\right]-1\right\} \tag{1}$$

where q is the unit charge, n is the ideality factor, k is the Boltzmann's constant, T is the absolute temperature, R_s is the series resistance and I_S is the saturation current defined by

$$I_s = AA^*T^2\exp\left(-\frac{q\phi_b}{kT}\right) \tag{2}$$

where A is the device area, A* is the Richardson's constant and Φ_b is the zero bias barrier height of the diode. The variation of lnI with V of p-SnS/n-WO_3:Sb heterojunction at room temperature (Figure 2) was found to be linear suggesting that the thermionic emission is the dominant conduction mechanism at low bias voltage. The diode ideality factor 'n' was calculated from the slope of the forward bias lnI-V plot through the relation:

$$n = \left(\frac{q}{kT}\right)\left(\frac{dV}{d\ln I}\right) \tag{3}$$

n=1 for an ideal diode. However, n has usually a value greater than unity. On the other hand, the barrier height can be obtained from the following expression:

$$\phi_b = \left(\frac{kT}{q}\right)\ln\left(\frac{AA^*T^2}{I_s}\right) \tag{4}$$

The values of the ideality factor and barrier height were found to be n = 15.383 and Φ_b = 0.520 eV at room temperature. Figure 3 shows lnI-lnV curve of p-SnS/n-WO_3:Sb heterojunction performed at room temperature. At low bias voltage region (region I), current was linearly dependent on voltage V exhibiting an Ohmic behavior due to its slope was nearly equal 1. On the other hand, at intermediate voltage region (region II), a non- linear region appeared according to thermionic emission (TE) mechanism. As applied bias voltage higher than 1.0 V (region III), the strong injection of space charge limited current (SCLC), with its slope nearly 2.5, was observed. The value of R_s was calculated from the forward bias I-V data using the Cheung's method [4]. R_s value can be obtained from the following equation:

IOP Conf. Series: Materials Science and Engineering **383** (2018) 012006 doi:10.1088/1757-899X/383/1/012006

$$\frac{dV}{d\ln I}=\left(\frac{nkT}{q}\right)+IR_s \qquad (5)$$

Experimental dV/d(lnI)-I at room temperature is presented in Figure 4. Equation (5) should give a straight line for the data of downward curvature region in the forward bias I-V characteristics. Thus, a plot of dV/d(lnI)-I will give R_S as the slope and nkT/q as the y-axis intercept. The I-V, lnI-V, lnI-lnV and dV/d(lnI)-I plots of p-SnS/n-WO_3:Sb heterojunction at different temteratures are shown in Figure 5, 6, 7 and 8, respectively. The obtained n, Φ_b and R_S were found to be strongly temperature dependent. Ideality factor greater than 1 has been attributed to particular distribution of interfacial layer and interface states between SnS and WO_3:Sb. As can be seen from Figure 9, the value of ideality factor showed that there was a deviation from TE theory in the current conduction mechanism. The temperature dependence of n suggested that the current conduction mechanism was controlled by the surface states were more pronounced because of the spatial distribution of the interfacial layer of SnS and WO_3:Sb. A new approach explaining the tunneling enhancement of recombination via deep centers in the space charge region or at junction interface was reported for the forward current transport in CdS/Cu(In,Ga)Se_2, CdS/CdTe and Cu_2O/ZnO solar cells [5, 6]. Rearranging equation (1) yields the relationship

$$n\ln J_s=\left(\frac{E_a}{kT}\right)+n\ln J_{00} \qquad (6)$$

Thus, a plot of $n\ln J_s$ versus 1/kT should yield a straight line where the slope provided activation energy (E_a). The E_a deduced from Figure 10 was 1.565 eV which smaller than energy gap of WO_3 (E_g = 2.76 eV). Thus, it is reasonable to suggest that the E_a corresponds to the potential barrier as seen by hole. Furthermore, tunneling of holes from the bulk of SnS into the interface states and subsequent recombination with electrons available in WO_3:Sb layer led the temperature dependence of the diode ideality factor as

$$n=\left(\frac{E_{00}}{kT}\right)\coth\left(\frac{E_{00}}{kT}\right) \qquad (7)$$

where E_{00} is the characteristic tunneling energy measuring the amount of tunneling contribution to the recombination process. From Figure 9, the large value of E_{00} about 99.5 meV suggested that tunneling enhancement recombination via deep centers at junction interface provided analytical expression for the forward current transport in our diode.

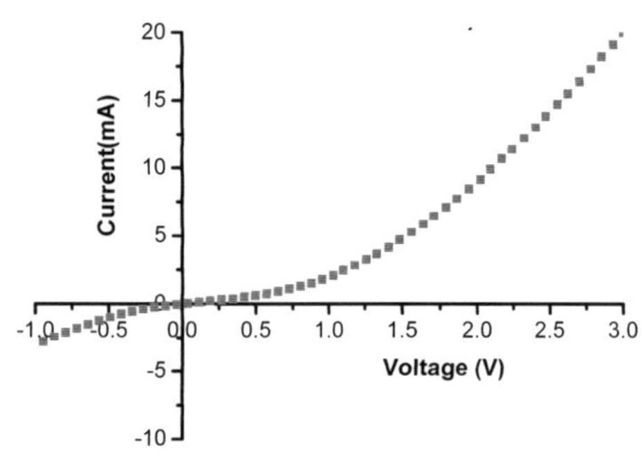

Figure 1. I-V plot of p-SnS/n-WO_3:Sb heterojunction measured at 300 K.

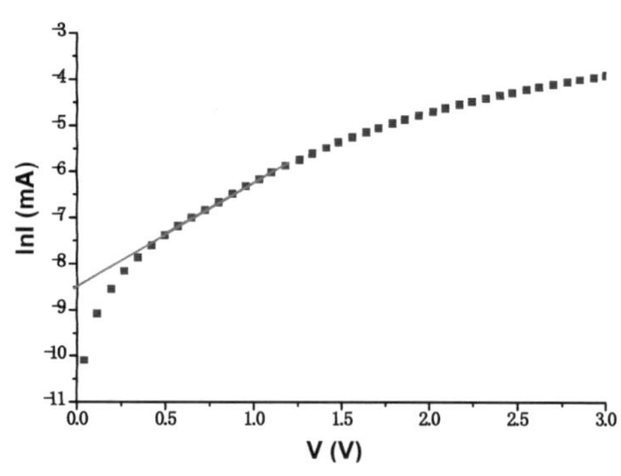

Figure 2. lnI – V plot of p-SnS/n-WO_3:Sb heterojunction measured at 300K.

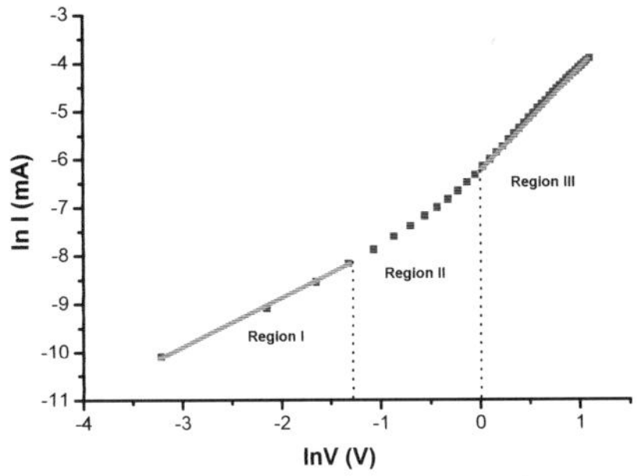

Figure 3. lnI –lnV plot of p-SnS/n-WO_3:Sb heterojunction at 300 K.

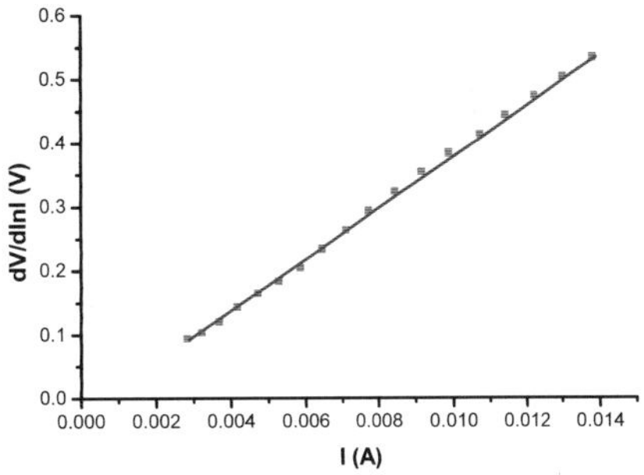

Figure 4. dV/d(lnI)-I plot of p-SnS/n-WO_3:Sb heterojunction at 300 K.

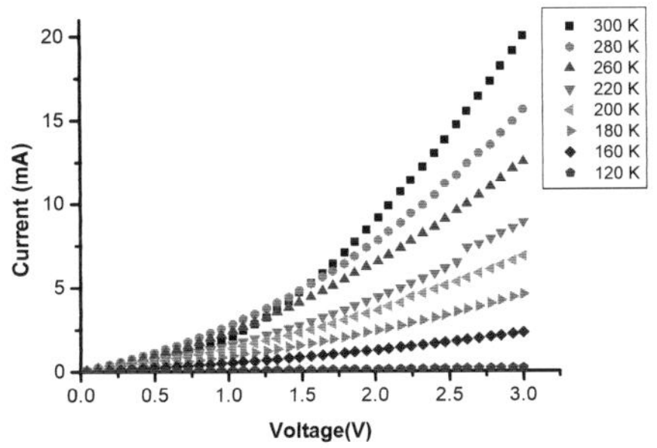

Figure 5. I-V plots of p-SnS/n-WO_3:Sb heterojunction at various temperatures.

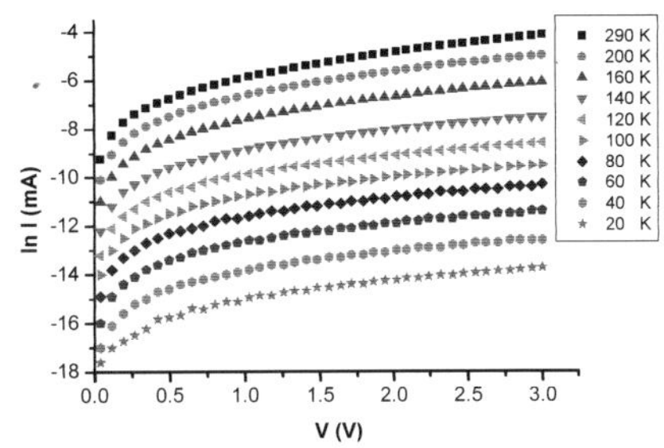

Figure 6. lnI-V plots of p-SnS/n-WO_3:Sb heterojunction at various temperatures.

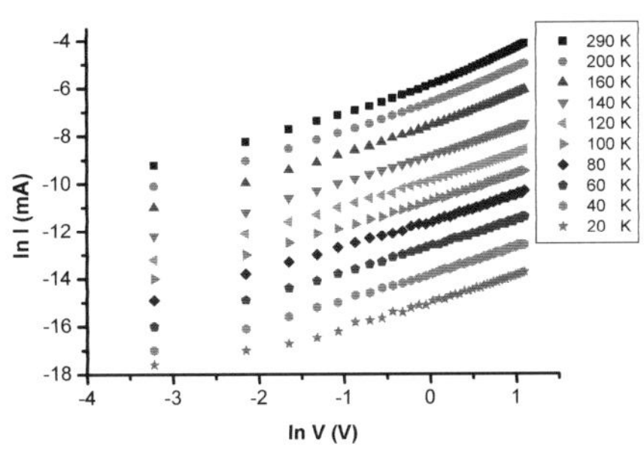

Figure 7. lnI-lnV plot of p-SnS /n-WO_3:Sb heterojunction at various temperatures.

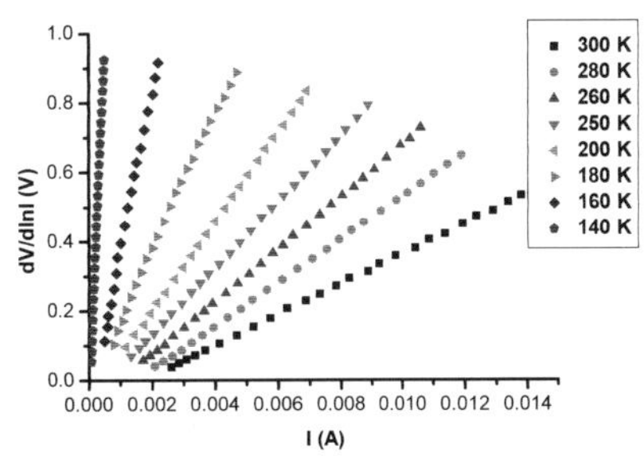

Figure 8. dV/d(lnI)-I plots of p-SnS/n-WO_3:Sb heterojunction at various temperatures.

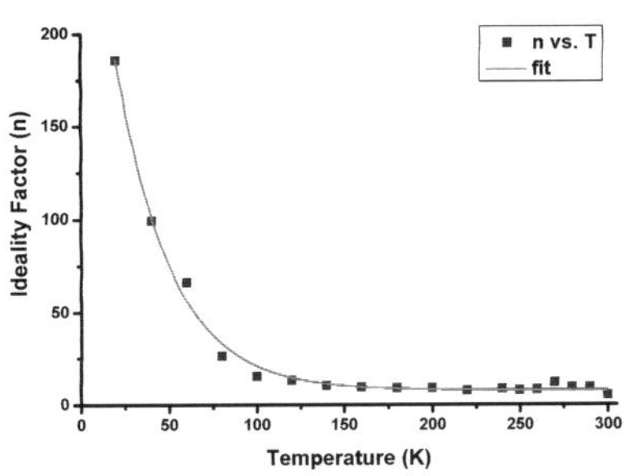

Figure 9. n vs. T of p-SnS/n-WO_3:Sb heterojunction.

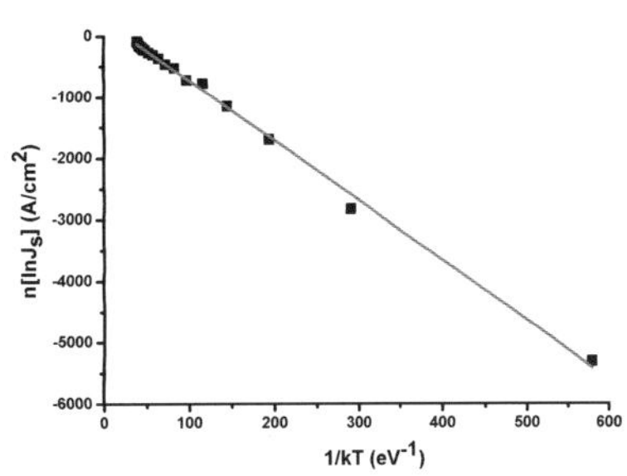

Figure 10. nlnJ_s vs. 1/kT of p-SnS/n-WO_3:Sb heterojunction.

In the simple Schottky barrier theory, the capacitance of the barrier varies with applied reverse

IOP Conf. Series: Materials Science and Engineering **383** (2018) 012006 doi:10.1088/1757-899X/383/1/012006

voltage V in such a way that a straight line is obtained if $1/C^2$ is plotted against V as show in Figure 11. This relation is given by the expression:

$$\frac{1}{C^2} = \left(\frac{2}{\varepsilon_s q N_d A^2}\right)\left(V_{bi} - \frac{kT}{q} - V\right) \tag{8}$$

where V_{bi} is the built-in potential, A is the effective area of the diode, ε_S is the dielectric constant of the semiconductor (ε_S = 12.1 for WO_3) and N_d is the concentration of ionized donor. Electron concentration of WO_3:Sb layer around 4.46×10^{14} cm^{-3} at 300 K evaluated from the slope of $1/C^2$-V plot. Based on these data, energy band diagram for p-SnS/n-WO_3:Sb heterojunction could be proposed as shown in Figure 12.

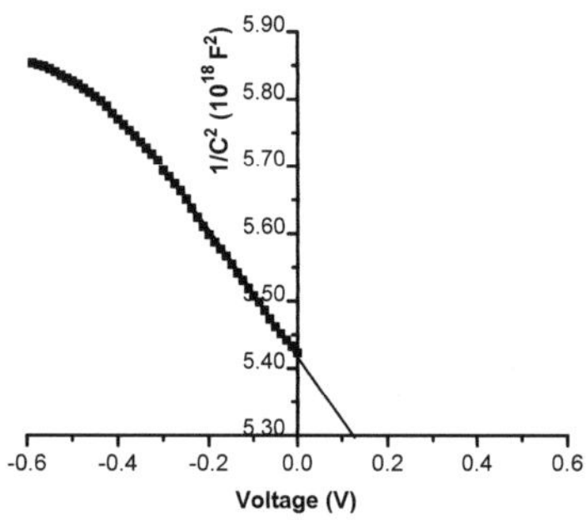

Figure 11. $1/C^2$ – V plot of p-SnS/n-WO_3:Sb heterojunction at 300 K.

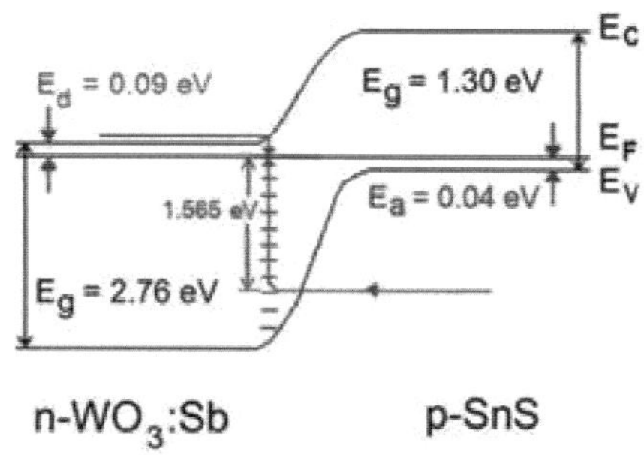

Figure 12. Proposed energy band diagram of p-SnS/n-WO_3:Sb heterojunction.

4. Conclusions

The p-SnS/n-WO_3:Sb heterojunction were successfully prepared by thermal evaporating SnS thin films on WO_3:Sb ceramic pellet substrate. The electrical conduction in the diode was explained by thermionic emission at intermediate bias voltage and by space charge limited current at high bias voltage. The temperature depencence of the saturation current and ideality factor are well described by tunneling enhanced recombination model occuring at junction interface with the activation energy and the characteristic energy values as about 1.565 eV and E_{00} = 99.5 meV, respectively. The variation of junction capacitance with reverse bias voltage was studied and the electron concentration of WO_3:Sb ceramics of 4.46×10^{14} cm^{-3} were determined from the slope of the $1/C^2$ - V plot.

References

[1] Diaz-Reyes J, Castillo-Ojeda R, Galvan-Arellano M and Zaca-Moran O 2013 *Adv. Cond. Matter Phys.* **ID 591787** p1

[2] El-Nahass M M, Zeyada H M, Aziz M S and El-Ghamaz N A 2002 *Opt. Mater.* **20** p159

[3] Rhoderick E H and Williams R H 1988 *Metal-Semiconductor Contacts* (Oxford University Press, 2nd ed) pp 21-58

[4] Cheung S K and Cheung N W 1986 *Appl. Phys. Lett.* **49(2)** p 85

[5] Bayhan H and Kavasoglu A S 2005 *Solid State Electron.* **49** p 991

[6] Jeong S H, Song S H, Nagaich K, Cambell A and Aydil E S 2011 *Thin Solid Films* **519** p 6613

IOP Conf. Series: Materials Science and Engineering **383** (2018) 012007 doi:10.1088/1757-899X/383/1/012007

Effect of source temperature on the electrical transport, microstructural and optical properties of tin monosulfide thin films prepared by close spaced sublimation method

Ng Wongcharoen* and T Gaewdang

Physics Department, King Mongkut's Institute of Technology Ladkrabang, Bangkok 10520, Thailand

*Email: ngamnit.wo@kmitl.ac.th

Abstract. Tin monosulfide (SnS) thin films were prepared by close spaced sublimation method at three different source temperatures of 550, 575 and 600°C. Structural, textural and compositional features were investigated by XRD, SEM, EDS and Raman spectroscopy. From absorption spectra, direct band gap increased from 1.18 to 1.23 eV with increasing source temperature. Electrical resistivity, Hall constant, carrier concentration and mobility values of the films prepared at source temperature of 600°C were 0.54 Ω.cm, 40.76 cm^3/C, $1.53x10^{17}$ and 75.48 $cm^2/(V.s)$, respectively. The temperature-dependent conductivity was performed in the temperature range of 50-300 K. It was shown that three types of conduction mechanisms can be expected: thermionic emission passing through grain boundary at high temperature range (240-300 K), the Mott variable-range hopping (Mott-VRH) at low temperature range (135-235 K) and the Efros-Shklovskii variable-range hopping (ES-VRH) at very low temperature range (80-130 K).

1. Introduction

In the past years thin films of SnS have attracted much attention because of their potential applications in the fabrications of photovoltaic, solar cells and optoelectronic devices [1]. SnS is a potential absorber of solar cell because of its high optical absorption coefficient above the photon energy threshold of 1.3 eV, which is close to the optimal band gap. Moreover, SnS shows intrinsic p-type conductivity with carrier concentrations in the range 10^{16}-10^{18} cm^{-3}, which is suitable for absorber of solar cell. In addition, both Sn and S are abundant in nature and non-toxic. The latter characteristics are very important for photovoltaic industry with the currently available CdTe and $CuInSe_2$ materials, although they do also exhibit high efficiencies. SnS thin films can be prepared by a variety of methods [2]. Among these methods, close spaced sublimation method is proposed low consumed materials, high growth rate, ease to control stoichiometric composition and ease to extend to large scale production [3, 4]. In this study, we prepared SnS thin films on glass substrate by close spaced sublimation method in vacuum. Influence of different source temperatures on structural, optical and electrical properties of SnS films was investigated.

Content from this work may be used under the terms of the Creative Commons Attribution 3.0 licence. Any further distribution of this work must maintain attribution to the author(s) and the title of the work, journal citation and DOI.
Published under licence by IOP Publishing Ltd

2. Materials and methods

SnS thin films were prepared on glass substrates by close spaced sublimation method using a home-made apparatus. Three different source temperatures (T_s) as 550, 575 and 600°C were used to prepare SnS thin films. The deposition time was 10 min. The glass substrate temperature was fixed at lower than source temperature around 100°C by PID temperature controller. The crystal structure of these films was checked by X-ray diffraction technique with a Brucker D 8 diffractometer using CuK$_\alpha$ radiation. Surface morphology was examined by LEO1455VP scanning electron microscope. The elemental composition was analyzed by using EDS. Raman spectroscopy was performed by Renishaw inVia micro-Raman microscope. The optical absorption spectra were measured with a UV-Vis spectrophotometer in the 400-1000 nm wavelength range. Electrical properties of the films were evaluated by Hall effect and resistivity measurements in the van der Pauw configuration at room temperature. The temperature-dependent conductivity of SnS thin films was measured by two probe method. The cryogenic system with a He closed cycle allows us to measure electrical conductivity of the films from 50 to 300 K.

3. Results and discussion

3.1. Structural analysis

The XRD patterns of SnS films deposited at three different source temperatures of 550, 575 and 600 °C are shown in Figure 1. All samples belonged to orthorhombic structure corresponding to file JCPDS 73-1859 with preferred orientation (400) plane. The variation of lattice parameters with source temperature is shown in Figure 2. The parameter "a" increased significantly attain a maximum value at source temperature of 575°C and above this source temperature, parameter "a" was rather constant. The small variation of lattice parameter "a" is probably the result of the strain especially in direction perpendicular to substrate. The parameter "b" and "c" are more or less constant with the increase of source temperature. The biggest texture coefficient value [5] about 2.11 was observed in the films prepared at source temperature of 600°C. Figure 3 shows SEM images of SnS films grown at different source temperatures. The films contained rectangular plate-like form with dimension around 12.5 μm in length and 1.5 μm in thickness was observed in the films prepared at source temperature of 600°C.

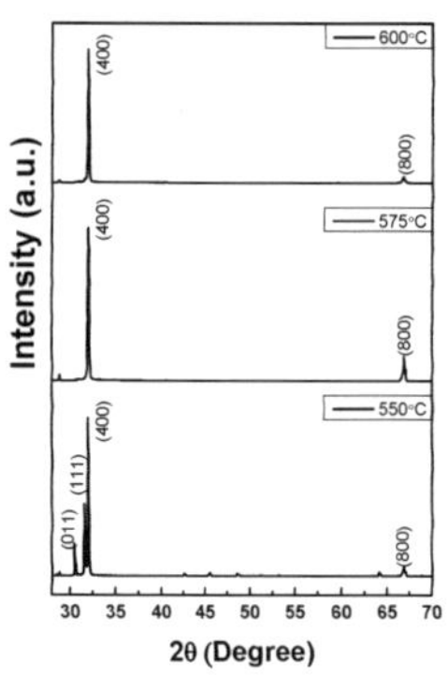

Figure 1. XRD patterns of SnS thin films prepared at different source temperatures.

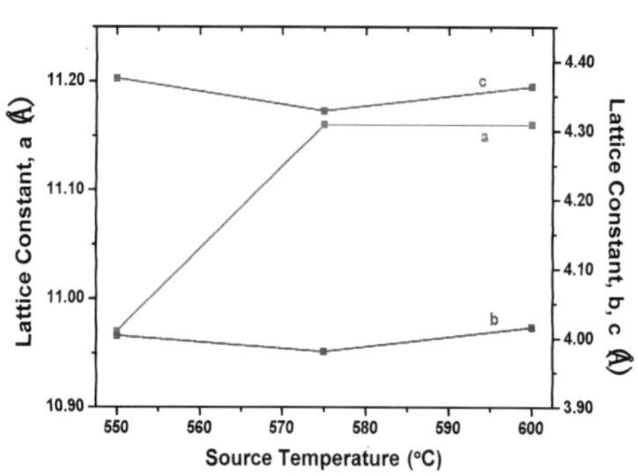

Figure 2. Variation of lattice constant values as a function of source temperature.

From EDS analysis, the relative atomic percentages of Sn and S in the films prepared with source temperature 600°C were 51.21% and 48.79 %, respectively. For all three typical films, only one Raman peak located around 220 cm^{-1} was presented in Figure 4. This Raman peak was attributed to LO phonon mode of SnS phase [6]. Regarding to the peak intensity, it suggested that the films prepared at source temperature of 600°C revealed the best crystal quality due to the strongest peak intensity.

Figure 3. SEM images of SnS thin films prepared at different source temperatures.

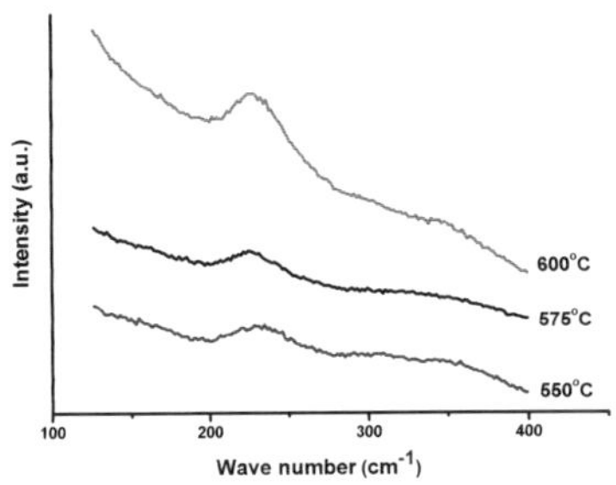

Figure 4. Raman spectra of SnS thin films prepared at different source temperatures.

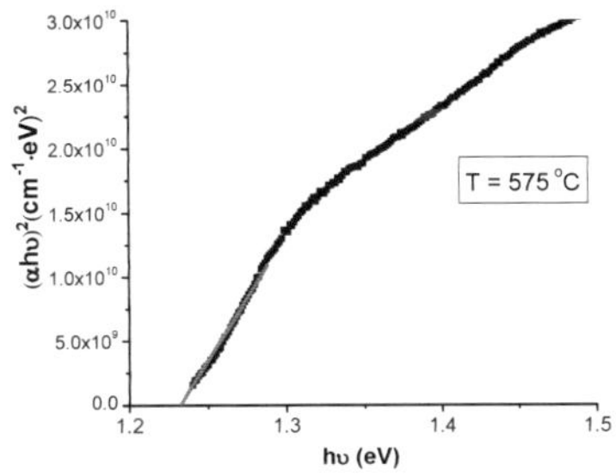

Figure 5. Plot of $(\alpha h\nu)^2 = A(h\nu - E_g)$ of SnS thin films prepared at source temperature of 600°C.

3.2. Optical properties

From Figure 5, the plot of $(\alpha h\nu)^2 = A(h\nu-E_g)$ of SnS films prepared at source temperature 600°C revealed the straight-line nature indicating that the optical transition in the films is direct allowed transition and the $h\nu$-axis interception resulting the band gap value. The band gap was found to increase from 1.18 to 1.23 eV when the source temperature increased from 550°C to 600°C.

3.3. Electrical properties

Electrical properties of the films were evaluated by Hall effect and resistivity measurements in the van der Pauw configuration. The positive sign of the Hall coefficient confirmed p-type conductivity in all studied films. The lowest resistivity and highest carrier concentration values were observed on the films prepared at source temperature of 600 °C. Resistivity, Hall constant, carrier concentration and mobility values of the films were 0.54 Ω.cm, 40.76 cm^3/C, 1.53×10^{17} and 75.48 $cm^2/(V.s)$, respectively. The obtained results are in agreement with the reported literatures [1,3,7]. Figure 6 (a) shows the Arrhenius plot of $\ln(\sigma)$ and 1000/T for the films prepared with source temperature of 600°C. As seen in Figure 6 (a), no single mechanism can be fitted in the whole curve of conductivity. At high temperature, the electrical conductivity of the sample is dominated by thermal activation passing through grain boundaries. The temperature dependence of the electrical conductivity of the films was firstly analyzed using Seto's model as the following relation [8]:

$$\ln\left(\sigma T^{1/2}\right) = -\frac{E_a}{k}\left(\frac{1}{T}\right) + \ln\sigma_0 \quad (1)$$

where σ_0 is pre-exponential factor, k is Boltzmann's constant and E_a is the activation energy. This model related to grain boundaries which trap carriers and the barrier height value appeared. As shown in Figure 6 (b), a good linear relationship can be observed in the Arrhenius plot in the high temperature range (240-300 K). The important parameters deduced from Seto's model were listed in Table 1: L is crystallite size, E_b is barrier height, N_a is acceptor concentration, E_t is trap energy level,

N_t is trap concentration and L_D is Debye length. Figure 6 (c) shows the plot of $\ln(\sigma T^{1/2})$ vs $T^{-1/4}$. The plot was found to be linear at temperature range of 135-235K. This suggested that the conduction is controlled by Mott variable-range hopping (Mott-VRH) instead of the Seto's model at high temperature. The Mott's parameters can be extracted from the relation [8]:

$$\ln\left(\sigma T^{1/2}\right) = -\left(\frac{T_{0,Mott}}{T}\right)^{1/4} + \ln \sigma_{0,Mott} \tag{2}$$

The important parameters deduced from Mott-VRH model were tabulated in Table 2: $T_{0,Mott}$ is a characteristic temperature coefficient, $\sigma_{0,Mott}$ is pre-exponential factor, $N_0(E_F)$ is density of state at the Fermi level, $R_{hop,\ Mott}$ is average hopping distance, and $W_{hop,\ Mott}$ is average hopping energy. At very low temperature range (80-130 K), the measured data were satisfied the Efros-Shklovskii variable-range hopping (ES-VRH). In this electrical conduction mechanism region, the electron interaction caused the density of state at Fermi level is separated and became to the coulomb gap (Δ_C). The conductivity in ES-VRH model is given as [8]

$$\ln(\sigma T) = -\left(\frac{T_{0,ES}}{T}\right)^{1/2} + \ln \sigma_{0,ES} \tag{3}$$

Figure 6 (d) shows the plot of $\ln(\sigma T)$ vs $T^{-1/2}$. The important parameters deduced from ES-VRH model were listed in Table 3: $T_{0,ES}$ is a characteristic temperature coefficient, $\sigma_{0,ES}$ is pre-exponential factor, $R_{hop,\ ES}$ is average distance and $W_{hop,\ ES}$ is average hopping energy.

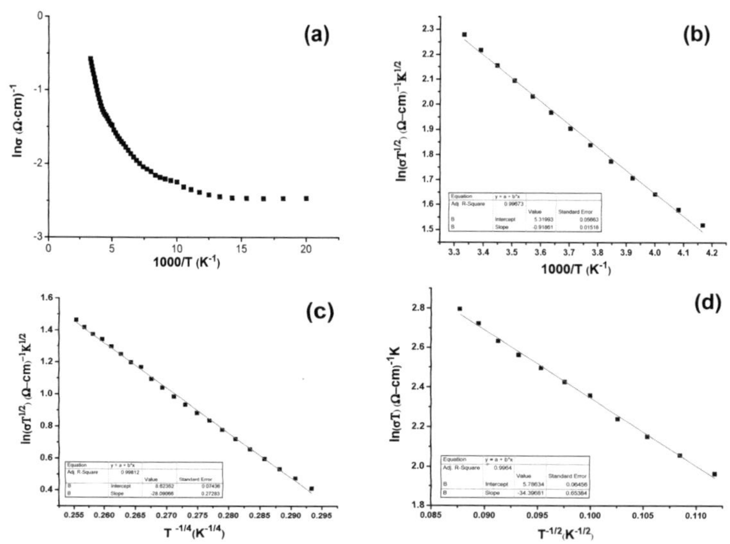

Figure 6. $\ln(\sigma)$ vs 1000/T plot, (b) $\ln(\sigma T^{1/2})$ vs 1000/T plot, (c) $\ln(\sigma T^{1/2})$ vs $T^{-1/4}$ plot and (d) $\ln(\sigma T)$ vs $T^{-1/2}$ plot of SnS thin films prepared at source temperature of 600°C.

Table 1. Seto's parameters at high temperature of the films prepared at source temperature 600°C.

Temperature range	E_g (eV)	ε_r	L (nm)	E_b (eV)	N_a (cm^{-3})	E_t (eV)	N_t (cm^{-2})	L_D (nm)
240-300	1.23	10.65	47.30	0.0792	1.67×10^{17}	0.527	7.90×10^{11}	9.07

Table 2. Mott's parameters at low temperature of the films prepared at source temperature 600°C.

Temperature range	$\sigma_{0,Mott}$ (S/cm)	$T_{0,Mott}$ (K)	ξ_{Mott} (nm)	$N_0(E_F)$ ($cm^{-3}\cdot eV^{-1}$)	$R_{hop,Mott}$ (nm)	$\frac{R_{hop,Mott}}{\xi_{Mott}}$	$W_{hop,Mott}$ (meV)
135-235	5.56×10^{3}	6.23×10^{5}	0.806	6.39×10^{20}	2.29	2.84	31.06

Table 3. ES's parameters at very low temperature of the films prepared at source temperature 600°C.

Temperature range	$\sigma_{0,ES}$ (S/cm)	$T_{0,ES}$ (K)	$R_{hop,ES}$ (nm)	ξ_{ES} (nm)	$\frac{R_{hop,ES}}{\xi_{ES}}$	$W_{hop,ES}$ (meV)	Δ_C (eV)
80-130	3.25×10^{2}	1.18×10^{3}	39.10	46.56	0.84	15.20	3.76×10^{-6}

4. Conclusions

Influence of source temperature on microstructure, optical and electrical properties of SnS films prepared by close spaced sublimation method in vacuum was investigated. XRD analysis results indicated that all studied films belonged to orthorhombic structure. Raman spectra of the all three typical films presented the LO phonon mode. The band gap was found to increase from 1.18 to 1.23 eV when the source temperature increased from 550 to 600°C. The lowest resistivity and highest carrier concentration values were observed on the films prepared at source temperature of 600 °C. In the temperature range of 50-300 K, it was shown that three types of conduction mechanisms can be expected in the films prepared at source temperature of 600°C: thermionic emission across grain boundary at high temperature range (240-300 K), the Mott variable-range hopping (Mott-VRH) at low temperature range (135-235 K) and the Efros-Shklovskii variable-range hopping (ES-VRH) at very low temperature range (80-130K).

Acknowledgments

This work is supported by King Mongkut's Institute of Technology Ladkrabang (KMITL) under grant number A118-0360-005.

References

[1] Kumar K K, Manohari A, Dhanapandian S and Mahalingam T 2014 *Mater. Lett.* **131** p 167
[2] Bashkirov S A, Gremenok V F and Ivanov V A 2011 *Semiconductor* **45(6)** p 749
[3] Yanuar Y, Guastavino F, Llinares C, Djessas K and Massc G 2000 *J. Mater. Sci.* **19** p 2135
[4] Zhan X, Shi C, Shen X, Yao M and Zhang Y 2012 *Adv. Mater. Res.* **590** p 148
[5] Pan L L, Li G Y and Lian J S 2013 *Appl. Surf. Sci.* **274** p 365
[6] Yang Y B, Dash J K, Xiang Y, Wang Y, Shi J. Dinolfo P H, Lu T M and Wang G C 2016 *J. Phys. Chem. C* **120** p 13199
[7] Vidal J, Lany S, d'Avezac M, Zunger A, Zakutayev A, Francis J and Tate J 2012 *Appl. Phys. Lett.* **100** 032104
[8] Baqiah H, Ibrahim N B, Abdi M H and Halim S A 2013 *J Alloys Compd.* **575** p 198

Design of a porous cathode structure of high performance Li-ion battery

Jihwan Song[1,*], Junhyung Kim[2] and Dongchoul Kim[2]
[1]Department of Mechanical Engineering, Hanbat National University, Daejeon 34158, Korea
[2]Department of Mechanical Engineering, Sogang University, Seoul 04107, Korea

*E-mail: jsong@hanbat.ac.kr

Abstract. The cathode materials of Li-ion battery are suffering from huge loss in the specific capacity as the C-rate or cycle increases due to the unusually low diffusivity of Li-ion in specific phase. Although the porous cathode has shown the improved performance, it still has not been clearly explained whether the porosity really increases the specific capacity and how much high rate capability we can obtain. Here we report the importance of the design of porous cathode, which may provide an ultrahigh or even lower than that of the simple sphere-shaped cathode. Numerical simulations show the improved specific capacity with the porosity and the effect of porosity becomes significant as the C-rate increases. As the C-rate increases from 2.5 to 40C, the loss of specific capacity becomes only 20% with the 50% porosity, while that is about 60% with the simple sphere-shaped cathode. In addition, that decreases during total cycles from 30% to only about 2% with the 50% porosity under 40C. Interestingly, the porous cathode does not guarantee the improved specific capacity even porosity increases unless the pores are evenly distributed. The presented analysis provides important issues to design the ultrahigh performance of cathode.

1. Introduction

Although the cathode materials provide the outstanding performance of Li-ion batteries (LIBs) based on their high energy density, high capacity, simple synthesis, low cost [1-5], they are still suffering from a severe problem, such as a loss of specific capacity under the high C-rate or high cycle [6,7]. It has revealed that the loss of specific capacity originates from the various reasons, such as the structural transformation of cathode [8], the low electronic conductivity [8], and the low diffusivity of Li-ion in the cathode material [9,10]. There have been various attempts to resolve the loss of specific capacity [11]. For instance, it has been turned out that the morphology of concave or notch-like should avoid at the cathode to reduce the structural transformation of cathode [12]. In addition, the hollow structure has been proved effective to buffer the structural transformation and to reduce the loss of specific capacity. To improve the electronic conductivity, the coating of high conductive materials on the cathode has been suggested [13-15]. The improved electronic conductivity with high conductive materials have ensured the high rate capability and high cycle stability. Meanwhile, since the typical cathode materials generate the phase that has unusually low diffusivity of Li-ion during their utilization, [9-11, 13, 14], there have been many efforts in order to overcome the low diffusivity of Li-ion in the cathode materials. Especially, the various porous structures in cathode have shown the great improvement in the rate capability and the cycle stability [11,13,14]. Although it has been shown that

Content from this work may be used under the terms of the Creative Commons Attribution 3.0 licence. Any further distribution of this work must maintain attribution to the author(s) and the title of the work, journal citation and DOI.
Published under licence by IOP Publishing Ltd

the porosity in cathode have the improved specific capacity, rate capability, and cycle stability than those without the porosity, however, it still has not been clearly explained whether the porosity in cathode really increases the specific capacity and how much high rate capability and high cycle stability we can obtain from a systematic analysis.

Here we report the importance of the design of porosity in cathode, which may provide an ultrahigh specific capacity with the porosity or even worsened specific capacity as the porosity increases. The specific capacity and the loss of it were systematically evaluated with various porosity, distribution of pores, C-rate, cycle. Numerical simulations show that the specific capacity can be improved as the porosity in cathode increases and the effect of porosity becomes significant as the C-rate increases. The loss of specific capacity is considerably reduced with respect to the increase of both of C-rate and cycle as the porosity in cathode increases. Interestingly, the porous cathode does not guarantee the improved specific capacity as the porosity increases unless the pores are evenly distributed.

2. Model and methods

In order to evaluate the performance of LIB with porous cathode, a three-dimensional (3D) dynamic model based on the phase field approach was developed. The phase field approach has been demonstrated its reliability and effectiveness in previous studies related with evolutions of nano- and micro-structures [16-18]. The schematic illustration of design of porous cathode are shown in Figure 1. The cathode consists of the microsphere structures, Li-ion, and electrolyte (Figure 1a). When LIB is discharged, Li-ion diffuses into the microstructure of cathode and the performance of LIB (*e.g.*, specific capacity and rate capability) can be determined by the Li-ion concentration in the cathode. As Li-ions diffuse into the cathode, the phases of cathode material are transformed according to the concentration of Li-ion and the diffusivity of Li-ion of some phases could be considerably low, which would reduce the total amount of Li-ion in the cathode during the discharging process and induce a low specific capacity and rate capability. For instance, the cathode made with vanadium pentoxide (V_2O_5), which is considered in this study, has the transformation of phases from α, ε, δ to γ phases as the concentration of Li-ion increases [6]. During the discharging process, the diffusion of Li-ion into the cathode is interrupted by the ε phase because the diffusivity of Li-ion is significantly low in ε phase, which is about 10 times lower than that of other phases (*i.e.*, α, δ, and γ). The porosity in the cathode could reduce the amount of cathode where Li-ion cannot reach due to ε phase and provide a full utilization of cathode (Figure 1b). As shown in Figure 1c, the distribution of pores also can be an important parameter in designing the microstructure of cathode.

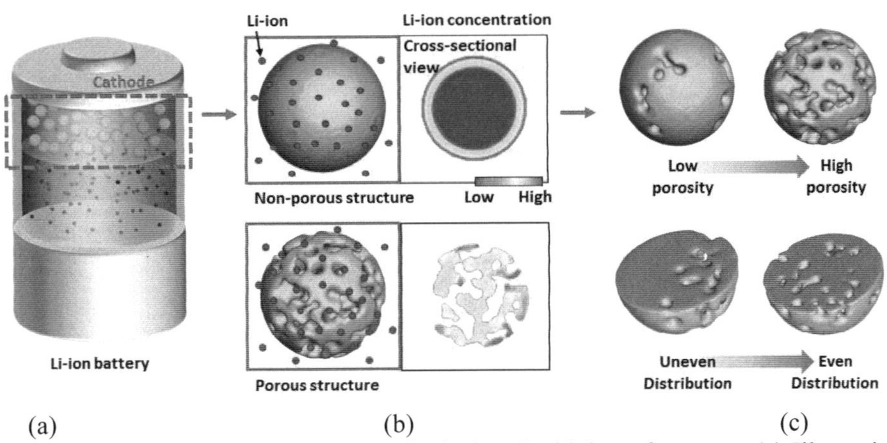

Figure 1. Design of porous structure in the cathode for ultrahigh performance. (a) Illustration of the cathode in Li-ion battery consisting of microsphere structure, Li-ion, electrolyte. Scale bars, 500 nm. (b) Li-ion concentration and e phase fraction in the cathode according to the design of porosity. (c) Design of porous cathode with respect to the porosity and the distribution of pores.

The diffusion flux of Li-ion into the cathode is represented by $\boldsymbol{J}_d = -D\nabla c$, where D and c are the diffusivity and the concentration of Li-ion, respectively. Since we assume that V_2O_5 is a cathode material, the diffusivity of Li-ion in the cathode depends on the internal phases (*e.g.*, α, ε, δ, γ, and ω) and the phases are determined by the Li-ion concentration in cathode. The diffusivities at each phase are defined according to the Li-ion concentration [19]. The Li-ions are supplied as a source at the cathode surface and it can be defined with current density as $\boldsymbol{S} = i_n / F$ [20], where F is Faraday constant. The governing equation with the flux and source of Li-ion is constructed by $\partial c/\partial t = \nabla\cdot(D\nabla c) + \boldsymbol{S}$, and then normalized with the characteristic time t_c, length l_c, and diffusivity $D_c = l_c^2 / t_c$ for computational convenience. The characteristic time and length set to be $t_c = 0.5$ s and $l_c = 5.0$ nm, respectively. The normalized diffusivity, $D_n = D/D_c$, is defined the characteristic diffusivity. A semi-implicit Fourier spectral method is employed for high space resolution and fast computation. In the Fourier space, the periodic boundary condition is assigned in all of directions. The domain of cathode was constructed by 500 nm of diameter of the microsphere. The porosity of cathode is designed by the distribution of sphere-shaped pores. The sphere-shaped pore has 60 nm of diameter and the porosity is varied according to the total number of pores in the cathode. The simulations are performed until the completion of discharge. The fully discharged state is assumed by that the fraction of ω phase in the cathode becomes 1% of total volume of cathode because LIB commonly stops operation when the ω phase is generated to avoid the increase of irreversible ω phase. The fully charged state is assumed by that the concentration of Li-ion at the surface of cathode becomes zero because the chemical reaction for charging process is terminated when Li-ion does not exit at the cathode surface.

3. Results and discussions

To analyze the performance of cathode with porosity, the difference in the porosity of cathode are systematically investigated. The numerical simulations are performed using a simulation platform for battery materials (iBat alpha; Korea Institute of Science and Technology).

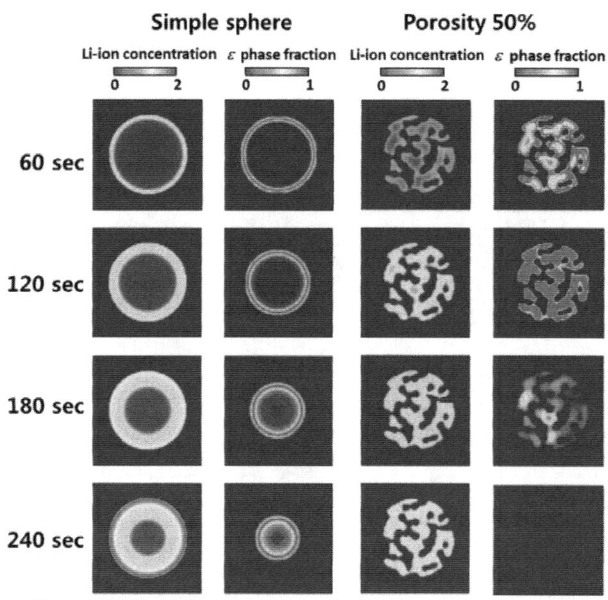

Figure 2. Cross-sectional images of time-dependent Li-ion concentration and ε phase fraction in the simple sphere-shaped cathode and the cathode with 50% porosity under 10C.

Figure 2 shows the cross-sectional images of time-dependent Li-ion concentrations and the fraction of ε phase during the discharging process, where the simple sphere-shaped cathode and the 50 % porosity of cathode are considere. The C-rate is set to be 10C. In the simple sphere-shaped cathode, Li-ion is stagnated at the out side of ε phase due to its unusaully low diffusivity. The ε phase still remains until the discharging process is finished. As a result, the cathode has an unutilized part and it lead to the loss of specific capacity. On the other hand, since the 50% porosity provides the significantly increased surface area of cathode and reduced diffusion pathway of Li-ion into the cathode, the cathode with porosity is less affected by the low diffusivity of ε phase. Indeed, the ε phase is transformed to the next phase that has usual diffusivity of Li-ion before the discharging process is finished and it is not observed at the final stage. Thus, the porosity in the cathode allows the full utilization of cathode.

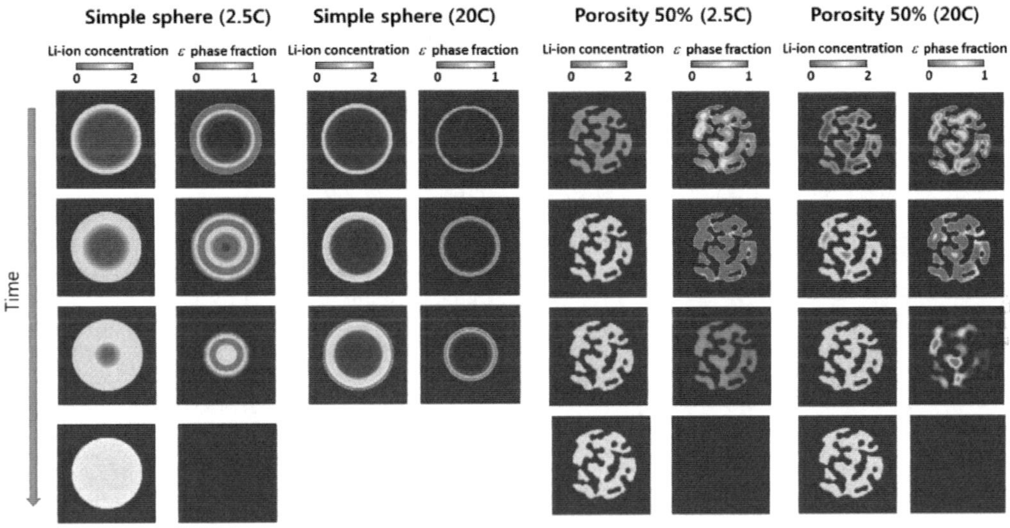

Figure 3. Cross-sectional images of time-dependent Li-ion concentration and ε phase fraction in the simple sphere-shaped cathode and the cathode with 50% porosity with C-rate 2.5 and 20C.

Additionally, the effect of porosity on the specific capacity was investigated with respect to the C-rates, where the C-rate increases from 2.5 to 20 (Figure 3). As shown in Figure 3, at the low C-rate (*i.e.*, 2.5C), Li-ion can diffuse into the cathode even without the porosity. Since the supplying rate of Li-ion at 2.5C is sufficiently small, it allows the transformation of ε phase to the next phase before the discharging process is finished. As a result, the cathode even with the simple sphere-shape can be fully utilized with the less affection of low diffusivity of ε phase. If the supplying rate of Li-ion is to be much bigger as the C-rate increases (*i.e.*, 20C), Li-ion is rapidly stagnated at the outside of ε phase. The ε phase remains until the discharge is finished and it result in the insufficient utilization of cathode. On the other hand, Li-ion can diffuse more effectively into the cathode with 50% porosity. Due to the significantly increased surface area and reduced pathway of diffusion, the porous cathode is lees affected by the ε phase than the simple sphere-shaped cathode under the same C-rate. As a result, the cathode with porosity can be fully utilized even under high C-rate condition.

A quantitative analysis on the specific capacity with different porosity in cathode was performed (data not shown). The specific capacity mostly increases as the porosity in cathode increases and the effect of porosity becomes significant as the C-rate increases. The cathode with the 50% porosity shows the 220% of improved specific capacity at the 40 C-rate condition. In addition, the high rate capability of the cathode was observed with porosity. For example, the loss of specific capacity is dramatically reduced to be only about 20 % when the porosity in cathode is 50%, while the simple sphere-shaped cathode shows about 60% loss of specific capacity. About 300% of reduction in the loss of specific capacity when the porosity is considered in the cathode means that the design of porosity in the cathode could resolve the severe loss of specific capacity of LIBs according to the increase of C-rate.

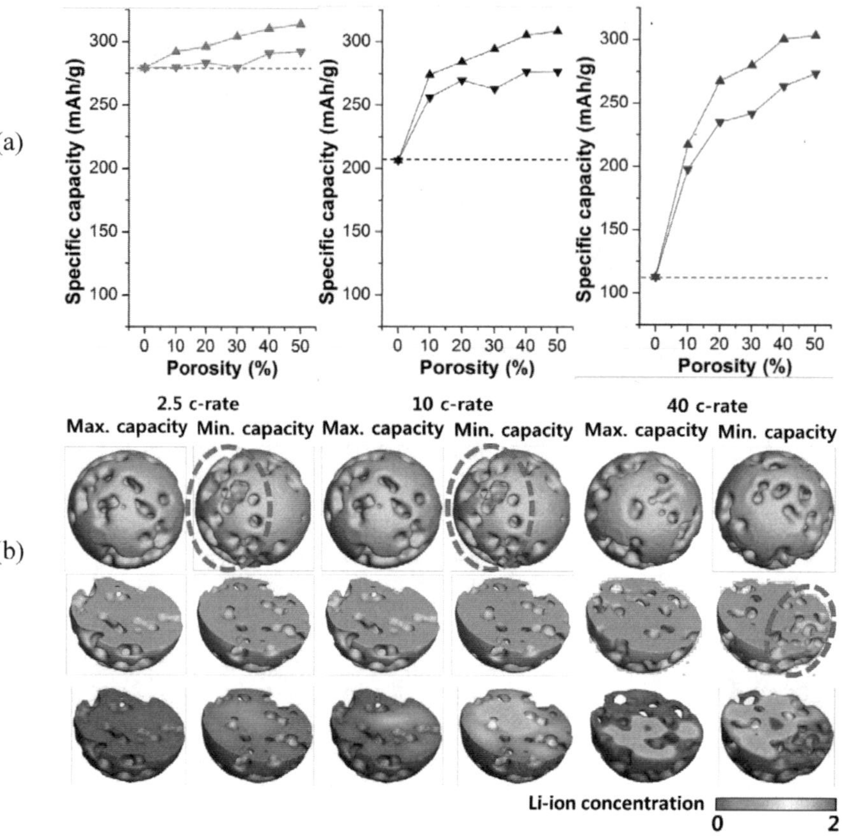

Figure 4. Effect of pore distribution on the specific capacity of cathode. (a) Maximum (triangle) and minimum (reverse triangle) specific capacities among the 15 different kinds of porous cathode at each porosity with 2.5 (red), 10 (black), and 40 (blue) C-rate condition, respectively. (b) Image of the 30% porosity in the cathode and Li-ion concentration for maximum and minimum specific capacity. Red circle, pores-concentrated region.

Next, the performance according to the distribution of pores in the cathode is investigated with 2.5, 10, and 40 C-rate conditions, respectively. 15 different kinds of porous cathode are designed for each porosity. In same porosity, 15 different kinds of cathode has the difference in only distribution of pores, not total number of pores. All of pores are randomly distributed in every porous structure. Among the specific capacities of each porous cathode, the maximum (*i.e.*, triangle) and minimum (*i.e.*, reverse triangle) specific capacities are presented in Figure 4a. Red, black, and red lines are the specific capacities at the 2.5, 10, and 40 C-rate, respectively. Dashed lines represent the specific capacity of simple sphere-shaped cathode at each C-rate. Simulation reveals that the porous cathode have the considerable variation of the specific capacity according to the distribution of pores, which is up to 16% of variation. Interestingly, the specific capacity can be decreased irrespective of c-rate even with the porosity in cathode unless the pores are evenly distributed. Qualitatively, the cathode with well-distributed pores has higher Li-ion concentration than the cathode with uneven distribution of pores (Figure 4b). This result can be explained by that the pores-concentrated region (*i.e.*, red circles in the morphological image of cathode) results in the restrictive use of cathode material. Namely, Li-ion is extremely intercalated at the pores-concentrated region due to the excessively decreased pathway of diffusion into the cathode and the ε phase is generated much earlier than other regions. Li-ion is increasingly stagnated where ε phase is generated due to significantly low diffusivity of Li-ion and thus, the discharging process is terminated without full utilization of the cathode. It means that the design of porosity in cathode might become important to achieve ultrahigh performance of LIBs because the design of porosity can provide the ultrahigh specific capacity of cathode or even worsened specific capacity as the porosity increases.

4. Conclusion

In summary, we have reported the importance of the design of porosity in cathode, which may provide an ultrahigh specific capacity with the porosity or even worsened specific capacity as the porosity increases. Simulation represents that the cathode with the porous cathode can have higher specific capacity than the simple sphere-shaped cathode and the effect of porosity on specific capacity becomes great as the C-rate increases. The rate capability and the cycle stability are improved with the porosity in cathode. As the C-rate increases from 2.5 to 40, the specific capacity loss is reduced to be only about 20% with 50% porosity in the cathode, while the simple sphere-shaped cathode shows about 60% loss of specific capacity. In addition, the loss of specific capacity during total cycles is only about 2% when the porosity in cathode is 50%, while the simple sphere-shaped cathode shows about 30% loss of specific capacity. Interestingly, the porous cathode might have the degenerated specific capacity than the simple sphere-shaped cathode unless the pores are evenly distributed. We expect that the presented analysis on the structural design of cathode can provide an efficient way for the ultrahigh performance Li-ion battery as an advanced power source.

References

[1] Strelcov E, Cothren J, Leonard D, Borisevich A Y and Kolmakov A 2015 *Nanoscale* **7** 3022-27.
[2] Delmas C, Brethes S and Menetrier M 1991 *J. Power Sources* **34** 113-18
[3] Zhang J, Xiang Y J, Yu Y, Xie S, Jiang G S and Chen C H 2004 *J. Power Sources* **132** 187-194.
[4] Mai L Q, Xu X, Xu L, Han C H and Luo Y Z 2011 *J. Mater. Res.* **26** 2175-85
[5] Wang Y, Takahashi K, Lee K and Cao G Z 2006 *Adv. Funct. Mater.* **16** 1133-44
[6] Whittingham M S 2004 *Chem. Rev.* **104** 4271-301
[7] Jo M, Hong Y S, Choo J and Cho J 2009 *J. Electrochem. Soc.* **156** A430-A434
[8] Liu Z L, Yu A S and Lee J Y 1998 *J. Power Sources* **74** 228-33
[9] Wang L, He X M, Sun W T, Wang J L, Li Y D and Fan S S 2012 *Nano Lett.* **12** 5632-36.
[10] Okubo M, Hosono E, Kim J, Enomoto M, Kojima N, Kudo T, Zhou H S and Honma I 2007 *J. Am. Chem. Soc.* **129** 7444-52
[11] Sun C W, Rajasekhara S, Goodenough J B and Zhou F 2011 *J. Am. Chem. Soc.* **133** 2132-35
[12] Malave V, Berger J R, Zhu H Y and Kee R J 2014 *Electrochem. Acta* **130** 707-17
[13] Zhu D, Liu H, Lv L, Yao Y D and Yang W Z 2008 *Scripta Mater.* **59** 642-45
[14] Zheng Z M, Wang Y, Zhang A, Zhang T R, Cheng F Y, Tao Z L and Chen J 2012 *J. Power Sources* **198** 229-35
[15] Shi Y, Chou S L, Wang J Z, Wexler D, Li H J, Liu H K and Wu Y P 2012 *J. Mater. Chem.* **22** 16465-470
[16] Han H S, Song J, Hong J, Kim D and KangT 2014 *Anal. Chem.* **86** 6160-65
[17] Yang M S, Jeong S, Kang T and Kim D 2015 *J. Phys. Chem. C* **119** 6148-51
[18] Song J and Kim D 2013 *J. Appl. Phys.* **114**
[19] McGraw J M, Bahn C S, Parilla P A, Perkins J D, Readey D W and Ginley D S 1999 *Electrochim. Acta* **45** 187-96
[20] Zhang X C, Sastry A M and Shyy W 2008 *J. Electrochem. Soc.* **155** A542-A552

IOP Conf. Series: Materials Science and Engineering **383** (2018) 012009 doi:10.1088/1757-899X/383/1/012009

Evaluation of the fine blanking tool for the precise clutch cam of automobile seat lifter pumping device

Jong-Deok Kim[1,*], Kil-Sang Yoon[1], Ok-Rae Kim[1] and Si-Tae Won[2]

[1]Dies & Molds Technology Group, Korea Institute of Industrial Technology, 156, Gaetbul-ro, Yeonsoo-gu, Incheon, 21999, Republic of Korea
[2]Department of Mechanical System Design Engineering, Seoul National University of Science and Technology, 232 Gongneung-ro, Nowon-gu, Seoul, 01811, Republic of Korea

*E-mail: jdk@kitech.re.kr

Abstract. In this paper, the multi-step forming process was studied with the fine blanking (FB) tool and the FB experiments for the precise clutch cam of automobile seat lifer pumping device (PCC). In order to form the PCC, the multi-step compound FB tooling system was taken; the three forces were directly gained from the rams of the FB press and the other force was gained from the auxiliary hydraulic device in the FB press. The FB tool was optimally designed through the FB simulation and fabricated. From the FB experiments with the FB press (Mori Iron Co., Ltd, FB400), the 5 samples of the PCC were taken and measured. All samples fulfilled the tolerance conditions.

1. Introduction

FB is a press working process which makes it possible to produce precise finished components with the clean cutting surface over the whole material thickness in a single stroke. Therefore the FB part has the advantages of the tight dimensional tolerances, no fractures and cracks in the shear surface, no secondary machining.

Generally, the FB process can be used with the conventional press working process in order to form the complex part with good shear surface and in the majority of cases, the progressive tooling system was taken because of good efficiency and productivity. But if the progressive tooling system was applied in case of forming the very precise part, it cannot be obtained because of the tolerance stack in relation to the material feeding. The Daesung Fine Tec. Co., Ltd. being one of the famous Korean companies in the field of fine blanking technology, had developed the forming of the PCC with the three stations progressive FB tooling system: the 1st station - the pilot piercing and semi piercing of diameter 9 mm, the 2nd station - semi piercing of diameter 14 mm, the 3rd station - FB of external form (referring to figure 1). But the total defective proportion was over 30% because of dimension error due to the tolerance stack. Therefore in this paper, the multi-step compound FB tooling system was applied in order to form the PCC as shown in figure 1. The three forces were directly gained from the rams of FB press and the other force was gained from the auxiliary hydraulic device (accumulator) in the FB press.

2. Optimal design of FB tool for PCC

Content from this work may be used under the terms of the Creative Commons Attribution 3.0 licence. Any further distribution of this work must maintain attribution to the author(s) and the title of the work, journal citation and DOI.
Published under licence by IOP Publishing Ltd

2.1. Analysis of the PCC for FB

The PCC (Material: JIS SNCM220, Thickness: 5mm) was constituted of the web width 2.5 mm. It had the high difficulty rating for FB because the web width is small compared to its thickness [1]. Table 1 shows the mechanical properties of the material.

Table 1. Mechanical Properties of PCC Material (SNCM220).

Yield strength(MPa)	Tensile strength (MPa)	Elongation (%)
349	486.5	32.5

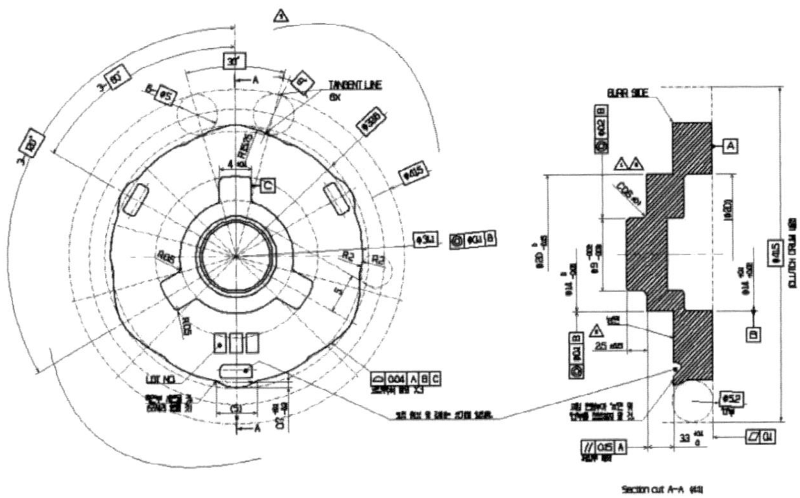

Figure 1. PCC drawing.

2.2. Design of FB tool

At first, the structure of the compound FB tool for the PCC was constructed and the strip layout was designed as shown in figure 2. And then FB forces were calculated according to Franz Birzer's formula [2]. The total forces were 214.92 tons. The FB press with more than 214.92 tons was required, 400 ton hydraulic FB press (Mori Iron Co., Ltd, FB400) was chosen for the FB experiment. The clearance between punch and die of the FB tool was decided to be 0.02mm according to external form [3].

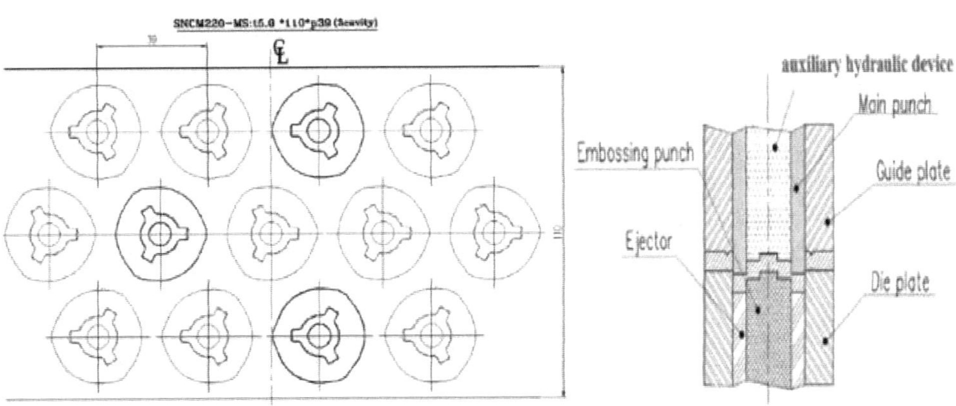

Figure 2. Strip layout and structure of compound FB tool for the PCC.

The V-ring of the FB tool clamps the material strip during the shearing process. It was only designed on the guide plate of the FB tool in order to save the tooling cost. Figure 3 shows the V-ring designed on the guide plate [4].

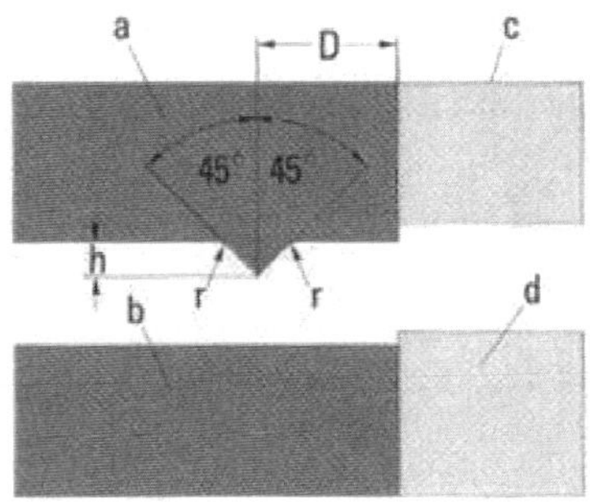

a : guide plate
b : die plate (insert)
c : main punch
d : ejector
D : 2.5 mm
h : 0.8 mm
r : 0.2 mm

Figure 3. Shape of V-ring on the guide plate.

2.3. FB simulation

The FB simulation was conducted using the commercial FEA software Deform 3D in order to predict the fine blanked shear surface, the stress and strain distribution for the PCC. The simulation model was set to the 1/3 model in consideration of a part's symmetry.

The FB simulation was performed at each step as shown in figure 4. The clamping process by the V-ring indenter (1st step), the semi piercing process of diameter 9mm (2nd step), the semi piercing process of diameter 14mm (3rd step) and then the FB process of external form (4th step) were simulated step by step. Some parts of large deformation were set to locally make less mesh with the mesh window function of the software Deform 3D. The numbers of the mesh element were set to 288,985. The shear speed was set to 20mm/sec and the time increment was 0.001 seconds [5].

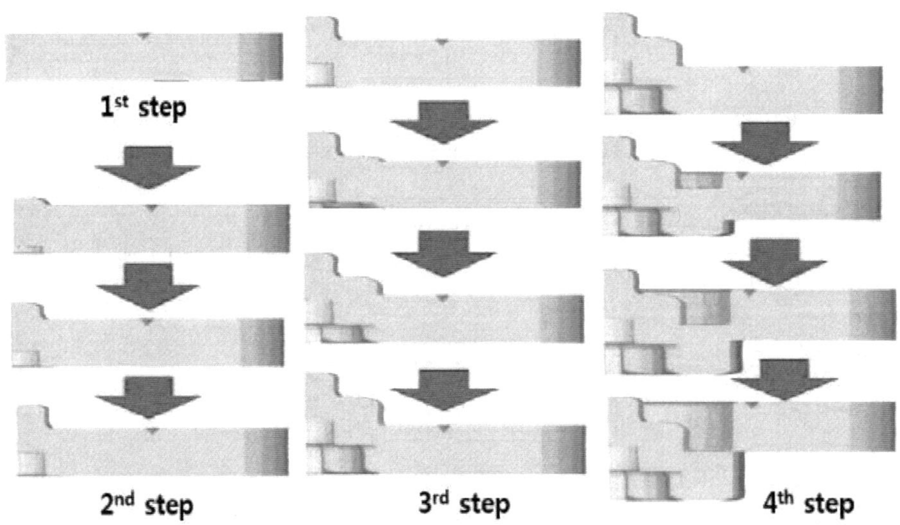

Figure 4. FB simulation procedure.

The shear surface of the PCC was estimated and then the stress distribution and the strain distribution on the PCC were predicted as shown in figure 5. The stress on the PCC was so evenly distributed in the shear surface (1,870MPa) without stress concentration.

3. FB experiments

IOP Conf. Series: Materials Science and Engineering **383** (2018) 012009 doi:10.1088/1757-899X/383/1/012009

The compound FB tool for the PCC was fabricated as shown in figure 6 and the FB experiments were performed on a 400 ton hydraulic FB press (Mori Iron Works Co., Ltd. FB 400, Japan). The conditions of the experiments were as follows; The shear force, the V-ring force, and the counter force were each set to 130 tons, 90 tons and 20 tons, respectively, while the shear speed was set to 20mm/sec. The samples from the FB experiments were obtained as shown in figure 7. Their geometric and dimensional accuracy and the surface quality satisfied the target as shown in Table 2.

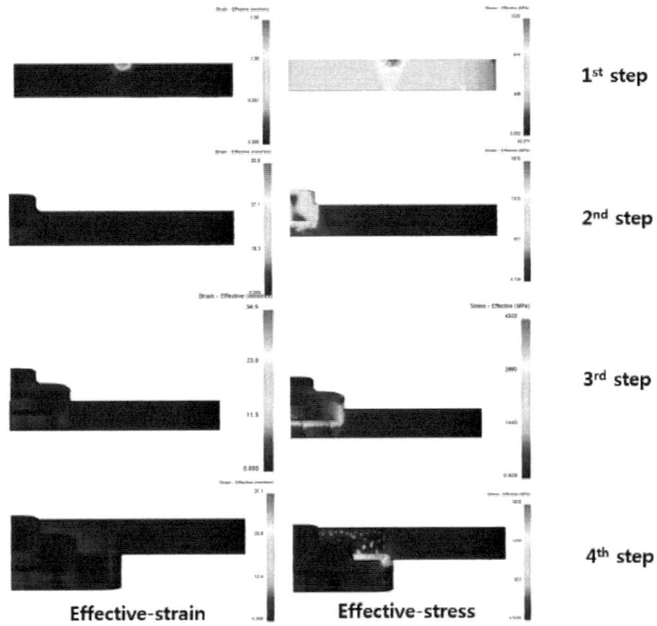

Figure 5. Distributions of the stress and strain on the PCC.

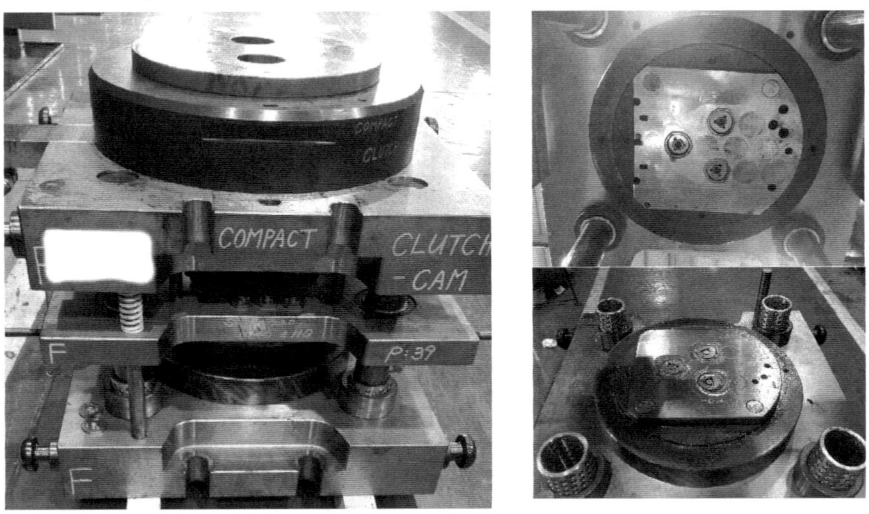

Figure 6. Fabricated compound fine blanking tool.

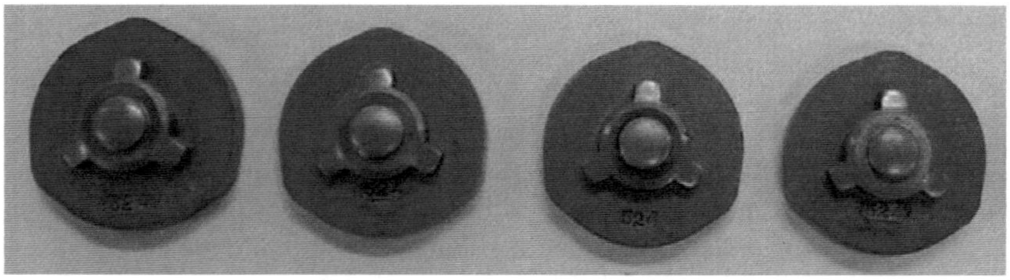

Figure 7. Samples from the FB experiments.

Table 2. Measured data of 5 samples.

items		target	Measured data
Rate of shear surface		80% ↑	92 %
External semi piercing (ø9)	dimensional tolerance	-0.02/-0.08 mm	8.966 mm
	roundness	0.2	0.013
External semi piercing (ø14)	dimensional tolerance	0/-0.06 mm	13.96 mm
	roundness	0.1	0.068
Height of semi piercing (2.5)	dimensional tolerance	± 0.15 mm	2.634 mm

4. Conclusions

In this study, the FB process for the PCC was evaluated by the simulations and the experiments. The following conclusions were obtained.

(1) The effective stresses on the PCC were uniformly distributed and there was no stress concentration.

(2) The PCC samples from the FB experiments had the shear surface with good quality and satisfied the geometric and dimensional accuracy of target.

References

[1] K. Lange, F. Birzer, P. Hoefel, A. Muhoty, H. Singer, Edelstahlwerke Buderus AG, Feintool AG Lyss, Hoesch Hohenlimburg GmbH, Kaltwaltzwerk Brockhaus GmbH 1997 *Cold forming and Fineblanking – A handbook on cold precessing, material properties, component design* pp 141-65.

[2] Franz Birzer 1997 Forming and fineblanking / cost-effective manufacture of accurate sheetmetal parts Landesberg/Lech, verlag moderne Indusrie

[3] W. Blatter 1991 Introduction into the technology fine blanking tools KAITECH

[4] Taylan Altan 1988 *Metal Forming Handbook* Shuler GmbH, Berlin, Heidelberg, New York, Springer-Verlag, Chapter 4

[5] Jong-Deok Kim, Heung-Kyu Kim, Young-Moo Heo, Sung-Ho Chang 2012 A study on the effect of V-ring position on the die roll height in fine blanking for special automobile seat recliner gear, *Advanced Materials Research* Vols. **383-390**, pp 7122-27

IOP Conf. Series: Materials Science and Engineering **383** (2018) 012010 doi:10.1088/1757-899X/383/1/012010

Surface enhancement approach for FDM rapid prototypes by organically modified montmorillonite nanoparticles

V Francis* **and P K Jain**

CAD/CAM lab, Mechanical Engineering Discipline, Indian Institute of Information Technology, Design and Manufacturing, Jabalpur, 482005, India

*Email: f.vishal@iiitdmj.ac.in

Abstract. Fused deposition modelling (FDM) is one of the widely used rapid prototyping techniques for fabricating complex shaped objects from CAD data. Minimum wastage of material, simple manufacturing process and economical desktop like machine have made the technique a major contender in the field of manufacturing. However, due to layered manufacturing process, the surface of FDM parts lacks quality. In the present research work, the FDM built parts were surface treated with organically modified montmorillonite (OMMT) nanoparticles. The post-treated specimens were characterised by XRD, FTIR, SEM and UV-vis analysis which revealed the successful deposition of nanoparticles on the surface of the FDM built specimens. Effect of post-treatment parameters on surface hardness, dimensional accuracy and average surface roughness was investigated using Taguchi technique. It was found that the embedment of OMMT nanoparticles increased the hardness of FDM parts due to its ceramic attributes and the combined effect of nanoparticles/chemicals reduced the surface roughness providing a smooth surface. Moreover, the process does not marred the dimensional accuracy of the fabricated parts. Further, multi objective optimisation technique was used to find the optimum post-treatment parameters. The present approach can be utilised for improving the surface integrity of FDM parts.

1. Introduction

Rapid prototyping (RP)/3D printing has emerged as a promising candidate for fabrication of intricate objects with ease [1]. Initially the technique was used for prototyping purposes, however due to its inevitable advantages various sectors have adopted the technique for fabrication of end use products [2]. FDM is one of the widely utilised RP techniques to fabricate complex parts with polymeric materials [3]. A polymer material generally acrylonitrile butadiene styrene (ABS), in the form of filament is used in FDM process. The filament is heated and extruded through a nozzle on the platform, where it solidifies and bonds with the adjacent deposited raster's. The process is repeated layer by layer and the final parts is fabricated resembling the computer aided design (CAD) model. The process gives freedom to design any complex shaped structures. However, due to the layered manufacturing, the surface of FDM built parts have inferior surface properties compared to other polymer processing techniques [4, 5]. The surface issues associated with FDM limits its potential for end use applications. To overcome this limitation several attempts have been made to post-treat the FDM built parts. Various techniques such as micro machining, parametric optimisation and chemical treatment have been used for post-treating the FDM parts [4, 6, 7]. Each process has its own benefits and shortcoming. However, there is inadequate literature available reporting the utilisation of nanoparticles for surface

Content from this work may be used under the terms of the Creative Commons Attribution 3.0 licence. Any further distribution of this work must maintain attribution to the author(s) and the title of the work, journal citation and DOI.
Published under licence by IOP Publishing Ltd

improvement of FDM parts. Utilising nanoparticles for surface treatment can increase the application domain of the FDM process by overcoming the limitation of surface integrity.

In the present research work, the FDM parts were post-treated with OMMT nanoparticles via chemical route. The post-treated specimens were characterised by XRD, FTIR, SEM and UV-vis analysis. The effect of OMMT/nanoclay content, dipping time and layer thickness were investigated using Taguchi technique on the surface properties of FDM parts namely, surface hardness, average surface roughness and variation in dimensional accuracy. It has been found that the proposed methodology can effectively improve the hardness and surface finish of the parts without marring the dimensional accuracy. The experimentation was planned according to L_{18} orthogonal array. Further, multi objective optimisation (grey relational analysis) was carried to optimise the coupled effect of the post-treatment process parameters on all the three responses namely, hardness, average roughness and dimension deviation. The present approach can be utilised for improving the surface integrity of FDM parts.

2. Experimental

2.1. Materials, sample preparation and post-treatment

Acrylonitrile butadiene styrene (ABS) was used to print the specimens using FDM 3D printer. The layer thickness of 0.25 and 0.3 mm was used with 230° C nozzle temperature and 110° C bed temperature. The travel feed rate of 80 mm/sec and 100% infill was used in the investigation. The raster orientation was kept at 0° for all the specimen preparations. For average surface roughness and dimensional analysis, specimens of 25 x 10 x 1.2 mm were fabricated. For hardness testing specimens were prepared according to ASTM D 2240. For post-treatment of the fabricated specimens, first OMMT nanoparticles were dissolved in acetone in varying proportions containing 1, 3 and 5 wt% of OMMT. Further the specimens were dipped in the prepared solution for 30, 50 and 70 seconds. After chemical treatment the specimens were heat treated at 60° C for the removal of the solvent. The experiments were planned according to L_{18} orthogonal array.

2.2. Characterization

The post-treated specimens were characterised using XRD, FTIR, SEM and UV-vis techniques. The morphology of the top surface of treated specimens were investigated using SEM analysis and internal structure of OMMT nanoparticles after treatment was analysed by XRD technique. FTIR was used to investigate molecular components of the post-treated specimens. UV-vis spectroscopy was used to observe the UV absorbance of post-treated ABS specimens in order to see the scope for outdoor applications of ABS specimens which generally undergo photo degradation. The characterisation was performed for specimens treated with 3 wt% OMMT content and subjected to 30 seconds of dipping which was selected randomly to observe the presence of nanoparticles on the post-treated specimens.

2.3. Testing

Hardness testing was done according to shore D scale. First the fabricated samples were tested and hardness values were recorded using a durometer. Then the samples were subjected to post-treatment and again the readings were recorded. Similar methodology was adopted for roughness testing and dimensional analysis. Three readings were taken and the average values were used for analysis purpose. For average roughness (R_a) testing 0.8 mm cut off length was used. The dimensional variation was tested in z direction as maximum deviation in FDM prototypes is observed in this direction compared to x and y directions.

2.4. Statistical analysis

Taguchi technique was used to investigate the individual effect of post-treatment parameters on the surface properties of the specimens. Orthogonal array was used to design the experiments. Three levels of nanoparticle content and dipping time were taken and two levels of layer thickness were considered. L_{18} orthogonal array was selected based on the number of factors and their levels. For optimisation of hardness larger the better type category of S/N ratio was used and for roughness and dimensional deviation smaller the better S/N ratio was used. For multi optimisation of the responses, grey relational analysis was used. The S/N ratios were normalised in the range of 0 to 1. Further the normalized values were used to calculate the grey relational coefficient. These coefficient were further used to calculate the grey relational grade and accordingly the ranks were assigned.

3. Results and discussion

The presence of OMMT nanoparticles on post-treated ABS specimens were observed by XRD results which showed a peak at 4.75° 2θ value. This is the characteristic peak of OMMT nanoparticles [8, 9]. A broad peak was also present around 20° 2θ value which is due to the ABS amorphous structure [10] (Figure 1). However, it was also evident from the figure that no delamination of OMMT platelets occurred since the gallery between the platelets was not increased from 1.85 nm. This signifies that the nanoparticles were embedded between the polymeric chains and not exfoliated. On observing the FTIR data it was confirmed that the nanoparticles were successfully deposited on the polymer (Figure 2). A sharp peak was observed at 1018 cm^{-1} due to the presence of Si-O-Si bond in the nanoparticle [11]. The presence of organic modifier in the nanoparticles was observed by the peak at 1488 cm^{-1}wich is due to the C-H vibrations of modifier [12]. The band at 702, 918 and 2237 cm^{-1} were due to the styrene, butadiene and acrylonitrile components of the ABS polymers [13]. The UV-vis spectroscopy of nanocomposite solution used for post-treatment was carried along with the post-treated specimen to identify the UV absorbance for both the cases. Figure 3 demonstrate the UV absorbance graph for both. It can be seen that the absorbance demonstrated by the post treated specimens resembles the absorbance of the solution used for post-treating. Also it was found that the UV absorbance was increased for the post treated specimens compared to the pristine ABS polymer (Figure 4). This can aid in reducing the photo degradation of ABS polymer. On observing the SEM images of the post-treated specimens the platelet structure was evident (Figure 5). The same morphology can be seen in the SEM image of OMMT nanoparticles. It was also clear that the nanoparticles were homogeneously distributed over the observed region. The SEM image of pristine ABS is much smoother.

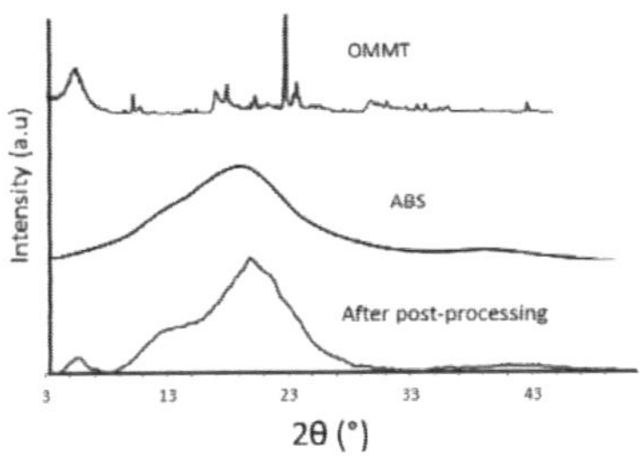

Figure 1. XRD plot for OMMT, ABS and post-processed specimens.

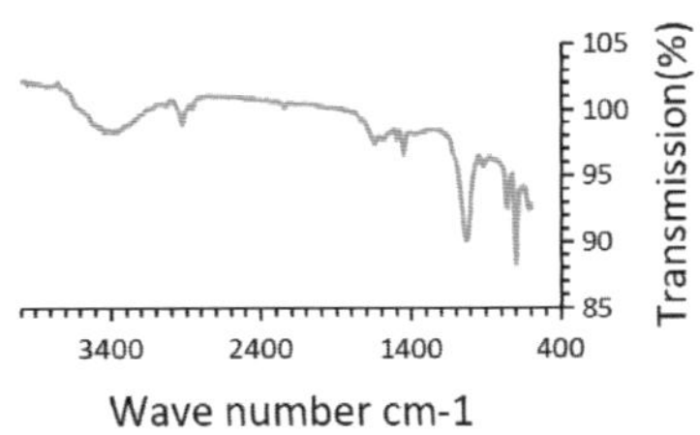

Figure 2. FTIR data for the post-treated specimens.

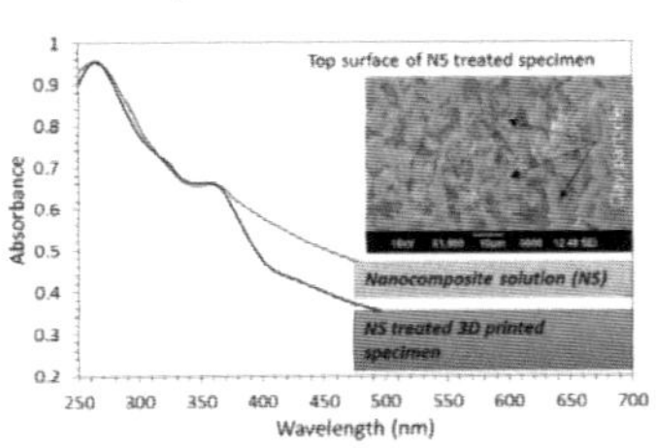

Figure 3. UV-vis spectroscopy of nano solution used for post-treatment and post-treated specimens.

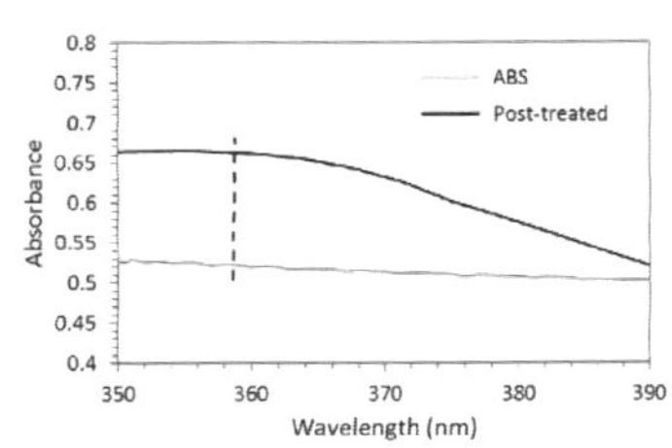

Figure 4. UV-Vis data demonstrating UV absorbance of pristine ABS and post-treated specimens.

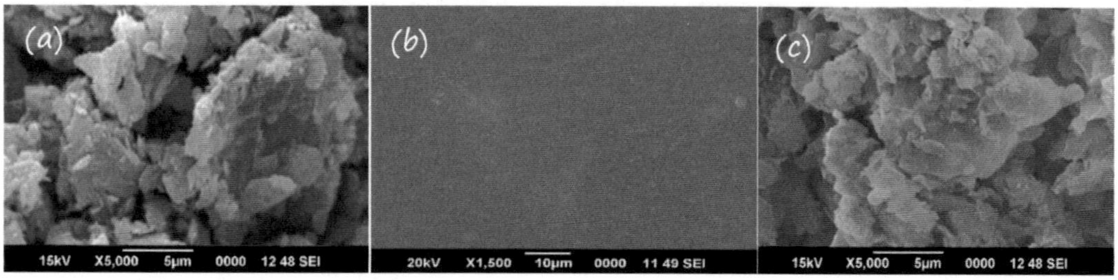

Figure 5. SEM images of (a) OMMT nanoparticles, (b) pristine ABS and (c) Post-treated ABS.

The characterisation of post-treated specimens revealed that OMMT nanoparticles were present on the surface of the post-treated specimens. This will aid in improving the surface properties of the FDM built parts.

In order to investigate the effect of each parameters on the responses Taguchi technique was used. Table 1 demonstrates the average values of the experimental results obtained. It is clear from the experimental results that the surface hardness enhanced by incorporation of OMMT on the surface of FDM parts. Maximum of 9.7% improvement was observed by the post-treatment compared to un-treated specimens. The increase is due to the ceramic attribute of OMMT nanoparticles which resisted indentation compared to pristine polymeric surface. Surface finish was also improved because of the nanoparticle/chemical treatment. Significant reduction in average surface roughness was achieved by the coupled effect of nanoparticles and acetone solvent [14]. Maximum of 94% reduction was observed compared to un-treated specimens. The dimensional variation was marginal compared to the ABS parts without post-treatment. It was found that the dimensional deviation was less than 3% which signifies that the post-treatment approach does not negatively affects the dimension of FDM specimens. However, reduction in thickness was observed after the post-treatment in some cases. The reduction in thickness is favourable as it helps to bring the dimensions closer to the dimension of the CAD model. Since in FDM parts positive deviation is commonly observed in Z height partially because of swelling of the filaments and partially due to the rapid and uneven cooling of the filaments deposited. Also the overlapping of the raster may cause the positive deviation in the z height.

Table 1. Results of experimental runs.

Exp. No.	Layer thickness (mm)	OMMT content (wt %)	Dipping time (sec.)	Hardness (Shore D)	Surface roughness (µm) (R_a)	Dimension (mm)
1	0.30	1	30	69.5	0.870	1.31
2	0.30	1	50	72.0	0.564	1.34
3	0.30	1	70	71.0	0.647	1.35
4	0.30	3	30	73.5	0.790	1.31
5	0.30	3	50	72.5	0.711	1.37
6	0.30	3	70	73.0	1.346	1.38
7	0.30	5	30	71.0	1.283	1.33
8	0.30	5	50	69.5	1.469	1.37
9	0.30	5	70	69.0	1.403	1.43
10	0.25	1	30	75.0	1.893	1.47
11	0.25	1	50	75.0	0.819	1.44
12	0.25	1	70	69.5	1.113	1.43
13	0.25	3	30	75.5	1.048	1.45
14	0.25	3	50	72.0	1.284	1.43
15	0.25	3	70	71.0	1.102	1.53
16	0.25	5	30	73.5	1.277	1.51
17	0.25	5	50	73.0	1.134	1.50
18	0.25	5	70	71.5	1.679	1.51
Un-treated (0.25 mm layer thickness)				70	9.61	1.46
Un-treated (0.30 mm layer thickness)				67	11.25	1.34

Figure 6 shows the main effect plot for the responses. From the figure the optimal parameters settings can be extracted. For hardness the optimal level setting is 0.25 mm layer thickness, 3 wt% OMMT content and 30 seconds dipping time. The most influential parameter was the dipping time as it governed the amount of OMMT particles deposition. However, on increasing the dipping time surface hardness decreases. This can be due to the clustering effect of nanoclay [15]. For surface roughness the optimal parameters were 0.25 mm layer thickness, 5 wt% OMMT content and 50 seconds dipping time. OMMT content was the most influential parameter for surface roughness. Lower nanoparticle content demonstrated better results. On increasing the nanoparticle content the roughness increased. However, this increase is very insignificant compared to the reduction in roughness obtained in un-treated specimens. The optimal setting for dimensional accuracy is 0.3 mm layer thickness 5 wt% OMMT content and 50 seconds dipping time. Layer thickness was the most influential parameter for marinating the dimensional accuracy compared to the CAD model. Also the dimension (thickness) of CAD model selected plays an important role since during tool path generation for material deposition the thickness in decimal is rounded off towards positive level. Thereby, producing positive deviation in 3D printed model.

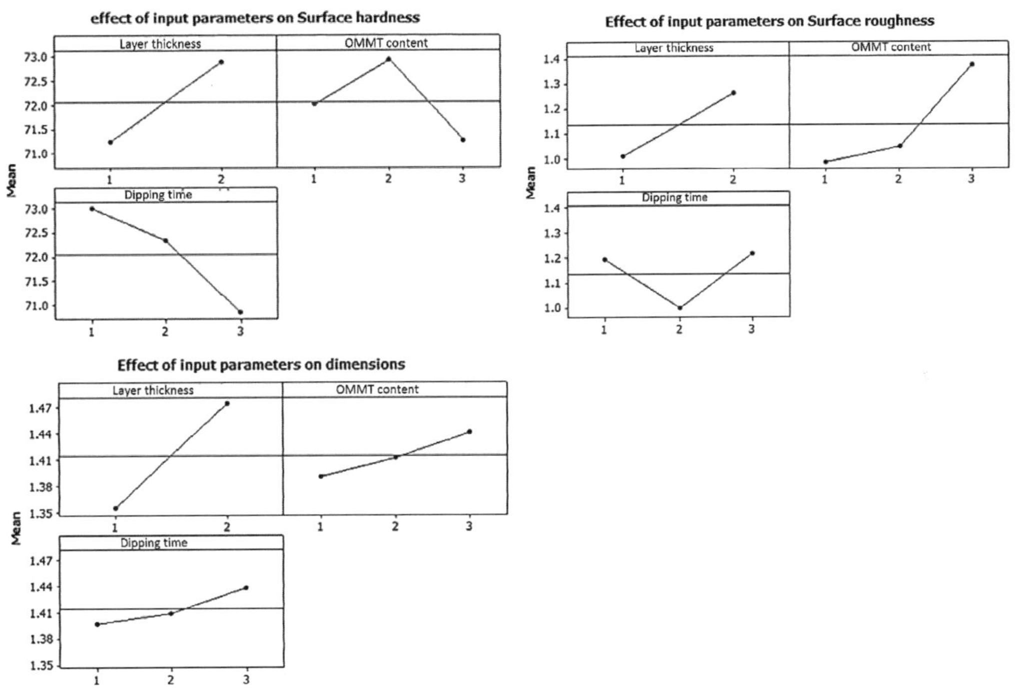

Figure 6. Main effect plot for the responses.

It should be important to note that these settings will optimize the responses individually. However, for practical conditions all the responses may come into picture and it is vitally important that all the responses should be considered for optimization. Since Taguchi method is based on the single response criterion. Therefore, grey relational analysis technique is adopted for multi response optimization [4]. The S/N ratio obtained from Taguchi method were normalized between 0 to 1 values for grey relational analysis. Further the normalized values were used to calculate the grey relational coefficient. These coefficient were further used to calculate the grey relational grade and accordingly the ranks were assigned as demonstrated in Table 2. The higher grey relational grade represents that the corresponding experimental result is closer to the ideally normalized value. Ranks are assigned according to the grade. First rank is assigned to the experiment run having highest grade value.

Table 2. Grey relational coefficient with their grades and ranks.

Exp. No.	Grey Relation coefficient			Grey relation grade	Rank
	Hardness (Shore D)	Surface roughness (μm) (R_a)	Dimension (mm)		
1	0.352159618	0.437812699	0.333333333	0.374435217	18
2	0.48671024	0.333333333	0.369236499	0.396426691	16
3	0.42280863	0.360590141	0.382774576	0.388724449	17
4	0.626371866	0.409264479	0.333333333	0.456323226	14
5	0.526142252	0.382053692	0.412703077	0.440299673	15
6	0.572107997	0.639682149	0.429303703	0.547031283	8
7	0.42280863	0.608845233	0.356530936	0.462728266	13
8	0.352159618	0.704800179	0.412703077	0.489887624	11
9	0.333333333	0.668998769	0.53452363	0.512285244	10
10	0.871391954	1	0.659893509	0.843761821	1
11	0.871391954	0.419487566	0.561468379	0.617449299	7
12	0.352159618	0.532699374	0.53452363	0.473127541	12
13	1	0.505913467	0.591056134	0.698989867	3
14	0.48671024	0.609322832	0.53452363	0.543518901	9
15	0.42280863	0.528084438	1	0.650297689	5
16	0.626371866	0.605989003	0.855052062	0.69580431	4
17	0.572107997	0.541607081	0.79673836	0.636817813	6
18	0.452619392	0.834621944	0.855052062	0.714097799	2

Since the experimental design is orthogonal, it is then possible to separate out the effect of each process parameter on the grey relational grade at different levels. Table 3 shows the effect of individual parameter based on grey relational grade. The optimal settings was found to be 0.25 m layer thickness, 3 wt% OMMT content and 30 seconds dipping time. Layer thickness is the most influencing factor followed by OMMT content and dipping time.

Table 3. Effect of individual parameter based on grey relational grade.

Factors	Average grey relational grade			Main effect	Rank
	Level 1	Level 2	Level 3		
Layer thickness	0.452015742	0.652651671*	-	0.20063593	1
OMMT content	0.51565417	0.556076773	0.585270176*	0.069616007	2
Dipping time	0.588673784*	0.520733333	0.547594001	0.067940451	3
Total mean value of grey relational grade 0.552333706					
*Optimal levels by grey relational grades					

4. Conclusions

OMMT nanoparticles were successfully embedded on the surface of FDM specimens. The approach presented provides an efficient way to enhance the surface properties of the FDM parts which can increasing their application domain. XRD, FTIR and SEM analysis showed the presence of embedded nanoparticles. UV-vis spectroscopy demonstrated that post-treatment can improve the UV absorbance of FDM built ABS polymer. 9.7% increase in surface hardness was achieved along with 94% reduction in surface roughness. Dimensional deviation of less than 3 % was achieved. The effect of parameters on individual responses according to Taguchi technique showed that dipping time, OMMT content and layer thickness were the most influential parameters for surface hardness, surface roughness and dimensional accuracy respectively. Further multi objective optimization revealed that layer thickness is the most influential parameter followed by OMMT content and dipping time. The optimal parametric values were 0.25 mm layer thickness, 5 wt % OMMT content and 30 seconds dipping time.

References

[1] Chua CK, Leong KF, Lim CS. Rapid Prototyping: principles and applications. *Rapid prototyping*. 2003

[2] Guo N, Leu MC. Additive manufacturing: Technology, applications and research needs. *Front Mech Eng*. 2013;8(3): pp 215-43

[3]. Ning F, Cong W, Qiu J, Wei J, Wang S. Additive manufacturing of carbon fi ber reinforced thermoplastic composites using fused deposition modeling. *Compos Part B*. 2015;80: pp 369-78

[4] Sood AK, Ohdar RK, Mahapatra SS. Improving dimensional accuracy of Fused Deposition Modelling processed part using grey Taguchi method. *Mater Des*. 2009;30(10): pp 4243-52

[5] Taufik M, Jain PK. A Study of Build Edge Profile for Prediction of Surface Roughness in Fused Deposition Modeling. *J Manuf Sci Eng*. 2016;138(6): p 61002

[6] Galantucci LM, Lavecchia F, Percoco G. Quantitative analysis of a chemical treatment to reduce roughness of parts fabricated using fused deposition modeling. *CIRP Ann - Manuf Technol*. 2010;59(1): pp 247-50

[7] Kuo CC, Mao RC. Development of a Precision Surface Polishing System for Parts Fabricated by Fused Deposition Modeling. *Mater Manuf Process*. 2016 Jun 10;31(8):pp 1113-18

[8] Sinha Ray S, Okamoto M. Polymer/layered silicate nanocomposites: A review from preparation to processing. *Prog Polym Sci*. 2003;28(11): pp 1539-641

[9] Francis V, Jain PK. Experimental investigations on fused deposition modelling of polymer-layered silicate nanocomposite. *Virtual Phys Prototyp*. 2016;11(2): pp 109-21

[10] Wang Z, Wang J, Li M, Sun K, Liu C. Three-dimensional Printed Acrylonitrile Butadiene Styrene Framework Coated with Cu-BTC Metal-organic Frameworks for the Removal of Methylene Blue. *Sci Rep*. 2014;4(5939): pp 1-7

[11] Pourabas B, Raeesi V. Preparation of ABS/montmorillonite nanocomposite using a solvent/non-solvent method. *Polymer*. 2005;46(15): pp 5533-40

[12] Bhagabati P, Chaki TK, Khastgir D. Panoptically exfoliated morphology of chlorinated polyethylene (CPE)/ethylene methacrylate copolymer (EMA)/layered silicate nanocomposites by novel in situ covalent modification using poly(ε-caprolactone). *RSC* adv.2015;5(48): pp 38209-22

[13] Skorski MR, Esenther JM, Ahmed Z, Miller AE, Hartings MR. The chemical, mechanical, and physical properties of 3D printed materials composed of TiO2-ABS nanocomposites. *Sci Technol Adv Mater*. 2016;17(1): pp 89-97

[14] Cunico MWM, Cunico MM, Cavalheiro PM, Carvalho J de. Investigation of additive manufacturing surface smoothing process. Rapid Prototyp J. 2017;23(1): pp 201-08

[15] Lam CK, Cheung HY, Lau KT, Zhou LM, Ho MW, Hui D. Cluster size effect in hardness of nanoclay/epoxy composites. *Compos Part B Eng*. 2005;36(3): pp 263-69

IOP Conf. Series: Materials Science and Engineering **383** (2018) 012011 doi:10.1088/1757-899X/383/1/012011

Zr-based metal-organic framework with dual BrØnsted acid-base functions

W Pratanpornlerd and S Bureekaew*

School of Energy Science and Engineering, Vidyasirimedhi Institute of Science and Technology, Rayong 21210, Thailand

*E-mail: sareeya.bureekaew@gmail.com

Abstract. Design and synthesis of metal-organic frameworks (MOFs) with multi-functionality have attracted a lot of attention since the different functional groups can enhance the cooperative effect providing properties superior to those found in mono-functional MOFs. The efficiency and selectivity of the reactions can be improved by the simultaneous activation of multi-functional catalysts. Herein, a series of BrØnsted acid-base bifunctional UiO-66 bearing acidic sulfo- and basic amino- functional groups were successfully prepared *via* a simple one-pot solvothermal method. PXRD patterns of the bifunctional frameworks were similar to that of simulated UiO-66. The existence of sulfo-and amino- functional groups was investigated by IR and H^1-NMR. The bifunctional frameworks showed high thermal stability up to 300°C examined by TGA. By varying the molar ratios of ligands, the proportions of acidic and basic moieties in the framework were successfully tuned. The amount of functional groups anchoring in resulting frameworks was determined by ^{1}H-NMR. This bifunctional UiO-66 was tested for CO_2 capture.

1. Introduction

Metal-organic frameworks (MOFs), also called porous coordination polymers (PCPs), are porous crystalline solids assembled from metal ions and organic ligands. MOFs have generated considerable interest due to their high surface area, the diverse design for pore sizes and shapes, and the possibility for chemical functionalization [1, 2]. To use MOFs as materials for specific applications, MOFs were functionalized generally in three strategies, e.g., trapping active guests into the framework cavities, modulating metal ions in the framework to obtain coordinatively unsaturated metal sites (CUSs), and employing organic ligands which have pendant functionalized groups [3].

MOFs have been demonstrated in various applications such as gas storage, gas separation, sensing and catalysis [1-3]. CO_2 capture using MOFs as adsorbents is one interesting studied area [4]. Many MOFs were investigated for CO_2 capture and separation from flue gas mixtures which is released into the atmosphere (1 atm). The CO_2 partial pressure (<0.15 atm) is relatively low in such conditions. The strong interaction between CO_2 and MOF frameworks is essential to increase the CO_2 uptake [4]. Linear CO_2 has a permanent electrical quadrupole moment that can enhance strong interactions with polar molecules by the quadrupole-dipole interaction. The introduction of polar functional groups into the frameworks such as amino- (NH_2), sulfo- (SO_3H), and carboxylic- (COOH) is one promising strategy to increase CO_2 and MOF interaction aiming for efficient CO_2 capture [5-7]. Amino and amine-derivative functional MOFs have demonstrated a great performance of CO_2 capture evaluated

Content from this work may be used under the terms of the Creative Commons Attribution 3.0 licence. Any further distribution of this work must maintain attribution to the author(s) and the title of the work, journal citation and DOI.
Published under licence by IOP Publishing Ltd

from the CO_2 heat of adsorption [8-10]. For instance, ethylenediamine, pendant alkylamine species, was incorporated into the CUSs of $H_3[(Cu_4Cl)_3(BTTri)_8]$ (H_3BTTri = 1,3,5-tris(1H-1,2,3-triazol-5-yl)benzene)). It showed a higher value of the CO_2 heat of adsorption (90 kJ mol^{-1}) compared with a non-functionalized framework (21 kJ mol^{-1}) [8].

Recently, multi-functional MOFs bearing more than one functional group have attracted much attention. Having multi-functional groups in the frameworks can provide synergistic behaviors which contribute to some properties superior to those found in mono-functional MOFs [11]. For example, the catalytic performance of USTC-253–TFA (USTC = University of Science and Technology of China, TFA = trifluoroacetate), a sulfone functionalized MOF, for the CO_2 cycloaddition of epoxides was investigated [12]. USTC-253-TFA contains different kinds of active sites, i.e., sulfone groups, Brønsted acid from bridging OH groups, and Lewis acid from CUSs (introduced by TFA). TFA has been used to generate structural defects in MOFs causing Lewis acid sites [13-15]. The Lewis acid CUSs and sulfone functional groups exhibit strong interaction with CO_2 molecules. Also, Brønsted and Lewis acid sites were used to activate the epoxides. The cooperative manners of the mentioned functional groups in USTC-253-TFA contributed to improved CO_2 storage capacity at 298 K (64.6 cm^3 g^{-1}) compared to un-functional EL-MIL-53 (27.5 cm^3 g^{-1}) and sulfone only functional USTC-257 (47.8 cm^3 g^{-1}) [12]. Also, USTC-253-TFA exhibited high CO_2 conversion (81.3% with propylene oxide), and catalytic recyclability [12].

In the present work, a series of Brønsted acid-base bifunctional UiO-66 (UiO = University of Oslo) bearing acidic sulfo- and basic amino- functional groups were successfully prepared *via* a simple one-pot solvothermal method. UiO-66 ($Zr_6O_4(OH)_4(1,4\text{-bdc})_6$, 1,4-bdc = 1,4-benzenedicarboxylate) is Zr based MOFs assembled from hexanuclear $[Zr_6O_4(OH)_4]^{12+}$ metal ions and twelve 1,4-bdc ligands [16]. The three-dimensional framework of UiO-66 possesses two different kinds of cages, a large octahedral cage (pore size of ~12 Å) encompassed with eight smaller tetrahedral cages (pore size of ~8 Å). UiO-66 has many advantages to design and synthesize MOFs for particular applications due to its high thermal and chemical stability [16]. Sulfo- and amino- functional groups were selected because the polar functional groups are expected to interact strongly with CO_2 molecules [5]. Moreover, the presence of such antagonistic acid-base functional groups in the framework aims the study at the synergistic effects of CO_2 capture.

2. Experimental

2.1. Materials

Zirconium (IV) chloride ($ZrCl_4$, Merck, ≥98%), 2-aminobenzene-1,4-dicarboxylate (NH_2-bdc, TCI, >98%), sodium 4-carboxy-2-sulfobenzoate (SO_3Na-bdc, TCI, >98.0%), 1,4-benzenedicarboxylate (1,4-bdc, Sigma-Aldrich, 98%) acetic acid (Sigma-Aldrich, 99.8%), dimethylformamide (DMF, Qrec, 99.8%), and deionized water (d.i. H_2O, 18.2 MΩ).

2.2. Preparation of parent MOFs UiO-66-SO_3H and UiO-66-NH_2

UiO-66-SO_3H was prepared following the previous report [17]. Note that UiO-66-SO_3H consists of mixed ligands: 1,4-bdc and SO_3Na-bdc. 789 mg (3.3 mmol) of $ZrCl_4$, 122 mg (0.7 mmol) of 1,4-bdc, and 726 mg (2.7 mmol) of SO_3Na-bdc were dissolved into 45 ml of DMF in a 100 ml glass reactor. Then, 10 ml of concentrated acetic acid was added. The mixture was heated at 120°C for 30 h, yielding a white powder. The product was washed three times with DMF and acetone. The resulting solids were activated by water and acetone under refluxing and then dried at 100°C for 5 h in a vacuum oven. The actual mole percent of sulfonic functional groups in UiO-66-SO_3H was 50% as determined by ^{1}H-NMR.

UiO-66-NH_2 was synthesized by adding 315 mg (1.35 mmol) of $ZrCl_4$ and 240 mg (1.35 mmol) of NH_2-bdc, 0.02 ml of water and 50 ml of DMF into a 100 ml glass reactor [18]. The mixture was heated at 120°C for 24 h, resulting in a yellow powder. The product was washed three times with DMF and acetone. The resulting powder was activated in refluxing water and then dried at 100°C for 5 h in a vacuum oven.

2.3. Preparation of UiO-66-SO_3H-NH_2
A series of UiO-66-SO_3H-NH_2 were synthesized by dissolving 466 mg (2 mmol) of $ZrCl_4$, various molar ratios of SO_3Na-bdc and NH_2-bdc in 38 ml of DMF. Then, 8 ml of concentrated acetic acid was added. The molar ratios of SO_3H-bdc and NH_2-bdc were denoted as UiO-66-(*x*)SO_3H-(*100-x*)NH_2, where *x* was the mole percent of SO_3H-bdc. The amounts of SO_3Na-bdc and NH_2-bdc linkers were varied for UiO-66-(*25*)SO_3H-(*75*)NH_2 = 134 mg (0.5 mmol) of SO_3Na-bdc, 270 mg (1.5 mmol) of NH_2-bdc, UiO-66-(*50*)SO_3H-(*50*)NH_2 = 268 mg (1 mmol) of SO_3Na-bdc, 180 mg (1 mmol) of NH_2-bdc, and UiO-66-(*75*)SO_3H-(*25*)NH_2 = 402 mg (1.5 mmol)of SO_3Na-bdc, 90 mg (0.5 mmol) of NH_2-bdc. The mixtures were heated at 120°C for 30 h obtaining a pale-yellow powder. The product was washed three times with DMF and acetone. The resulting solids were activated in refluxing water and dried at 100°C for 5 h in a vacuum oven.

2.4. Characterizations
The powder X-ray diffraction (PXRD) patterns were obtained from a Bruker D8 Advance X-ray powder diffractometer equipped with a Cu sealed tube (λ = 1.54178 Å) at a scan rate of 0.02 deg s^{-1}. Thermogravimetric analysis (TGA) experiments were examined by a Linseis STA PT1600 under nitrogen gas with a flow rate of 100 cm^3 min^{-1}. The samples were heated at the rate of 5°C min^{-1} up to 700°C. Quantitative analysis of linker was measured by proton nuclear magnetic resonance (^{1}H-NMR) using a Bruker Advance III HD (600 MHz). Before ^{1}H-NMR measurement, 20 mg of MOF was digested in 600 μL of a 1 M NaOH-D_2O solution [19]. Attenuated total reflectance Fourier transform infrared spectra (ATR-FTIR) were recorded on a Perkin Elmer Universal-ATR. The samples were measured over the wavenumber range 400–4000 cm^{-1} with 4 cm^{-1} resolution and 32 scans. CO_2 sorption isotherms of the parent UiO-66 and the bifunctional frameworks measured at 298 K were obtained from a BELSORP-mini II. Before gas sorption measurement, MOF samples were activated again at 100°C for 5 h under vacuum conditions.

3. Results and discussion
Structure and purity of UiO-66-SO_3H-NH_2 samples with various ligand ratios were confirmed by PXRD analysis. The PXRD patterns of UiO-66-(*25*)SO_3H-(*75*)NH_2, UiO-66-(*50*)SO_3H-(*50*)NH_2, and UiO-66-(*75*)SO_3H-(*25*)NH_2 were similar to that of the simulated parent UiO-66 as shown in Figure 1 [16]. The results indicated that all synthesized UiO-66-SO_3H-NH_2 frameworks shared the same structure with the parent UiO-66.

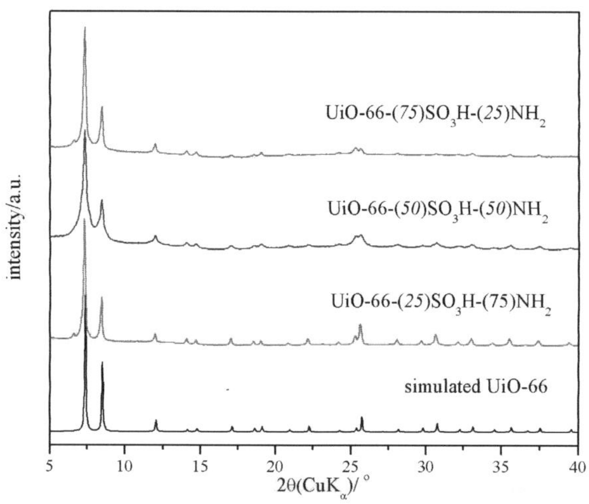

Figure 1. PXRD patterns of simulated UiO-66 (black), as-synthesized UiO-66-(*25*)SO_3H-(*75*)NH_2 (red), UiO-66-(*50*)SO_3H-(*50*)NH_2 (blue), and UiO-66-(*75*)SO_3H-(*25*)NH_2 (green).

The coexisting of SO_3H-bdc and NH_2-bdc in UiO-66-SO_3H-NH_2 was determined by IR and ^1H-NMR. Figure 2 shows the IR spectrum of UiO-66-(*50*)SO_3H-(*50*)NH_2 which presented the characteristic stretching peaks of N-H (primary amine) at 3471 and 3358 cm^{-1}, O=S=O at 1253 and 1167 cm^{-1} and S-O at 1074 and 1024 cm^{-1} [17, 20, 21]. These results confirmed the existence of acid and base functional groups in the bifunctional frameworks.

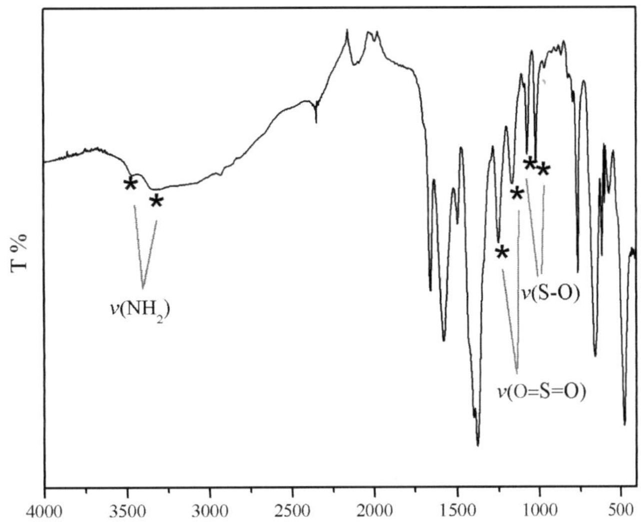

Figure 2. IR spectrum of as-synthesized UiO-66-(*50*)SO_3H-(*50*)NH_2 possessing sulfo- and amino-functional groups.

The ^1H-NMR spectra of the digested UiO-66-SO_3H-NH_2 in NaOH/D_2O solution also confirmed the coexistence of acidic SO_3H-bdc and basic NH_2-bdc ligands as shown in Figure 3. The proportions of acid and base moieties in the bifunctional frameworks were successfully tuned. The SO_3H-bdc and NH_2-bdc peaks of all UiO-66-SO_3H-NH_2 samples were measured to calculate the actual percentages of acid and base moieties in each bifunctional framework. Table 1 presents percentages of SO_3H functions from the molar ratio of ligands and ^1H-NMR results after synthesis. The ^1H-NMR results show that all bifunctional frameworks have slightly less percentage of acidic moieties than the original molar ratio of ligands.

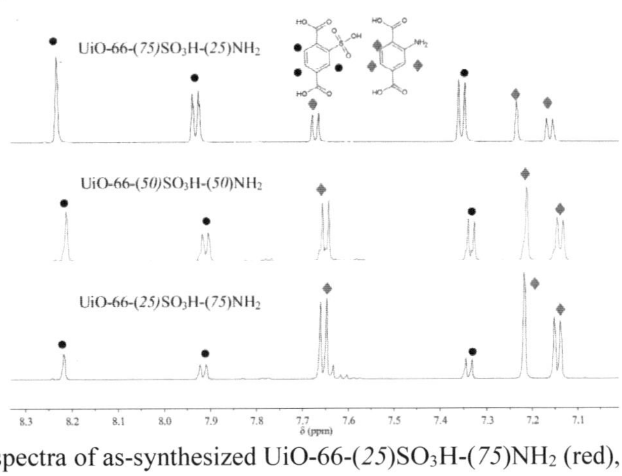

Figure 3. ^1H-NMR spectra of as-synthesized UiO-66-(*25*)SO_3H-(*75*)NH_2 (red), UiO-66-(*50*)SO_3H-(*50*)NH_2 (green), and UiO-66-(*75*)SO_3H-(*25*)NH_2 (blue).

Table 1. The percentages of SO_3H-bdc from the molar ratio of ligands at preparation and ^{1}H-NMR results of UiO-66-(*25*)SO_3H-(*75*)NH_2, UiO-66-(*50*)SO_3H-(*50*)NH_2, and UiO-66-(*75*)SO_3H-(*25*)NH_2.

MOF	The percentage of SO_3H-bdc	
	Molar ratio of ligands	^{1}H-NMR results
UiO-66-(*25*)SO_3H-(*75*)NH_2	25	17
UiO-66-(*50*)SO_3H-(*50*)NH_2	50	41
UiO-66-(*75*)SO_3H-(*25*)NH_2	75	68

Thermal stability of UiO-66-(*50*)SO_3H-(*50*)NH_2 was examined by TGA. Figure 4 shows the TGA profile with two steps of weight loss. In the first step, the weight loss of 23.3% for temperatures up to 225 °C corresponded to the evaporation of the water solvent. The second weight loss starting from 300 °C was due to the decomposition of the framework. This result confirmed that the bifunctional framework was thermally stable up to 300°C.

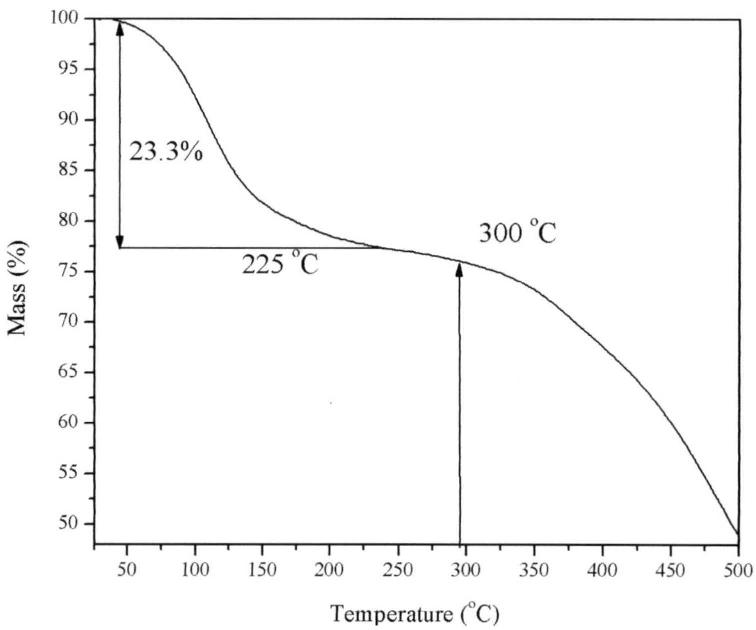

Figure 4. TGA of UiO-66-(*50*)SO_3H-(*50*)NH_2 which has thermal stability up to 300°C.

The UiO-66-SO_3H-NH_2 frameworks were tested for CO_2 capture. Figure 5 and Table 2 show CO_2 sorption isotherms and storage capacity of UiO-66-NH_2, UiO-66-SO_3H (50% of sulfonic groups), UiO-66-(*25*)SO_3H-(*75*)NH_2, UiO-66-(*50*)SO_3H-(*50*)NH_2, and UiO-66-(*75*)SO_3H-(*25*)NH_2 at 298 K. UiO-66-NH_2 exhibited CO_2 storage capacity of 43 $cm^3\ g^{-1}$, lower than reported in the literature (c.a. 67 $cm^3\ g^{-1}$) [22]. UiO-66-(*25*)SO_3H-(*75*)NH_2 showed the highest CO_2 storage capacity (57 $cm^3\ g^{-1}$) while UiO-66-SO_3H exhibited the lowest capacity (14 $cm^3\ g^{-1}$).

IOP Conf. Series: Materials Science and Engineering **383** (2018) 012011 doi:10.1088/1757-899X/383/1/012011

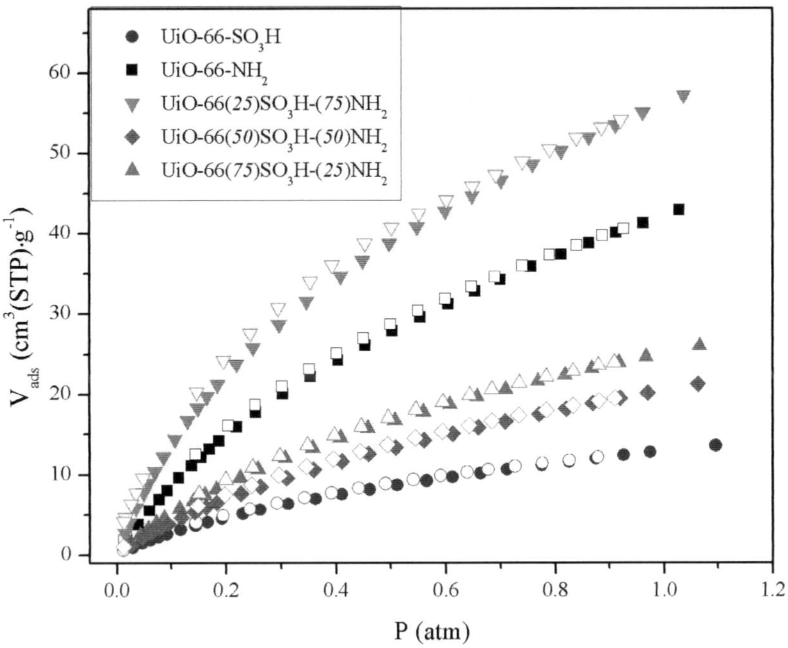

Figure 5. CO_2 sorption isotherms at 298 K, adsorption (filled circles) and desorption (open circles) of UiO-66-SO_3H (blue), UiO-66-NH_2 (black), UiO-66-(*25*)SO_3H-(*75*)NH_2 (pink), UiO-66-(*50*)SO_3H-(*50*)NH_2 (green), and UiO-66-(*75*)SO_3H-(*25*)NH_2 (red).

Table 2. CO_2 storage capacity of UiO-66-SO_3H, UiO-66-NH_2, UiO-66-(*25*)SO_3H-(*75*)NH_2, UiO-66-(*50*)SO_3H-(*50*)NH_2, and UiO-66-(*75*)SO_3H-(*25*)NH_2 at 298 K.

MOF	CO_2 storage capacity (cm^3 g^{-1})
UiO-66-SO_3H	14
UiO-66-NH_2	43
UiO-66-(*25*)SO_3H-(*75*)NH_2	57
UiO-66-(*50*)SO_3H-(*50*)NH_2	21
UiO-66-(*75*)SO_3H-(*25*)NH_2	26

4. Conclusion

A series of Brønsted acid-base bifunctional UiO-66 MOFs carrying acidic sulfo- and basic amino functional groups were successfully synthesized *via* a one-pot solvothermal procedure. PXRD results confirmed that the structures of the bifunctional frameworks shared the same structure as the parent UiO-66. The existence of sulfo- and amino- ligands in the bifunctional frameworks was examined by IR and ^{1}H-NMR. The proportions of acid and base moieties were successfully tuned by changing molar ratios of the ligands before synthesis. TGA analysis showed the bifunctional framework had high thermal stability up to 300°C. The bifunctional UiO-66 with ~17% of SO_3H (25% original molar ratio) showed improvement of CO_2 storage capacity compared with the parent MOFs, acidic and basic UiO-66.

References

[1] Rowsell J L C and Yaghi O M 2004 *Microporous Mesoporous Mater.* **73** pp 3–14
[2] Furukawa H, Cordova K E, O'Keeffe M and Yaghi O M 2013 *Science.* **341** 1230444
[3] Zhou H C, Long J R and Yaghi O M 2012 *Chem. Rev.* **112** pp 673-4
[4] Sabouni R, Kazemian H and Rohani S 2014 *Environ. Sci. Pollut. Res.* **21** pp 5427-49
[5] Li J-R, Ma Y, McCarthy M C, Sculley J, Yu J, Jeong H-K, Balbuena P B and Zhou H-C 2011 *Coord. Chem. Rev.* **255** pp 1791-23
[6] Torrisi A, Bell R G and Mellot-Draznieks C 2010 *Cryst. Growth Des.* **10** pp 2839–41
[7] Chen J, Li H, Zhong M and Yang Q 2017 Capture *Chem. – An Asian J.* **12** pp 2291–8
[8] Demessence A, D'Alessandro D M, Foo M L and Long J R 2009 *J. Am. Chem. Soc.* **131** pp 8784-6
[9] Liu J, Thallapally P K, McGrail B P, Brown D R and Liu J 2012 *Chem. Soc. Rev.* **41** pp 2308-22
[10] An J, Geib S J and Rosi N L 2010 *J. Am. Chem. Soc.* **132** pp 38-9
[11] Huang Y-B, Liang J, Wang X-S and Cao R 2017 *Chem. Soc. Rev.* **46** pp 126-57
[12] Jiang Z-R, Wang H, Hu Y, Lu J and Jiang H-L 2015 *ChemSusChem* **8** pp 878-85
[13] Wu H, Chua Y S, Krungleviciute V, Tyagi M, Chen P, Yildirim T and Zhou W 2013 *J. Am. Chem. Soc.* **135** pp 10525-32
[14] Cai G and Jiang H-L 2017 *Angew. Chemie Int. Ed.* **56** pp 563-67
[15] Fang Z, Bueken B, De Vos D E and Fischer R A 2015 *Angew. Chemie Int. Ed.* **54** pp 7234-54
[16] Cavka J H, Jakobsen S, Olsbye U, Guillou N, Lamberti C, Bordiga S and Lillerud K P 2008 J. *Am. Chem. Soc.* **130** pp 13850-51
[17] Foo M L, Horike S, Fukushima T, Hijikata Y, Kubota Y, Takata M and Kitagawa S 2012 *Dalt. Trans.* **41** pp 13791-94
[18] Vermoortele F, Ameloot R, Vimont A, Serre C and De Vos D 2011 *Chem. Commun.* **47** pp 1521-23
[19] Chavan S M, Shearer G C, Svelle S, Olsbye U, Bonino F, Ethiraj J, Lillerud K P and Bordiga S 2014 *Inorg. Chem.* **53** pp 9509-15
[20] Shen L, Liang S, Wu W, Liang R and Wu L 2013 *Dalt. Trans.* **42** pp 13649-57
[21] Andriamitantsoa R S, Wang J, Dong W, Gao H and Wang G 2016 *RSC Adv.* **6** pp 35135-43
[22] Cmarik G E, Kim M, Cohen S M and Walton K S 2012 *Langmuir* **28** pp 15606-13

Evaluation of laser Induced Breakdown Spectroscopy for analysis of annealed Aluminum Germanium alloy at different temperatures

W. A. Farooq[1,*], Walid Tawfik[1,2], M. Atif [1,3], M S Alsalhi[1], H.Y.Zahran[4,5], A. F. Abd El-Rehim[4,5], I.S. Yahia[5] and Mansoor Sarfraz[6]

[1]Department of Physics and Astronomy, College of Science, P.O.Box 2455, King Saud University Riyadh Saudi Arabia
[2]Department of Environmental Applications, NILES National Institute of Laser, Cairo University Cairo, Egypt
[3]National Institute of Laser and Optronics, Nilore, Islamabad, Pakistan
[4]Physics Department, Faculty of Education, Ain Shams University, P.O. Box 5101, Heliopolis 11771, Roxy, Cairo, Egypt
[5]Physics Department, Faculty of Science, King Khalid University, P.O. Box 9004, Abha 61413, Saudi Arabia
[6]Sustainable Energy Technologies Centre, King Saud University, Riyadh, Saudi Arabia

*E-mail: wafarooq@hotmail.com

Abstract. Laser Induced Breakdown Spectroscopy (LIBS) technique has been applied to study the annealing effect of Al-Ge alloy at 150, 175, 240, 280, 325 ^{0}C. LIBS spectra of all samples prepared at these annealing temperatures were recorded. The signal intensity of the Al and Ge spectral lines increases with increase in annealing temperature. This increase in the signal intensity is due to structural modification of Al-Ge sheets surface with annealing temperatures. This variation in intensity verifies the assessment of annealed materials which can be performed with LIBS technique.

1. Introduction

The Laser Induced Breakdown Spectroscopy (LIBS) technique is a suitable analytical tool for analyzing various types of sample such as solid, liquid and gas [1-15]. It is useful for on line, real time measurement and it is also suitable for field application in inaccessible environments owing to its advantages that include in situ, quick, and remote elemental analysis [1,2]. The technique utilizes atomic emission spectra directly collected from plasmas generated by the interaction of high-energy laser pulses and the target material. The line emission radiated from the plasma presents a spectral signature indicating the presence of constituents of the sample. The line emission intensity from the observed atomic emission spectra can be used for quantitative evaluation of the elemental composition.

LIBS technique has been used to measure the qualitative and quantitative composition of solid, molten materials and powder samples [16-18]. It has also been used to measure the composition of the molten materials. Only limited work has been carried out on LIBS application to thin film and thin

Content from this work may be used under the terms of the Creative Commons Attribution 3.0 licence. Any further distribution of this work must maintain attribution to the author(s) and the title of the work, journal citation and DOI.
Published under licence by IOP Publishing Ltd

sheets [19-21]. Recently, LIBS has been used to measure the elemental composition of thin sheets of copper. An interesting application of LIBS in variation of sample surface in polymer sheets has been reported [22].

In the present paper, we aimed to use LIBS technique for investigation of annealing effect at different temperatures on Al-Ge alloy thin sheets. Variation in the signal intensity of Al-Ge alloy samples annealed at different temperatures demonstrates the suitability of LIBS technique for assessment of level of annealing of the material. In addition impurities in the sample can also be identifies at the same time.

2. Experimental procedure

2.1. Samples

Al-Ge alloys preparation

Al-1.0 wt. % Ge alloy was prepared from pure elements 99.99% Al and Ge by melting in graphite crucible together with calcium chloride $CaCl_2$ flux to prevent oxidation. The ingot was homogenized at 823K for 48 hours. The ingot in the shape of rod was swaged into sheets of 2 mm in thickness for analysis with LIBS. The samples were annealed for 2 h at 773 K then quenched in iced water to obtain samples with the supersaturated - solid solution structure. Sheets of Al-Ge alloy samples were aged at temperature range (423- 598K) for 2h then quenched in iced water before applying the measurements.

2.2. The LIBS system

The schematic diagram of the LIBS experimental setup used in the present investigation is described elsewhere in details [16]. Briefly, the LIBS experiment was carried out with spectrolaser-7000 (Laser Analysis Technologies, Australia). The spectrolaser-7000 is an integrated system consists of excitation laser, optical spectrographs, sample chamber, focusing optics and data acquisition system equipped with charge coupled detector (CCD). The excitation laser is a high power (5-300 mJ variable) Nd:YAG laser having a 7 ns pulse duration at the fundamental wavelength of 1064 nm and operates in the Q-switched mode with a pulse repetition rate of 10 Hz. It also has the ability to focus on a fresh region of the sample through a short focal length lens for each successive laser pulse. In the present study, 100 mJ energy of the laser has been used with a varied delay time between 0.2 to 1μs (gate delay between the laser firing and the CCD detector). The laser radiation was focused by a convex lens with a focal length of about 48 mm onto a target to ablate material from the sample surface and produce a plasma cloud. A 600 μm diameter optical fiber was used to collect the radiation emitted by the generated plasma cloud. This collected light was then delivered to a high-resolution (0.1 nm FWHM) spectrometer equipped with a gated CCD camera. Each LIBS spectrum was recorded in the range of 190 – 1100 nm and an average of 10 points was taken to reduce the error due to laser fluctuation.

3. Results and discussion

The observed LIBS spectrum of Al-Ge sheets without annealing from 200 nm to 900 nm is shown in Figure 1(a). The emission from LIBS plasma was recorded at a delay of t_{gate} = 1000 ns with respect to firing time of the laser at the target material and at a gate width of ΔT_{gate} = 2000 ns. Figure 1(a) reveals that the spectral emission lines from neutral atoms of aluminum and germanium dominate the LIBS spectrum over the whole spectral range from 200- 900 nm. The other impurities present in the samples are listed in Table 1. Actually, this is expected because Al and Ge are the main contents of the samples. Abundances of aluminum ions rather than germanium ions in the observed LIBS plasma were recognized due to the lower ionization potential of aluminum (IPAl = 5.98 eV) compare to germanium (IPGe = 7.89 eV). The observed aluminum transitions of 394 nm, and 396 nm have strong lines intensities and large self-absorption as expected since these represent aluminum resonance lines from ground states with high transition probabilities. The proper choice of the experimental conditions is therefore essential to overcome the effect of high number density and reduce the optical thickness of the plasma in a viable approach to minimize self-absorption. In doing so, the reduction of spectral lines intensity mitigates population number density of the species in the lower energy level of the transition. There are two ways to decrease the population number densities of low-lying energy levels: (A) By decreasing the laser pulse energy, the overall number of ablated species reduces, or (B) by studying the plasma during its early time of plume expansion (increased plasma temperature). Usually, the choice (A) is not very useful because the reduction of the laser energy not only reduces the number of ablated atoms but it also reduces the plasma temperature. On the other hand, the increase of temperature via the reduction of the gate delay (Condition B) is an efficient way to reduce the self-absorption as the temperature strongly reduces during the expansion of the laser-produced plasma. Thus, the optimized recording conditions are obtained when the observation time is early enough to limit the self-absorption and large enough to get a satisfactory signal-to-noise ratio. These conditions were achieved at 100mJ of laser energy, gate delay of t_{gate} = 1000 ns and gate width of Tgate = 2000 ns as mentioned earlier.

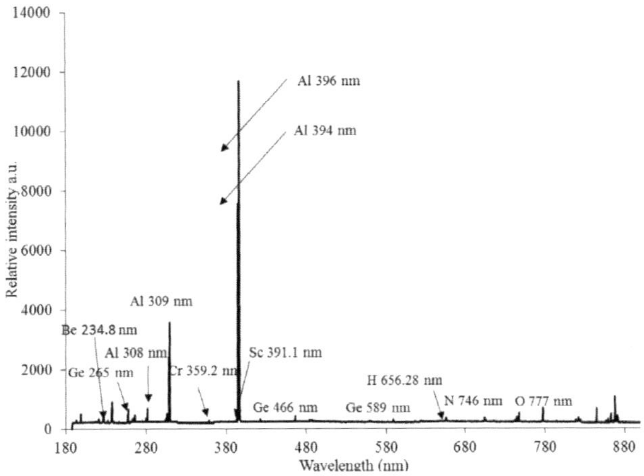

Figure 1(a). LIBS spectrum from 200-900 nm for Al-Ge sample without annealing.

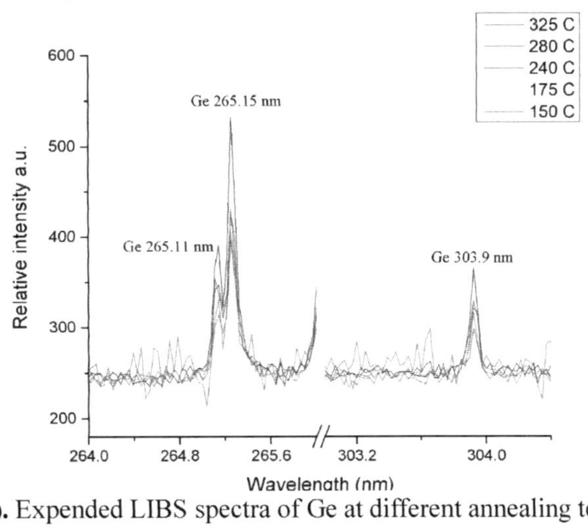

Figure 1(b). Expended LIBS spectra of Ge at different annealing temperatures.

Variations in the intensities of Ge lines at 265.15 nm, 446.35 nm and 285.16 nm observed in the LIBS spectra at different annealing temperatures are depicted in Figure 1(b). Similar effect in the Aluminum lines at wavelengths, 308.21 nm and 309.28 nm at different annealing temperatures was also observed which is shown in Figure 1(c).

Table 1. Spectroscopic data for trace elements.

Wavelength	Element	Elower	Eupper	A
194.295	Be I	21978.27	73429.32	1.30E+05
234.86	Be I	0	42565.4	5.60E+07
359.348	Cr I	0.00	27820.23	1.50E+07
391.181	Sc I	168.33	25724.67	1.80E+07
423.804	Sc I	16210.84	39799.98	7.10E+06
656.28	H I	82259	97492.3	4.40E+06
742.395	N I	83285.5	96751.7	6.00E+05
777.195	O I	73767.8	86631	3.70E+06
868.028	NI	83364.6	94881.8	2.50E+07

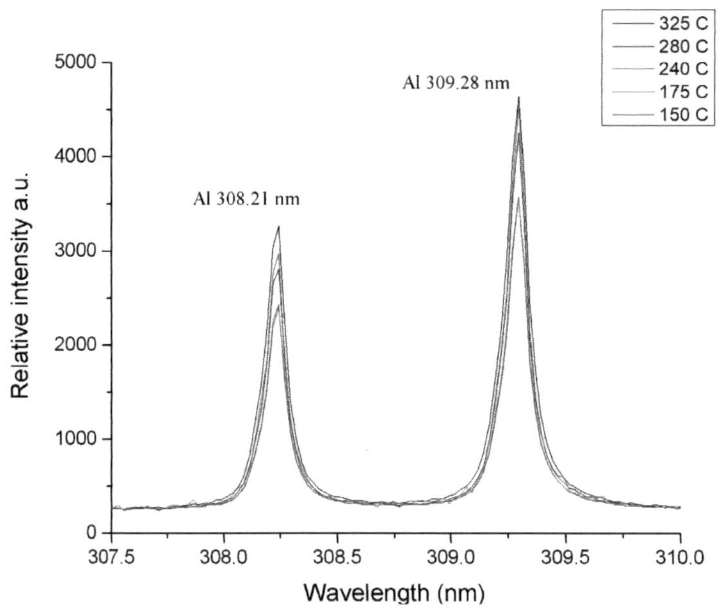

Figure 1(c). Al-Ge LIBS spectra of Al lines at different annealing temperatures.

When a sheet is annealed, the crystal structure is changed and as the annealing temperature increases, the energy band gap of the material decreases [23, 24]. Due to decrease in band gap less energy is required to excite and ionize the material. Since the energy in LIBS experiment was kept constant, therefore LIBS spectra of the sheet samples at higher annealing temperature are showing high intensity as compared to the samples annealed at lower temperatures.

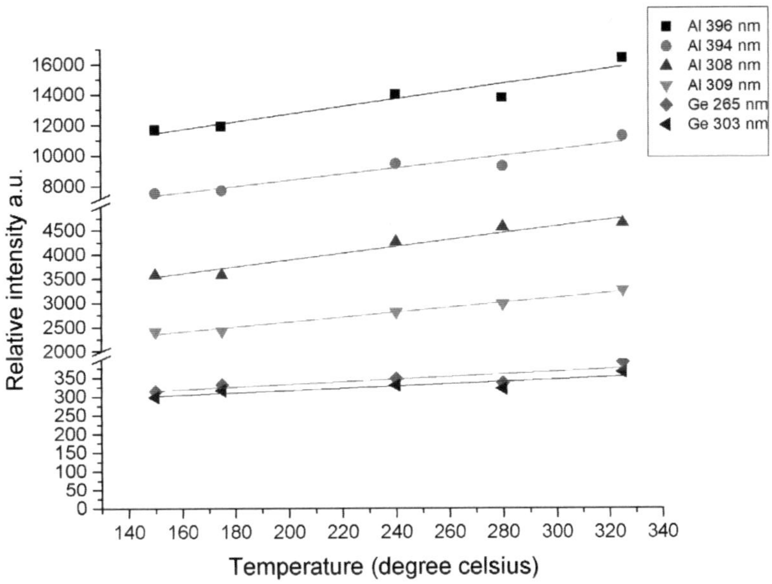

Figure 2. Variation of Al 309.28 nm line intensity with annealing temperature.

Variation in intensity of Aluminum lines at 396, 394, 309.28, 308.21 nm and Ge lines at 265, 303 nm in the LIBS spectra of Al-Ge samples annealed at different temperatures are depicted in Figure 2. This Figure reveals that the intensity variation in Al and Ge lines are linear with increasing temperature. This behavior indicates the structural modification due to laser interaction with materials, which may depends on the type of materials [25, 26] and on the thermal effects of the material. Such structural modifications due to thermal effects were discussed in the ref [27-29]. These results demonstrate that LIBS technique can be applied to assess the annealing effects in materials.

4. Conclusions

LIBS technique has proved to be a very good technique to study annealing effect in the form of variation in signal intensity of the respective material. It has been successfully applied to annealed Al-Ge alloy samples. It is demonstrated that signal intensity increases linearly with increasing annealing temperatures. This happens due to recrystallization of material with annealing temperatures. Moreover, impurities in the sample can be identified at the same time from LIBS spectra.

Acknowledgments

The authors would like to extend their sincere appreciation to the Deanship of Scientific Research at King Saud University for its funding of this research through the Research Group Project No. RG -1435-059.

References

[1] J. P. Singh and S. N. Thakur, Elsevier Science, ISBN: 978-0-444-51734-0, (2007)
[2] L. J. Radziemski, and T.R. Loree, *Plasma Chemistry and Plasma Processing*, **1(3)**, pp 281-93 (1981)
[3] D. Anglos, S. Couris and C. Fotakis, *Applied Spectroscopy*, **51 (7)**, pp 1025-30 (1997)
[4] J. M. Vadillo, J. J. Laserna, *Spectrochim. Acta B* **59** pp 147-61 (2004)
[5] C. Lopez-Moreno, S. Palanco, J. J. Laserna, F. DeLucia Jr, A. W. Miziolek, J. Rose, R. A. Walters and A. I. Whitehouse, *J. Anal. At. Spectrom*, **21(1)**, pp 55-60 (2006)
[6] G. Bazalgette Courrages-Lacoste, B. Ahlers and F. R. Parez, SpectrochimicaActa Part A:

Molecular and Biomolecular Spectroscopy, **68(4)**, pp 1023-28 (2007)

[7] K. J. Grant, G. L. Paul, and J.A. O'Neill, *Applied Spectroscopy*, **45**, pp 701-05 (1991)

[8] Nazar Farid, Shazia Bashir and Khaliq Mahmood, *Phys. Scripta*, **85**, 015702 (2012)

[9] Matthieu Baudelet, C. C. Christina, Willis, Lawrence Shah, and Martin Richardson Optics Express 18, pp 7905-10 (2010)

[10] W. A. Farooq, F. N. Al_Mutairi, A. E. M. Khater, A. S. Al_Dwayyan, M. S. AlSalhi, and M. Atif, **112**, pp 874-80 (2012)

[11] M. Sabsabi and P. Cielo, *Appl. Spectrosc*, **49**, pp 499-507 (1995)

[12] P. Inakollu, T. Philip, A. K. Rai, F. Y. Yueh, and J. P. Singh, *Spectrochim. Acta B*, **64** pp 99-104 (2009).

[13] W. T. Y. Mohamed, *Progress in Physics*, **2**, pp 87-92 (2007)

[14] W A Farooq, W. Tawfik, Z. A. Alahmed, K Ahmad and J. P. Singh, *Journal of Russian Laser Research*, **35**, pp 252-62 (2014)

[15] W. A. Farooq, F. N. Al-Mutairi, Z. Alahmed, *Optics and Spectroscopy*, **115**, pp 241-48 (2013)

[16] Kaleem Ahmad, Walid Tawfik, Wazirzada A. Farooq, Jagdish P. Singh, *Appl. Phys. A*, **117(3)**, pp 1315-22 (2014)

[17] M. M. Sarfraz, W. A. Farooq, Mohammad A Al-Eshaikh, Ahmed N, Laser Phys. 23 055701 (8pp) (2013)

[18] M. M. Suliyanti, M. Pardede, TjungJie Lie and Koo Hendrik Kurniawan, *Journal of the Korean Physical Society*, **58**, pp 1129-34 (2011)

[19] Walid Tawfik, W. Aslam Farooq and Zeyad A. Alahmed M. M. Sarfraz, K. Ahmad, Fahrettin Yakuphanoglu, IEEE Explore, 978-1-4673-6195-8/13/$31.00 (2013)

[20] Jung-Hwan In, Chan-Kyu Kim, Seok-Hee Leea, Sungho Jeong, *J. Anal. At. Spectrom*, **28**, pp 473-81 (2013)

[21] N. E. Widjonarko, J. D. Perkins, J. E. Leisch, P. A. Parilla, C. J. Curtis, D. C. Ginley, J.J. Berry, Rev SciInstrum, 81(7), 073103 (2010)

[22] Walid Tawfik, W. Aslam Farooq, and Z. A. Alahmed, *Journal of the Optical Society of Korea* **18**, pp 50-54 (2014)

[23] Gandhimathinathan Saroja, Veerapandy Vasu, Nagayasamy Nagarani, *Open Journal of Metal*, **3**, pp 57-63 (2013).

[24] Mohd Rafie Johan, Mohd Shahadan Mohd Suan, Nor Liza Hawari, Hee Ay Ching, *Int. J. Electrochem. Sci*, **6**, pp 6094-104 (2011)

[25] W. Aslam Farooq, M. Atif, Syed Mansoor Ali, AmanullahFatehmulla, and M. Aslam *Optics and Spectroscopy*, *117*, pp 386-91 (2014)

[26] W. A. Farooq, S. Mansoor Ali, J. Muhammad, S. Danish Ali, and M. Atif, *Optics and Spectroscopy*, **116**, pp 151-56 (2014)

[27] W. D. Callister, "Materials science and engineering; an introduction", John Wiley & Sons. (2000)

[28] Y. Kaneno, H. Inoue and N. Inakazu, Text. and Microstr, 14-18, pp 709-14 (1991)

[29] A. Kikuchi, Y. Iijima, K. Inoue, *Physica C*, **372**, pp 1321-24 (2002)

IOP Conf. Series: Materials Science and Engineering **383** (2018) 012013 doi:10.1088/1757-899X/383/1/012013

Conceptual metadata model for sensor data abstraction in IoT environments

Yoosang Park, Jongsun Choi and Jaeyoung Choi*

School of Computer Science and Engineering, Soongsil University, Seoul 06978 KOREA

*E-mail: choi@ssu.ac.kr

Abstract. Sensor data abstraction is necessary to provide users with context-aware services. Sensor data abstraction mechanism in context-aware system usually consists of collecting data, converting, and context reasoning. For this mechanism in IoT environments, sensor data is used that is described in the pairs of key-value set or digit value. However, environmental data cannot be sufficiently formed since these sensors data are provided with these values. For example, it may need metadata of contexts and things description. In this paper, we propose a new conceptual metadata model for sensor data abstraction in IoT environments. The proposed model provides sensor data and their metadata as low-level context, which is a part of basic context for presenting given environment, to context-aware system. In the experiments, we describe a procedure to generate low-level context for sensor data abstraction based on the proposed model and to provide this information to the context-aware system.

1. Introduction

Context-aware system provides users with context-aware services based on context information. At this point, context refers to evaluated state information that applies current circumstances, users, and their surroundings. One of the straightforward examples is described as follows: "a certain sensor device sensed temperature as warm state."

This context can be presented from sensor data processing. Procedures for processing sensor data normally consist of three major steps: collecting sensor data, converting this collected data into state information, and reasoning this state information for application services. These steps are known as sensor data abstraction [1,2]. The main purpose of sensor data abstraction is to provide understandable content to users, and related systems as appropriate data form, which is context. Sensor data abstraction helps to accumulate sensor data with any relevant piece of meaningful information. However, this sensor data abstraction cannot be done by individual sensor data. Because the users and systems have no hints for choosing sensor data state, and they need guidelines for making decisions of this sensor data. Sensor devices should be given more information to deal with these problems, for example, profile, device specification, their characteristics. And they are generally called as metadata. However, sensor devices in IoT environments do not provide these metadata with sensor data, and it contains key-value based dataset or sensed digit values. In these circumstances, metadata model for converting sensor data into state information is necessary. In this paper, we propose a new conceptual metadata model for sensor data abstraction in IoT environments. The proposed model consists of two major data fields, which are STATICMETADATA and ACTIVEMETADATA.

Content from this work may be used under the terms of the Creative Commons Attribution 3.0 licence. Any further distribution of this work must maintain attribution to the author(s) and the title of the work, journal citation and DOI.
Published under licence by IOP Publishing Ltd

STATICMETADATA carries fixed data, for example, profile, device specification, and device characteristics. And ACTIVEMETADATA carries changeable values such as access information, entity name, types of sensed values, and actual sensor data values. Through the proposed model, sensor data can be formalized for sensor data abstraction with detail contents.

2. Related work

The main concept of context was mentioned by the research of Schilit and Theimer [3], and later Abowd and Dey [4] introduced context-awareness. Until now, they are still important ideas for context related researches. Perera et al. [5] and Maarala et al. [6] represented and outlined processing models and systems for sensor data abstraction. And they are organized by three major procedures: sensor data acquisition, converting sensor data into state information, and reasoning this state information for application services. Bettini et al. [7] summarized data abstraction levels as three degrees with their individual purposes and representation forms as low-level context information, high-level context, and situational relationships.

Meanwhile, for systems to process these contexts, researches for representing context are mainly conducted with three major representing forms as key-value based, mark-up scheme based, and ontology-based. Firstly, key-value based context representation is mentioned by a research of Nambi et al. [8] and their main purpose was designing knowledge base. Sensor data acquisition in this research to build the knowledge base, is done by collecting data by key-value forms. Key-value based representation appears to be a pair of an identifier and sensed value. They considered sensed value as raw data. Secondly, Mark-up scheme based context representation is mentioned by a research of Knappmeyer et al. [9] and its main aim was to transmit context to remote sites. For this situation, ContextML was suggested. Mark-up scheme based context representation holds information such as profile and sensed values in a structured way. Finally, ontology-based context representation is mentioned by a research of Kayes et al. [10] and their main aim was to make decisions for each patient in health facilities. Ontology-based context representation is able to describe RDF-based entities as an identifier with each patient state and they are considered individually. RDF format supports universal description methods for those who describe any documents representing states.

By the researches above we followed, key-value based context representation which contains the identifier and sensed value and is the simplest representing way. However, it is difficult to contain additional information which can lead to reasoning techniques because ontology procedure needs much more information than key-value based context representation that only contains the identifier and raw data in a restricted format.

3. Proposed metadata model

3.1. Context levels in sensor data abstraction

Context levels in sensor data abstraction are arranged as three levels: low-level context, high-level context, and situational relationships followed by the research of Bettini et al. [7] in section 2. Figure 1 shows that context levels in sensor data abstraction and outlines for each level. Details for each level description are as follows.

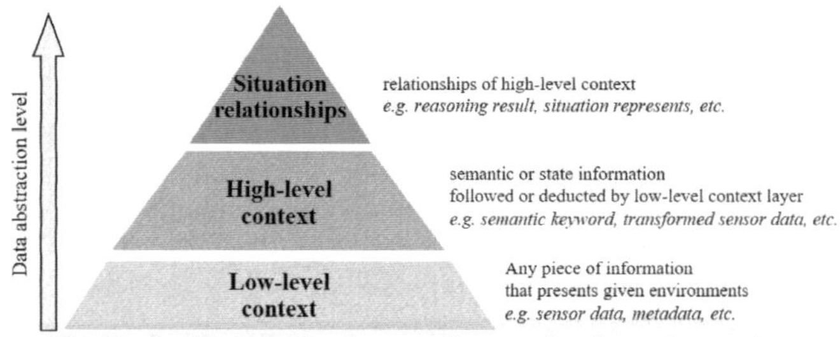

Figure 1. Context level according to the data abstraction.

3.1.1. Low-level context

Low-level context is the bottom level of the sensor data abstraction and it contains any part of information that represents given environments, for example, sensor data, metadata, etc. Sensor data is sensed value measured by a certain sensor device and represents key-value set or digit values. Metadata refers to data about data, and any object information in real-world, such as profile, device specification, etc. Metadata is typically described in a structured format such as XML or JSON. There are two main input data in low-level context; sensor data and metadata. Output data is a formed entity that is considered input data in high-level context for the next step.

3.1.2. High-level context

High-level context level carries semantic information or state information. In this level, output data from the low-level context is considered as input data, and then they are processed to create state information based on the structured sensor data converting scheme. Output data is formed as entity which contains each identifier in the real world and converted sensor data into state information. State information is described as keywords which are able to be accessed by context-aware system.

3.1.3. Situation relationships

Situation relationships level carries the top-notch context in this paper, and it is directly connected to services in real-world applications. To provide these services, entities at this level are formed as triplets of {ENTITY, PREDICATE, ENTITY}. With this representation, the context indicates certain entities and a relationship between them, so that it is able to express surroundings for each state information processed by the guideline and its condition is defined by the users.

3.2. Elements for low-level context in sensor data abstraction

3.2.1. Metadata

Metadata is commonly used for data about data. In this paper, metadata also indicates objects in the real world, so that metadata also carries a set of descriptions; sensor device, profile, and access information. Figure 2 shows the summarized structure of the metadata and details for STATICMETADATA. Metadata contains two major fields: STATICMETADATA and ACTIVEMETADATA. STATICMETADATA field holds fixed descriptions such as device profile, vendor information, and sensing unit, etc. ACTIVEMETADATA holds changeable data which is information for accessing device, sensed value and their type information.

3.2.2. Sensor data acquisition model

Collected sensor data is assembled to ACCESSINFORMATION field in ACTIVEMETADATA which contains access information to the sensor device, and sensed value type descriptions, etc. ACCESSINFORMATION field is for accessing sensor devices, which has two main types of connection: manually access and automatically access. Manually access is about to create sensor thread that handles connection procedure and its address. Automatic access happens in IoT-based sensor devices with their hub or gateway. For this IoT sensor, in general, access information is only required to the identifier and its sensor data. And the sensor data from the hub, is provided in web services or publish/subscribe based communication protocols.

IOP Conf. Series: Materials Science and Engineering **383** (2018) 012013 doi:10.1088/1757-899X/383/1/012013

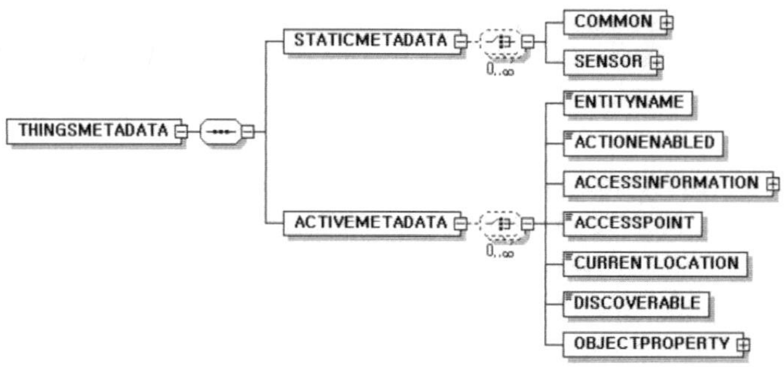

Figure 2. Proposed metadata scheme overview.

Collected sensor data is processed with three children fields of ACCESSINFORMATION field: VALUETYPE, BASISVALUE, and GATHEREDVALUE. VALUETYPE is used for describing data types of collected data and an equation for converting sensor data, BASISVALUE is used for describing comparison information by user's choice, and GATHEREDVALUE is used for describing collected data and their processed result. For example, given a temperature sensor device from ANALOG DEVICE named TMP36, and a user wants to define the state between warm and cold by 26°C. Table 1 shows the details for which description used for among these fields.

Table 1. Example for describing metadata with TMP36 temperature sensor.

Field		Valuetype	Basisvalue	Gatheredvalue
Generic element	Typename	Temp26	Temp26	Temp26
	Transform	true	Warm, Cold	Warm
	Equation	(((x1*5)/1024)-0.5)*100	26.0	26.3
	Comparison	Greater	Greater	Greater
	Valuelength	1	1	1
	Valuelist	x1	var_temp	var_temp
Has type	Type	singletype	singletype	singletype
	Singletype	var_temp	var_temp	var_temp
	Complextype	N/A	N/A	N/A

For users, they tend to describe VALUETYPE and BASISVALUE for the decision making on the certain temperature degree. VALUETYPE is related to raw data description which contains the Celsius converting formula in the Equation section of VALUETYPE. And user chooses the 26°C information in the Equation section of BASISVALUE. The user wants to figure out Warm state when temperature is over 26°C. Since the Transform section of VALUETYPE is marked as true, "Warm" or "Cold" expression in the Transform section of BASISVALUE is used when the temperature information is changed. The operator "greater" is described in every Comparison section, that means processed result value will be followed whether true or false with this operator. Collected and processed sensor data will be put in GATHEREDVALUE along with the VALUETYPE and BASISVALUE contents. Through this mechanism, entity will contain the identifier, which is sensor name, and state information with collected data which is "Warm" state.

4. Experiments

We describe processing elements that are explained in section 3. Procedures for generating low-level context with the proposed model is shown in figure 3.

Figure 3 shows the conceptual procedure to compose low-level context to high-level context. In figure 3, sensor devices are presented that communicate with the collector. The collector handles accessing devices, merges sensor data and their metadata, and returns entities which is the input data for high-level context.

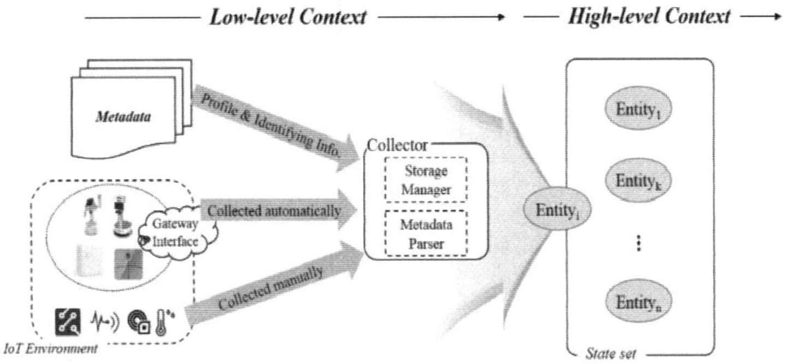

Figure 3. Procedures for generating low-level context with the proposed model.

Step 1. Initialize. Initialize step performs creating objects for the collector's use from the described metadata input. At this point, validation check for the input data also proceeds.

Step 2. Data acquisition & Combine. In this step, the collector collects sensor data through the access information described in ACTIVEMETADATA field. For the IoT-based sensor devices, each connected hub will transmit the sensor data with the identifier in the form of key-value based context representation, and the collector classifies their identifier and sensor data type. Both cases will do the same combining procedure that merges sensor data and metadata described in section 3 and prepare to create the object.

Step 3. Generate state information. After Step 2 is done, each object contains sensor data and metadata. Metadata contains converting scheme for the sensor data with the equation form. State information is generated by the equation and the operator.

Step 4. Return output. When step 3 is done, each object contains sensor data and metadata with GATHEREDVALUE field filled. This output is considered as the entity and the object is transmitted to a context pool which is presented state set in figure 3.

5. Conclusion

In this paper, we proposed a conceptual metadata model for the sensor data abstraction. A key idea of this model is merging sensor data and metadata in IoT environments, and the proposed model consists of elements: sensor data, metadata, and their sensor data acquisition model. With the proposed model, low-level context can be formed in structured format and it can be used for higher abstraction procedures, which are high-level context and situational relationships described in this paper.

Acknowledgments

This work was supported by the Industrial Convergence Core Technology Development Program (No. 10048474) funded by the Ministry of Trade, Industry & Energy(MOTIE), Korea and was supported by Basic Science Research Program (No. 2016R1C1B2009744) through the National Research Foundation of Korea(NRF) funded by the Ministry of Science and ICT(MSIT).

References

[1] Atzori L, Iera A, and Morabito G 2010 *Computer Networks* **54** 15 pp 2787-2805
[2] Li X, Eckert M, Martinez J F, and Rubio G 2015 *Sensors* **15** 8 pp 20570-607

[3] Schilit B N and Theimer M M 1994 *IEEE Network* **8** 5 pp 22-32
[4] Abowd G, Dey A, Brown P, Davies N, Smith M and Steggles P 1999 *Handheld and ubiquitous computing* Springer Berlin/Heidelberg pp 304-07
[5] Perera C, Zaslavsky A, Christen P, and Georgakopoulos D 2014 *IEEE communications Surveys & Tutorials* **16** 1 pp 414-54
[6] Maarala A I, Su X, and Riekki J 2017 *IEEE Internet of Things Journal*, **4** 2 pp 461-73
[7] Bettini C, Brdiczka O, Henricksen K, Indulska J, Nicklas D, Ranganathan A, and Riboni D 2010 *Pervasive and Mobile Computing* **6** 2 pp 161-80
[8] Nambi S A U, Sarkar C, Prasad R V, and Rahim A 2014 *Internet of Things(WF-IOT) on IEEE World Forum* pp 575-80
[9] Knappmeyer M, Kiani S L, Frà C, Moltchanov B, and Baker N 2010 *Wireless Pervasive Computing(ISWPC) on 5th IEEE International Symposium* pp 367-72
[10] Kayes A S M, Rahayu W, Dillon T, Chang E, and Han J 2017 *OTM Confederated International Conference* Springer pp 132-53

IOP Conf. Series: Materials Science and Engineering **383** (2018) 012014 doi:10.1088/1757-899X/383/1/012014

A control method to save machine energy in production

Wei Weng*, Yifan Yang and Shigeru Fujimura

Graduate School of Information, Production and Systems, Waseda University
Kitakyushu, Japan

*Email: wengwei@toki.waseda.jp

Abstract. Nowadays, energy saving is one of the most talked about issues in our life, it is also increasingly important in the manufacturing industry. This research considers the dynamic flexible flow shop scheduling (DFFS) problem, which is an extended version of the classical flow-shop scheduling problem. A flexible flow shop has multiple stages with multiple machines at each stage for processing multiple products. Previous research on DFFS aimed to achieve just-in-time production, or reducing difference between the actual completion time and the due date of each job. However, little research has been made on energy saving of machines in production. To address such a need, this paper proposes a method that dynamically turns on and off machines so as to reduce energy consumption while achieving JIT production. The proposed method has been tested on different environments, and the results show that it is high performing for both JIT production and energy saving.

1. Introduction

Energy and environmental issues are one of the most talked about issues in the 21st century. As a result of climate change, regulation of carbon dioxide emissions has been imposed globally, which has put much pressure on the manufacturing industry to reduce energy (particularly electric energy) [1]. Environmental protection, solar, wind and other new energy sources are being widely used in life and production. Sustainable development is becoming a trend to promote energy saving. With the need for energy efficiency and low-carbon production, it is a big challenge for high energy consumption enterprises to reduce energy use [2]. Therefore, effective energy and utility management is a key factor to enhance the competitive advantage of organizations, and to promote green and sustainable practices [3]. The factory as a major energy consumer, energy conservation is particularly important.

Current studies have been highlighting the importance of energy efficiency on the environmental impact of manufacturing processes. They primarily focus on the evaluation of manufacturing processes based on operating state and components of machine tools [4], thermodynamics approaches [5], and empirical modelling [6].

What's more, cutting parameters optimization for material removal processes is also one of the important energy-saving strategies using by most research. Anderberg et al. analyzed the relationship between cutting parameters, machining costs, and energy consumption in a CNC machining environment [7]. Diaz et al. analyzed the effect of material removal rate and workpiece material on cutting power and energy consumption [8]. Mativenga and Rajemi proposed a method for selecting optimum cutting parameters for minimum energy consumption in turning a cylindrical steel billet [9]. These methods have been developed which enable consideration of energy consumption for the selection of optimal cutting parameters. Based on machine level, this paper presents a method called

Content from this work may be used under the terms of the Creative Commons Attribution 3.0 licence. Any further distribution of this work must maintain attribution to the author(s) and the title of the work, journal citation and DOI.
Published under licence by IOP Publishing Ltd

dynamic machine on/off method which can dynamic turn on and turn off machines when needed in order to save energy consumption of machines.

Similar to machine level, at shop floor level, most of the research related to energy efficient shop floor scheduling have shown improvements in reducing energy consumption for manufacturing. Multi-objective scheduling models have been proposed to optimize both makespan and energy consumption [10], [11] presented a general multi-objective mixed-integer programming formulation for flow shop scheduling problem that considers makespan, peak power demand, and carbon footprint. The proposed scheduling problem considers the operation speed as an independent variable, which can be changed to affect the peak load and energy consumption. Bruzzone et al. [12] presented an energy-efficient scheduling method to modify the original timetable of the jobs in FFS in order to reduce the shop floor's peak power demand while some level of tardiness and makespan increase are accepted.

The just-in-time (JIT) production and flexible flow shop (FFS) type of manufacturing system are selected as the research object for the following reasons:

With the increase in inventory costs, more and more factories tend to JIT production rather than reduce the maximum completion time of all jobs. The objective of JIT production is to reduce both inventory costs of jobs completed before due dates and delay costs of jobs completed after due dates. The due date is the date that the factory expects to deliver the product to the customer. There is a complex manufacturing environment called FFS which is a further development of the classical flow-shop scheduling. FFS has multiple stages with multiple machines at each stage for processing multiple products. All jobs must go through all the stages to become finished products, which means that there are many different flow lines (routes) available for each job to go through the system. There is a previous study [13] concentrating on JIT production for FFS problem. It detailed three distributed-intelligence approaches to achieve the JIT objective. We add the proposed dynamic machine on/off method in the best-performing approach among the previous study, which is named stage to stage (S2S) feedback approach. This research intends to minimize both total inventory costs and delay costs of all jobs and total energy consumption of all machines.

The remainder of this paper is organized as follows. We give the problem description in chapter 2. After that, we present the dynamic machine on/off method proposed in this paper in chapter 3, and give the computational simulation results and the conclusions in chapters 4 and chapter 5 respectively.

2. Problem description

This paper considers the dynamic FFS problem in which the manufacturing environment consists of multiple stages with multiple machines at each stage. All jobs must go through all stages in sequence in order to become a product. If job j is finished before its due date D_j, it has an earliness E_j. On the other hand, if job j is finished after its due date D_j, it has a tardiness T_j. Let C_j denote the completion time of job j, then the earliness of a job completed before its due date is given as $E_j = max(0, D_j - C_j)$, and the tardiness of a job completed after its due date is given as $T_j = max(C_j - D_j, 0)$. Both early jobs and tardy jobs have penalties but earliness penalties are always smaller than tardiness penalties. Each stage has a waiting queue before it, with jobs inside the queue processed in an Earliest-Due-Date (EDD) sequence.

The objective function is equation (1)

$$Min\begin{bmatrix} \sum_{j=1}^{J}\left(E_j \times EP_{p(j)} + T_j \times TP_{p(j)}\right) \\ + \sum_{m=1}^{M}\left(tp_m \times MP_m + ti_m \times MI_m + nt_m \times MS_m\right) \end{bmatrix} \quad (1)$$

where $p(j)$ is product type of job j and J is the total number of jobs, E_j is earliness of job j and $EP_{p(j)}$ is unit penalty of earliness, T_j is tardiness of job j and $TP_{p(j)}$ is unit penalty of tardiness, tp_m is sum of processing time of jobs operated on machine m, MP_m is energy cost per unit time for machine m in processing mode, ti_m is sum of idling time for machine m, MI_m is energy cost per unit time for

IOP Conf. Series: Materials Science and Engineering **383** (2018) 012014 doi:10.1088/1757-899X/383/1/012014

machine m in idle mode, nt_m is number of times for turning on and turning off machine m, MS_m is energy cost for turning on or turning off machine m.

Besides, the following conditions are assumed in this paper:

1. One machine can only process one job at a time.
2. The processing of each job is nonpreemptive.
3. Machines do not break down.
4. Machines have the same energy cost per unit time when they are in the same mode.
5. The processing time for each product on each machine is different but fast machines process all types of products faster than slow machines.
6. The waiting queue of each stage is unlimited.

3. Proposed method

We propose a method that dynamically turns off some machines without causing jobs to be tardy in order to save energy and turns on them again when they need to be used for reducing job tardiness.

We assume that each machine has three modes: processing mode, idle mode and off mode [14]. Machines are in the processing mode when they are processing jobs and will switch to the idle mode automatically after they finish the jobs. The energy cost per unit time for the processing mode is bigger than that for the idle mode. There is also an energy cost for either turning on or turning off a machine. The proposed method aims to reduce the total energy consumption by dynamically turning on and off machines. In the following, we introduce how to dynamically turn off and turn on machines respectively.

3.1. Design of turning off machines

As mentioned before, machines are in the processing mode when they are processing jobs and will switch to the idle mode automatically after they finish the jobs. Since jobs come dynamically, the system does not always need all the machines to work when the system is not busy. We focus on these idle machines when trying to turn off a machine. The procedure of turning off the machines is shown in figure 1.

At each stage, when a machine in the processing mode finishes a job, we use the estimated completion time of each job provided in the S2S feedback approach to check if any job in the waiting queue will be tardy if we turn off the machine. If no job is found to be tardy, we will compare idle energy consumption with machine on/off energy consumption, if the idle energy consumption is greater, we will compare the average processing time of the last job in the stage waiting queue with the waiting time of the last job in queue of current stage. The average processing time (APT) is calculated by equation (2)

$$APT = \frac{\sum_{m=1}^{M_s} P(j,m)}{M_s}, \tag{2}$$

where M_s is the number of machines of each stage, $P(j,m)$ is the processing time of job j operated on machine m. The waiting time of a job means the time duration from this job enters the waiting queue of current stage to it is processed by machine. If the former is larger than the latter, the idle machine will be turned off. Each stage repeats this procedure until it finds at least one condition in figure 1 is not satisfied.

3.2. Design of turning on machines

The procedure of turning on the machines is shown in figure 2. When a new job enters the waiting queue of a stage, if there are some machines that are turned off at the stage and any of the waiting jobs is found to be tardy based on the estimated completion time provided by S2S in the stage, we will compare the average processing time of the last job in waiting queue of current stage with the waiting time of the last job in queue of current stage. After comparison, we will turn on the fastest off machine using the procedure shown in figure 2. Since jobs arrive dynamically, machines are turned off and on dynamically at each stage.

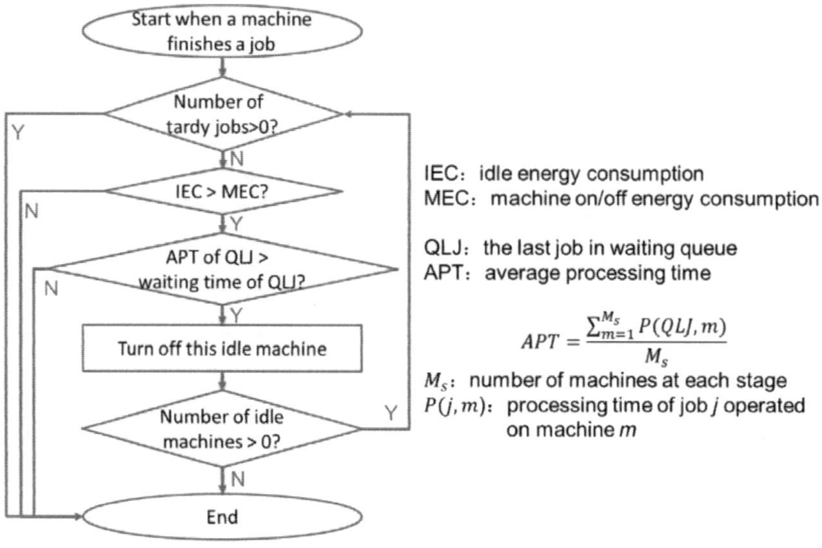

Figure 1. Design of turning off machines.

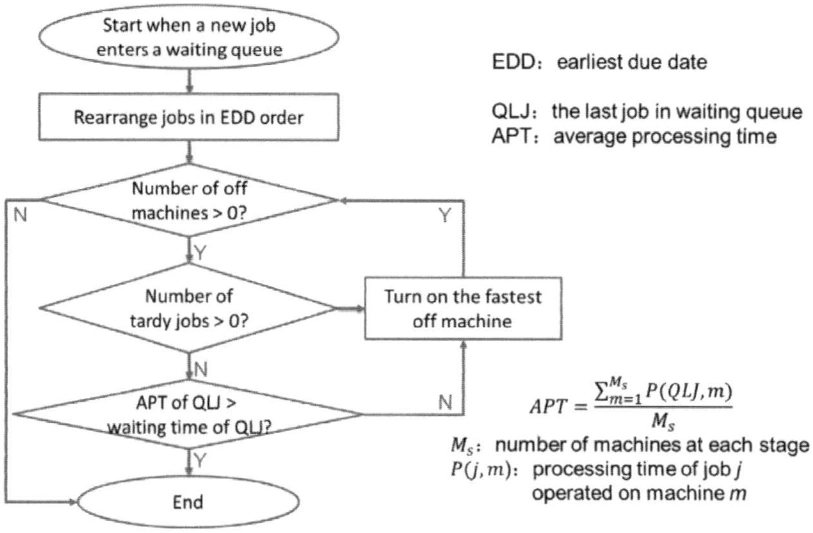

Figure 2. Design of turning on machines.

4. Computational simulations

4.1. Simulation environment

Table 1 describes parameters design of simulation environment. The type of products is designed to be 1 and 5, the number of stages is designed to be 3, 5 and 10. The total number of machines in the system is designed to be 5, 10 and 20. Since each factor has more than one design, the number of combinations is equal to 18 (2*3*3=18). Then we let the problem instance for each feasible combination equals to 3, so the total number of problem instances is 48. When there are 10 stages, the number of machines must be at least 10, so there are 6 infeasible instances because the number of machine is 5 in system is unacceptable.

In order to evaluate the results in a fair way, each instance has 2200 jobs, of which the first 200 jobs are used to warm up the system to a stable state, so we focus on only the remaining 2000 jobs for getting the results.

Table 1. Design of simulation environment.

Type of Products: 1, 5
Number of Stages: 3, 5, 10
Number of Machines: 5, 10, 20
Number of combinations: 18
Problem instance for each feasible combination: 3
Total number of problem instances: 48

These designs are intended to test whether the proposed dynamic machine on/off method can work well under all kinds of situations. In addition, we set the energy costs as follows after refer to a previous paper [9]:

- Energy cost per unit time for a machine in processing mode: 1;
- Energy cost per unit time for a machine in idle mode: 0.4;
- Energy consumption for turning on or turning off a machine: 0.6.

4.2. Simulation results

In order to show the effect of this dynamic machine on/off method, Tables 2-4 give the average results of all instances in different designs.

In the tables, *Total Cost* is sum of *Power Consumption Cost* and *Earliness & Tardiness Cost*, which means the total penalty including energy consumption, earliness and tardiness of all jobs. The data of tables are calculated by equation $result = \frac{r-r_0}{r_0} \times 100\%$, where r is the result calculated by using machine on/off method and r_0 is the result without using method.

Table 2. Average results of different products.

		Power Consumption Cost	*Earliness & Tardiness Cost*	*Total Cost*
P	1	-31.57%	-0.05%	-18.39%
	5	-24.10%	-0.08%	-15.67%

Table 3. Average results of different stages.

		Power Consumption Cost	*Earliness & Tardiness Cost*	*Total Cost*
S	3	-38.65%	-0.07%	-20.96%
	5	-25.58%	-0.05%	-15.20%
	10	-25.48%	-0.07%	-16.90%

Table 4. Average results of different machines.

		Power Consumption Cost	*Earliness & Tardiness Cost*	*Total Cost*
M	5	-12.54%	-0.03%	-6.30%
	10	-20.12%	-0.06%	-12.42%
	20	-41.27%	-0.07%	-27.54%

4.3. Result analysis

As revealed by tables 2-4, both energy consumption and total earliness and tardiness of all jobs are becoming better after we use dynamic machine on/off method, regardless of different factors or different types, so does the *Total Cost*, which means the total penalty including energy consumption, earliness and tardiness of all jobs.

Table 2 shows the results become better when *P* increases from 1 to 5. It is probably because the dynamic machine on/off method can effectively reduce both job earliness and tardiness, the total cost will become better with the increase in the number of products. Table 3 shows that when *S* increase, the results become worse. The reason is that a job need more time to be processed to a product when the number of stage increase. When new order comes constantly, the number of waiting jobs will increase rapidly, which lead to the increase number of tardy jobs. Table 4 shows that the results become better when M increases. This is because there are more machines that can process jobs and machines are turned off from slow to fast, jobs have more chances to be assigned to faster machine.

5. Conclusions

This research aims at saving energy consumption for JIT production by using a dynamic machine on/off method based on a previously proposed S2S feedback approach. The results show that all problem instances'objective function values become better, thereby proving the effectiveness of the proposed method.

Acknowledgments

The authors acknowledge the support of JSPS Grants-in-Aid for Scientific Research (KAKENHI) Grant Number 15K16296.

References

[1] Park, C.W.; Kwon, K.S.; Kim,W.B.; Min, B.K.; Park, S.J.; Sung, I.H. Energy consumption reduction technology in manufacturing—A selective review of policies, standards, and research. Int. J. Precis. Eng. Manuf. 2009, 10, pp 151–73

[2] Tong, Y.; Li, J.; Li, S.; Li, D.; 2016. Research on Energy-Saving Production Scheduling Based on a Clustering Algorithm for a Forging Enterprise. Sustainability 2016, 8, p 136

[3] Ngai, E.W.T.; Ng, C.T.D.; Huang, G. Energy Sustainability for Production Design and Operations. Int. J. Prod. Econ. 2013, 146, pp 383–85

[4] Avram, O.I., Xirouchakis, P., 2011. Evaluating the use phase energy requirements of a machine tool system. J. Clean. Prod. 19 (6e7), pp 699-711

[5] Gutowski, T., Dahmus, J., Thiriez, A., 2006. Electrical Energy Requirements for Manufacturing Processes. 13th CIRP International Conference on Life Cycle Engineering, pp 623-28. Leuven, Belgium

[6] Draganescu, F., Gheorghe, M., Doicin, C.V., 2003. Models of machine tool efficiency and specific consumed energy. J. Mater. Process. Technol. 141, pp 9-15

[7] Anderberg, S.E., Kara, S., Beno, T., 2010. Impact of energy efficiency on computer numerically controlled machining. Proc. Inst. Mech. Eng. Part B J. Eng. Manuf. 224, pp 531-41

[8] Diaz, N., Redelsheimer, E., Dornfeld, D., 2011. Energy Consumption Characterization and Reduction Strategies for Milling Machine Tool Use. 18th CIRP International Conference on Life Cycle Engineering, pp 263-67. Braunschweig, Germany

[9] Mativenga, P.T., Rajemi, M.F., 2011. Calculation of optimum cutting parameters based on minimum energy footprint. CIRP Ann. - Manuf. Technol. 60 (1), pp 149-52

[10] He, Y., Liu, F., Cao, H., Li, C., 2005. A bi-objective model for job-shop scheduling problem to minimize both energy consumption and makespan. J. Cent. South Univ. 12 (Suppl. 2), pp 167-71

[11] Fang, K., Uhan, N., Zhao, F., Sutherland, J.W., 2011. A new approach to scheduling in manufacturing for power consumption and carbon footprint reduction. J. Manuf. Syst. 30, pp 234-40

[12] Bruzzone, A.A.G., Anghinolfi, D., Paolucci, M., Tonelli, F., 2012. Energy-efficient scheduling for improving manufacturing process sustainability: a mathematical model for flexible flow shops. CIRP Ann. - Manuf. Technol. 61, pp 459-62

[13] Weng W, Fujimura S. Distributed-intelligence approaches for weighted just-in-Time production. IEEJ Transactions on Electrical and Electronic Engineering 2010; 5: pp 560-68

[14] Min Dai, Dunbing Tang, Adriana Giret, Miguel A. Salido, W.D. Li. Energy-efficient scheduling for a flexible flow shop using an improved genetic-simulated annealing algorithm. Robotics and Computer-Integrated Manufacturing 29 (2013) pp 418-29

IOP Conf. Series: Materials Science and Engineering **383** (2018) 012015 doi:10.1088/1757-899X/383/1/012015

Trajectory tracking with ANFIS and uncalibrated vision system for uncertain robotic manipulators

Aung Myat San*, Wut Yi Win and Saint Saint Pyone

Department of Mechatronic Engineering, Mandalay Technological University, Myanmar

*Email: agmyatsanmc@googlemail.com

Abstract. This paper describes a new approach for tracking the trajectory of uncertain robotic manipulators using ANFIS (Artificial Neuro-Fuzzy Inference System) and uncalibrated vision system. The main emphasis of this work is on the ability to estimate the positioning accuracy and repeatability of a low-cost robotic arm with unknown parameters under uncalibrated vision system. In this study, captured image data are collected from two fixed-cameras vision system; installed on the top and lateral sides of the robot, respectively. For training and validating purposes, the robot is manoeuvring within its workspace using forward kinematics. The tracking system is trained using ANFIS with subtractive clustering method in MATLAB. Extensive simulations were carried out to illustrate the effectiveness of the proposed visual tracking method for LabVolt R5150 manipulator in our laboratory. Observing the simulation results, the performance of the proposed approach is efficient for using the vision-based learning system as visual feedback in uncertain robotic manipulator.

1. Introduction

In practical usage, a robot is a mechanical device which performs automated physical tasks, either according to direct human supervision, a pre-defined program, or a set of general guidelines using artificial intelligence techniques. The positioning problem of robot manipulators using visual information has been an area of research over the last 40 years. Attention to this subject has drastically grown in recent years. The visual feedback loop can solve many problems that limit applications of current robots: automatic driving, long range exploration, medical robotics, aerial robots, etc.

This paper focuses about mapping visual sensory information for robot control. Recently, the three-dimension (3-D) vision systems for robot applications have been popularly studied. Okada et al. [1] used multi-sensors for the 3-D position measurement. Winkelbach et al. [2] combined one camera with one range sensor to find the 3-D coordinate position of the target. Huang [3] addressed a 3-D position control for a robot arm utilizing two-CCD vision geometry and inverse kinematics. Zhou et al. [4] used position sensitive detector (PSD) for high-precision parallel kinematic mechanisms (PKMs) in order to allow them to accurately achieve their desired poses. Dallej et al. [5] developed 3D pose visual servoing for cable driven parallel robots.

In this paper, the positioning problem of articulated robot manipulators is addressed under two fixed cameras configurations. The main contribution is the development of a new pose-independent learning method for the robotic end-effector positioning using two uncalibrated fixed cameras and

Content from this work may be used under the terms of the Creative Commons Attribution 3.0 licence. Any further distribution of this work must maintain attribution to the author(s) and the title of the work, journal citation and DOI.
Published under licence by IOP Publishing Ltd

forward kinematics. The objective concerning the control is defined in terms of cartesian coordinates which are deduced from visual information.

The paper is organized by four sections. In section 2, the system approach is presented along with the forward kinematic analysis of the R5150 robotic, the learning of ANFIS technique and Kalman approach of motion estimation. In section 3, implementation of ANFIS training system, simulation tests and results are presented. Finally, the paper is concluded with the observed results in section 4.

2. System approach

In this section, the system approach is introduced including forward kinematic analysis for the robotic simulation, ANFIS learning for the Cartesian coordinates, and Kalman filtering for estimating the motion of the robot in occlusion cases. As shown in figure 1, the overall system approach is implemented along with two fixed imaging devices of view limits [0, 1024, 0, 1024].

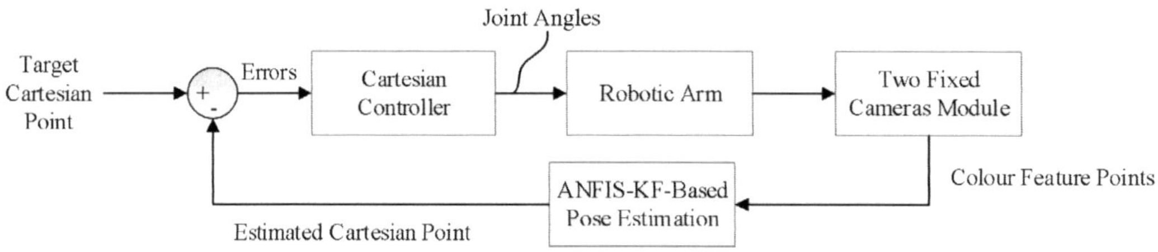

Figure 1. Overall system approach.

2.1. Forward kinematic analysis of the robot

The first step in modelling robot LabVolt R5150 is determining D-H parameters. It is used in robotics, where a robot can be modelled as a number of related solids (segments) where the D-H parameters are used to define the relationship between the two adjacent segments.

Table 1. D-H parameters.

	Link	d (m)	a (m)	α	Joint limits (min, max)
1	Base	0.2555	0	90°	338° (-185°, 153°)
2	Shoulder	0	0.19	0	181° (-32°, 149°)
3	Elbow	0	0.19	0	198° (-147°, 51°)
4	Wrist Pitch	0	0	90°	185° (-5°, 180°)
5.	Wrist Roll	0.115	0	0	360° (-360°, 360°)

The observed robot has 3-DOF three links with a 2-DOF wrist mechanism. In this work, the location of the end of the first three links is tracked using one colour feature point. For specifying D-H parameters of the robot, conventional D-H method is used. Using D-H parameters defined in the previous steps in table 1, robot model was created in MATLAB software using the Robotic and Vision Toolbox [6]. Robot model in addition to previously determined D-H parameters contains physical parameters which is using in the calculation of the dynamics movement.

The link transformation matrix between coordinate frames {i-1} and {i} has equation (1) [7];

$$ {}^{i-1}T_i = \begin{bmatrix} \cos\theta_i & -\sin\theta_i.\cos\alpha_i & \sin\theta_i.\sin\alpha_i & a_i.\cos\theta_i \\ \sin\theta_i & \cos\theta_i.\cos\alpha_i & -\cos\theta_i.\sin\alpha_i & a_i.\sin\theta_i \\ 0 & \sin\alpha_i & \cos\alpha_i & d_i \\ 0 & 0 & 0 & 1 \end{bmatrix} \quad (1) $$

Since a coloured feature point is attached to the end of the first three links of the robot, we need to calculated the first three basic transformation matrices to get the location of the feature point. The final homogeneous transformation matrix for locating the feature point is got from the product of three basic transformation matrices as equation (2):

$$^{0}T_{3} = {}^{0}T_{1}.{}^{1}T_{2}.{}^{2}T_{3} = \begin{bmatrix} n_x & o_x & a_x & p_x \\ n_y & o_y & a_y & p_y \\ n_z & o_z & a_z & p_z \\ 0 & 0 & 0 & 1 \end{bmatrix} = \begin{bmatrix} Q^{3\times3} & & & P^{3\times1} \\ 0 & 0 & 0 & 1 \end{bmatrix} \tag{2}$$

where,

$$Q = \begin{bmatrix} \cos(\theta_1)\cos(\theta_2+\theta_3) & -\cos(\theta_1)\sin(\theta_2+\theta_3) & \sin(\theta_1) \\ \sin(\theta_1)\cos(\theta_2+\theta_3) & -\sin(\theta_1)\sin(\theta_2+\theta_3) & -\cos(\theta_1) \\ \sin(\theta_2+\theta_3) & \cos(\theta_2+\theta_3) & 0 \end{bmatrix}$$

$$P = \begin{bmatrix} \cos(\theta_1)\left(a_3\cos(\theta_2+\theta_3)+a_2\cos(\theta_2)\right) \\ \sin(\theta_1)\left(a_3\cos(\theta_2+\theta_3)+a_2\cos(\theta_2)\right) \\ d_1+a_3\sin(\theta_2+\theta_3)+a_2\sin(\theta_2) \end{bmatrix}$$

From the position matrix *P*, we get the location of the feature point as the robotic motion. In this work, the forward kinematics of the robot is used to simulate and drive the robot for learning ANFIS networks and driving the robot to a specified trajectory or location.

2.2. ANFIS training

In this work, the visual positioning is trained using ANFIS as well as robotic forward kinematics and multi-view geometry. The first step was to test the working space of the robotic arm in vision. In order to do so, a m-file have been created in MATLAB based on the direct kinematics of the robotic arm and the perspective geometries of two cameras. In this work, two pin-hole cameras are installed on the top and lateral sides of the robot, respectively. The cameras' focal length and view limits are identical as 0.002 and [0, 1024, 0, 1024], respectively. In this research, mapping the visual sensory information into the Cartesian coordinates is performed using ANFIS method and MATLAB Fuzzy Logic toolbox.

In MATLAB, "*genfis2*" generates a Sugeno-type FIS structure using subtractive clustering. *genfis2* is generally used where there is only one output; hence here it has been used to generate initial FIS for training the ANFIS. On the other hand, "*genfis2*" achieves this by extracting a set of rules that simulates the data values. In order to determine the number of rules and antecedent membership functions, "*subclust*" function has been used by the rule extraction methods. Further it uses the linear least squares estimation to determine each rule's consequent equations.

However, ANFIS itself is only suitable for single output system. For a system with multiple outputs, ANFIS will be placed side by side to produce a Multiple ANFIS (MANFIS) [8]. The number of ANFIS required depends on the number of required output. In this research, three cartesian coordinate (XYZ) points have to be outputted as ANFIS outputs (i.e., X, Y and Z).

2.3. Generating train data and check data

For a feature point, a red-coloured feature point is placed at the end of the three links. Therefore, the feature point will be moved as long as the arm is moved. In this research, Corke's RVC v9.8 MATLAB toolbox [6] is used for simulation of robot and two cameras.

The data for ANFIS training process is generated by manoeuvring the robot in its operational workspace and acquiring image data from two cameras. This is so called as motor babbling phase in which the robot is driven in joint space. The displacement between two consecutive joint elements was limited to 30 degrees among their maximum and minimum ranges. To get the required data, the manipulator is manoeuvred within its workspace using forward kinematics, and the end-effector image coordinates are acquired from two cameras. For the neuro-fuzzy model in this work, data set are 588 cartesian points analytically obtained using forward kinematics, and feature points captured by two cameras in motor babbling phase. They are halved for training and validation of the ANFIS models; 294 and 294, respectively. The image coordinates of these cameras are [u_1, v_1] and [u_2, v_2]. And position matrix using forward kinematics is [p_x, p_y, p_z]. So, there are four inputs and one Cartesian point output for individual ANFIS; i.e. [u_1, v_1, u_2, v_2, p_x], [u_1, v_1, u_2, v_2, p_y] and [u_1, v_1, u_2, v_2, p_z].

2.4. Tracking an object on cameras' image plane using discrete Kalman filter

Kalman filter is a recursive solution to the discrete-data linear filtering problem. The Kalman filter has been used extensively for tracking in interactive computer graphics. This research focuses on tracking an object on camera's image plane to compensate the losses of image feature data due to lighting or occlusion. Therefore, a simple discrete Kalman filter is implemented in the system.

The state model of discrete Kalman filter is defined as equation (3) and (4).

$$\text{Estimate: } x_{k+1} = Ax_k + w_k \tag{3}$$

$$\text{Measurement: } z_k = Hx_k + v_k \tag{4}$$

where, w_k and v_k are white noises denoting the process and measurement noises, respectively. Parameters of the proposed discrete Kalman filter using equation (3) and (4) are as follows:

$$A=\begin{bmatrix}1&1&0&0\\0&1&0&0\\0&0&1&1\\0&0&0&1\end{bmatrix},\ H=\begin{bmatrix}1&0&0&0\\0&0&1&0\end{bmatrix},\ Q=\begin{bmatrix}1&0&0&0\\0&1&0&0\\0&0&1&0\\0&0&0&1\end{bmatrix},\ R=0.01\begin{bmatrix}50&0\\0&50\end{bmatrix},\ \hat{x}_0=\begin{bmatrix}0\\0\\0\\0\end{bmatrix},\text{ and } P_0=\begin{bmatrix}100&0&0&0\\0&100&0&0\\0&0&100&0\\0&0&0&100\end{bmatrix}.$$

3. Simulation tests and results for trajectory tracking

For trajectory tracking purpose, MATLAB Simulink model is built as shown in figure 2. In which, the robot is driven in joint trajectory control from one joint coordinate to another (e.g., from [π/6, π/6, 0, 0, 0] to [π/4, π/2, π/4, 0, 0] as shown in figure 3). The comparative results of forward-kinematic-based and ANFIS-estimated trajectories are shown in figure 4. RMSE errors are 0.0014903, 0.0032865 and 0.0032303 for X, Y and Z, respectively. And Kalman filtering results for estimation the feature points in cameras' image planes are shown in figure 5.

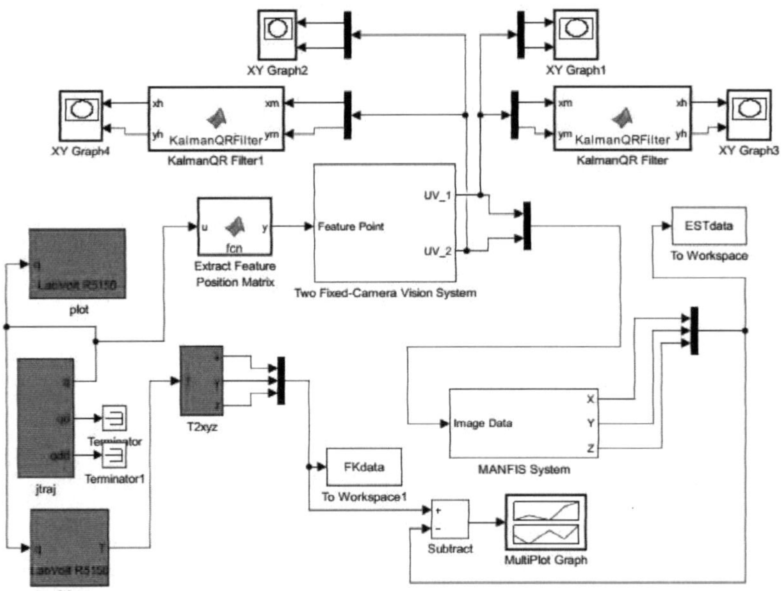

Figure 2. Simulink of the trajectory tracking for the robotic arm.

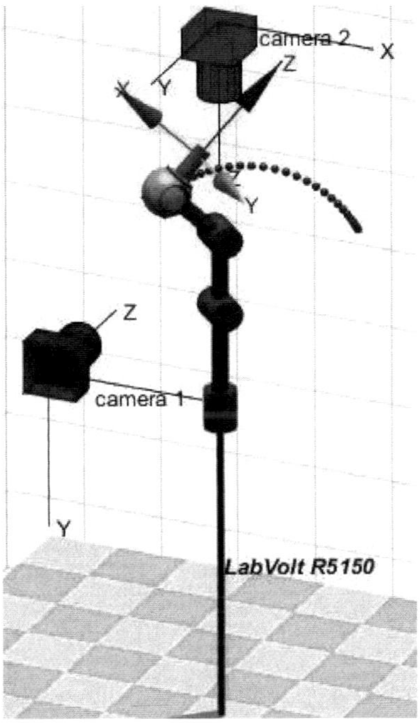

Figure 3. Simulation of the trajectory.

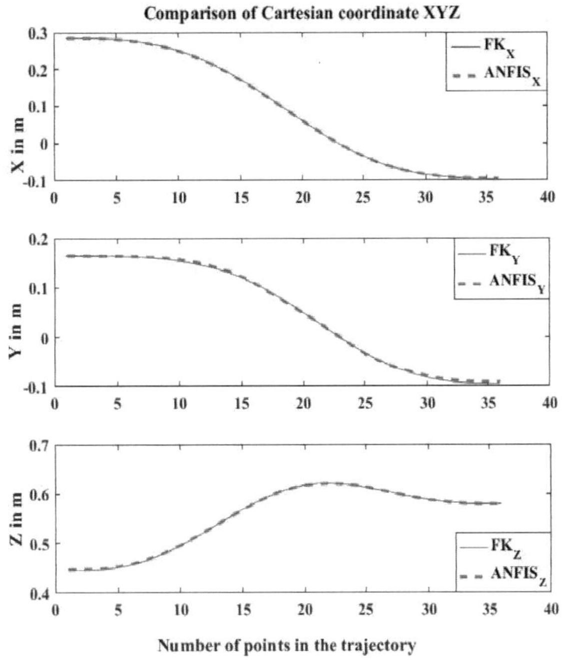

Figure 4. Comparative result for the trajectory.

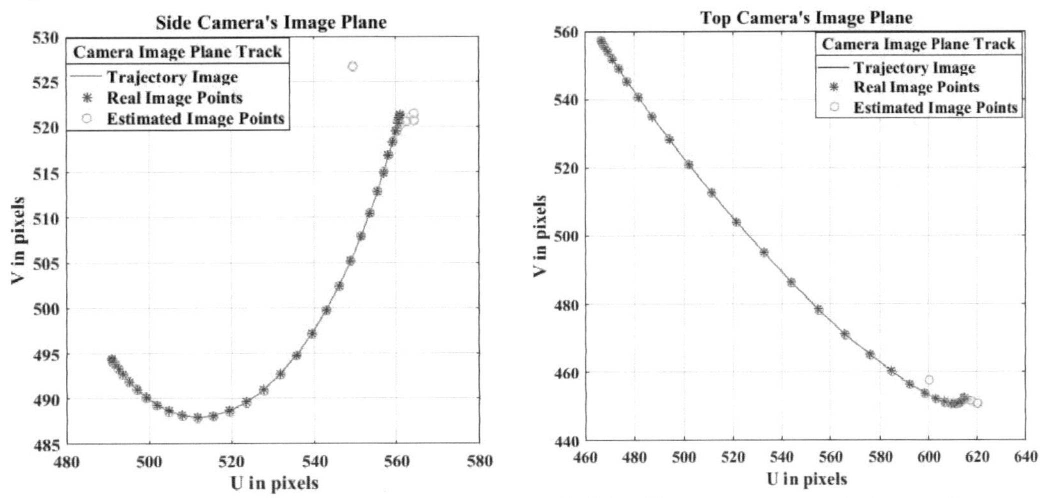

(a) Tracking on side camera's image plane. **(b)** Tracking on top camera's image plane.

Figure 5. KF-based tracking estimations on cameras' image planes for the trajectory.

The proposed ANFIS-KF tracking method gives good estimation of the position of the robotic end-effector. The learning ability of ANFIS makes the kinematic and dynamic knowledge of the robot flexible under uncalibrated vision system. Since the learning is based upon Cartesian coordinates, it is easy to implement the controller designs and human-machine interfaces better. Moreover, the implementation of Kalman filter makes the system reliable in tracking concept. Therefore, the proposed ANFIS-KF approach for the trajectory tracking has better in flexibility, user-friendly manner and computational concepts over conventional techniques.

IOP Conf. Series: Materials Science and Engineering **383** (2018) 012015 doi:10.1088/1757-899X/383/1/012015

4. Conclusion

An ANFIS-based visual trajectory tracking approach using two uncalibrated cameras is proposed in this paper. The idea of using forward kinematic equations and two cameras for generating training data for ANFIS led to a nearly accurate training of the ANFIS network. The simulation experiments show that the location of the robotic arm can be trained in ANFIS using two uncalibrated cameras, and the tracking approach makes the robotic vision control systems more flexibly in control designs and more user-friendly in learning the robotic motion in Artificial Intelligence than conventional techniques. The proposed ANFIS-based approach of tracking the pose of the robot is very useful in obtaining the position of the robotic arm as it can work as a control algorithm. The Cartesian-coordinate-based learning can be used in robotic calibration, visual servoing and Cartesian controller. The authors are planning to use the pose tracking using MANFIS and uncalibrated cameras for the visual servoing of the robot in future.

Acknowledgments

The author would like to thank our teachers at the department for many interesting and informative discussions. The author would also like to thank the anonymous reviewers of the original version of this paper who provided comments and remarks that have been invaluable in helping us understand what we are trying to do as well as significantly improving the paper. And he also would like to thank his family and friends who have helped him throughout these years.

References

[1] K Okada, M Kojima, S Tokutsu, T Maki, Y Mori and M Inaba 2007 *Multi-cue 3D object recognition in knowledge-based vision-guided humanoid robot system* (*IEEE/RSJ Int. Conf. Intell. Robot Syst.*) (CA, USA) pp 3217-22

[2] S Winkelbach, S Molkenstruck and F M Wahl *2006 Low-cost laser range scanner and fast surface registration approach* (DAGM Symp. Pattern Recognit.) (Berlin, Germany)

[3] C H Huang, C S Hsu, P C Tsai, R J Wang and W J Wang 2011 *Vision Based 3-D Position Control for a Robot Arm* (*IEEE Control Systems*) pp 1699-1703

[4] E Zhou, M Zhu, A Hong, G Nejat and B Benhabib 2015 *Line-Of-Sight Based 3D Localization of Parallel Kinematic Mechanisms* (*Internal Journal on Smart Sensing and Intelligent System* vol 8 no 2) pp 1172-83

[5] T Dallej, M Gouttefarde, N Andreff, M Michelin and P Martinet 2011 *Towards vision-based control of cable driven parallel robots* (*IEEE Int. Conf. Intell. Robots and Systems IROSS'11*) (San Francisco, United States) pp 2855-60

[6] Corke P 2011 *Robotic, Vision and Control*, ed Bruno S, Oussama K and Frans G (Berlin: Spring Verlag)

[7] J J Denavit and R S Hartenberg 1955 *A Kinematics Notation for Lower-Pair Mechanisms Based on Artices* (*ASME Journal of Applied Mechanics* vol 22) pp 215-21

[8] Jang S R, Sun C T and Mizutani E 1997 *Neurl-Fuzzy and Soft Computing* (Prentice Hall)

IOP Conf. Series: Materials Science and Engineering **383** (2018) 012016 doi:10.1088/1757-899X/383/1/012016

A study on slim optical fingerprint sensor for fake fingerprint detection on mobile environment

Young-Hyun Baek*, Byunggeun Kim, Seok-han Kim and Jong-woo Ahn

Research Institute, UNION COMMUNITY Co., Ltd. Seoul, South Korea

*Email: neural76@unioncomm.co.kr

Abstract. We propose a slim optical fingerprint sensor for fake fingerprint detection on mobile environment. The sensor proposed in this paper is compact size and slim thickness. Also, it can be used by only the power of smart phone without battery or external power. This sensor has a new lens design and optical path to minimize distortion. This sensor shows detailed images of ridges and valleys characteristic of fingerprints and has a characteristic of being resistant to moisture. The proposed sensor can detect various kinds of fake fingerprint (Paper, rubber, silicone, etc.). Simulations show that the slim optical fingerprint sensor has an effective performance on the results of the fake fingerprint detection rate.

1. Introduction

In the recent years, biometric technologies have gained increasing attention in the various field. The fingerprint recognition is a well-known biometric technology. There various techniques have been used in the optical, capacitive, radio frequency, ultrasound, and thermal methods [1-5]. Among them, the most common sensors are optical and capacitive fingerprint method. Optical fingerprint sensor is to offer greater durability and reliability, providing lower cost and longer life. Capacitive fingerprint sensor is to offer lower durability and providing higher cost.

In this paper, we propose a slim optical fingerprint sensor for mobile environment. This paper is organized as follows. In section 2, we proposed the slim optical structure and design. section 3 presents the simulations results. And section 4 is conclusions.

2. Proposed slim optical fingerprint sensor design

2.1. Optical fingerprint sensor

Figure 1 shows the Optical fingerprint sensor structure. The light is reflected at the fingerprint (as they pass through(valleys) or reflect off objects(ridges)). The sensor is to offer greater durability and reliability, providing lower cost and longer life, but they cannot be miniaturized due to optical characteristic (the distance between the prism and the image sensor) [3-6].

2.2. Fingerprint recognition system

This section provides a basic introduction to fingerprint recognition systems and their main parts, including a brief description of the most widely used techniques and algorithms. Figure 2 shows main modules of a fingerprint recognition system. Data capture(fingerprint sensing), in which the fingerprint of an individual is acquired by a fingerprint scanner to produce a raw digital representation. Next

Content from this work may be used under the terms of the Creative Commons Attribution 3.0 licence. Any further distribution of this work must maintain attribution to the author(s) and the title of the work, journal citation and DOI.
Published under licence by IOP Publishing Ltd

preprocessing, in which the input fingerprint is enhanced and adapted to simplify the task of feature extraction. Fingerprint feature extraction, in which the fingerprint is further processed to generate discriminative properties, also called feature vectors. Final step matching, in which the feature vector of the input fingerprint is compared against one or more existing templates. The templates of approved users of the biometric system, also called clients, are usually stored in a database. Clients can claim an identity and their fingerprints can be checked against stored fingerprints.

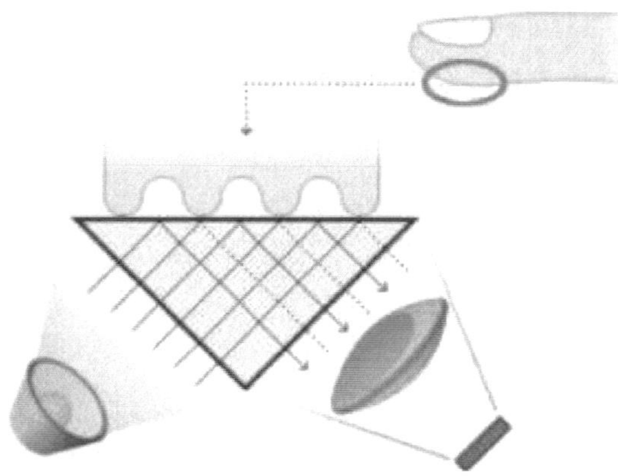

Figure 1. Optical capture method.

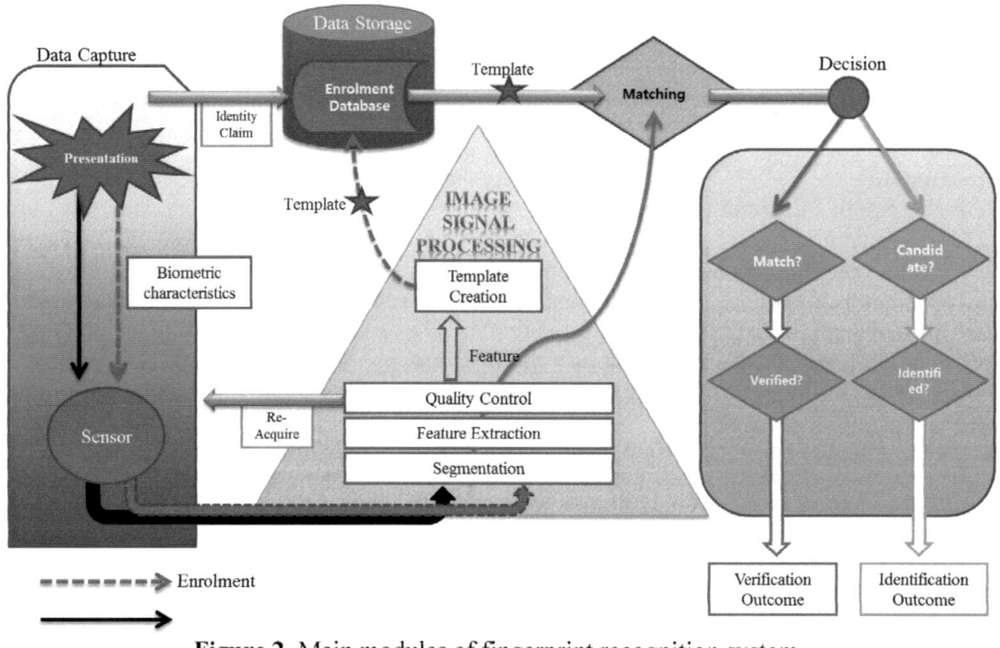

Figure 2. Main modules of fingerprint recognition system.

2.3. *Definition of fake fingerprint*

The fake fingerprint generally refers to a fake access by an illegal user into fingerprint biometric system by using a fake fingerprint reproducing that one of an unauthorized user. These artifacts are made from various materials like silicon, clay, rubber, paper, etc(figure 3). The whole process of creation of fake fingerprints can be done with or without co-operation and get easy access to highly authenticated society. We can countermeasure against such attacks using liveness check.

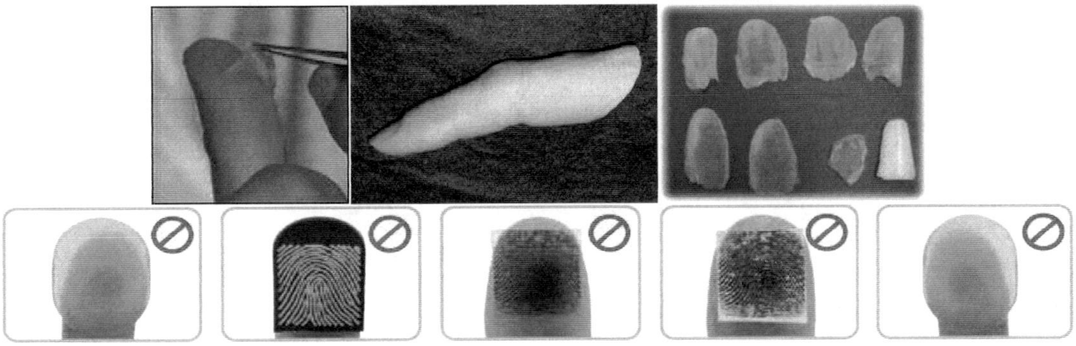

(a) Silicon (b) Rubber (c) Film (d) Paper (d) Gelatin

Figure 3. A kind of fake fingerprint.

2.4. Proposal slim optical lens design

The figure 4 is the slim optical design output applying the spherical lens to satisfy the standard FBI PIV Spec. In order to satisfy the PIV must be able to decompose the line of 0.05mm prism fingerprint surface. 0.05mm line of the prism surface is reduced through the image formation lens system to the image sensor there is in early value to be dismantled should enter the one line to a pixel of the image sensor. But consider that the line of machining tolerances and tolerances enclosures, an assembler 1line differential is 1 pixel is put in danger. It is most desirable to enter the 1 line to 2 pixel. If 1 line is designed to fall into the 3 pixel image. This is because the larger the angle of incidence to the micro-lens over each pixel in the image CMOS sensor. The prism surface is at least 0.05mm to be suitable for the more than 2.8um, 5.6um or less, when incident on the image sensor through the lens system. At this time, the ratio of the line is between 0.112 times 0.056 times. To enter the 1 line 21 pixel image differs, so the vertical direction is 25mm cis shall be imaged by 2.8mm. Figure 4 is applying a spherical lens design.

The prism is set to start because it should be separated so that they are at the wet. In addition, the surface of the fingerprint image without being affected by the external light source should be uniform.

Ability to separate water and silicon in all areas of active area:

- Minimum angle of 59.9 ° to distinguish water from silicon
- The angle of the active area is 60.2 °(figure 5), which is greater than 59.9 °.

2.5. Slim fingerprint internal configuration

Figure 6 shows the configuration of the proposed slim fingerprint sensor.

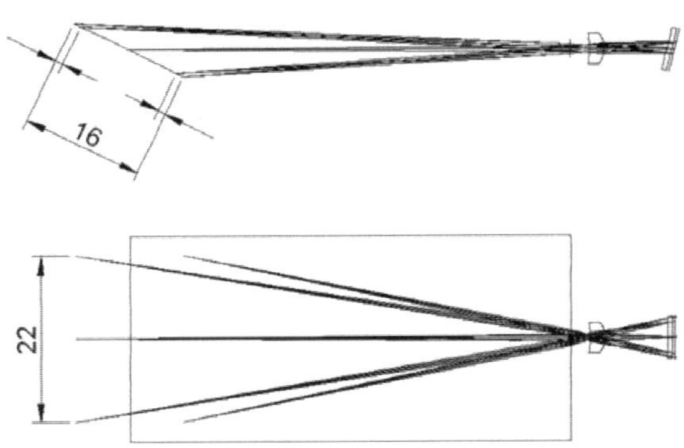

Figure 4. New spherical lens design (Active area in prism : 22 *16mm).

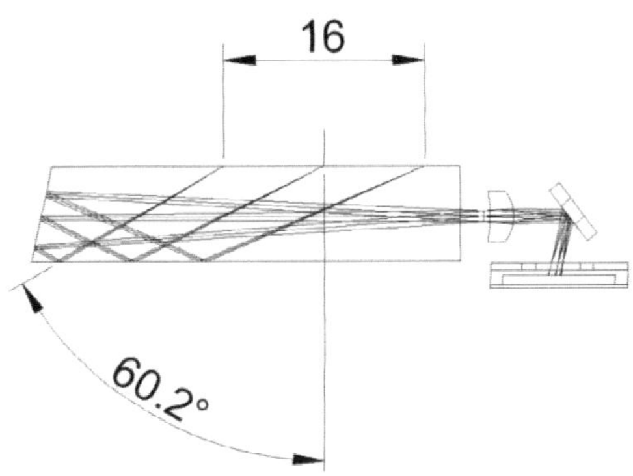

Figure 5. The prism design in slim moisture fingerprints.

IOP Conf. Series: Materials Science and Engineering **383** (2018) 012016 doi:10.1088/1757-899X/383/1/012016

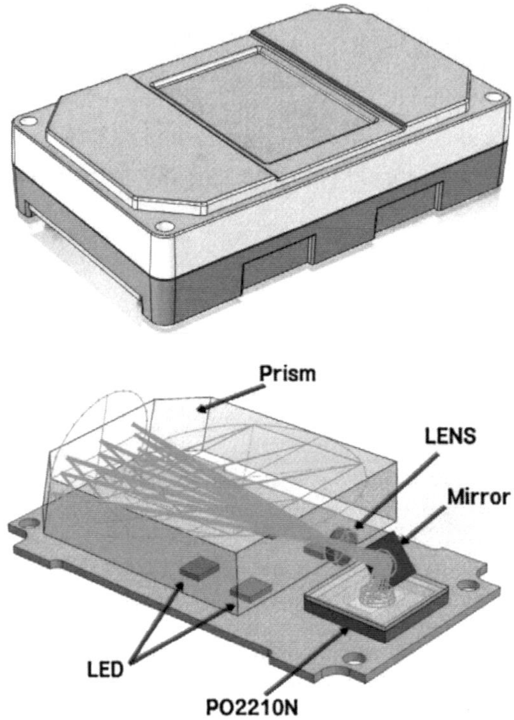

Figure 6. The device design in slim optical fingerprints.

3. Simulations

This section focuses on evaluating the proposed sensor. We start by describing the test optical simulation data. Then, the experiment results of the proposed slim optical sensor is presented for fake fingerprint detection rate.

Table 1 shows the comparison results when the result shows the high performance in all fake feature type. As can be seen in table 1, the fake fingerprint detection a 97.25% accuracy in the make fake fingerprint 100 set.

Table 1. The comparison results (100set).

	Fake Detection(detect/try)	Detection rate(%)
Real Finger	**0/100**	**Pass**
Paper	100/100	100%
Rubber	98/100	98%
Film	100/100	100%
Silicon	91/100	91%

4. Conclusion

The performance of the slim optical sensor with mobile environment is shown in figure 7. As a result, we succeeded in overcoming the limitation of optical design and implementing optical type fingerprint sensor with slim structure. The proposed system tested on the real finger and fake fingers. And The table 1 results show that the fake fingerprint detection accuracy has been 97.25% which proves that our system has a performance. In future works, we will need to variety of material fake fingerprint testing.

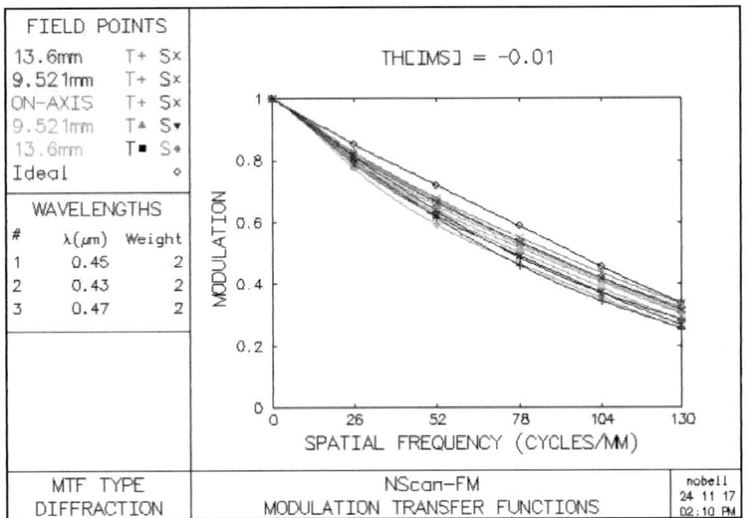
FIELD POINTS
13.6mm T+ Sx
9.521mm T+ Sx
ON-AXIS T+ Sx
9.521mm T▲ S▼
13.6mm T■ S◆
Ideal ◇
WAVELENGTHS
λ(μm) Weight
1 0.45 2
2 0.43 2
3 0.47 2
TH[IMS] = -0.01
MODULATION
SPATIAL FREQUENCY (CYCLES/MM)
MTF TYPE
DIFFRACTION
NScan-FM
MODULATION TRANSFER FUNCTIONS
nobell
24 11 17
02:10 PM

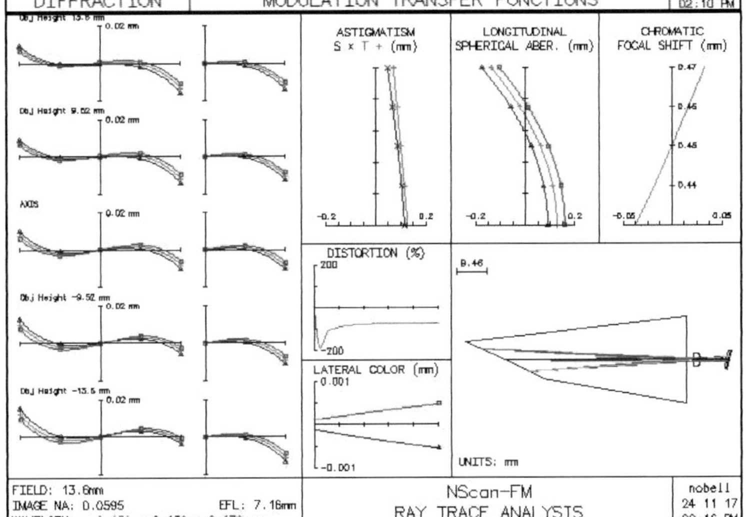
ASTIGMATISM
S x T + (mm)
LONGITUDINAL
SPHERICAL ABER. (mm)
CHROMATIC
FOCAL SHIFT (mm)
DISTORTION (%)
LATERAL COLOR (mm)
UNITS: mm
FIELD: 13.6mm
IMAGE NA: 0.0595
EFL: 7.16mm
NScan-FM
RAY TRACE ANALYSIS
nobell
24 11 17

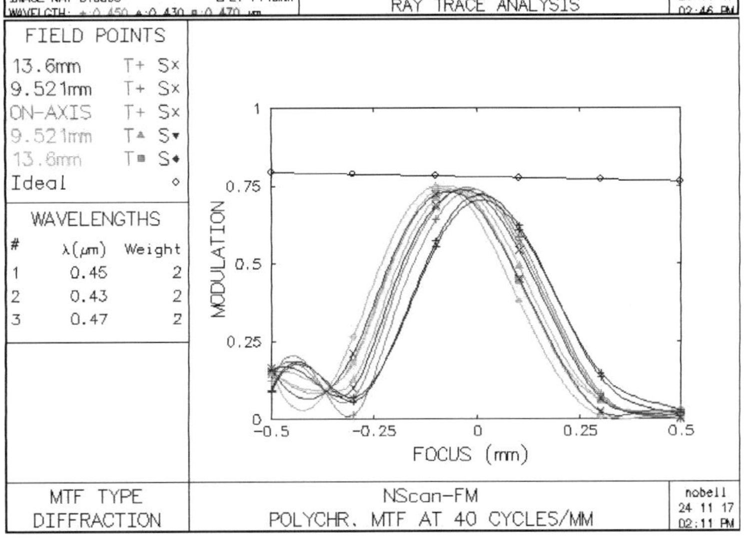
FIELD POINTS
13.6mm T+ Sx
9.521mm T+ Sx
ON-AXIS T+ Sx
9.521mm T▲ S▼
13.6mm T■ S◆
Ideal ◇
WAVELENGTHS
λ(μm) Weight
1 0.45 2
2 0.43 2
3 0.47 2
MODULATION
FOCUS (mm)
MTF TYPE
DIFFRACTION
NScan-FM
POLYCHR. MTF AT 40 CYCLES/MM
nobell
24 11 17
02:11 PM

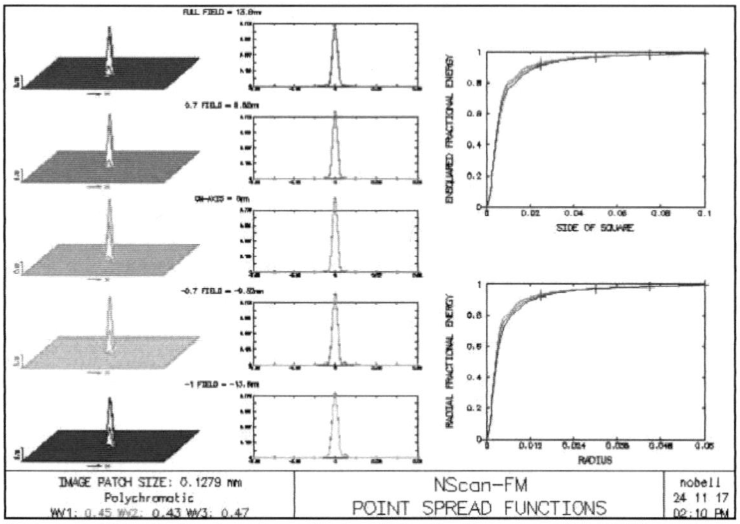

Figure 7. The Optical system design / Optical system simulation result.

Acknowledgments
This research was supported Advanced Technology Center (ATC) grant funded by the Korean government (Ministry of Trade, Industry and Energy) (No. 10051484).

References

[1] A. Al-Ajlan, "Survey on fingerprint liveness detection", In International Workshop on Biometrics and Forensics, pp 1-5, 2013
[2] D., Maio, D., Jain, A.K., Prabhakar, S.: "Handbook of Fingerprint Recognition", Springer-Verlag, New York, 2003
[3] Qinzhi Zhanga, Hong Yan, "Fingerprint classification based on extraction and analysis of singularities and pseudoridges", Pattern Recognition, vol.**37**, pp 2233–43, 2004
[4] J. W. Lee, D. J. Min, J. Y. Kim and W. C. Kim, "A 600-dpi Capacitive Fingerprint Sensor Chip and Image-Synthesis Technique", IEEE J. of Solid-state circuits, vol. **34**, no. 4, (1999) April, pp 469-75
[5] S. -M. Jung, J. -M. Nam, D. -H. Yang and M. K. Lee, "A CMOS Integrated Capacitive Fingerprint Sensor with 32-bit RISC Microcontroller", IEEE Journal of Solid-state Circuit, vol. **40**, no. 8, (2005), pp 1745-50
[6] S. -J. Kim, K. -H. Lee, S. -W. Han and E. Yoon, "A CMOS Fingerprint System-on-a-Chip With Adaptable Pixel Networks and Column-Parallel Processors for Image Enhancement and Recognition", Journal of Solid-state Circuit, vol. **43**, no. 11, (2008), pp 2558-67, 2008

IOP Conf. Series: Materials Science and Engineering **383** (2018) 012017 doi:10.1088/1757-899X/383/1/012017

Subcarrier selection for efficient CSI-based indoor localization

Yu Taso[1], Shih-Chun Yeh[2], Yu-You Liang[2,*], Chu-Hsuan Wang[2] and Shih-Hau Fang[2]

[1]Academia Sinica, Taiwan
[2]Yuan-Ze University, Taiwan

*Email: y.y.liang.hchs@gmail.com

Abstract. Indoor positioning systems have received increasing attention for supporting location-based services. In recent Wi-Fi networks, the rich information in the physical layer, known as channel state information (CSI), has been recognized an effective positioning characteristic rather than traditional received signal strength. However, the positioning performance depends on a very high-dimensional CSI due to all pairs of transceiver antenna, which may incur over-fitting problems. This paper proposes a subcarrier-selection approach based on information theoretic learning to compensate for over-fitting problems in CSI-based localization systems. After equalizing the histogram of CSIs, the proposed algorithm computes the information gain of each subcarrier and forms a new low-dimensional subset of CSIs to reduce the complexity and to decrease possible over-fitting caused by redundant CSIs. We demonstrate the effectiveness of the proposed algorithm through experiments. On-site experimental results demonstrate that the proposed approach outperforms traditional feature selection schemes.

1. Introduction

Indoor localization has become a crucial concern for various mobile applications in pervasive computing environments [1]. The massive deployment of Wi-Fi access points (APs) has made Wi-Fi a suitable technology for developing such indoor location systems. Numerous studies have addressed indoor location estimation by using existing Wi-Fi infrastructure. Among the various indoor Wi-Fi positioning systems, the fingerprinting-based approach, in which the user's location is estimated by matching receive signal strength (RSS) with the radio map, is one of the most feasible solutions [2].

In recent Wi-Fi systems with the IEEE 802.11a/g/n standard, the Orthogonal Frequency Division Multiplexing (OFDM) and Multiple-input and multiple-output (MIMO) technology have be used, where data are modulated on multiple subcarriers in different frequencies and transmitted among different pairwise antenna to increase throughput [3-4]. Thanks to the development of firmware and open sources, the wireless MIMO channel response can be extracted from the receivers in the format of channel state information (CSI) [5-7].

CSI reveals a set of channel measurements depicting the amplitudes and phases of every subcarrier. Recently, CSI-based fingerprinting approaches, which uses PHY layer fine grained information, have been proposed for indoor localization. Instead of receive signal strength (RSS), these approaches exploit CSI as new positioning features, with the location granularity to improve the accuracy [8-14].

Content from this work may be used under the terms of the Creative Commons Attribution 3.0 licence. Any further distribution of this work must maintain attribution to the author(s) and the title of the work, journal citation and DOI.
Published under licence by IOP Publishing Ltd

However, the positioning performance depends on a very high-dimensional CSI due to all pairs of transceiver antenna, which may incur over-fitting problems.

The studies in [15] proposed FIFS, which leverage both the frequency and spatial diversity and uses a correlation filter with the probability positioning algorithm. That is, FIFS represents the location by aggregating the power of all subcarriers. However, although the summation over all subcarriers can significantly reduce the dimensions, it does not consider the detail information within subcarriers. A similar approach, namely, MIMO, was presented in [16-17], in which the amplitude and phase value were subtracted for subsequent subcarrier to generate the location fingerprint. Although this approach effectively extracts the relatively robust fingerprints owing to the subtraction process, the high-dimension still maintains.

This paper presents a robust CSI-based indoor localization using histogram equalization (HEQ) [18] and information theoretic learning (ITL) [19] to reduce the required dimensions. Specifically, we propose a subcarrier-selection approach based on information theoretic learning to compensate for over-fitting problems. There are two steps of the proposed algorithm. First, we normalize CSIs using the HEQ technique by converting the coefficients into a reference to enhance the system robustness. Second, after equalizing the histogram of CSIs, the proposed algorithm computes the information gain of each subcarrier based on ITL and forms a new low-dimensional subset of CSIs to reduce the complexity and to decrease possible over-fitting caused by redundant CSIs. We demonstrate the effectiveness of the proposed algorithm through experiments. On-site experimental results demonstrate that the proposed approach provide compare or even better performance with a relatively lower-dimension subcarrier subset.

The rest of the paper is structured as follows. Section II illustrates the proposed localization algorithm. Section III presents the experimental setup and results. Finally, the conclusion is given in Section IV.

2. Proposed algorithm

To enhanced data throughput, Wi-Fi networks used OFDM (orthogonal frequency division multiplexing) and MIMO (multiple input multiple output) techniques in the 802.11a/g/n standard, where data are modulated on multiple channels in different frequencies and are transmitted simultaneously among multiple antenna pairs. In these systems, channel response can be extracted from the receivers in the format of channel state information (CSI), as indicated in figure 1.

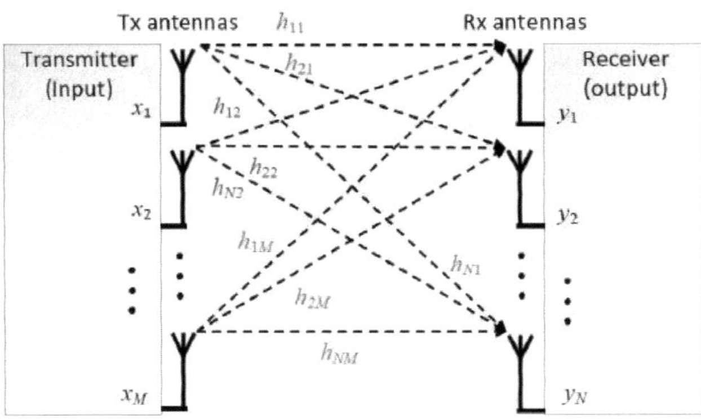

Figure 1. Main idea of CSI in a MIMO system.

Because CSI reports each channel state for each antenna pair, we assume that a complete measured CSI signal in the MIMO wireless network is given as equation (1):

$$\mathrm{h} = [h_{11},\cdots,h_{1M},\cdots,h_{N1},\cdots,h_{NM}]^T \quad (1)$$

To localize a user, a typical fingerprinting location algorithm estimates the likelihood of real-time CSI observations for all candidate locations and selects the one having maximum likelihood as the result. Let R be the entire set of reference locations; the CSI-fingerprinting localization problem can be formulated as equation (2)

$$r^* = arg \max_{r \in \Re} P(h|\wedge_r) \tag{2}$$

where Λ_r is the probabilistic model of the r-th reference location, which can be obtained on the basis of the training data during the offline stage [20]. A challenging issue in CSI- based localization is the very high-dimensional CSI. In this study a mechanism is designed to select the important CSIs as positioning feature.

First, we normalize the CSIs using the histogram equalization (HEQ) technique (i.e., converting the coefficients into a reference). HEQ provides a transformation that converts the probability density function of an original variable into a reference probability density function. This study uses a typical Gaussian as the reference. Then, the proposed approach ranks CSIs in descending order of their InfoGain values which are calculated as equation (3):

$$InfoGain(h_i) = I(D) - I(h_i)\,,\ i = 1,\cdots, NM \tag{3}$$

where I(D) is the information incurred in the entire training database D (entropy of the reference locations without h_i, and I(h_i) computes the conditional entropy of the location given h_i. This approach calculates the discriminative ability for each subcarrier and the top are considered the best. As the above equation indicates, the feature with larger information gain is assigned a higher rank, representing more contribution in indoor localization.

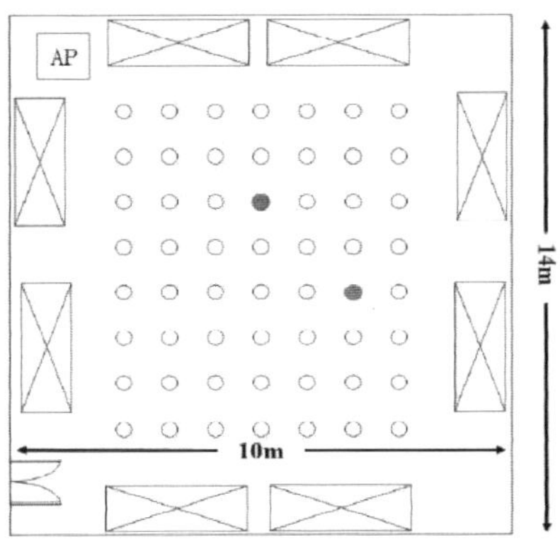

Figure 2. Part of the fifth floor plane of the test-bed building, where we had performed the experiments.

3. Performance evaluation

3.1. Experimental setup

This study involved experiments conducted on the fifth floor of the telecommunications building at Yuan-Ze University. Figure 2 shows where the experiments were performed. The size of the test-bed was 14 (m) × 10 (m). One access point (ASUS RT-N12E) with two transmitted antennae (M=2) was deployed at the corner. After installing the Linux 802.11n driver, which is built on the Intel Wi-Fi Wireless Link 5300 MIMO radios, the gain and phase of the signal path between a single transmit–receive antenna pair is available in the measured device (Lenovo x200). In the experiments, the laptop measured CSIs using three receiver antenna (N=3), and thus there were 6 links available as positioning features. Each reported channel state contains 30 subcarrier channel groups. Each channel state matrix entry is a complex number and only the magnitudes were used in this study.

Figure 3 shows the realistic RSS and CSI measurements over 100s at two different locations. In figures 3(b) and 3(d), CSIs are measured over 100s at a fixed location, indicated by a red and blue circles in figure 2, respectively. The wider curve represents the higher time variation of an antenna pair. Similarly, figures 3(a) and 3(c) show the measured RSSs at the sane locations. Figure 3 clearly reveals the changing CSI patterns measured from two different locations. It indicates the possibility of determining the location on the basis of CSI. Moreover, the figures verify that CSI provides fine-grained information, for each subcarrier and each Tx-Rx pair, whereas RSS provides coarse-grained information only.

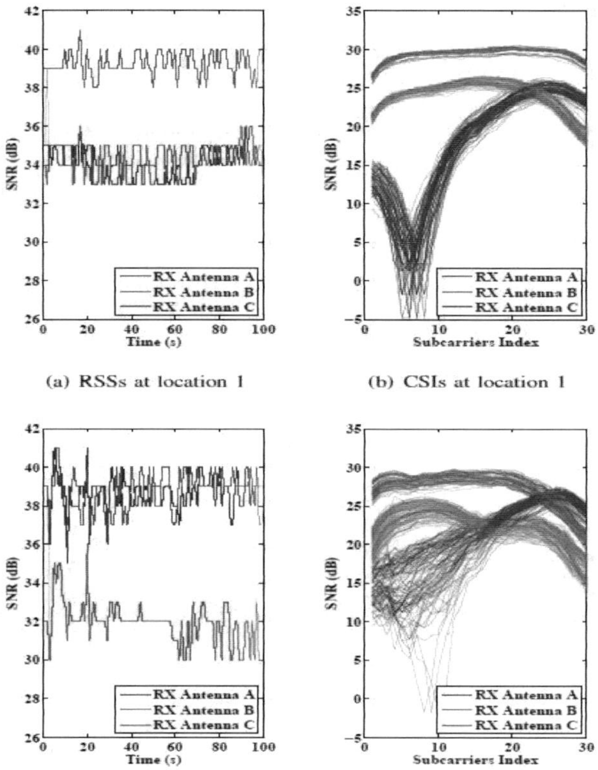

(a) RSSs at location 1 (b) CSIs at location 1

(c) RSSs at location 2 (d) CSIs at location 2

Figure 3. The measured RSSs and CSIs over 100s at two different locations.

3.2. Performance evaluation

Figure 4 displays the mean positioning error versus the number of subcarriers based on six algorithms, including RSS, CSI, HEQ, MIMO, FIFS, and the proposed ITL algorithm. The HEQ approach indicates the direct normalization of CSI, and the RSS and FIFS methods are constant because the performance is independent of the number of subcarriers. This figure first shows that the error generally decreases as the number of channels increases. Moreover, this figure indicates the effectiveness of ITL, which considerably reduce the error while using the lowest number of subcarriers. This shows that ITL can select the useful location-related information among CSIs more effectively.

Next, figure 5 compare the performance of different sub- carrier selection methods on mean positioning error. In this figure, the Maxmean method selects the most stronger CSIs (in terms of the maximal mean magnitude over time) while the Minvar approach selects the most stable ones (in terms of the minimal variance over time). This figure shows that the proposed ITL algorithm outperforms traditional selection approaches. Using 40 subcarriers, ITL reduces the mean localization errors by 40% and 22%, respectively, compared to Maxmean and Minvar. The trend is similar using 50 subcarriers. The results again demonstrate that ITL constructs a more suitable subset of subcarriers according to information gain, thus achieving more accurate location estimations with fewer numbers of CSIs. The computed information gain of CSIs is shown in figure 4.

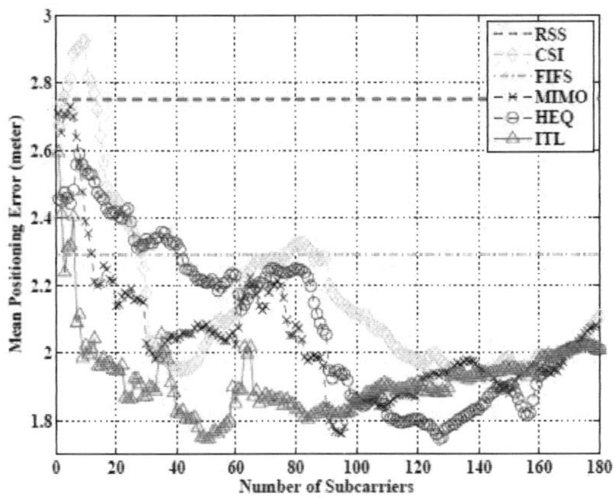

Figure 4. Performance comparison between different positioning algorithms.

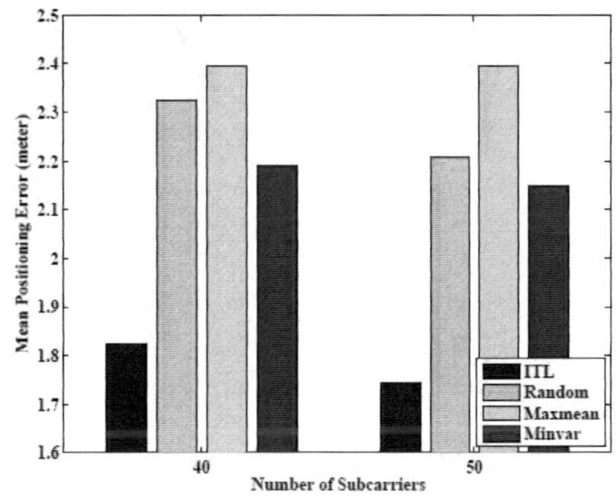

Figure 5. Performance comparison between different subcarrier selection algorithms.

4. Conclusions

This paper proposes a subcarrier-selection approach based on ITL to compensate for over-fitting problems in CSI-based localization systems. After equalizing the histogram of CSIs, the proposed algorithm computes the information gain of each subcarrier and forms a new low-dimensional subset of CSIs to reduce the complexity and to decrease possible over-fitting caused by redundant CSIs. We demonstrate the effectiveness of the proposed algorithm through experiments. The experimental results demonstrate that ITL constructs a more suitable subset of subcarriers according to information gain, thus achieving more accurate location estimations with fewer numbers of CSIs.

References

[1] Y. Gu, A. Lo, and I. Niemegeers, "A Survey Of Indoor Positioning Systems For Wireless Personal Networks," *IEEE Communications Surveys & Tutorials*, vol. **11**, no. 1, pp 13–32, 2009

[2] A. S. Paul and E. A. Wan, "RSSI-Based Indoor Localization And Tracking Using Sigma-Point Kalman Smoothers," *IEEE J. Select. Topics Signal Process*, vol. **3**, pp 860–73, Oct 2009

[3] A. J. Paulraj, D. A. Gore, R. U. Nabar, and H. Bolcskei, "An Overview Of MIMO Communications - A Key To Gigabit Wireless," *Proc. IEEE*, vol. **92**, pp 198–218, Feb 2004

[4] P. Xia, S. Zhou, and G. B. Giannakis, "Adaptive MIMO-OFDM Based On Partial Channel State Information," *IEEE Trans. Signal Process.*, vol. **52**, pp 202–13, Jan 2004

[5] D. Halperin, W. Hu, A. Sheth, and D. Wetherall, "Predictable 802.11 Packet Delivery From Wireless Channel Measurements," *SIGCOMM Comput. Commun. Rev.*, vol. **41**, no. 4, pp 159–70, 2011

[6] D. Halperin, W. Hu, A. Sheth, and D. Wetherall, "Tool Release: Gathering 802.11n Traces With Channel State Information," *SIGCOMM Comput. Commun. Rev.*, Jan 2011

[7] S. Sen, R. R. Choudhury, B. Radunovic, and T. Minka, "Precise Indoor Localization Using PHY Layer Information," in *ACM HotNets*, pp 18:1–18:6, 2011

[8] K. Wu, J. Xiao, Y. Yi, M. Gao, and L. Ni, "FILA: Fine-Grained Indoor Localization," in *INFOCOM*, pp 2210–18, March 2012

[9] Z. Yang, Z. Zhou, and Y. Liu, "From RSSI To CSI: Indoor Localization Via Channel Response," *ACM Comput. Surv.*, vol. **46**, no. 2, pp 25:1– 25:32, November 2013

[10] K. Wu, J. Xiao, Y. Yi, D. Chen, X. Luo, and L. Ni, "CSI-Based Indoor Localization," *IEEE Trans. Parallel Distrib. Syst.*, vol. **24**, no. 7, pp 1300–09, July 2013

[11] Z.-P. Jiang, W. Xi, X. Li, S. Tang, J.-Z. Zhao, J.-S. Han, K. Zhao, Z. Wang, and B. Xiao, "Communicating Is Crowdsourcing: Wi-Fi Indoor Localization With CSI-Based Speed

Estimation," *JCST*, vol. **29**, no. 4, pp 589–604, 2014

[12] M. Kotaru, K. Joshi, D. Bharadia, and S. Katti, "SpotFi: Decimeter Level Localization Using Wi-Fi" in *SIGCOMM*, pp 269–82, 2015

[13] F. Wen and C. Liang, "Fine-Grained Indoor Localization Using Single Access Point With Multiple Antennas," *IEEE Sens. J.*, vol. **15**, pp 1538–44, March 2015

[14] X. Wang, L. Gao, S. Mao, and S. Pandey, "CSI-Based Fingerprinting For Indoor Localization: A Deep Learning Approach," *IEEE Trans Veh Technol*, vol. PP, no. 99, pp 1–1, 2016

[15] J. Xiao, K. Wu, Y. Yi, and L. Ni, "FIFS: Fine-Grained Indoor Fingerprinting System," in *ICCCN*, pp 1–7, July 2012

[16] Y. Chapre, A. Ignjatovic, A. Seneviratne, and S. Jha, "CSI-MIMO: Indoor Wi-Fi Fingerprinting System," in *Proc. IEEE LCN*, pp 202– 09, Sept 2014

[17] Y. Chapre, A. Ignjatovic, A. Seneviratne, and S. Jha, "CSI-MIMO: An Efficient Wi-Fi Fingerprinting Using Channel State Information With MIMO," *Pervasive and Mobile Computing*, vol. **23**, pp 89–103, July 2015

[18] S. H. Fang, C. H. Wang, and Y. Tsao, "Compensating For Orientation Mismatch In Robust Wi-Fi Localization Using Histogram Equaliza- tion," *IEEE Trans. Veh. Technol.*, vol. **64**, pp 5210–20, Nov 2015

[19] J. C. Principe, *Information Theoretic Learning: Renyis Entropy and Kernel Perspectives.* Springer Verlag, 2010

[20] S. H. Fang, W.H. Chang, Y. Tsao, H. C. Shih, and C. W. Wang, "Channel State Reconstruction Using Multilevel Discrete Wavelet Transfrom for Improved Fingerprinting-Based Indoor Location," *IEEE Sensors Journal*, 2016

IOP Conf. Series: Materials Science and Engineering **383** (2018) 012018 doi:10.1088/1757-899X/383/1/012018

Agrizero: an O2O adversarial framework and triangulation for agricultural forecasting

Yenming J. Chen*

Department of Logistics Management, National Kaohsiung University of Science & Technology, Kaohsiung, Taiwan

*Email: yjjchen@ccms.nkfust.edu.tw

Abstract. This study proposes an online to offline (O2O) learning framework, ArgiZero, based on three components: a generative adversarial network, an online auction market, and offline simulated agents (digital twins: buyers, farmers, natures, and markets). The generative time series and digital twins are massively generated in a manner of Monte Carlo but with extremely efficient algorithms. The goal of the generator is to produce time series that are statistically indistinguishable with the records from auction market. The goal of the discriminator is to develop a triangulation method based on semi-modeless assimilation to separate generated from actual time series. Most farmers believe agriculture is impossible to be planned because uncertainty, crowding, crisis, and risk cannot be foreseen and accommodated. This AgriZero framework alleviates the challenges through the techniques of Bayesian deep learning and data assimilation as well as the mega power of GPU computation. Thanks to Bayesian hierarchical estimation, which is akin to deep learning but more sophisticate and longer history. We are able to estimate human behaviour in the agents of buyers and farmers, natural disaster in the agents of natures, and price fluctuations in the agents of markets. The framework has been validated by a large amount of records in vegetable auctions of Taiwan and USA. The hierarchical Bayesian estimation and Monte Carlo Markov Chain particle filters used in hidden Markov model are appreciated during the massive construction of the most probable digital twins. The feature space mapping in wavelet time series, Bayesian deep learning in recurrent neural network, kernel induced Hamiltonian dynamics ABC, and hybrid SDE-kernel based forecasting for time series analysis embodies the particle generator in the GAN structure. We also apply time series clustering, RNN, bagging and boosting, and semi-modeless assimilation to assist the performance of the triangulated discriminator.

1. Introduction

Challenges exist in forecasting cultivation yields and crop prices. Market prices may fluctuate without obvious reasons. Unlike the business practices in the industry, no actual demand is specified for a farmer. In other words, no customers place orders for agricultural produce. Price is determined completely in an auction market through a demand-supply-price relation. In the agriculture business, no one promises to buy produce with promised prices. Instead, farmers guess future trends of prices through a history of transactions. Without the help of scientific management, farmers mostly gamble in low bet efficiency. Because all farmers look at the same price history, they may make the same "optimal planting decision" in variety and quantity. This excessive over-supply usually makes produce

Content from this work may be used under the terms of the Creative Commons Attribution 3.0 licence. Any further distribution of this work must maintain attribution to the author(s) and the title of the work, journal citation and DOI.
Published under licence by IOP Publishing Ltd

price plunge. Whenever a loss is encountered, the government is obligated to allocate tax money to subsidize the losses, and consumers must buy produce with high prices.

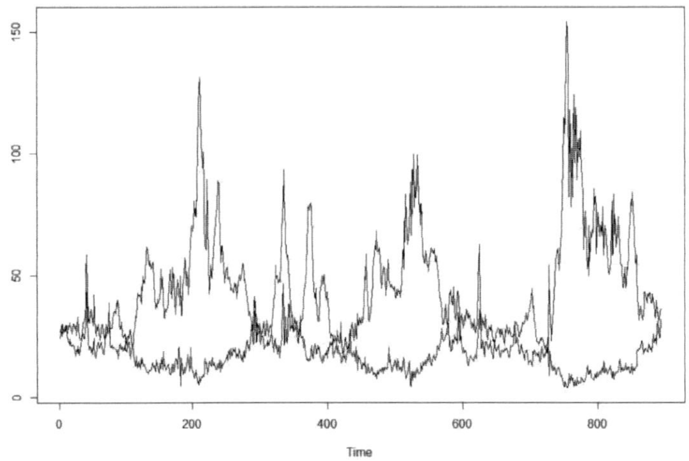

Figure 1. Transaction quantities and price for celery in three consecutive years.

By inspecting figure 1, the motivation for our agriculture research is well understood. The graph shows the transaction quantity and price for celery in three consecutive years. The supply quantities oscillate yearly and prices response reversely. Apparently, the most profitable timing cannot be at the high and low peaks of production. An experienced farmer usually uses their intuition to guess a right timing and quantity for their cultivation plan. Traditionally, because of multiple constraints imposed by the natural characteristics of crops, such as weather, temperature, and scale of economies, people mistakenly believe that they have limited freedom in making the optimal decisions for cultivation. Without the possibility of control, allowing for the contribution of forecast becomes unconventional. This mis-understanding is probably the reason why prior studies in cultivation planning are few. However, almost all farmers, regardless of whether they are doing in a right way or not, implicitly apply newsvendor theory to their planting quantity decisions. According to their intuition, farmers usually plant an increased number of crops if they envision an increase in the probability of high demand (demand >> supply) and higher profit (price >> cost). If this intuition is toward the right direction, scientific approaches can definitely help farmers avoid mistakes and improve their managerial yields.

We propose an online to offline (O2O) learning framework, ArgiZero to tackle the challenges in agriculture business. Basically, the generative time series and digital twins are massively generated in a manner of Monte Carlo but with extremely efficient algorithms.

2. Literature

This framework is related to three streams of literature. We are mostly concerned with business and practical aspects of the agriculture supply chain as well as with the mechanism to make the system sustainable. To make the research practical, big data analytics and deep learning in generative models are investigated.

This framework exploits machine learning in stochastic modelling. Deep generative models recently received tremendous attention because of their superior performance. However, parameter estimation for latent variables within the models could be challenged due to highly non-linearity and chaotic state trajectories. [1] proposed a generative adversarial net in which two models are trained iteratively. The two models form a minimax two-player game such that a generative model is trained to mimic given data and the other discriminative model is trained to evaluate the performance of the target generative model. This approach effectively replaces the ordinary mean square error metric used traditionally. [2] developed deep convolutional to replace the inner product operator in the middle layers for application of anomaly detection, classification, and clustering. [3] summarized the algorithms of unsupervised deep learning for time-series feature extraction. [4] reviewed the deep learning evolution in the form of literature chronicles. On the other hand, the deep generative models can help the MCMC sampling more efficient. [5] facilitated neural dynamics of recurrent networks for the sampling of stochastic computation and suggested that the stochastic computation is closer to the mechanism of biological brain.

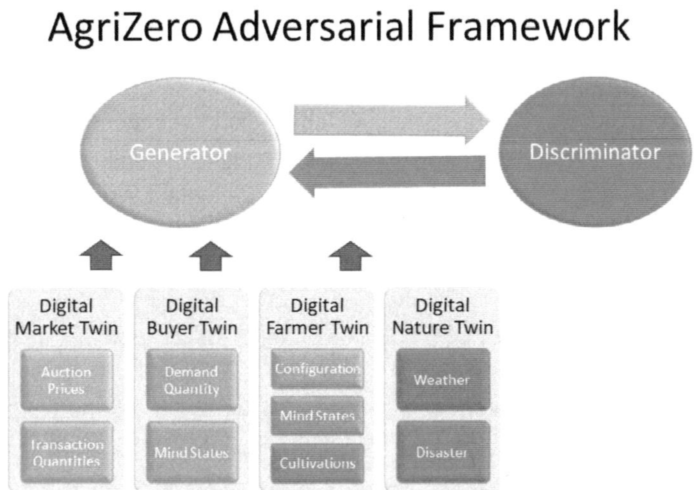

Figure 2. AgriZero adversarial framework.

3. Proposed framework

Our AgriZero framework incorporates big data estimation techniques and price dependent demand model into an integrated decision model, as shown in figure 2. Therefore, our model is different from others in three perspectives: 1) the performance of training is not fixed by a metric like *mse* but it will be discriminated by another learning agent, which devote to challenge the former's decisions, 2) the long consequence of a decision will firstly reflect to a virtual market which can embed all kinds of factors, and 3) the reinforcement of decision logic can be immediately tested by the online transaction results.

This framework uses partially observed Markov process (POMP) models as a fundamental computation platform for the generative agents. The models are also known as hidden Markov models (HMM) or state space models, which are powerful for time series analysis. We focus on Monte Carlo statistics, where the process and noises are nonlinear and non-Gaussian. The underlying algorithms include sequential Monte Carlo MLP, Hamiltonian MLP, nonlinear iterated filtering, particle MCMC, and approximate Bayesian computation (ABC). As shown in figure 3, the prediction of celery price and quantity through the learning of POMP in the three years are promising.

Triangular data assimilation used in the AgriZero framework refers to the quantitative methods by which the observations of variables are combined with system behaviours to provide estimates of internal states and key parameters. The discriminator can easily to use the framework of triangular data assimilation to boost the performance of generative agents. General data assimilation applies sparse observed information to estimate system internal states by making assumptions based on the fact that the system behaviours are typically governed by a set of pre-defined equations or statistical models. A suggested treatment can therefore be derived once all internal states are clear.

The estimation of the unmeasured demands depicted in the left panel of figure 4 is not accurate with a single measuring source. True demand should be presented in the solid line, but the estimation falls to the dashed lines due to the lack of sufficient information. However, through other sources, as shown in the right panel of figure 4, the estimation becomes more accurate. New evidences can be derived from observations that result from the common causes.

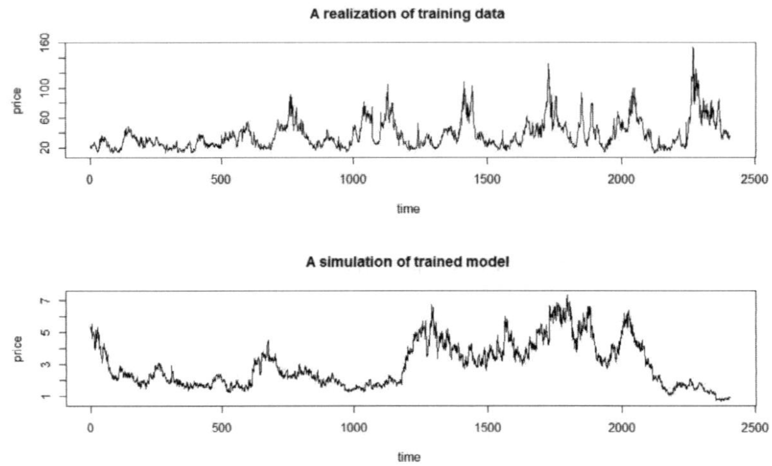

Figure 3. The learning results of POMP and prediction of celery price and quantity in three years.

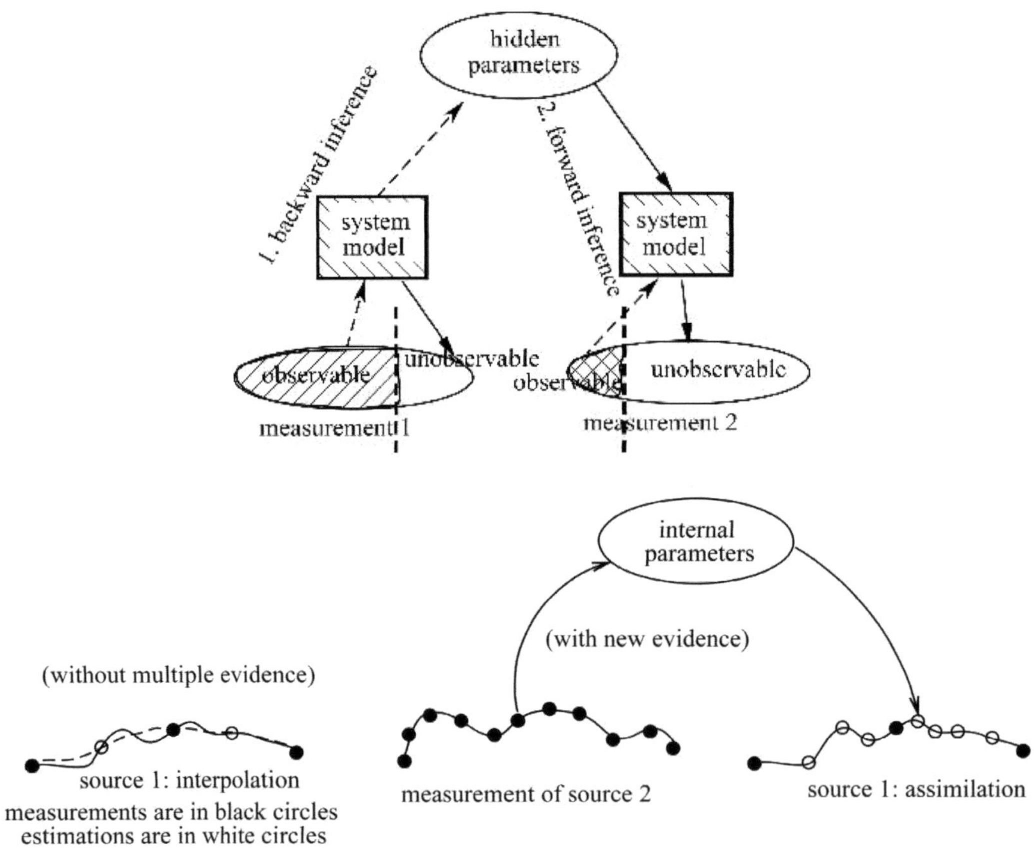

Figure 4. Triangular Assimilation.

4. Results

Solving a parameter estimation problem is highly complicated and mostly nonlinear and combinatorial. Combining recent advances in machine learning and stochastic achievements offer an excellent means to achieve the goal. Bagging and boosting has been proven effective in machine learning. The algorithm of adaBoost classification was reported to have excellent capability. In the classification work, the adaBoost classification algorithm successfully combines the merits of multiple algorithms to tackle the challenges. In this framework, the machine learning work is supported mainly by the high-efficiency MCMC and deep learning.

To estimate the joint probability densities, the hierarchical Bayesian learning and MCMC methods were used in the algorithm. The MCMC methods represent a broad range of integration techniques and can approximate a multidimensional integral by generating samples through an ergodic Markov chain, thereby ensuring that the generated samples converge to the desired distribution based on the ergodic theorem. We apply the hierarchical Bayesian estimation and Monte Carlo Markov Chain particle filters during the massive construction of the most probable digital twins. The feature space mapping in wavelet time series, Bayesian deep learning in recurrent neural network, kernel induced Hamiltonian dynamics ABC, and hybrid SDE-kernel based forecasting for time series analysis embodies the particle generator in the GAN structure. figure 5 shows the result of forecasting by applying time series clustering, RNN, bagging and boosting, and semi-modeless assimilation.

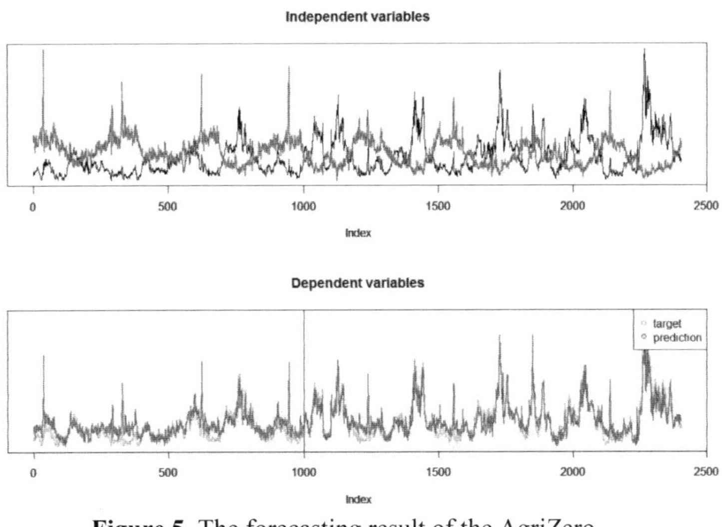

Figure 5. The forecasting result of the AgriZero.

To conduct high-performance computation algorithms, we apply contemporary GPU, in particular NVIDIA GPU. Because NVIDIA is among the most dominant in the graphics hardware market, their proprietary CUDA programming interface has become the industry standard to distributing thread tasks to parallel programming cores with the scalable ability.

To further overcome misclassification errors, the AdaBoost algorithm combines a collection of weak classifiers (around 500 to 1000) to approach the Bayes classifier iteratively. To improve the classification rate, training data with high misclassification rates are emphasized and boosted in the next iteration. The final classifier is generally obtained by collecting votes from all the classifiers, although it could also be a simple linear combination of the classifiers.

5. Conclusions

Our AgriZero framework is novel in helping investigate the effects of business and managerial insights in terms of the optimal control theory, O2O adversarial framework, and triangular assimilation. We established a modeling work for cultivation planning and will focus on model identification as

IOP Conf. Series: Materials Science and Engineering **383** (2018) 012018 doi:10.1088/1757-899X/383/1/012018

well as Markov time series prediction. This study contributes to the field of operations management in a holistic view of time series analysis in government data.

With our AgriZero framework, farmers may increase their profits by using modern management technologies. For whatever reasons that agriculture management research has been ignored for a long time, we wish to raise the possibility of increasing the efficiency of crop production and reducing the waste through the help of modern AI technologies.

References

[1] Goodfellow, Ian, Jean Pouget-Abadie, Mehdi Mirza, Bing Xu, David Warde-Farley, Sherjil Ozair, Aaron Courville, Yoshua Bengio. 2014. Generative adversarial nets. Advances in neural information processing systems. pp 2672-80

[2] Yuan, Chao, Amit Chakraborty. 2016. Deep convolutional factor analyser for multivariate time series mod- eling. Data Mining (ICDM), 2016 IEEE 16th International Conference on. IEEE, pp 1323-28

[3] L¨angkvist, Martin, Lars Karlsson, Amy Loutfi. 2014. A review of unsupervised feature learning and deep learning for time-series modeling. Pattern Recognition Letters 42 pp 11-24

[4] Schmidhuber, Ju¨rgen. 2015. Deep learning in neural networks: An overview. Neural networks pp 61 85-117

[5] Buesing, Lars, Johannes Bill, Bernhard Nessler, Wolfgang Maass. 2011. Neural dynamics as sampling: a model for stochastic computation in recurrent networks of spiking neurons. PLoS computational biology 7(11) e1002211

Improved wavelet cerebellar model articulation controller for precision positioning of piezo-driven stage

Van-Phuong Ta[1,2,*] **and Xuan-Kien Dang**[2]

[1]HCMC University of Technology And Education, Vietnam
[2]Hochiminh University of Transport, Vietnam

*Email: phuongtv@hcmute.edu.vn

Abstract. This paper proposes an improved wavelet cerebellar model articulation control system (IWCMACS) to control precision positioning of the linear piezoelectric motor (LPM). The proposed system combines the improved wavelet cerebellar articulation model controller (IWCMAC) with the estimated bound compensator controller to obtain high-precision positioning for applications requring accuracy within micrometer or nanometer range. In this research, the momentum and propotional factors are added into learning algorithm to escape from local minima points and speed up convergence of the system. The Lyapunov-Like Lema theory is used to estimate error bound to guarantee the stable of the system. The experimental results for the piezo-priven stage conclude the effectiveness and applicability of the proposed control system for model-free nonlinear.

1. Introduction

Together with development of nanotechology, linear piezoelectric motors with their superior properties such as small dimension, high force at low speed, high-holding force, silence and nanometer displacement resolution have been more popular in many industrial applications and attracted much attention of researchers [1-3]. However, the dynamic equation of the LPM is complex and nonlinear [4] so the model-based controllers can't obtain favorable performance and robust for the entrire domain of working [5]. To overcome this problem, many ingelligent controller have been developed for controlling the LMPs in the recent years such as sliding mode controllers [6], neural network [7], CMAC and wavelet cerebellar model articulation controller (WCMAC) [8-10]. However, the local minima problems is not treated and discussed in these researches.

In this reseach, the improved wavelet cerebellar model articulation controller (IWCMAC) combines with the estimated bound compensator controller are used to control piezo-driven stage to obtain precision positioning. Where, the IWCMAC is used to imitate the ideal controller which is affected by uncertainties and disturbances to minimize error function and avoid local minima. Futhermore, the estimated bound compensator controller is used to attenuate effects of disturbances and noises to the stability of the system.

This rest of the paper is organized as follows: Section 2 describes the precision positioning system and proposed control system. Section 3 describes the structure of proposed control sytem, which consists of the training and avoiding local minima method. Section 4 provides the experimental results and discussion. Section 5 is conclusion and future research.

Content from this work may be used under the terms of the Creative Commons Attribution 3.0 licence. Any further distribution of this work must maintain attribution to the author(s) and the title of the work, journal citation and DOI.
Published under licence by IOP Publishing Ltd

2. The precision positioning system dynamic description

The structure of the precision positioning system is reprented in figure 1. The system has main parts such as, XY stage base controled by LPMs, driver AB5, distance sensors, 1711 PCI card, PC with matlab software.

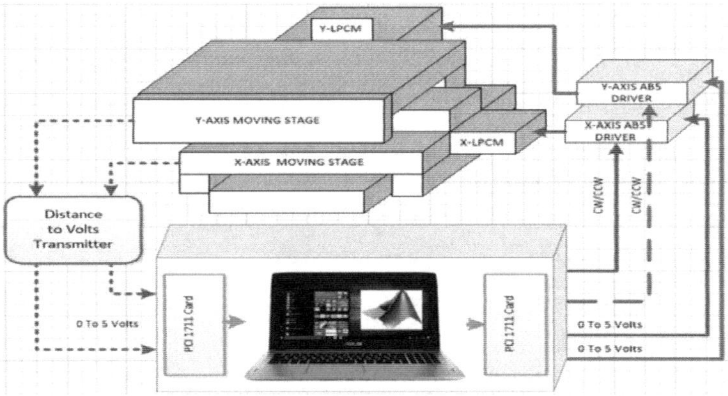

Figure 1. The structure of the precision positioning system.

The dynamic equation of LPMs including hysteresis and stiffness behavior is described in detail in [7] and has following equation (1).

$$\ddot{d}_a = -\frac{D_a}{M_a}\dot{d}_a + \frac{K_{Ea}}{M_a}u_a - \frac{F_{Ha} + F_{La}}{M_a} \tag{1}$$

where a is x, y for the x-axis, y-axis respectively; M_a is the effective mass of the moving stage; $\ddot{d}_a$, $\dot{d}_a$ and d_a are acceleration, velocity and displacement, respectively; D_a is the friction coefficient; F_{La} is an unknown external disturbance force; K_{Ea} is the voltage-to-force coefficient; u_a is the control volt of the LPMs; F_{Ha} is the hysteresis frictional force given as equation (2) [7].

$$F_{Ha} = \alpha b + \beta\frac{db}{dt} + \gamma\dot{d}_a \tag{2}$$

where b is the average deflection of the elastic bristles between two contact surfaces of rigid bodies; α, β, γ are positive constants depending on the bristle stiffness, bristle damping, and viscous damping-coefficient respectively.

In genearal, the parameters of the LPMs such as D_a, F_{La} and especially F_{Ha} can't be obtained or measurable exactly. Therefore, from the point of view of control system, these terms are considered as uncertainties or disturbances and the controllers must be designed to compensate the difference between nominal and practical value. So the dynamic equation of the LPMs can be rewritten as equation (3)

$$\ddot{\mathbf{d}} = \mathbf{F_0(d)} + \mathbf{G_0(d)u} + \mathbf{UD(d)} \tag{3}$$

where $\mathbf{F_0(d)} = -\frac{D_{a0}}{M_{a0}}$, $\mathbf{G_0(d)} = \frac{K_{Ea0}}{M_{a0}}$, u are nominal nonlinear vector, control gain and the control voltage of the LPMs, respectively. $\mathbf{UD(d)} = -\frac{F_{Ha0} + F_{La0}}{M_{a0}} + f(D, F_{Ha}, F_{La}, t)$ denote nominal hysteresis frictional force, nominal external force, uncertainties and disturbances which are assumed to be bounded, $\mathbf{d} = [d, \dot{d}]^T$ is the state vector. The tracking error is defined as equation (4)

$$e_\theta(t) = d_d(t) - d(t) \tag{4}$$

Assuming that, an error sliding surface is described as equation (5)

$$s(t) = \dot{e}_\theta(t) + K_1 e_\theta(t) + K_2 \int_0^t e_\theta(\tau) d\tau. \tag{5}$$

where: K_1 and K_2 are nonzero positive constants.

In case the parameters of the system are definitely defined and the external torque is measurable, a feedback linearization ideal controller can be designed as equation (6) [9]

$$u_{ideal} = \frac{1}{\mathbf{G}_0(\mathbf{d})}\left[\ddot{\theta}_d - \mathbf{F}_0(\mathbf{d}) - \mathbf{UD}(\mathbf{d}) + K_1\dot{e}_\theta + K_2 e_\theta\right] \tag{6}$$

The error dynamic equation is obtained by substituting (6) into (3)

$$\dot{s} = \ddot{e}_\theta + K_1\dot{e}_\theta + K_2 e_\theta = -\mathbf{F}_0(\mathbf{d}) - \mathbf{G}_0(\mathbf{d})\mathbf{u} - \mathbf{UD}(\mathbf{d}) + \ddot{\theta}_d + K_1\dot{e}_\theta + K_2 e_\theta = 0. \tag{7}$$

The convergence of error dynamic equation (7) is guaranteed under condition that the constants K_1 and K_2 are chosen to correspond to the Hurwitz polynomial coefficients. The ideal controller in (6), however, can't be obtained since **UD(d)** is immeasurable or exactly known in the practical applications.

To deal with the uncertainties and disturbances, **UD(d)**, a proposed IWCMAC control system is depicted in figure 2. This proposed controller combines a main controller u_{IWCMAC} and a estimated bound compensator controller u_{CC}. Where u_{IWCMAC} is developed to imitate the ideal controller and escape from stucking in a local minima, u_{CC} is designed to dispel the effects of uncertainties and disturbances during learning. The total proposed control system has following equation (8).

$$u = u_{ideal} - u_{IWCMAC} + u_{CC} \tag{8}$$

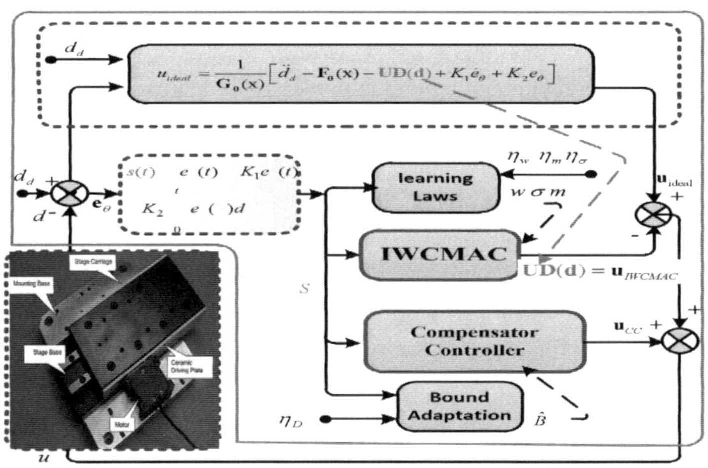

Figure 2. The structure of the proposed control system.

3. The improved wavelet cerebellar model articulation controller

3.1. The structure of WCMAC

The WCMAC with its fast learning and good generalization capability places an important role to learn uncertainties **UD(d)** to minimum error sliding surface. The structure of the WCMAC is depicted in figure 3 includes input Space S, association memory space A, receptive field space R, weight memory space W and output spaces O. The signal propagation in the WCMAC is presented as follows [9].

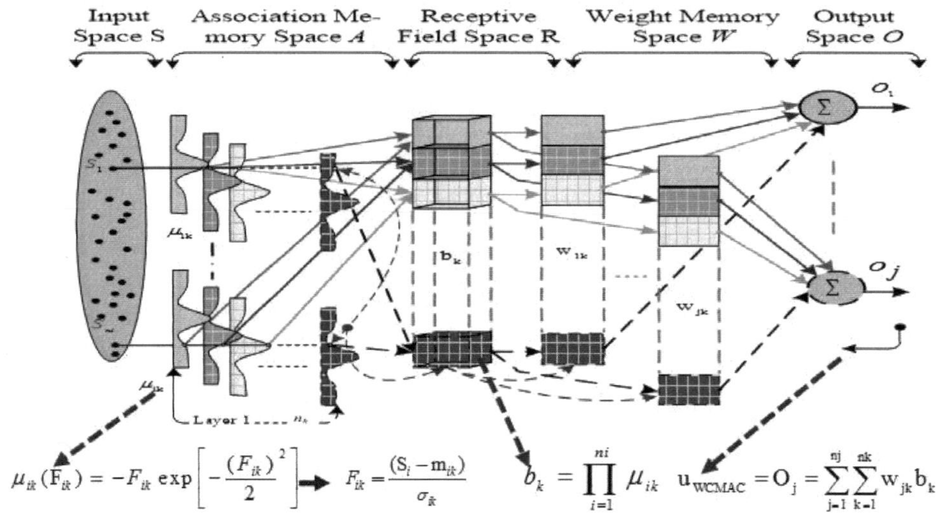

Figure 3. The structure of the WCMAC.

3.2. The online learning rule

The varying load torque and the unknown lumped uncertainties **UD(d)** are learned by the WCMAC $\mathbf{u}_{WCMAC}$ with approximation error ε as equation (9).

$$D(x) = \frac{1}{\mathbf{G_0}(x)}\mathbf{u}_{WCMAC}(\mathbf{S}, w_{kj}, m_{ik}, \sigma_{ik}) + \varepsilon = \sum_{j=1}^{nj}\sum_{k=1}^{nk} w_{jk} \prod_{i=1}^{n} \mu_{ik}(S_i) \tag{9}$$

The error function and the learning algorithm and compute parameters of the controller are described in detail in [9]. However, the weakness of this algorithm is that it may stuck in local minima points [11].

To advoid stucking in a local minima, the standard back propagation algorithm is modified by adding momentum term ρ and propotional term υ for calculating new parameters of the controller [12]. The momentum term places an important role in preventing the network trap in to local minima and the propotional term speed up the convergence as the activation having flat slope. So the new update rules of the IWCMAC as equation (10), (11) and (12).

$$\Delta w_{kj}(t) = \Delta w_{kj}(t) + \rho\Delta w_{kj}(t-1) + \upsilon(d - d_d) \tag{10}$$

$$\Delta m_{kj}(t) = \Delta m_{kj}(t) + \rho\Delta m_{kj}(t-1) + \upsilon(d - d_d) \tag{11}$$

$$\Delta \sigma_{kj}(t) = \Delta \sigma_{kj}(t) + \rho\Delta \sigma_{kj}(t-1) + \upsilon(d - d_d) \tag{12}$$

3.3. The estimated bound compensator controller

The most useful property of IWCMAC is to learn linear or nonlinear functions through learning capability. In (9), the approximation error ε is assumed to be bounded by a positive B. if error bound B is assumed to be a constant during the operation then the compensator controller is designed

IOP Conf. Series: Materials Science and Engineering **383** (2018) 012019 doi:10.1088/1757-899X/383/1/012019

following the sliding mode theory to compensate for the effect of the approximation error due to uncertainties and disturbances as equation (13)

$$u_{CC} = -\mathbf{G}_0^{-1}(\mathbf{d})\hat{B}\,\mathrm{sgn}(s) \tag{13}$$

However, this bound parameter is immeasurable in practical applications. Therefore, a error bound estimation is developed to estimate this bound and given by estimation law as equation (14) [10].

$$\dot{\hat{B}} = \eta_B \|s\|_1 \tag{14}$$

In conclusion, the proposed system IWCMAC is represented in (9), the parameters $\hat{w}$, $\hat{m}$ and $\hat{\sigma}$ of which are adjusted and apdated online (10)-(12). Moreover, the estimated bound compensator controller u_{CC} is given in (13) with bound estimation law as (14). The IWCMAC control system can be guaranteed to be stable in the Lyapunov-Like Lemma sense[13].

4. Experimental results

An image of the experimental equipment of the precision positioning system is showed in figure 4. To illustrate the superior of the proposed control system, the experimental results of the proposed control system and the proportional–integral-derivative (PID) controller are provided to further demonstrate the effectiveness of the proposed control system over PID.

The parameters of the nominal model of the drive system are given as:

$M_{a0} = 5kg$, $D_{a0} = 66[N.\mathrm{sec}/m]$, $K_{Ea0} = 3[N/Volt]$, $\eta_w = 0.005$, $\eta_m = 0.005$, $\eta_\sigma = 0.005$, $\eta_D = 0.001$, $K1 = 0.041$, $K2 = 0.055$, $\eta_k = 11$, $\upsilon = 0.05$ $\rho = 0.96$ $K_P = 0.074, K_I = 0.0002, K_D = 0.0001$

$S = [-1.0\ -0.8\ -0.6\ -0.4\ -0.2\ 0.0\ \ 0.2\ \ 0.4\ \ 0.6\ \ 0.8\ \ 1.0]$

Sample time = 0.01s

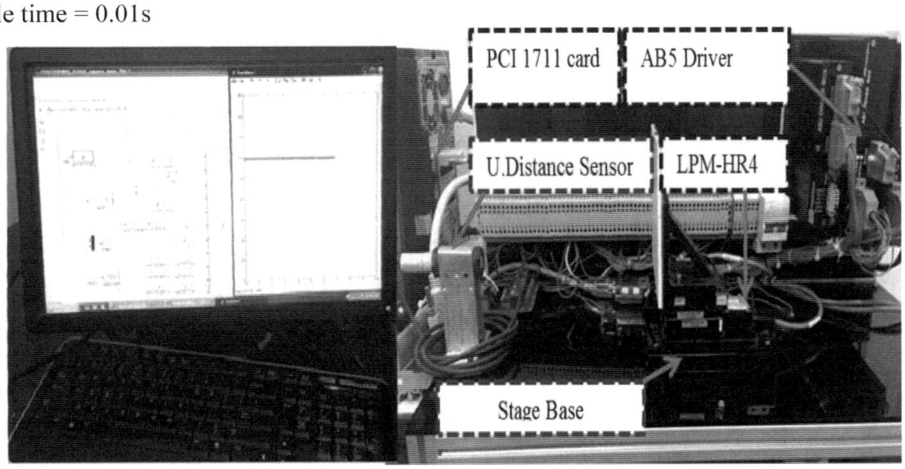

Figure 4. Image of practical control system.

The experimental results of the PID controller and the IWCMAC due to sinusoidal and periodic step commands are depicted in figure 5 and figure 6, respectively.

IOP Conf. Series: Materials Science and Engineering **383** (2018) 012019 doi:10.1088/1757-899X/383/1/012019

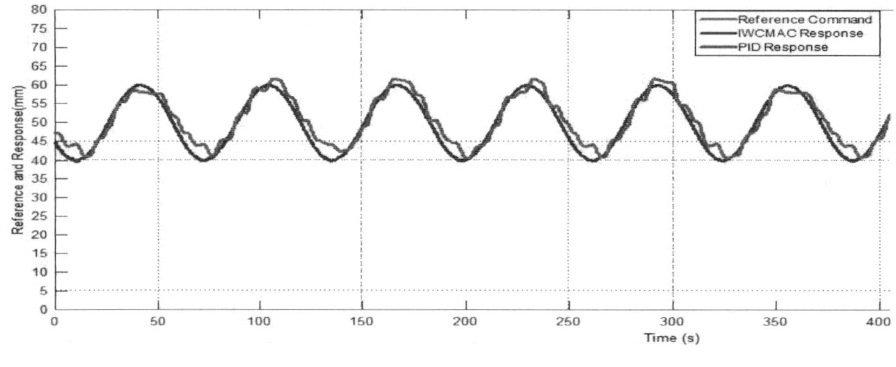

(a)

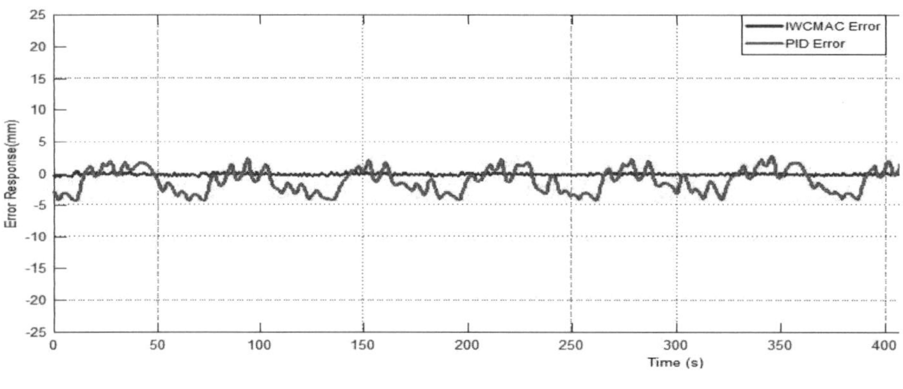

(b)

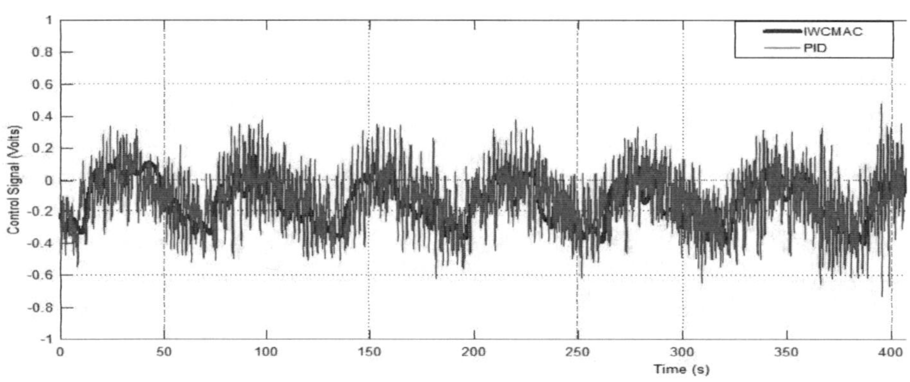

(c)

Figure 5. Experimental results of PID and IWCMAC controller due to sinusoidal command. (a) Tracking response. (b) Error response. (c) Control signal.

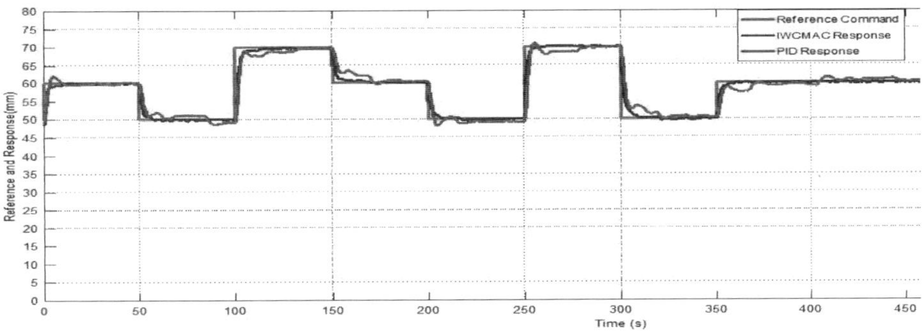

(a)

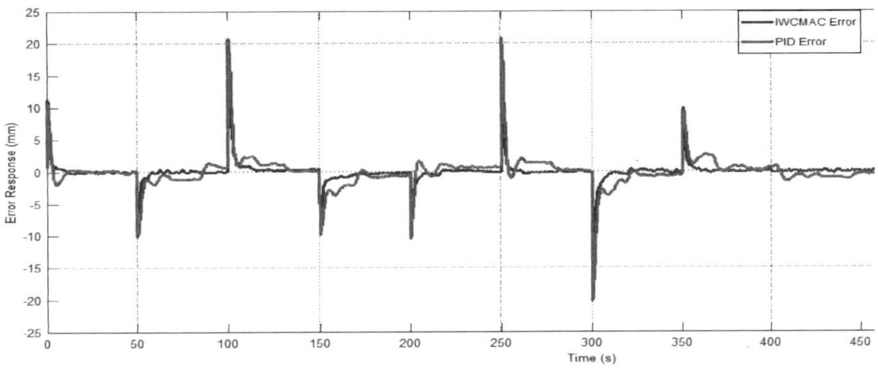

(b)

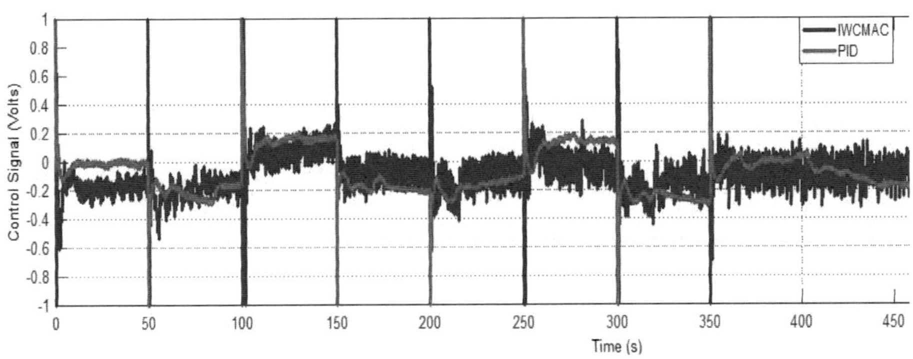

(c)

Figure 6. Experimental results of PID and IWCMAC controller due to periodic step command. (a) Tracking response. (b) Error response. (c) Control signal.

The experimental results due to sinusoidal and periodic step commands show that the IWCMACS can work well in realtime applications. Comparing to PID controller, the better performances are obtained such as small convergence error and time, small voltage control signal at steady state and stability for a long time. These results once again confirm the effectiveness of the proposed system.

5. Conclusion and future works

In this paper, the IWCMAC system was proposed to control the precision positioning for piezo-driven stage. The proposed control system is composed by the IWCMAC and the estimated bound compensator controller. The parameters of the proposed IWCMAC system are tuned on-line and the system's stability has been proven in the Lyapunov-Like Lemma sense. In addition, the effectiveness of the proposed control system has proved by the experimental results. Furthermore, the nominal nonlinear vector and control gain $\mathbf{F}_0(\mathbf{d})$ and $\mathbf{G}_0(\mathbf{d})$ non-essential to know exactly. If there are differences between nominal and practical parameters, they belong to uncertainties, $\mathbf{UD(d)}$. Consequently, the proposed control system is available for model-free nonlinear control designs. However, to enhance the performance index, the robust controller should be investigated and added to guarantee the robustness of all the system in working space.

References

[1] Grossard M, Boukallel M, Chaillet N, and Rotinat-Libersa C 2011, Modeling and robust control strategy for a control-optimized piezoelectric microgripper, *IEEE/ASME Trans. Mechatronics,* **16**, p 674

[2] Kenton B.J and Leang K.K 2012, Design and control of a three-axis serial kinematic high-bandwidth nanopositioner, *IEEE/ASME Trans. Mechatronics*, **17**, p 356

[3] Sente P.A, Labrique F.M, and Alexandre P.J 2012, Efficient Control of a Piezoelectric Linear Actuator Embedded Into a Servo-Valve for Aeronautic Applications, *IEEE Transactions on industrial electronics,* **59**, p 1971

[4] Al-Ghanimi A, Zheng J, and Man Z 2016, A Fast Non-Singular Terminal Sliding Mode Control Based on Perturbation Estimation for Piezoelectric Actuators Systems, *International Journal of Control*, **3**, p 480

[5] Mo J.S, Qiu Z.C, Wei J.Y, and Zhang X.M 2017, Adaptive positioning control of an ultrasonic linear motor system, *Robotics and Computer-Integrated Manufacturing*, **44** p 156

[6] Li Y and Xu Q 2010, Adaptive Sliding Mode Control With Perturbation Estimation and PID Sliding Surface for Motion Tracking of a Piezo-Driven Micromanipulator, *IEEE transactions on control systems technology*, **18**, p 798

[7] Lin F.J, Lee S.Y, and Chou P.H 2016, Intelligent integral backstepping sliding-mode control using recurrent neural network for piezo-flexural nanopositioning stage, *Asian Journal of Control*, **18**, p 456

[8] Wen C.M and Cheng M.Y 2013, Development of a recurrent fuzzy CMAC with adjustable input space quantization and self learning rate for control of a dual-axis piezoe micromotion stage, *IEEE Trans. Ind. Electron ,* **11**, p 5105

[9] Van-Phuong Ta, Xuan-Kiên Dang, and Thanh Quyen Ngo 2017, Adaptive Tracking Control Based On CMAC for nonlinear Systems, *Proceedings of the International Conference on System Science and Engineering, p* 494

[10] Lin C M and Li H Y 2012, A Novel Adaptive Wavelet Fuzzy Cerebellar Model Articulation Control System Design for Voice Coil Motors, *IEEE transactions on industrial electrionics*, **59**, (2012)

[11] Dai Q and Liu N 2012, Alleviating the problem of local minima in Backpropagation through competitive learning, *Elsevier Journal, neurocomputing*

[12] Burse K, Manoria M, Vishnu P S K 2010, Improved Back Propagation Algorithm to Avoid Local Minima in Multiplicative Neuron Model, *International Journal of Electrical, Computer, Energetic, Electronic and Communication Engineering*, **4**

[13] Wai R J, Lin C M, and Peng Y F 2004, Adaptive Hybrid Control for Linear PiezoelectricCeramic Motor Drive Using Diagonal Recurrent CMAC Network, *IEEE Transaction Neural Network*, **15**

IOP Conf. Series: Materials Science and Engineering **383** (2018) 012020 doi:10.1088/1757-899X/383/1/012020

Novel condition for exponential stability of linear system with mixed non-differential time-varying delay with feedback control

Thongchai Botmart[1] and Patarawadee Prasertsang[2,*]

[1]Faculty of Science, Department of Mathematics, 40002 Khon Kaen, Thailand
[2]Faculty of Science and Engineering, Department of General Sciences, 47000 Sakon Nakhon, Thailand

*Email: patarawadee.s@ku.th

Abstract. The problem of exponential stability for linear system with mixed time-varying delay is investigated. The stability of linear system are proposed with feedback control. Based on the constructing of improved Lyapunov-Krasovskii functionals combined with Leibniz-Newton's formula and by utilizing the integral inequality, new sufficient condition for the stabilization of the systems is first established in terms of LMIs. A numerical example is given to illustrate that the effectiveness of the proposed feedback control.

1. Introduction

The stability analysis of dynamic systems with time delays has been one of the most attention in the field of control theory, because time-delay systems occur in various areas including chemical engineering systems, neural networks, biological systems etc. The stability analysis for linear time-delay systems has been discussed widely and various approaches to such problems have been proposed and the references therein. Time-varying delays systems have been an interesting topic in recent year such as interval time-varying delays [1-2], distributed time-varying delay [3], mixed time-varying delays [4] and so on. The stability analysis and control of time-varying delay systems have received considerable attention widely for the last few decades. Some delay-dependent conditions for designing stabilizing feedback control [4-9] and intermittent feedback control [4,5], [9,10] are studied in various system. Both of them have been considered see in [4] which is more advantageous in the field of the stability of the control theory. In this paper, the problem of exponential stabilization of linear systems with mixed time-varying delay with feedback control is studied. The time delay is a continuous function belonging to a given interval, which means that the lower and upper bounds for the time-varying delay are available, but the delay function is not necessary to be differentiable. By utilizing the construction of improved Lyapunov-Krasovskii functionals combined with Leibniz-Newton's formula and using the technique of dealing with some integral terms, new delay-dependent sufficient conditions for the exponential stabilization of the those systems are first established in terms of LMIs without introducing any free-weighting matrices. A numerical example is given to demonstrate the effectiveness of the obtained results.

2. Network model and mathematical preliminaries

Content from this work may be used under the terms of the Creative Commons Attribution 3.0 licence. Any further distribution of this work must maintain attribution to the author(s) and the title of the work, journal citation and DOI.
Published under licence by IOP Publishing Ltd

Let us consider the linear system as equation (1)

$$\dot{x}(t) = Ax(t) + Bx(t-h(t)) + C\int_{t-k_1(t)}^{t} x(s)ds + U(t),\ t \ge 0\ , \tag{1}$$

$$x(t) = \phi(t),\ t \in [-\tau_{\max}, 0],\ \tau_{\max} = \max\{h_2, d, k_1, k_2\},$$

where $x(t) = (x_1(t),...,x_n(t))^T \in \Re^n$ is the state vector; A, B and C are known real constant matrices, $U(t) \in \Re^m$ is the control input. In order to stabilize the origin of linear system as equation (1) by means of the state feedback controller $U(t)$ satisfying as equation (2),

$$U(t) = D_1u(t) + D_2u(t-d(t)) + D_3\int_{t-k_2(t)}^{t} u(s)ds,\ \forall t \ge t_0, \tag{2}$$

where $D_i, i = 1,2,3$ are given matrices of appropriate dimensions, $u(t) = Kx(t)$ and K is a constant matrix control gain, and n is a non-negative integer. Then, substituting $U(t)$ into linear system as equation (1),

$$\dot{x}(t) = Ax(t) + Bx(t-h(t)) + C\int_{t-k_1(t)}^{t} x(s)ds + D_1u(t) + D_2u(t-d(t)) + D_3\int_{t-k_2(t)}^{t} u(s)ds. \tag{3}$$

It is clear that, if the zero solution of linear system as equation (3) is globally exponentially stable, the exponential stability of the controlled linear system as equation (1) is achieved. The time-varying delay functions $h(t), d(t),\ k_1(t)$ and $k_2(t)$ satisfy the conditions as equation (4).

$$0 \le h_1 \le h(t) \le h_2,\ 0 \le d(t) \le d,\ 0 \le k_1(t) \le k,\ 0 \le k_2(t) \le k_2\ . \tag{4}$$

The initial condition function $\phi(t)$ denotes a continuous vector-valued initial function of $t \in [-\tau_{\max}, 0]$.

3. Exponential stability of linear delay system via feedback control

Let us denote

$$\|\phi(t)\| = \|x(0)\|, \|\varphi(t)\| = \sup_{-\tau_{\max} \le s \le 0} \|x(s)\|, K = -LP^{-1}, \gamma = \lambda_{\min}(P^{-1}), \beta = \frac{h_2 - h(t)}{h_2 - h_1}, N = N_1\|\phi(t)\|^2 + N_2\|\varphi(t)\|^2,$$

$$N_1 = \lambda_{\max}(P^{-1}) + \left[2h_2\lambda_{\max}(P^{-1}RP^{-1}) + h_2\lambda_{\max}(P^{-1}UP^{-1})\right]\left(\frac{1-e^{-2\alpha h_2}}{2\alpha}\right) + d\lambda_{\max}(P^{-1}L^TT^{-1}LP^{-1})\left(\frac{1-e^{-2\alpha d}}{2\alpha}\right),$$

$$N_2 = \left[2\lambda_{\max}(P^{-1}QP^{-1}) + 2h_2\lambda_{\max}(P^{-1}RP^{-1}) + h_2\lambda_{\max}(P^{-1}UP^{-1})\right]\left(\frac{1-e^{-2\alpha h_2}}{2\alpha}\right)$$
$$+ k_1\lambda_{\max}(P^{-1}SP^{-1})\left(\frac{1-e^{-2\alpha k_1}}{2\alpha}\right) + d\lambda_{\max}(P^{-1}L^TT^{-1}LP^{-1})\left(\frac{1-e^{-2\alpha d}}{2\alpha}\right) + k_2\lambda_{\max}(P^{-1}L^TW^{-1}LP^{-1})\left(\frac{1-e^{-2\alpha k_2}}{2\alpha}\right).$$

Theorem 1 For some given scalars $\alpha > 0$, the linear system as equation (3) with time-varying delay satisfying equation (4) are exponentially stable if there exist symmetric positive definite matrices $P > 0, Q > 0, R > 0,\ S > 0,\ \ U > 0, T > 0, W > 0$ and a matrix L appropriately dimensioned such that the following symmetric linear matrix inequality holds, equations (5)-(9):

$$\Sigma_1 = \Sigma - [0\ \ 0\ \ I\ \ -I\ \ 0]^T e^{-2\alpha h_2} U [0\ \ 0\ \ I\ \ -I\ \ 0] < 0, \tag{5}$$

$$\Sigma_2 = \Sigma - [0\ \ 0\ \ 0\ \ I\ \ -I]^T e^{-2\alpha h_2} U [0\ \ 0\ \ 0\ \ I\ \ -I] < 0, \tag{6}$$

$$\Sigma_3 = \begin{bmatrix} -0.5\left(e^{-2\alpha h_1} + e^{-2\alpha h_2}\right)R & 2k_1CP & k_2L^T & 2L^T \\ * & -2k_1e^{-2\alpha k_1}S & 0 & 0 \\ * & * & -k_2W & 0 \\ * & * & * & -2\,de^{2\alpha d} \end{bmatrix} < 0, \tag{7}$$

$$\Sigma_4 = \begin{bmatrix} -0.5P & 2k_1CP & d^2L^T \\ * & -2k_1e^{-2\alpha k_1}S & 0 \\ * & * & -d^2T \end{bmatrix} < 0, \tag{8}$$

where

$$\Sigma = \begin{bmatrix} \Sigma_{11} & \Sigma_{12} & \Sigma_{13} & \Sigma_{14} & \Sigma_{15} \\ * & \Sigma_{22} & 0 & \Sigma_{24} & 0 \\ * & * & \Sigma_{33} & \Sigma_{34} & 0 \\ * & * & * & \Sigma_{44} & \Sigma_{45} \\ * & * & * & * & \Sigma_{55} \end{bmatrix}, \tag{9}$$

$$\Sigma_{11} = P^T\left(A+\alpha I\right)+\left(A+\alpha I\right)^T P - D_1L - L^TD_1^T + 3e^{2\alpha l}D_2^{\ T}TD_2 + 2k_2e^{2\alpha k_2}D_3^{\ T}WD_3 + 2Q$$
$$+k_1S - 0.5\left(e^{-2\alpha h_1} + e^{-2\alpha h_2}\right)R,$$
$$\Sigma_{12} = PA^T - L^TD_1^{\ T},\ \Sigma_{13} = e^{-2\alpha h_1}R,\ \Sigma_{14} = BP,\ \Sigma_{15} = e^{-2\alpha h_2}R,$$
$$\Sigma_{22} = h_1^2R + h_2^2R + \eta^2U - 1.5P + 3e^{2\alpha d}D_2^{\ T}TD_2 + 2k_2e^{2\alpha k_2}D_3^{\ T}WD_3,\ \eta = h_2 - h_1$$
$$\Sigma_{24} = BP, \Sigma_{33} = -e^{-2\alpha h_1}Q - e^{-2\alpha h_1}R - e^{-2\alpha h_2}U, \Sigma_{34} = e^{-2\alpha h_2}U,$$
$$\Sigma_{44} = -2e^{-2\alpha h_2}U, \Sigma_{45} = e^{-2\alpha h_2}U, \Sigma_{55} = -e^{-2\alpha h_2}Q - e^{-2\alpha h_2}R - e^{-2\alpha h_2}U,$$

then, the linear system as equation (3) has an exponential stability. Moreover, the feedback control is equation (10),

$$u(t) = -LP^{-1}x(t) \tag{10}$$

and the solution $x(t,\phi)$ satisfies $\left\|x(t,\phi)\right\| = \sqrt{\frac{N}{\gamma}}e^{-\alpha t}, \forall t \geq 0.$

Proof Let $Y = P^{-1}$ and $y(t) = Yx(t)$. By using the feedback control equation (10), let us consider the following Lyapunov-Krasovskii functional as equation (11) show:

$$V\left(x(t)\right) = \sum_{i=1}^{9} V_i(t) \tag{11}$$

where as equation (12) show

$$V_1(t) = x^T(t)Yx(t), V_2(t) = \int_{t-h_1}^{t} e^{2\alpha(s-t)}x^T(s)YQYx(s)ds, V_3(t) = \int_{t-h_2}^{t} e^{2\alpha(s-t)}x^T(s)YQYx(s)ds,$$
$$V_4(t) = h_1\int_{-h_1}^{0}\int_{t+s}^{t} e^{2\alpha(\tau-t)}\dot{x}^T(\tau)YRY\dot{x}(\tau)d\tau ds, V_5(t) = h_2\int_{-h_2}^{0}\int_{t+s}^{t} e^{2\alpha(\tau-t)}\dot{x}^T(\tau)YRY\dot{x}(\tau)d\tau ds, \tag{12}$$
$$V_6(t) = \eta\int_{-h_2}^{-h_1}\int_{t+s}^{t} e^{2\alpha(\tau-t)}\dot{x}^T(\tau)YUY\dot{x}(\tau)d\tau ds,\ V_7(t) = \int_{-k_1}^{0}\int_{t+s}^{t} e^{2\alpha(\tau-t)}x^T(\tau)YSYx(\tau)d\tau ds,$$
$$V_8(t) = d\int_{-d}^{0}\int_{t+s}^{t} e^{2\alpha(\tau-t)}\dot{x}^T(\tau)K^TT^{-1}K\dot{x}(\tau)d\tau ds, V_9(t) = \int_{-k_2}^{0}\int_{t+s}^{t} e^{2\alpha(\tau-t)}x^T(\tau)K^TW^{-1}Kx(\tau)d\tau ds.$$

It is easy to check that equation (13)

$$\gamma \|x(t)\|^2 \le V(x(t)) \tag{13}$$

By taking the derivatives of $V_i(t), i = 1,...,9$ along the trajectories of linear system as equation (3), and then apply Jensen's inequality, Cauchy inequality and Leibniz-Newton formula, we have the following equations (14)-(22):

$$\begin{aligned}\dot{V}_1(t) \le\; & y^T(t)\left[PA + A^T P\right]y(t) + 2y^T(t)BPy(t-h(t)) - 2y^T(t)D_1Ly(t) + 2k_1e^{2\alpha k_1}y^T(t)CPS^{-1}PC^Ty(t) \\ & + \frac{1}{2}e^{-2\alpha k_1}\int_{t-k_1(t)}^{t} y^T(s)Sy(s)ds + 3e^{2\alpha d}y^T(t)D_2TD_2^Ty(t) + \frac{1}{3}e^{-2\alpha d}u^T(t-d(t))T^{-1}u(t-d(t)) \\ & + 2k_2e^{2\alpha k_2}y^T(t)D_3TD_3^Ty(t) + \frac{1}{2}e^{-2\alpha k_2}\int_{t-k_2(t)}^{t} u^T(s)W^{-1}u(s)ds,\end{aligned} \tag{14}$$

$$\dot{V}_2(t) = -2\alpha V_2 + y^T(t)Qy(t) - e^{-2\alpha h_1}y^T(t-h_1)Qy(t-h_1), \tag{15}$$

$$\dot{V}_3(t) = -2\alpha V_3 + y^T(t)Qy(t) - e^{-2\alpha h_2}y^T(t-h_2)Qy(t-h_2), \tag{16}$$

$$\dot{V}_4(t) \le -2\alpha V_4 + h_1^2\dot{y}^T(t)R\dot{y}(t) - e^{-2\alpha h_1}\left[y(t)-y(t-h_1)\right]^T R\left[y(t)-y(t-h_1)\right], \tag{17}$$

$$\dot{V}_5(t) \le -2\alpha V_5 + h_2^2\dot{y}^T(t)R\dot{y}(t) - e^{-2\alpha h_2}\left[y(t)-y(t-h_2)\right]^T R\left[y(t)-y(t-h_2)\right], \tag{18}$$

$$\begin{aligned}\dot{V}_6(t) \le\; & -2\alpha V_6 + \eta^2\dot{y}^T(t)U\dot{y}(t) - e^{-2\alpha h_2}\left[y(t-h(t))-y(t-h_2)\right]^T U\left[y(t-h(t))-y(t-h_2)\right] \\ & - e^{-2\alpha h_2}\left[y(t-h_1)-y(t-h(t))\right]^T U\left[y(t-h_1)-y(t-h(t))\right] \\ & -(1-\beta)e^{-2\alpha h_2}\left[y(t-h(t))-y(t-h_2)\right]^T U\left[y(t-h(t))-y(t-h_2)\right] \\ & - \beta e^{-2\alpha h_2}\left[y(t-h_1)-y(t-h(t))\right]^T U\left[y(t-h_1)-y(t-h(t))\right],\end{aligned} \tag{19}$$

$$\dot{V}_7(t) \le -2\alpha V_7 + k_1y^T(t)Sy(t) - e^{-2\alpha k_1}\int_{t-k_1(t)}^{t} y^T(s)Sy(s)ds, \tag{20}$$

$$\dot{V}_8(t) \le -2\alpha V_8 + d^2\dot{y}^T(t)L^TT^{-1}L\dot{y}(t) + 2e^{-2\alpha d}y^T(t)L^TT^{-1}Ly(t) - \frac{2}{3}e^{-2\alpha d}u^T(t-d(t))T^{-1}u(t-d(t)), \tag{21}$$

$$\dot{V}_9(t) \le -2\alpha V_9 + k_2y^T(t)L^TW^{-1}Ly(t) - e^{-2\alpha k_2}\int_{t-k_2(t)}^{t} y^T(s)W^{-1}y(s)ds. \tag{22}$$

By using the identity relation and then multiplying its by $2\dot{y}^T(t)$ gives as equation (23)

$$\begin{aligned}& -2\dot{y}^T(t)P\dot{y}(t) + 2\dot{y}^T(t)APy(t) + 2\dot{y}^T(t)BPy(t-h(t)) + 2\dot{y}^T(t)CP\int_{t-k_1(t)}^{t} y(s)ds \\ & - 2\dot{y}^T(t)D_1Ly(t) + 2\dot{y}^T(t)D_2u(t-d(t)) + 2\dot{y}^T(t)D_3\int_{t-k_2(t)}^{t} u(s)ds = 0\end{aligned} \tag{23}$$

Hence, according to equations (14) - (22) and adding the zero items of equation (12), we have equation (24)

$$\dot{V}(x(t)) + 2\alpha V(x(t)) \le \xi^T(t)\left[(1-\beta)\Sigma_1 + \beta\Sigma_2\right]\xi(t) + y^T(t)M_1y(t) + \dot{y}^T(t)M_2\dot{y}(t), \tag{24}$$

where Σ_1 and Σ_2 are defined as in equation (5) and equation (6), respectively, and

$\xi^T(t)=\left[y^T(t)\quad \dot{y}^T(t)\quad y^T(t-h_1)\quad y^T(t-h(t))\quad y^T(t-h_2)\right],$

$M_1=-0.5\left(e^{-2\alpha h_1}+e^{-2\alpha h_2}\right)R+2k_1e^{2\alpha k_1}CPS^{-1}PC^T+k_2L^TW^{-1}L+2e^{-2\alpha d}L^TT^{-1}L,$

$M_2=-0.5P+2k_1e^{2\alpha k_1}CPS^{-1}PC^T++dL^TT^{-1}L.$

By $(1-\beta)\Sigma_1+\beta\Sigma_2<0$ as in equation (24) holds if and only if $\Sigma_1<0$ and $\Sigma_2<0$ as in equation (5) and (6). Apply Schur complement, the inequalities $M_1<0$ and $M_2<0$ are equivalent to $\Sigma_3<0$ and $\Sigma_4<0$ as in equation (7) and equation (8), respectively. It follows from equations (5)-(9) and equation (24), we obtain equation (25)

$$\dot{V}(x(t))+2\alpha V(x(t))\le 0,\forall t\ge 0. \tag{25}$$

Integrating both sides of equation (25) from 0 to t, and using the condition equation (13), we have equation (26)

$$\|V(x(t))\|\le\sqrt{\frac{V(x(0))}{\gamma}}e^{-\alpha t},\forall t\ge 0. \tag{26}$$

Finally, by estimating $V(x(0))$, we have that $\|V(x(t))\|\le\sqrt{\frac{N}{\gamma}}e^{-\alpha t},\forall t\ge 0,$ which implies the linear system as equation (3) is globally exponential stable under the controller $U(t)$ as in equation (2), then exponential stability of the controlled linear system as equation (3) is achieved.

4. Numerical example

Example 1 Consider the linear feedback control (3) with the parameters

$$A=\begin{bmatrix}-0.3 & -0.1\\ 0 & 0.2\end{bmatrix},B=\begin{bmatrix}-0.3 & 0\\ -0.1 & -0.1\end{bmatrix},C=\begin{bmatrix}-0.1 & 0.1\\ 0.2 & -0.3\end{bmatrix},D_1=\begin{bmatrix}2 & 3\\ 3 & 5\end{bmatrix},D_2=\begin{bmatrix}4 & 2\\ 3 & 1\end{bmatrix},D_3=\begin{bmatrix}2 & 1\\ 1 & 1\end{bmatrix}, \tag{27}$$

where $h_1=0.1,h_2=0.4,d=0.3,k_1=0.1,k_2=0.3$ and $\alpha=0.1$. Then, using the MATLAB LMI control toolbox to solve in (5)-(9), we obtain the solution as follows:

$$P=\begin{bmatrix}3.0305 & 0.7147\\ 0.7147 & 0.6970\end{bmatrix},Q=\begin{bmatrix}0.0754 & 0.0584\\ 0.0584 & 0.0513\end{bmatrix},R=\begin{bmatrix}0.6304 & 0.2910\\ 0.2910 & 0.3079\end{bmatrix},U=\begin{bmatrix}0.86\ 20 & 0.1010\\ 0.10\ 10 & 0.2662\end{bmatrix},$$

$$T=\begin{bmatrix}0.0093 & -0.0344\\ -0.0344 & 0.1513\end{bmatrix},W=\begin{bmatrix}0.0565 & -0.0350\\ -0.0350 & 0.1104\end{bmatrix},S=\begin{bmatrix}1.5603 & 0.3950\\ 0.3950 & 0.4117\end{bmatrix},K=\begin{bmatrix}-0.0064 & 0.0317\\ 0.0266 & -0.1727\end{bmatrix}.$$

The numerical simulation of linear system as equation (3) with the parameters as equation (27) with time-varying delays $h(t)=0.1+0.3|\cos t|,k_1(t)=0.1|\sin t|$.

The initial condition $\phi(t)=[0.5\cos s,-0.5\cos s],\forall s\in[-0.4,0]$ and without feedback control is represented in figure 1, which shows that the linear system as equation (3) is unstable. Figure 2 shows the trajectories of $x_1(t)$ and $x_2(t)$ of linear system as equation (3) with time-varying delays $d(t)=k_2(t)=0.3|\cos t|$ via feedback control. The method proposed in [2] is not applicable for the case that the delay function is not necessary to be differentiable and the improved control input.

5. Conclusion

This study has investigated the stability of dynamical system with mixed time-varying delays via feedback control. Based on the constructing of an improved Lyapunov-Krasovskii functionals combined with Leibniz-Newton's formula and using the integral inequality technique of dealing with some integral terms, new delay-dependent sufficient condition for the exponential stabilization of the these systems are first established in terms of LMIs without introducing any free-weighting matrices.

The delay feedback controllers designed can guarantee the exponential stability of the dynamical system. A numerical example is provided to show the advantages of the present results.

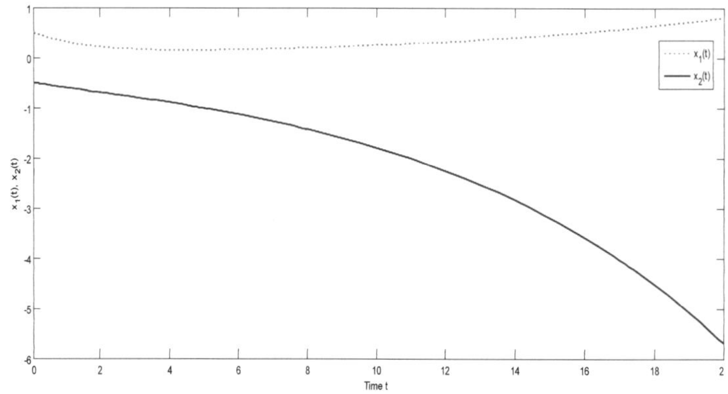

Figure 1. The trajectories of $x_1(t)$ and $x_2(t)$ of the uncertain linear system with interval time-varying delay and without feedback control activated.

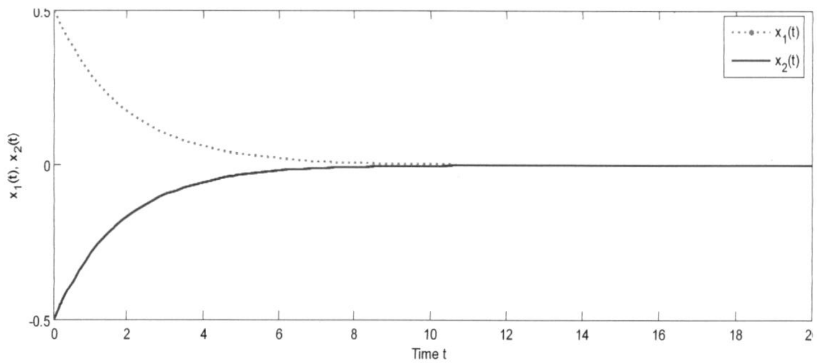

Figure 2. The trajectories of $x_1(t)$ and $x_2(t)$ of the uncertain linear system with interval time-varying delay and feedback control activated.

Acknowledgments
The first author was supported by Khon Kaen University. The second author was supported by Research and Academic Service Division and Faculty of Science and Engineering, Kasetsart University, Chalermprakiat Sakon Nakhon Province Campus.

References

[1] O.M. Kwon, M.J. Park, J. H. Park, S.M. Lee and E.J. Cha, Appl. Math. Comput. 224 (2013)
[2] J. Cheng, L. Xiong, Neurocomputing, 160 (2015)
[3] M. S. Ali and R. Saravanakumar, Appl. Math. Comput., 249 (2014)
[4] T. Botmart and P. Niamsup, Adv. Difference Equ., (2014)
[5] T. Huang and C. Li, J. Comput. Appl. Math., 234 (2010)
[6] M. V. Thuan and N. T. T. Huyen, Differential Equations and Control Processes, N 4 (2011)
[7] M.V. Thuan, V.N. Phat and H.M. Trinh, Appl. Math. Comput. , 218 (2012)
[8] Z. Zhang, C. Lin and B. Chen, J. Franklin Inst. 352 (2015)
[9] Q. Son and T. Huang, Neurocomputing, 154 (2015)
[10] P. Yang and X. Tang, Int. J. Biomath., 07(02) (2014)

Research on an ultra-low power thermoelectric-type anemometer

Chih-Hsiung Shen*, Po-Hsuan Lin and Shu-Jung Chen

Department of Mechatronics Engineering, National Changhua University of Education, NO. 2, Shi-Da Road, Changhua, Taiwan, R.O.C.

*Email: hilbert@cc.ncue.edu.tw

Abstract. Beyond the conventional hot-wire sensor, a new thermoelectric sensor with ultra-low power consumption without heating is proposed. Using the TSMC 0.35 um CMOS-MEMS process, the thermoelectric sensor is fabricated with 32 pair of central-symmetrical thermocouples positioned. When the fluid passes through the thermopile, the fluid will take away the heat, so that a temperature difference is generated between the cold end and hot end of the thermocouples. The temperature sensor is calibrated and senses the drop of temperature at the center of membrane during the measurement. For different flow velocities, it is interesting to find that the drop of temperature is verified by the output voltage of sensing circuit from thermopile and the same for temperature sensor which behaves as a function of flow velocity. The new approach for the anemometer of sensing flow velocity is realized by our proposed thermopile which is proved to be a practical technique with ultra-low power consumption.

1. Introduction

Thermal flow sensors plays an important role by means of heating and temperature measurement which are widely used to measure thermal mass-flow rate of fluids. Most of these sensors mainly include a heater, which is designed on a membrane and placed in the stream of flow to reduce the solid heat conduction and also the power of heating. The membrane for the heater is fabricated and floating while the substrate beneath is removed with thermally decoupling. As compared with sensors which the heater is directly placed on bulk silicon or ceramics, it leads a benefit to reduce the lower power consumption of heating and give a better performance with a faster response time and high sensitivity since both the solid thermal conductance and the heat capacity of the membrane is much lower [1-4].

However, these thermal flow sensors based on heat convection principle usually suffer the long-term drift of TCR and reading due to the continuous heating of hot-wire. Therefore, it is a great challenge to reduce the power consumption and temperature drifting issues and avoid of repeated time-consuming calibrations. It is necessary to develop a robust flow sensor design with high stability over the proper flow range and working temperature conditions. Several modulation of heating techniques for micro thermal flow sensors [5] have been presented and it is difficult to maintain a reliable reading during the dynamic operation and time-varying heating. The main reason for the instability of continuous heating is because the heater of thermistors is exposed to high temperatures over time, sometimes referred to as "aging," their resistivity can change over time. Temperature

Content from this work may be used under the terms of the Creative Commons Attribution 3.0 licence. Any further distribution of this work must maintain attribution to the author(s) and the title of the work, journal citation and DOI.
Published under licence by IOP Publishing Ltd

IOP Conf. Series: Materials Science and Engineering **383** (2018) 012021 doi:10.1088/1757-899X/383/1/012021

dramatically change may be thought of as a main factor of aging [5]. Besides, the cumulative exposure to high temperature that has the greatest influence on a thermistor component.

2. Convection-cooling and sensing principle

Based on the Newton's law of cooling, the hot-wire sensors operate by heat transfer to measure the heat dissipation from a heated element to a surrounding cooler fluid. It means the mathematical model of increasing fluid flow and forced convective cooling of the element can be described and a baseline calibration for the model is necessary for sensing applications. King's Law describes the coefficient of convection for the heat transfer from the sensing element to the flow in terms of the flow velocity and geometrical factors of sensor. These constants including complex combination of fluid thermal conductivity properties and flow geometry and usually it is derived from empirical experiments.

To build a model for the flow sensor with gain and lose energy by heating and dissipation of thermal conduction including the solid conductance and convection, it can be described in equation (1). We omit the thermal radiation because the working temperature is at the ambient temperature without the infrared radiation exchange between sensor and environment.

$$\left(\rho A_s \Delta x\right) s \frac{\partial T}{\partial t} = \phi_{path} - \phi_{convection} + I^2 R \tag{1}$$

Usually, there is a heating power of the heater in equation (1) which may come from the Joule's effect and the heat loss to the surrounding by solid conduction depending on the geometrical factors of the details of structure and also the material along the conduction path in equation (2). These formulations of effect and law have been rebuilt and modelled by numerous researchers. The differential form of analytical description for the heat transfer by Morris and Foss (2003) as a useful reference for the convective-cooling effect. The thermal behaviour of the sensor due to the heat loss under the flow through the membrane can be described in equation (3). Especially, it is for the small piece of heating element between the supported ends can be subdivided into short length segments, Δx, and each segment can be described by the thermal balance equation. For the Δx, the heating power is I^2R, our proposed ultra-low power thermoelectric flow sensor where $\emptyset_{path}$ equal to axial heat transfer,

$$\phi_{path} = -K_s (A_s \Delta x) \frac{\partial^2 T}{\partial x^2} \tag{2}$$

The convection heat transfer from the sensor to the flow is potential to the temperature difference and can be expressed as equation (3):

$$\phi_{convection} = h_n \pi D_s \Delta x \left[T_f - T_s \right] \tag{3}$$

where the subscript '*s*' refers to the sensor. The convection coefficient, h_n, is expressed in terms of the Nusselt number, Reynolds number and Prandtl number and the expression is as equation (4):

$$Nu = [0.42 Pr]_f^{0.26} + [0.57 Pr]_f^{0.33} [Re]_f^{0.45} \tag{4}$$

While $Nu = h D_s / k_f$ and $Re = \rho V D_s / \mu$ where the value of the Reynolds number is used to determine whether the flow is laminar or turbulent and D_s is the diameter of sensor. It is expected the thermal sensor installed in the steam of flow, the reduction of temperature on the membrane of sensor is induced from equation (3). For our proposed thermopile, a thermoelectric output voltage for the serial connected thermocouples is generated from Seebeck effect and given as equation (5):

$$V_{emf} = -S \Delta T \tag{5}$$

2.1. Thermoelectric type sensor

For the investigation of convective cooling, a thermopile is fabricated by using the standard CMOS TSMC 0.35 um process and after these processes the silicon substrate beneath the floating membrane is removed subsequently by TMAH. The following figure, figure 1 shows the design of central-symmetrical layout of thermoelectric elements of flow sensor.

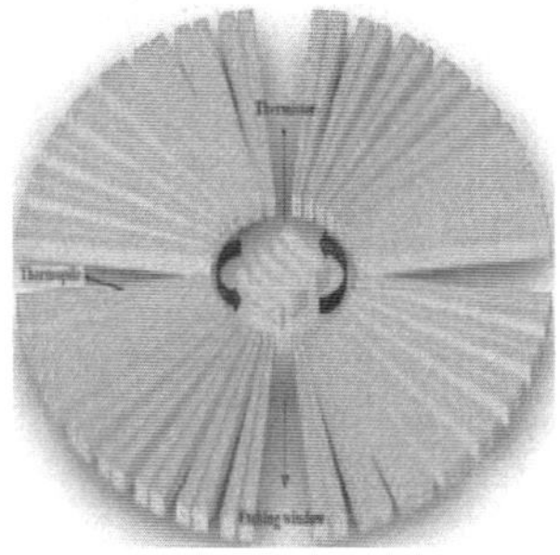

Figure 1. Illustration of structure drawing of proposed convective thermopile.

Figure 2. Photography of proposed thermopile after fabrication.

There are 32 pair of thermocouples placed around the center of membrane and each thermocouple is designed with two kinds of material, n+ Poly and metal in series which give a weak voltage according to the difference of temperature between the hot and cold junctions. We also consider a temperature sensor of thermistor situated at the centre of sensing area and the two terminals of resistor are connected with interconnection to the PAD. The material of thermistor is designed with p+ Poly and the resistance of sensor is 720 Ω at 25 °C. After the CMOS process, a series of post-processes including an isotropic RIE and wet etching processes to remove the silicon under the membrane are proceeded. The photography of microscope of final chip is shown in figure 2.

2.2. Sensing circuit of thermoelectric sensor

The circuit diagram of sensor signal processing composed of amplifier, ADC and data acquisition is shown in figure 3. This data acquisition system is designed with an embedded ARM based Cortex M0, NUC120 to control and communication with PC. For the temperature change of sensing area duo the flow, the output voltage of thermoelectric is weak and after amplification and low-pass filtering the signal is delivered to one of the channels of ADC, MCP3202. A chopper amplifier, AD8551 is used for its low offset and low temperature-drift. Without heating, the signal of temperature sensor is picked up to the second channel of ADC. After the ADC, the digital signals of two sensors are passed to the PC through the SPI interface of ADC.

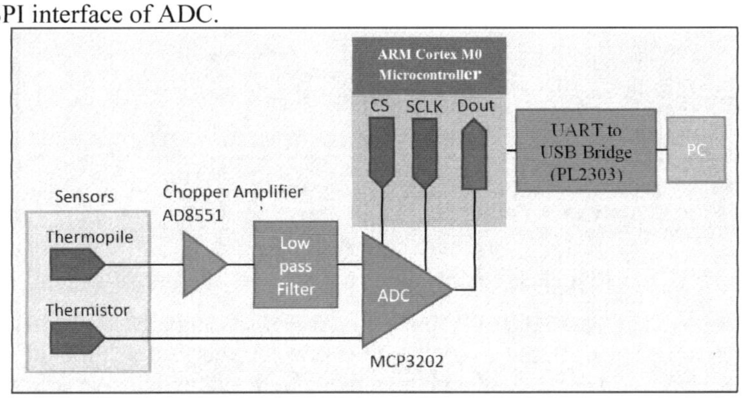

Figure 3. Architecture of sensing circuit for proposed ultra-low power thermoelectric-type anemometer.

3. Experiment and analysis

3.1. Experimental setup and calibration

A small wind tunnel is built and the flow velocity can be controlled under 0~5 m/s, while a standard flow meter is installed with outputs of flow velocity, flow temperature and humidity. The calibration curve is derived at room temperature 25 °C from the fitting function of data which are measured with thermoelectric sensor vs. flow velocities under different conditions of flow velocities, and shown in the figure 4. It is interesting to find the cooling of membrane temperature without heating due to the flow convection and the fitting curve shows a good behavior of squared function.

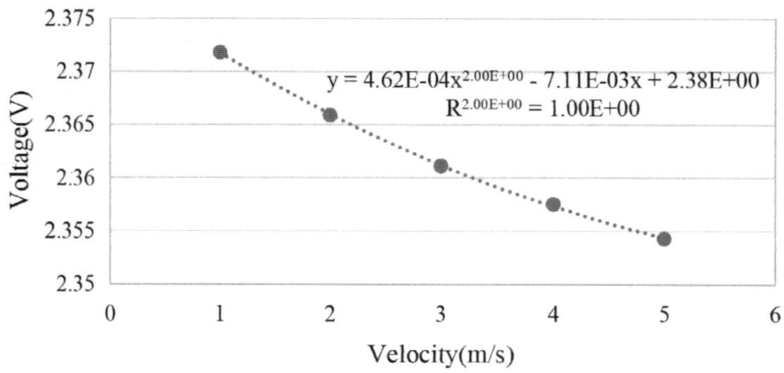

Figure 4. Measurement of output voltage vs. flow velocity.

3.2. Results and discussions

The convective cooling of sensor for flow measurement can be also observed from the thermistor at the center of membrane. From figure 5 the thermistor of p+ Poly shows a positive TCR and is calibrated for the temperature from 35C~80 °C and the resistance varies from 730~760 Ω. After the calibration of thermistor, it is used to verify the reduction of temperature of sensing area during the flow and the measurement is carefully proceeded under a pulse reading of bridge circuit while is biased with 10 ms and reset to zero bias for 990 ms. It is also verified in another research that the pulse reading will cause no extra side-effect during the investigation of flow measurement.

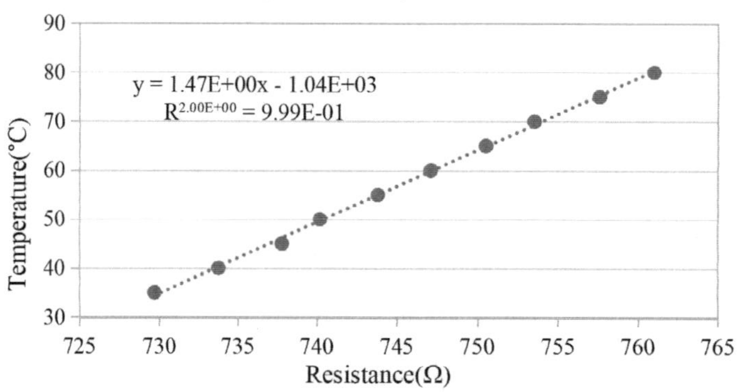

Figure 5. Temperature calibration of thermistor.

It is obvious to see that from figure 6, the temperature of sensing area is gradually getting lower as the flow velocity increases. Because of the precision of thermistor calibration, the change of temperature is observed and quite small and the range of temperature change is from 41.95 ~ 41.6 °C while the velocity of flow is controlled from 0 ~ 5 m/s.

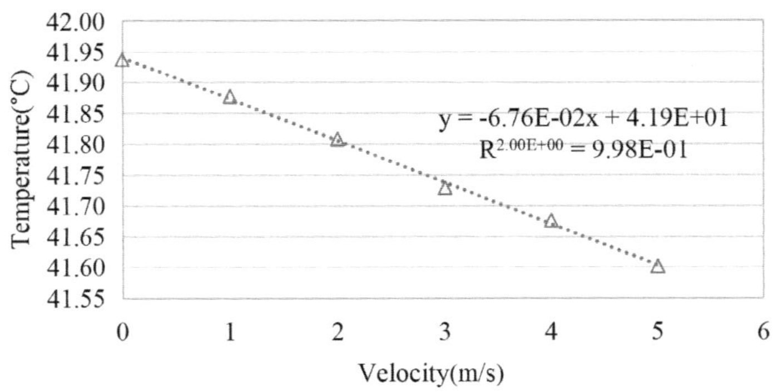

Figure 6. Investigation of temperature reduction vs. the increase of flow velocity.

4. Summary

A new thermopile with 32 pair central-symmetrical thermocouples is proposed in this work which is fabricated by using a standard TSMC 0.35 um CMOS process. The new approach of flow sensing technique without heating is proposed while the fluid passes through the thermopile, the fluid will take away the heat, so that a temperature difference is generated between the cold end and hot end of the thermocouple. The new approach for the anemometer of sensing flow velocity is realized and verified by our proposed thermopile which is proved to be a practical technique of flow sensing with ultra-low power consumption.

Acknowledgments

This research was supported in part by grant of making chips from National Chip Implementation Center (CIC). The authors would also like to thank the National Science Council of the Republic of china, Taiwan for financially supporting this research under Contract. MOST 106-2221-E-018-026.

References

[1] U. Schmid, R. Hoffmann, and G. Kroetz, "Ein mikromechanisch gefertigter hochdruckstabiler Massenflusssensor für Common-Rail(CR)-Einspritzsysteme," Tech. Mess., vol. **68**, no. 5, pp 208–14, May 2001.M. Ben Rabha, M.F. Boujmil, M. Saadoun, B. Bessaïs, Eur. Phys. J. Appl. Phys. (In press)
[2] U. Mescheder, Mikrosystemtechnik. Stuttgart, Germany: B.G. Teubner Verlag, 2000
[3] Z. Y. Tan, M. Shikida, M. Hirota, and K. Sato, "Characteristics of on-wall in-tube thermal flexible mass-flow sensors," in Proc. 19th IEEE MEMS, Istanbul, Turkey, 2006, pp 622–25
[4] R. Buchner, M. Maiwald, C. Sosna, T. Schary, W. Benecke, and W. Lang,"Miniaturised thermal flow sensors for rough environments," in Proc. IEEE MEMS, Istanbul, Turkey, 2006, pp 582–85
[5] O. Mrooz et al., "Aging of copper-nickel-cobalt manganite NTC thermistors [Cu0.1Ni0.8Co0.2Mn1.9O4]," 2002 23rd International Conference on Microelectronics. Proceedings (Cat. No.02TH8595), 2002, pp 375-78 vol.**1**

Analysis of metallic hemispherical shell vibration modes for coriolis vibratory gyroscope

Gopala Krishna Murthy Mittapally[1], Dinkar Dantala[2,*], IM Chhabra[1], Putha Kishore[2], N V N Rao[1] and K C Das[1]

[1]Research Center Imarat Hyderabad, India
[2]National Institute of Technology Warangal, India

*E-mail: dinakar.anu@gmail.com

Abstract. Gyroscope is an instrument widely used in the navigation systems for sensing inertial angular motion. One of the principle of Gyroscope is Coriolis Vibrating Gyroscope (CVG), which is simple and easy to operate. It is realized during the development of the CVG that it is most suitable inertial sensor for aerospace applications. In this paper, a novel method for high reliable gyroscope application is demonstrated to verify the 'Mode 2' vibration pattern of vibrating hemispherical metallic shell. A parallel plate capacitor arrangement is made by placing discrete metallic plates around the metallic shell to study the vibration mode pattern. The shell vibrating at this particular mode, flexes along the axial direction of capacitor. The air gap of the parallel plate capacitor changes as the shell vibrates. The 'Mode 2' vibration pattern from the shell is confirmed by studying the phase relation at different capacitor locations.

1. Introduction

Gyroscope is an instrument widely used in the navigation systems for sensing inertial angular motion about its input axis without any external reference [1]. Over the last few decades, a host of physical laws were utilized to develop operational gyroscopes, which are currently in operation [2]. Development of Coriolis Vibratory Gyroscope (CVG) addresses all the aerospace applications needs like aircraft, satellite, deep space missions, launch vehicles, missiles and Submarine. In this type of gyroscopes, a vibrating hemisphere is excited to one of its resonant vibration modes at the prescribed amplitude. When the structure rotates about a particular fixed axis of its body, Coriolis forces act on the vibrating mass element and excites another resonant mode [3]. The rate of exchange of energy can be measured in terms of change of amplitude and phase of the capacitive signals at different locations around the hemisphere.

In the present work, a parallel plate capacitor arrangement is made by placing discrete metallic plates around the metallic shell to study the vibration mode pattern.

2. Working principle

In a CVG, one of the resonant modes also referred as driving mode is excited to a prescribed amplitude [4]. When the device rotates about a particular fixed axis, the resulting Coriolis forces acting on the vibrating mass excites a different resonant mode [5-6]. The rate at which the energy is transferred to this mode is used to measure the rotation rate about the sensitive axis. In most cases, the natural frequency of the second mode is identical to that of the first mode, like in vibrating string,

Content from this work may be used under the terms of the Creative Commons Attribution 3.0 licence. Any further distribution of this work must maintain attribution to the author(s) and the title of the work, journal citation and DOI.
Published under licence by IOP Publishing Ltd

hemispherical shell and in the Foucault pendulum. The arctangent of the ratio of the mode amplitude changes in response to the input rotation rate.

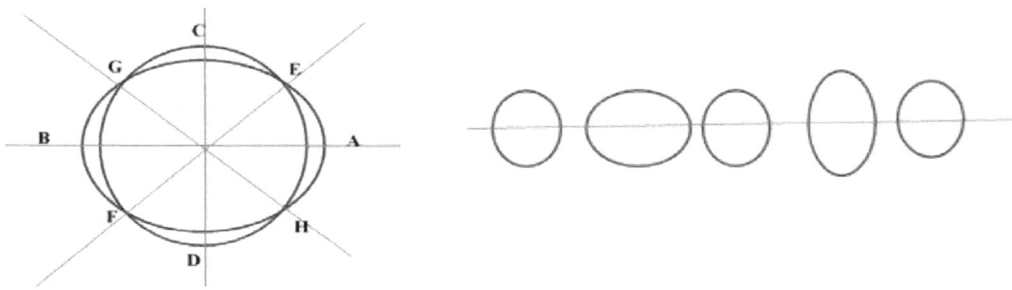

Figure 1. The mode pattern of the shell.

The CVG driven and readout mode amplitude satisfy coupled oscillator equations [7]. The equations are either same as two dimensional oscillator or they can be transformed in to desired two dimensional oscillator equations. When the hemispherical shell is excited to 'mode 2', the pattern looks as shown in the figure 1. The displacement of the shell is maximum at antinodal location and zero at nodal location. A parallel plate capacitor structure can be made with a shell and set of discrete metal pieces at nodal and antinodal locations.

A thin walled hemispherical shell is made to vibrate in its fundamental flexural mode by means of piezoelectric exciter. This mode pattern of the shell defines that in the first half cycle the shell deforms to its greatest ellipsoidal geometry and then returns to its spherical shape as shown in figure 1. In the next half cycle, a similar deformation takes place but spatially shifted by 90° in azimuth. This mode of vibration leads to four antinodes (A, B, C, D) with maximum deflection of the shell and four nodes (E, F, G, H) with zero deflection.

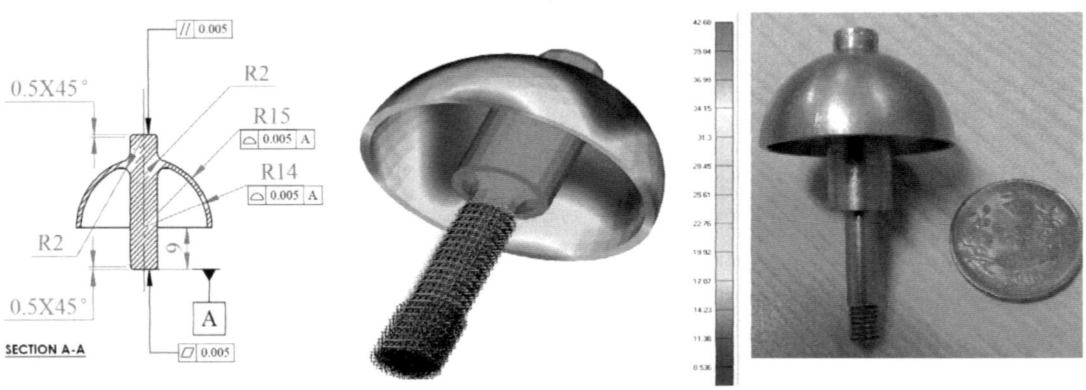

(a) (b) (c)

Figure 2. Designed Vibration sensor Shell(a) Dimensions, (b) Simulated model and (c) Photograph.

A thin wall of approximately 1mm aluminium hemispherical resonator with 30mm diameter is designed with high geometrical tolerances (±5μm). The dimensions of the designed metallic hemispherical shell is shown in figure 2(a). NASTRAN simulation software is used to understand various modes, their coupling and material selection to get required resonant frequency. Figure 2(b) shows the stress pattern. High performance CNC machine is used to realize the metallic shell as shown in figure 2(c).

3. Experiment

To detect the nodes and the antinodes of the mode 2 vibration, at least 8 capacitors are required, which are to be placed around the metallic shell at equal angular difference as shown in figure 3. Eight no. of copper pieces of uniform dimension of 5mm x 2mm are used as the electrode to form a parallel plate capacitor along with the metallic shell with air gap of 30μm. The capacitance will not change in case of the metallic shell has no vibration. The static capacitance of all the capacitors under no vibration is of the order of 10 – 12pF. When the shell vibrates, its equatorial rim flexes up to a maximum of ±5μm along the A, B and C, D directions alternately. Under this condition the air gap changes by 30±5μm, thus results change in capacitance value at these locations. This dynamic change in capacitance is between 15 – 20 fF. This small change in capacitance will be added to the capacitance under no vibration state. Hence total change in capacitance across anti nodal locations ranges from ±10.015pf to ±10.02pf and 12.015pf to ±12.02pf respectively. As the shell vibrates at its natural frequency 6.279 KHz, the change in capacitance also corresponds to same frequency. An operational amplifier with very high input impedance, high band width with good gain linearity (AD4000-4) is used to detect the change in capacitance and boost the signal with the gain of 20.

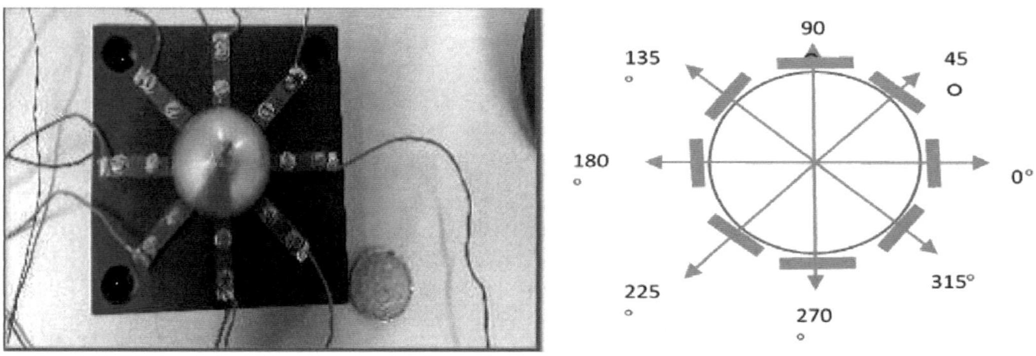

Figure 3. Arrangement of capacitors for designed shell.

4. Results and discussion

Figure 4(a) shows the electrical response of the vibrating shell at antinodal locations A, B, C, D. Since the Q factor of the aluminium shell is less (600-800), the signal damps rapidly. The vibration mode under investigation can be analysed by comparing the phase of the signals at these locations. When the mode under investigation is excited in the shell, capacitors at locations A and B experience a decrease in the air gap which leads to increase in the capacitance. At the same time capacitors at C and D experience decrease in capacitance as the air gap increases. These changes in the capacitances also result in similar change in phases at the locations A, B and C, D. These signals are recorded using the designed high input impedance ultra-low noise operational amplifier.

Mode 2 vibration on the hemispherical shell results in out of phase signal at A, C and B, D is shown in figure 4(b). From the phase information it is evident that the shell experiences a vibrational mode required for gyroscope operational applications. When the shell is under the vibration, maximum amplitude is experienced at antinodal locations and at the same time the nodal locations does not show any displacement, which results in no variation in capacitance. This can be seen by picking the capacitor signals at the locations E, F, G and H as shown in figure 4(c).

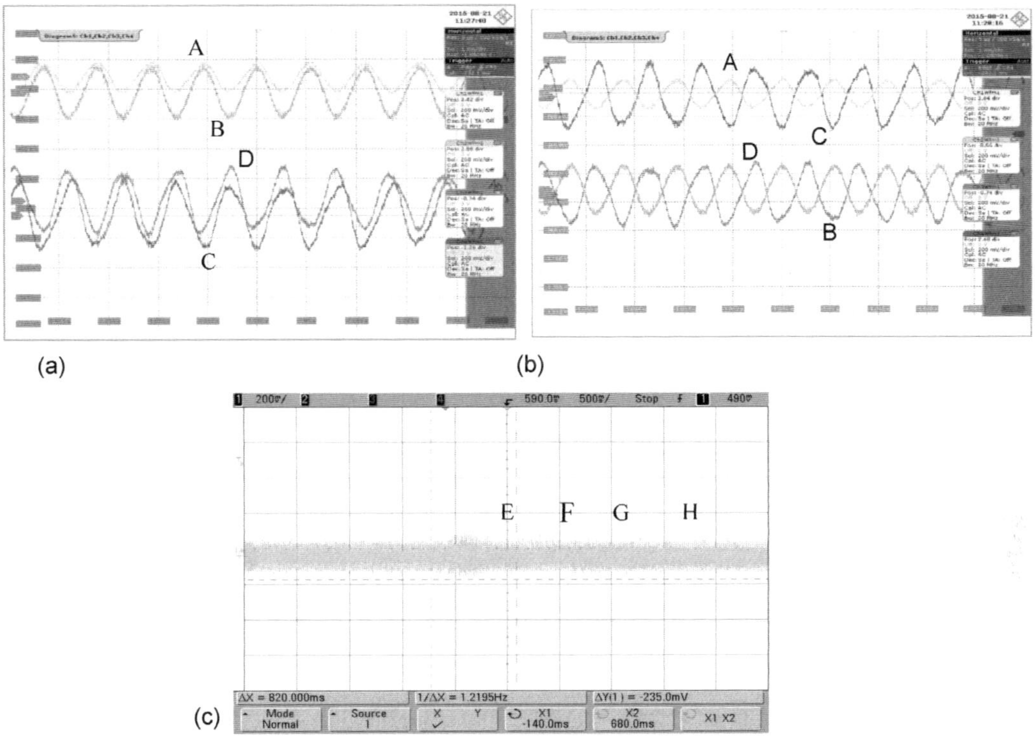

Figure 4. Capacitive response at anti nodal locations a) A, B & C, D b) A, C & B, D and c) E, F & G, H nodal locations.

5. Conclusion

A novel and low cost metallic hemispherical shell is designed and developed to measure and to analyse the vibrational modes for CVG. From the analysis of data, it is evident that the vibration mode imparted on to the shell is confined only at antinodal locations. Frequency and phase information of the signals reveal that the shell experiences 'mode 2' vibration. It is also clear from the data that the nodal locations do not show any vibration which implies that the shell is in the mode 2 alone.

Acknowledgments

Authors acknowledge the support and encouragement by Director and Technology Director, RCI, Hyderabad.

References

[1] D Lynch, A Matthews and G T V arty 1997Innovative mechanization to optimize inertial sensors for high or low rate operations *Proc. Symp. Gyro Technol. Stuttgart, Germany* **6** pp 16- 17

[2] D D Lynch, R .R. Savaya and J J Campanile Nov 2006 Hemispherical resonator gyro control *U. S. Patent* 2006/0248 953

[3] A Matthews and F J Rybak 1992 Comparison of hemispherical resonator gyro and optical gyros *IEEE Aerospace. Electron.Syst.Mag.* **7(5)** pp 40-46

[4] G I Dzhandzhgava, K. A Bakhonin, G M Vinogradov and A V Trebukhov 2010 Strapdown inertial navigation system based on a hemispherical resonator gyro *Gyros copy Navigat.* **1(2)** pp 91-97

[5] A Matthews and F J Rybak 1992 Comparison of hemispherical resonator gyro and optical

gyros", *IEEE Aerospace. Electron.Syst.Mag.* **17(5)** pp 40-46
[6] D M Rozelle 2009 The hemispherical resonator gyro: From wineglass to the planets *Proc.19th AAS/AIAA Space Flight Mech. Meeting* pp 1157-78
[7] A Matthews, G T Varty, C M Li and D D Lynch 1998 Vibratory rotation sensor with whole angle tracking *U.S.Patenet* **5 801 310**

IOP Conf. Series: Materials Science and Engineering **383** (2018) 012023 doi:10.1088/1757-899X/383/1/012023

Modelling and infrared radiation compensation for non-contact temperature measurement

Chih-Hsiung Shen*, Tsung-Dai Chang and Shu-Jung Chen

Department of Mechatronics Engineering, National Changhua University of Education, NO. 2, Shi-Da Road, Changhua, Taiwan, R.O.C.

*Email: hilbert@cc.ncue.edu.tw

Abstract. Beyond the conventional non-contact temperature measurement with ambient temperature compensation and our previous work of multi-sensors compensation, two compensation schemes are proposed and compared. An infrared radiation measurement module with several temperature sensors are built and calibrated with ambient compensation to investigate the dynamic temperature distribution under moving of module from one place to another. The new approach we proposed including the infrared radiation exchanges model between the target, sensor and the optical path thorough temperature monitoring and two fast temperature measurement schemes with dynamic compensation. After careful calibrations and verification of several experiment conditions, our models of two dynamic compensation schemes both show an excellent agreement with the measuring data. The experimental data of compensation scheme reach a stable reading value of target temperature down from 40 min to 2 min for the differential scheme compensation and 6 min for multi-sensors scheme compensation with temperature error around 0.2 °C.

1. Introduction

Temperature plays a significant role as an indicator of the condition both in manufacturing, quality control and medical diagnosis. We can divide temperature sensor into two types for contact temperature sensor and non-contact temperature sensor. Among different kinds of temperature measurement devices, thermopile-type and infrared thermometers for non-contact temperature measurement are highly developed sensors. Everything above the absolute zero produces infrared radiation, a phenomenon known as thermal radiation. So far, infrared sensing technology has been utilized successfully in industrial processing and research. Thermopile sensor and focusing lens are used to measure the temperature of high-voltage switch contactors as a security device [1]. Furthermore, freely-suspended micro channels with integrated thermopiles has been used to make a highly-sensitive thermal flow sensor with ambient temperature-gradient compensation [2]. Thermopile has the features of linear thermoelectric signal transduction and low sensing range limit make it attractive for many low-cost commercial application products [3-6].

Conventionally, the infrared thermometer is operated under a stable state. While the ambient temperature will change, it causes a drawback of measurement for many infrared instruments under this thermal non-equilibrium state. It should take a long time to wait for ambient temperature to achieve stability with infrared instruments. To overcome this drawback, we have to consider the thermal radiation between the target, surrounding and ambient under this non-equilibrium state.

Content from this work may be used under the terms of the Creative Commons Attribution 3.0 licence. Any further distribution of this work must maintain attribution to the author(s) and the title of the work, journal citation and DOI.
Published under licence by IOP Publishing Ltd

IOP Conf. Series: Materials Science and Engineering **383** (2018) 012023 doi:10.1088/1757-899X/383/1/012023

2. Theory and modelling

2.1. Principles of infrared temperature measurement

The source of thermal radiation is the thermal motion of charged particles. The thermal motion will produce thermal which will be converted into the energy form of electromagnetic radiation. That means thermal radiation can exchange heat without medium and without contact. There are many radiation laws that describe the basic characteristics of thermal radiation. The basic characteristics of blackbody radiation describes the heat radiation frequency depends on the temperature of the object. Planck's Law describes every object emits radiation at all times and at all wavelengths. Besides, Planck's Law mentions the relationship between emissivity and frequency of electromagnetic radiation emitted by a black body at different temperatures. The Stefan-Boltzmann Law states that the total amount of energy per unit area emitted by an object is proportional to the 4th power of the temperature. We can derive the radiation equilibrium relationship for non-contact temperature measurement based on these theories.

We propose two novel infrared compensation schemes, one for check the exchange of infrared radiation between the surrounding environment and sensor, the other for verifying the temperature difference of thermopile sensor under thermal non-equilibrium state. In the new proposed non-contact temperature measurement model, there are several temperature-monitoring sensors include the temperature of blackbody, surrounding environment and thermopile sensor at ambient temperature which are denoted as T_b , T_s and T_a . The infrared radiation exchanges between blackbody and thermopile sensor is denoted as Φ_{ba} and Φ_{ab} ,while the exchanges between surrounding environment and thermopile sensor is denoted as Φ_{sa} and Φ_{as} . The infrared radiation exchange relationship are shown in figure 1.

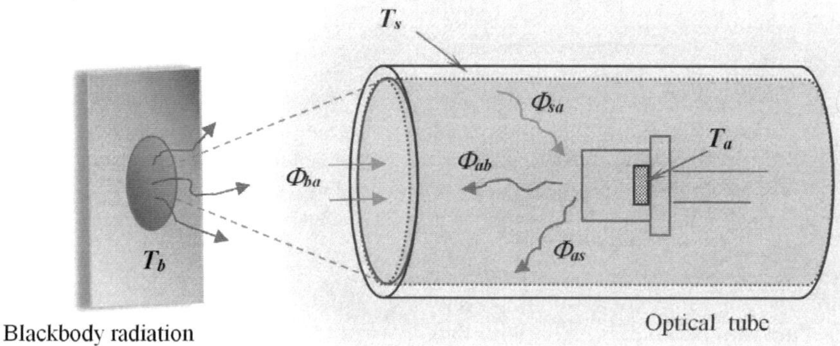

Figure 1. Infrared radiation exchange between target, environment and sensors.

2.2. Mathematical model of radiation exchange

To build the mathematical model, firstly an unbalanced temperature difference between surrounding and sensor is proposed under a dynamic ambient temperature change. We should pay attention to the dynamic process between steady-state to another steady-state. During the dynamic temperature measurement, we need to consider the change of the relative radiation between the blackbody, the thermopile sensor and the surrounding environment. We denote the total infrared radiation thermopile sensor received as $\Delta\Phi$ in equation (1), which will later processed and indicate the blackbody target temperature. The total infrared radiation $\Delta\Phi$ comprise of four major terms as Φ_{ba} , Φ_{sa} , Φ_{ab} and Φ_{as} . According to Stefan-Boltzmann law, we can express the relative radiation from equation (2) To equation (5).

$$\Delta\Phi = \Phi_{ba} + \Phi_{sa} - \Phi_{ab} - \Phi_{as} \tag{1}$$

$$\Phi_{ba} = A_{ba}\varepsilon_b * T_b^4 \tag{2}$$

$$\Phi_{sa} = A_{sa}\varepsilon_s * T_s^4 \tag{3}$$

$$\Phi_{ab} = A_{ab}\varepsilon_a * T_a^4 \tag{4}$$

$$\Phi_{as} = A_{as}\varepsilon_a * T_a^4 \tag{5}$$

where A represents the geometric factor is a parameter that characterizes the relative influence of various media in the space on the measurement results. It including the geometrical effects of area of sensor, field of view angles, cut on wavelength and transmission of optical lens in our measurement. Besides, ε represents the emissivity of different materials.

When the instrument is moved from one place to another, it is necessary to realize the dynamic temperature behaviours for thermopile sensor and surrounding environment. Before the time reach thermal equilibrium, the temperature gradient drift will cause thermal radiation exchange between thermopile sensor and surrounding environment which we denote $\Delta\Phi_{as}$ in equation (6). While there is at thermal equilibrium $\Delta\Phi_{as} \neq 0$, it will cause a temperature error ΔT_b between the measured value T_b and stable temperature T_{b0} as equation (7)

$$\Delta\Phi_{as} = \Phi_{as} - \Phi_{sa} = A_{as}\varepsilon_a * T_a^4 - A_{sa}\varepsilon_s * T_s^4 \tag{6}$$

$$\Delta T_b = T_{b0} - T_b \tag{7}$$

We can suppose that there is a correction factor K for the relationship between the temperature error ΔT_b as and the thermal radiation exchange $\Delta\Phi_{as}$. In that way, ΔT_b can be derived by Stefan-Boltzmann law as equation (8) and equation (9).

$$\Delta T_b = K(A_{as}\varepsilon_a * T_a^4 - A_{sa}\varepsilon_s * T_s^4) \tag{8}$$

$$= K_0\left[4(\overline{T})^3(T_a - T_s)\right] \tag{9}$$

where $K_0 = KA_{as}\varepsilon_a = KA_{sa}\varepsilon_s =$ constant and $\overline{T} = \dfrac{T_a + T_s}{2}$

From the above mathematical deduction that we can conclude the temperature error ΔT_b is proportional to temperature difference $(T_a - T_s)$. It is the basic principle for the multi-sensors scheme of dynamic compensation. Nevertheless, a more powerful compensation scheme, the differential scheme is proposed and described as follows. For a thorough description of thermal behaviour of thermopile sensor and environment, a thermal equilibrium equation is introduced as follows equation (10) and equation (11).

$$H\frac{\Delta T_a}{\Delta t} + G(T_a - T_s) = P = 0 \tag{10}$$

$$\frac{\Delta T_a}{\Delta t} \propto (T_a - T_s) \tag{11}$$

where H is the thermal sensor conductivity, G is the convection heat-transfer coefficient between thermopile sensor and surrounding environment. It is obvious to see that the temperature error ΔT_b is proportional to both the temperature difference $(T_a - T_s)$ and ΔT_a.

3. Experiment setup and results

3.1. Experimental setup and sensor calibration

The new infrared radiation measurement circuit which was built to investigate the thermal behaviour of infrared radiation exchanges between the target, thermopile sensor and surrounding environment is

IOP Conf. Series: Materials Science and Engineering **383** (2018) 012023 doi:10.1088/1757-899X/383/1/012023

shown in figure 2. This data acquisition system is designed with an embedded ARM based Cortex M0 micro-controller, NUC120, to control the circuit and communication with PC. A thermopile of TO-5 package is used with a filter 5~14 μm which receive the infrared radiation from a plastic Fresnel lens with a focal length 2.5 cm. The output signal of thermopile is delivered to an amplifier AD8551 with low offset and a 12-bit ADC MCP3202 follows to transfer it into the digital signal then we can get bias voltage of thermopile V_b and bias voltage of inner thermistor V_a . Besides of the thermopile, we put a SMD package thermistor attached to the optical tube which serves as a monitor of unbalanced temperature of surrounding environment. We use the embedded ARM microcontroller build in ADC to get the bias voltage of outer thermistor V_s .

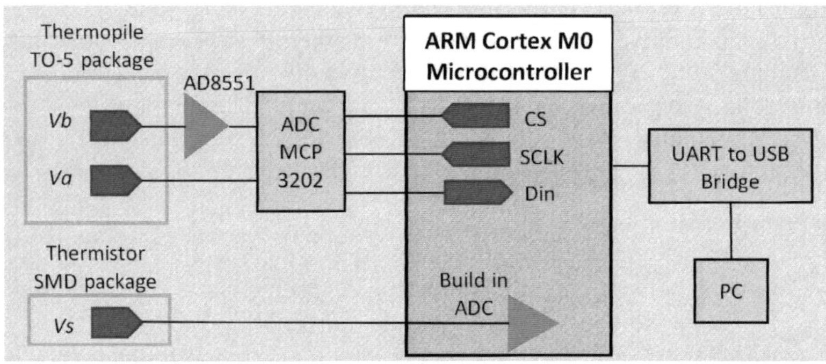

Figure 2. An infrared radiation measurement circuit compensation with sensor embedded.

Both of the temperature sensors including thermopile inner thermistor, outer thermistor and thermopile have to be calibrated before the testing of dynamic measurement. The thermopile inner thermistor is chosen as $30\,k\Omega$ and outer thermistor is chosen as $50\,k\Omega$ at 25 °C. The bias output voltages of thermopile inner thermistor and outer thermistor are calibrated from 20 °C to 45 °C inside a temperature controlled chamber. The thermopile is chosen as $50\,k\Omega$ at 25 °C and we measure the output signal after the amplifier AD8551.

During the thermopile calibration, we use the thermopile inner thermistor to compensate the ambient temperature difference to guarantee thermal radiation exchange only exist between blackbody and thermopile sensor. Then we adjust the standard blackbody with temperature from 20°C to 45 °C and measure the thermopile bias voltage under a stable environment. The calibration results of V_s , V_a and V_b vs. temperature are shown as figure 3.

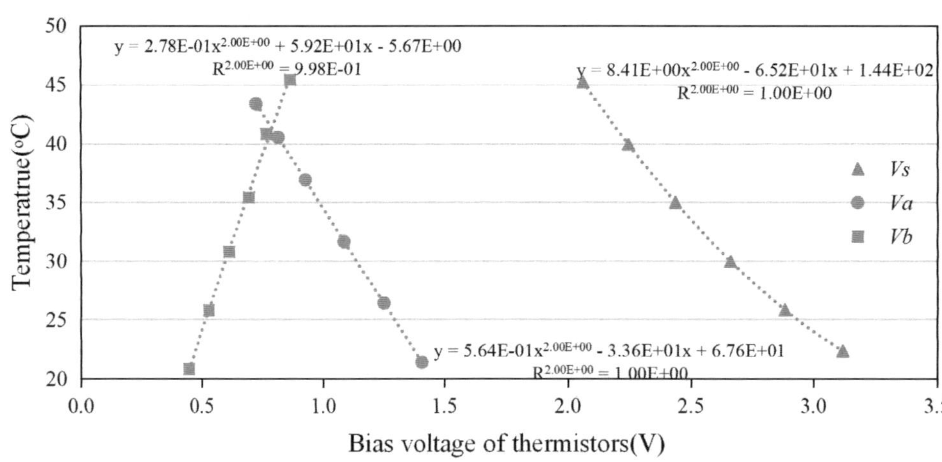

Figure 3. Calibration of the output voltage V_s , V_a and V_b vs. temperature.

After calibration of temperature sensor for blackbody, surrounding environment and thermopile sensor. We can proceed to the dynamic infrared radiation temperature measurement for moving of measurement module from a higher temperature chamber at 50 °C to room temperature and measure the blackbody target at 38 °C. The dynamic temperature behaviours of blackbody, surrounding environment and thermopile sensor are recorded for 50 minutes and shown as figure 4. One can see that the initial temperature T_a is higher than T_s. It is because the outer thermistor touch the surrounding cool air first and it will cool down more quickly.

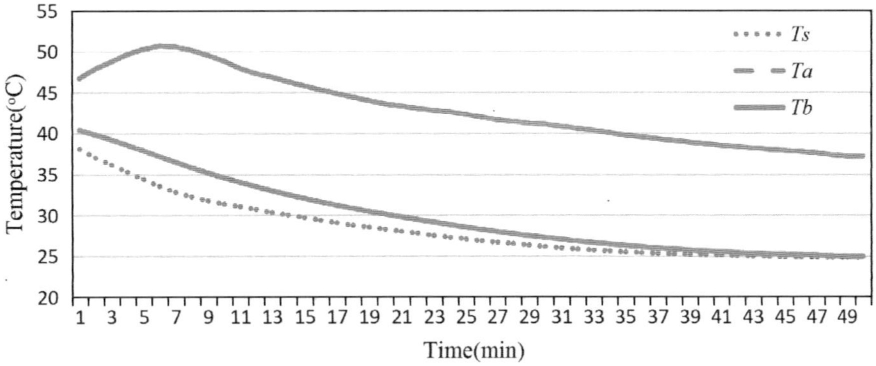

Figure 4. Dynamic temperature behaviours of T_s, T_a and T_b.

3.2. Results and discussions

After we get the dynamic temperature of different sensors, we can calculate the temperature difference between inner thermistor and outer thermistor $(T_a - T_s)$ and inner thermistor temperature difference ΔT_a to compare with unbalanced target temperature error ΔT_b at different time, respectively. The results is shown as figure 5. One can see that both data shows a good agreement relationship between ΔT_b vs. $(T_a - T_s)$ and ΔT_a. We can get two convert equation to estimate the target temperature error by $(T_a - T_s)$ and ΔT_a.

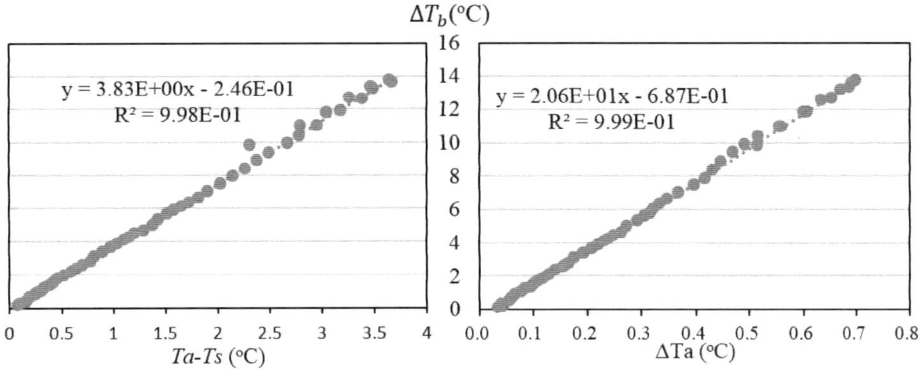

Figure 5. Calibration of target temperature error ΔT_b vs. $(T_a - T_s)$ and ΔT_a.

Consequently, we can use these two compensation schemes to compensate the dynamic unbalanced target temperature error. The comparison of measurement temperature error and compensate temperature by two compensation schemes is shown as figure 6. After calibrations and verifications , our compensation schemes provide an efficient way to reduce the conventional measuring waiting time for thermal equilibrium from 40 min down to 2 min for ΔT_a compensation and 6 min for

$(T_a - T_s)$ compensation with temperature error around 0.2 °C, which is proved to be a practical technique of non-contact temperature measurement.

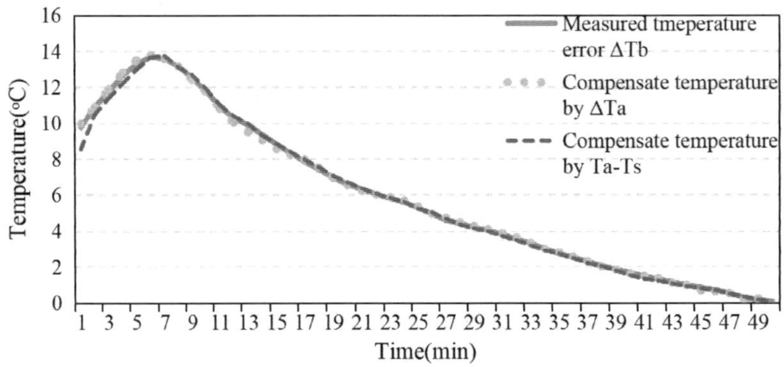

Figure 6. Comparison of measurement temperature error and compensate temperature by two compensation schemes of ΔT_a and $(T_a - T_s)$.

4. Conclusions

A non-contact temperature measurement model and two novel and high efficient infrared temperature compensation techniques are proposed in this research work. We calculate the temperature difference between internal and external thermistor to estimate the relative amount of radiation. After careful calibrations and verifications of several experiment conditions, our models of two dynamic compensation schemes both show an excellent agreement with the measurement data. The experimental data of compensation scheme reach a stable reading value of target temperature to 2 min for the differential scheme compensation and 6 min for multi-sensors scheme compensation with temperature error around 0.2 °C.

Acknowledgments

This research was supported in part by grant of making chips from National Chip Implementation Center (CIC). The authors would also like to thank the National Science Council of the Republic of china, Taiwan for financially supporting this research under Contract. MOST 106-2221-E-018-026.

References

[1] Jinbo Zhang, Shushu Liang, Yan Gao, Dan Lv, ICMTMA- international conference on measuring technology and mechatronics automation pp 1047-50, China(2010)

[2] M. Difkstra, TS.J Lammerink, MJ. de Boer, R.J. Wiegerink andM. Elwenspoek, IEEE 22nd 7. Int. Conf. On Micro Electro Mechanical Systems pp 479 - 82, Italy (2009)

[3] S. H. Tsang, A. H. Ma, K. S. Karim, A. Parameswaran, and A. M. Leung, Monolithically fabricated polymermems 3-axis thermal accelerometers design for automated wirebonder assembly, IEEE MEMS Conference, p 880 (2008)

[4] A. Schaufelbuhl, N. Schneeberger, U. Munch, M. Waelti, O. Paul, O. Brand, H. Baltes, C. Menolfi, Q. Huang, E. Doering, M. Loepfe, *IEEE J. Microelectromech.* Syst. 10, p 503 (2001)

[5] H. Baltes, O. Brand, J. G. Korvink, R. Lenggenhager and O. Paul, IMEMS- integrated micro electro mechanical systems by VLSI and micromaching, Proc. 24th Eur. Solid State Device Research Conference, ESSDERC, Edinburgh, UK, pp 11-15 Sept., (1994)

[6] Bing Shen, Zhongsheng Lai, A. M. Robinson and W. Allegretto, Rev. *Sci. Instrum.* 11, p 3528 (1994)

IOP Conf. Series: Materials Science and Engineering **383** (2018) 012024 doi:10.1088/1757-899X/383/1/012024

Development of a surface plasmon resonance acetone sensor for noninvasive screening and monitoring of diabetes

F Usman [1,3,*], J O Dennis[1] and F Meriaudeau[2]

[1]Department of Fundamental and Applied Sciences, Universiti Teknologi PETRONAS (UTP), Bandar Seri Iskandar, 31750 Tronoh, Perak, Malaysia
[2]Department of Electrical and Electronic Engineering, Universiti Teknologi PETRONAS (UTP), Bandar Seri Iskandar, 31750 Tronoh, Perak, Malaysia
[3]Department of Physics, Al-Qalam University Katsina, Nigeria

*Email: fahatu11@gmail.com

Abstract. Diabetes can cause many complications in the human body. The disease is commonly diagnoses invasively using blood glucose test. A good correlation between blood glucose and exhale breathe acetone has opened a way for non-invasive diagnosis through the detection of the breath acetone from the exhale breathe. However, the conventional means of detecting exhale breathe acetone (e.g. GC-MS) are bulky, require technical knowhow, do not offer real time measurement and found only at advanced institutions that can afford the equipment. This work is aimed at developing a real time, sensitive and selective optical based Surface Plasmon Resonance (SPR) breath acetone biosensor for non-invasive monitoring and screening of diabetes. The motivation of this work arises from the promising advantages of optical based biosensors over other biosensors. Additionally, a novel doped polyaniline (PANI) conducting polymer would be the SPR metallic layer while Chitosan would be the selective sensing layer. Mathematical modeling and SPR simulation has been conducted. The result has shown that the replacement of conventional gold or silver SPR layer with doped Polyaniline (PANI) SPR layer provides better promising sensitivity at mid infrared wavelength range. Therefore, additional inclusion of chitosan layer on the proposed SPR based biosensor would give rise to a sensitive and selective breathe acetone biosensor for easy monitoring and screening of diabetes.

1. Introduction

Diabetes has been listed among the most common life threatening diseases affecting humanity globally. The number of persons suffering from diabetes (type-1 and -2) is increasing due to population growth, aging, urbanization, increasing occurrence of obesity and physical inactivity [1, 2]. In order to hinder its rapid occurrence, Proper means of diagnosis is fundamental to hindering the disease. Currently diabetes is diagnosed through blood glucose test. Unfortunately, the method requires trained personnel, consumes time, laboratory restricted usage, invasive procedures etc.[2, 3]. Exhale breathe acetone has been identified as a good biomarker for diabetes that can provide a non-invasive detection of diabetes [2, 4, 5]. It has been stated that Acetone concentration in the human body is generally very low (0.1 ppm - 0.8 ppm), while it might be high in the case of metabolism disorders, including diabetes mellitus (DM) (1.8 ppm – 5.0 ppm) [6].

Content from this work may be used under the terms of the Creative Commons Attribution 3.0 licence. Any further distribution of this work must maintain attribution to the author(s) and the title of the work, journal citation and DOI.
Published under licence by IOP Publishing Ltd

Since the recognition of exhale breath acetone as the diabetes biomarker, a lot of researches have been conducted in a move to its ease detection [7-14]. The conventional means of acetone detection includes gas chromatography – mass spectrometry (GC-MS), selected ion flow tube mass spectrometry (SIFT-MS), proton transfer reaction mass (PTR-MS), high performance liquid chromatography (HPLC), ion mobility spectrometry (IMS), plus laser techniques like tunable diode laser absorption spectroscopy (TDLAS) and cavity ring down spectroscopy (CRDS) etc. They are capable of detecting trace of acetone vapour with better sensitivity and selectivity relatively. However, these methods rely on sophisticated instrumentation, complicated sample collection methods, expensiveness and present only at advanced medical institutions that can afford [15].

Fortunately, the detection of exhale breathe acetone via biosensors has been shown to feature many advantages over the conventional method. Easy and real time noninvasive detection is possible. In fact, it solves most of the problems associated with the other methods. Biosensors are generally devices that can respond to a physical stimulus (analyte) of biological interest and converts (transduce) the stimulus into a signal conveyed to another device (computer/detector) for interpretation [16-21]. Most of the exhale breathe acetone biosensors are fabricated using metal oxide semiconducting materials based on chemiresistive transduction method [4, 10, 11, 22-30]. However, metal oxide chemiresistive biosensors suffer from high operational temperature issue [15, 31, 32]. Contact resistance influence also results to unreliable selectivity[33]. Furthermore, the humidity influence is another issue of concern [23]. In addition, most of the fabrication processes were complicated; incorporating non-easy to find materials, difficult to process and synthesis materials and features poor biocompatibility.

Some of the advantages of optical biosensors over other types of biosensors are: greater sensitivity, electrical passiveness, freedom from electromagnetic interference, wide dynamic range, non-requirement of reference electrode, free from electrical hazards, high stability relatively, potential for higher-information content than electrical transducers, multiplexing capabilities etc[14, 24, 28, 34-38]. Surface Plasmon Resonance (SPR) based sensors has been identified to be a promising biosensor among optical biosensors [18, 19, 39]. It measures the amount of an analyte by exploiting the change in the refractive index at the interface of metal/heavily doped semiconductor-dielectric interface.

When a light source is incidented on a metal/heavily doped semiconductor-dielectric interface, after total internal reflection, a drop is observed on the reflected light at a particular angle called SPR angle. Presence of a biomolecule or any other material at the vicinity of the interface shifted the SPR angle. This shift is called dip. In order to achieve a cheapest, highly selective and sensitive breath acetone biosensor, this work is mainly aimed at developing an SPR based breath acetone sensor for diabetes' monitoring and screening with doped Polyaniline (PANI) as the SPR layer and chitosan polymer as the highly selective active sensing layer. As part of the fabrication process, the initial result from the mathematical modeling and simulation is presented in this paper.

2. Mathematical modelling and simulation

2.1. Mathematical modelling

The mathematical modeling of the SPR based breathe acetone biosensor started with the modeling of the surrounding of the biosensor. Penetration depth through doped PANI layer (δ_{PANI}) and dielectric (δ_d) are the most important in this regard. The penetration depth in the dielectric gives idea on surrounding sensitivity, while the penetration depth into doped PANI gives idea of the thickness of the doped PANI film required for the coupling of light. The equations for the penetration depth through doped PANI and dielectric (air) are (equation (1) and (2), respectively):

$$\delta_{PANI} = \frac{\lambda_0}{2\pi}\left[\frac{\varepsilon'_{PANI} + \varepsilon_d}{\varepsilon'^2_{PANI}}\right]^{1/2} \tag{1}$$

$$\delta_d = \frac{\lambda_0}{2\pi}\left[\frac{\varepsilon'_{PANI} + \varepsilon_d}{\varepsilon_d^2}\right]^{1/2} \tag{2}$$

Equations (3)-(7) were used for remaining part of the mathematical modeling. The values obtained by [40] and [41] at mid infrared wavelength range in table 1 were used throughout the modeling.

Table 1. Optical properties for the materials of the proposed PANI.

S/N	Material	Real dielectric constant	Imaginary dielectric constant	Refractive index
1	Doped PANI	-30.62	2	-5.5
2	Chitosan	2.25	0	1.5
3	Acetone	1.25	0	1.12
4	Air	1.00	0	1.00
5	Silica	2.1	0	1.45

$$\varepsilon_{eff} = \varepsilon'_{PANI}\left[\frac{1+2f_{ch}\left(\dfrac{\varepsilon_{ch}-\varepsilon'_{PANI}}{\varepsilon_{ch}+2\varepsilon'_{PANI}}\right)}{1-f_{ch}\left(\dfrac{\varepsilon_{ch}-\varepsilon'_{PANI}}{\varepsilon_{ch}+2\varepsilon'_{PANI}}\right)}\right] \tag{3}$$

$$\theta_{SPR} = \sin^{-1}\left[\sqrt{\frac{\varepsilon'_{PANI}\varepsilon_d}{(\varepsilon'_{PANI}+\varepsilon_d)\varepsilon_p}}\right] \tag{4}$$

$$\theta_{SPRB} = \sin^{-1}\left[\sqrt{\frac{\varepsilon_{eff}\varepsilon_d}{(\varepsilon_{eff}+\varepsilon_d)\varepsilon_p}}\right] \tag{5}$$

$$\theta_{SPRBA} = \sin^{-1}\left[\sqrt{\frac{\varepsilon_{eff}\varepsilon_{Ac}}{(\varepsilon_{eff}+\varepsilon_{Ac})\varepsilon_p}}\right] \tag{6}$$

$$\theta_{shift} = (\theta_{SPRBA}) - (\theta_{SPRB}) \tag{7}$$

where, ε'_{PANI} is the real part of the dielectric constant of doped PANI, ε_d is the relative permittivity of the dielectric (air), λ_0 is the free space wavelength, ε_p is the dielectric constant of the material of the prism, ε_{eff} is the effective dielectric constant of the doped PANI/chitosan bilayer, ε_{ch} is the dielectric constant of chitosan, f_{ch} is the fractional thickness of chitosan, θ_{SPR} is the SPR angle, θ_{SPRB} is the SPR Angle (doped PANI/chitosan bilayer), θ_{SPRBA} is the SPR Angle (bilayer plus Acetone), θ_{shift} is the SPR shift (which is correlated with the concentration of breathe acetone) and ε_{AC} is the dielectric constant of the analyte (Acetone).

2.2. SPR simulation

The simulation has been carried out using WinSpall SPR simulation software built based on Fresnel equations. Mid infrared wavelength (3000nm) and hemispherical prism options were also selected for the simulation in the case of the doped PANI. However, visible wavelength was used in simulating SPR curves for gold and silver that were aimed for comparison purpose.

3. Result and discussion

3.1. Theoretical calculations

Theoretically, the SPR phenomenon is expected to occur at 43^0 (table 2). This indicates wide detection range capability. The penetration depth also shows the need for increasing the thickness of doped PANI layer. Also, the penetration depth through the dielectric (2.68 μm) shows how sensitive the sensor could be. A very huge SPR shift is observed (6.41^0) after slightly changing the refractive index of the dielectric medium by exposing the biosensor to breathe acetone. This indicates promising high sensitivity of the proposed biosensor.

Table 2. Result of theoretical modeling of SPR breathe acetone biosensor.

Parameter	Value
δ_{PANI}	88 nm
δ_{d}	2.68μm
θ_{SPR}	42.77^0
θ_{SPRB}	42.84^0
θ_{SPRBA}	49.25^0
θ_{shift}	6.41^0

3.2. WinSpall simulation

The simulated result in figure 1(a) confirms the good sensitivity of this SPR based biosensors to slight change in the refractive index of breathe acetone. Similar to the modeling result, it shows an SPR shift with acceptable deviation (8^0). Also, the expected better sensitivity of this SPR based biosensor over conventional gold and silver based SPR based biosensors is highlighted in figure 1 (b). Its SPR curve using is sharper than Gold and silver ones. The sharpness generally indicates greatness in terms of sensitivity and accuracy.

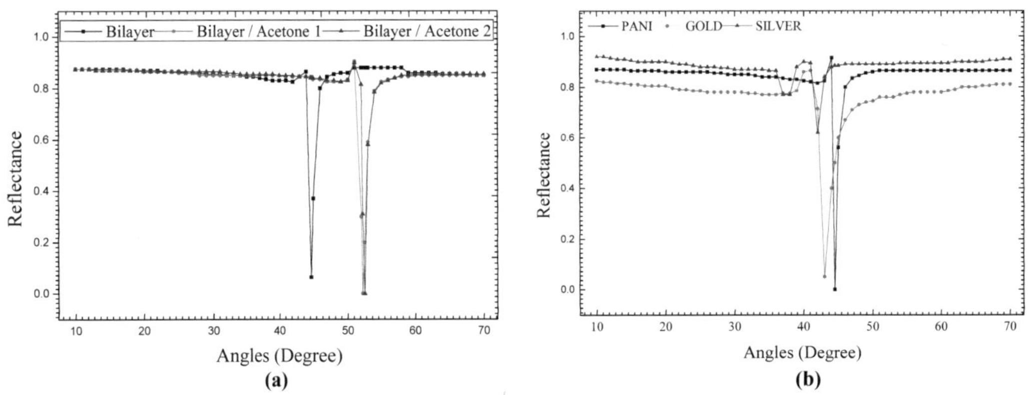

Figure 1. (a) SPR simulation of θ_{SPRB}, θ_{SPRBA} and θ_{shift} (b) Comparison of SPR angle between doped PANI, gold and silver layers.

4. Conclusion

The mathematical modeling and the SPR simulation of the doped PANI layer produces almost similar results. The shifts in SPR angle were 6.41^0 and 8^0 for the modelling and the simulation respectively. Slight change in dielectric constant of 0.1 order has shown to be detectable by the biosensor. Also, the doped PANI/Chitosan SPR curve was sharper than that of both the conventional gold and silver. Therefore, our proposed biosensor is expected to provide a reliable non-invasive monitoring and screening of diabetes.

Acknowledgments

Authors wishing to acknowledge financial support from GA scheme of Universiti Teknologi PETRONAS, Malaysia.

References

[1] S. Wild, G. Roglic, A. Green, R. Sicree, and H. King, "Global prevalence of diabetes: estimates for the year 2000 and projections for 2030 Diabetes Care 27: pp 1047-53," 2004
[2] M. Righettoni and A. Tricoli, "Toward portable breath acetone analysis for diabetes detection," Journal of breath research, vol. **5**, p. 037109, 2011
[3] P. Songthung and K. Sripanidkulchai, "Improving type 2 diabetes mellitus risk prediction using classification," 13th *Int.* Joint *Conf.* on Computer Science and Software Engineering (JCSSE), 2016, pp 1-6
[4] B. Bhowmik, V. Manjuladevi, R. Gupta, and P. Bhattacharyya, "Highly selective low-temperature acetone sensor based on hierarchical 3-D TiO 2 nanoflowers," IEEE Sensors Journal, vol. **16**, pp 3488-95, 2016
[5] S. Das, A. Bag, R. Kumar, and D. Biswas, "Fast Response (7.6 s) Acetone Sensing by InGaN/GaN on Si (111) at 373 K," IEEE Electron Device Letters, vol. **38**, pp 383-86, 2017
[6] A. Rydosz, K. Wincza, and S. Gruszczynski, "Microsystem in LTCC technology for the detection of acetone in healthy and diabetes breath," in ANDESCON, 2016 IEEE, 2016, pp 1-4
[7] A. Nooke, "Gas Detection by Means of Surface Plasmon Resonance Enhanced Ellipsometry," Dissertation, Berlin, Technische Universität Berlin, 2012
[8] B. Lakard, S. Carquigny, O. Segut, T. Patois, and S. Lakard, "Gas sensors based on electrodeposited polymers," Metals, vol. **5**, pp 1371-86, 2015
[9] S. Karthikeyan, H. M. Pandya, M. Sharma, and K. Gopal, "Gas sensors-a review," J. Environ. Nanotechnol, vol. **4**, pp 01-14, 2015
[10] Z. Jiang, R. Zhao, B. Sun, G. Nie, H. Ji, J. Lei, et al., "Highly sensitive acetone sensor based on Eu-doped SnO 2 electrospun nanofibers," Ceramics International, vol. **42**, pp 15881-888, 2016
[11] T. Hyodo, T. Kaino, T. Ueda, K. Izawa, and Y. Shimizu, "Acetone-Sensing Properties of WO3-Based Gas Sensors Operated in Dynamic Temperature Modulation Mode—Effects of Loading of Noble Metal and/or NiO onto WO3—," Sensors and Materials, vol. **28**, pp 1179-89, 2016
[12] G. W. Hunter, R. A. Dweik, D. B. Makel, C. C. Grigsby, R. S. Mayes, and C. E. Davis, "Portable Breath Monitoring: A New Frontier in Personalized Health Care," The Electrochemical Society Interface, vol. **25**, pp 63-67, 2016
[13] K. Toda, R. Furue, and S. Hayami, "Recent progress in applications of graphene oxide for gas sensing: A review," Analytica chimica acta, vol. **878**, pp 43-53, 2015
[14] J. Lou, Y. Wang, and L. Tong, "Microfiber optical sensors: A review," Sensors, vol. **14**, pp 5823-44, 2014
[15] Q. Zhang and D. Wang, "Room temperature acetone sensor based on nanostructured K 2 W 7 O 22," in SENSORS, 2016 IEEE, 2016, pp 1-3
[16] A. Biacore, "Biacore sensor surface handbook," GE Healthcare Bio-Sciences AB: Uppsala, Sweden, 2003
[17] G. Healthcare, "Biacore assay handbook," Bio-Sciences AB, Uppsala, Sweden, 2012
[18] L. Kapinos, R. Schoch, and R. Y. Lim, "Surface Plasmon Resonance (SPR)."
[19] M. Lesňák, F. Staněk, I. Hlavatý, J. Pištor, and J. Procházka, "SPR Method and Its Utilisation for Low Alcohols Concentrations Determination," Journal of Modern Physics, vol. **6**, p 363, 2015
[20] V. Perumal and U. Hashim, "Advances in biosensors: Principle, architecture and applications," Journal of Applied Biomedicine, vol. **12**, pp 1-15, 2014
[21] A. P. Turner, "Biosensors: sense and sensibility," Chemical Society Reviews, vol. **42**, pp 3184-96, 2013
[22] M. Esmaeili, G. Kiani, F. Shahriari Nogorani, and S. Boroomand, "Acetone sensing properties of hierarchical WO3 core-shell microspheres in comparison with commercial nanoparticles,"

International Journal of Nano Dimension, vol. **7**, pp 254-62, 2016

[23] A. Fioravanti, S. Morandi, and M. Carotta, "Chemoresistive gas sensors for sub-ppm acetone detection," Procedia Engineering, vol. **168**, pp 485-88, 2016

[24] J. Fraden, Handbook of modern sensors vol. **3**: Springer, 2010

[25] I. Fratoddi, I. Venditti, C. Cametti, and M. V. Russo, "Chemiresistive polyaniline-based gas sensors: A mini review," Sensors and Actuators B: Chemical, vol. **220**, pp 534-48, 2015

[26] V. Galstyan, E. Comini, I. Kholmanov, A. Ponzoni, V. Sberveglieri, N. Poli, et al., "A composite structure based on reduced graphene oxide and metal oxide nanomaterials for chemical sensors," Beilstein Journal of Nanotechnology, vol. **7**, pp 1421-27, 2016

[27] H. Jamalabadi and N. Alizadeh, "Enhanced Low-Temperature Response of PPy-WO 3 Hybrid Nanocomposite Based Gas Sensor Deposited by Electrospinning Method For Selective and Sensitive Acetone Detection," IEEE Sensors Journal, vol. **17**, pp 2322-28, 2017

[28] J. L. Santos and F. Farahi, Handbook of optical sensors: Crc Press, 2014

[29] Y. G. Song, Y.-S. Shim, S. Kim, S. D. Han, H. G. Moon, M. S. Noh, et al., "Downsizing gas sensors based on semiconducting metal oxide: Effects of electrodes on gas sensing properties," Sensors and Actuators B: Chemical, vol. **248**, pp 949-56, 2017

[30] Y. Wang, F. Liu, Q. Yang, Y. Gao, P. Sun, T. Zhang, et al., "Mesoporous ZnFe 2 O 4 prepared through hard template and its acetone sensing properties," Materials Letters, vol. **183**, pp 378-81, 2016

[31] A. Simo, K. Kaviyarasu, B. Mwakikunga, M. Mokwena, and M. Maaza, "Room temperature volatile organic compound gas sensor based on vanadium oxide 1-dimension nanoparticles," Ceramics International, vol. **43**, pp 1347-53, 2017

[32] R.-C. Wang, S.-N. Lin, and J.-y. Liu, "Li/Na-doped CuO nanowires and nanobelts: Enhanced electrical properties and gas detection at room temperature," Journal of Alloys and Compounds, vol. **696**, pp 79-85, 2017

[33] H. Bai and G. Shi, "Gas sensors based on conducting polymers," Sensors, vol. **7**, pp 267-307, 2007

[34] G. Gauglitz, "Direct optical sensors: principles and selected applications," Analytical and bioanalytical chemistry, vol. **381**, pp 141-55, 2005

[35] V. V. Tuchin, Handbook of optical sensing of glucose in biological fluids and tissues: CRC press, 2008

[36] X. Chen, C. K. Wong, C. A. Yuan, and G. Zhang, "Nanowire-based gas sensors," Sensors and Actuators B: Chemical, vol. **177**, pp 178-95, 2013

[37] R. Tabassum, S. K. Mishra, and B. D. Gupta, "Surface plasmon resonance-based fiber optic hydrogen sulphide gas sensor utilizing Cu–ZnO thin films," Physical Chemistry Chemical Physics, vol. **15**, pp 11868-874, 2013

[38] D. Ahuja and D. Parande, "Optical sensors and their applications," Journal of Scientific Research and Reviews, vol. **1**, pp 060-068, 2012

[39] J. Homola, I. Koudela, and S. S. Yee, "Surface plasmon resonance sensors based on diffraction gratings and prism couplers: sensitivity comparison," Sensors and Actuators B: Chemical, vol. **54**, pp 16-24, 1999

[40] K. Lee, S. Cho, S. H. Park, A. Heeger, L. Chan-Woo, and S.-H. Lee, "Metallic transport in polyaniline," Nature, vol. **441**, p 65, 2006

[41] M. N. Polyanskiy, "Refractive index database," https://refractiveindex.info. Accessed on 2017-10-03

IOP Conf. Series: Materials Science and Engineering **383** (2018) 012025 doi:10.1088/1757-899X/383/1/012025

Seismic responses comparison of RC building in consideration of plastic hinge models and properties

D R Teruna[1,*] **and H Wijaya**[2]

[1]Department of Civil Engineering, University of Sumatera Utara, Jln. DR. Mansyur Kampus USU Medan 20155, Indonesia
[2]Department of Civil and Construction Engineering, Swinburne University of Technology, Australia

*E-mail: danielteruna@usu.ac.id

Abstract. Assessment of reinforced concrete building capacity under seismic excitation has been recognized to be highly important for the past decades especially in Indonesia. Previous studies showed failures of RC building occurred in columns with bad confinement. The confinement in RC column has significant role in providing ductility and shear capacity for seismic loads. For this aim, a five-story RC building having different confinement on column members are investigated under seismic excitation. Three set of ground motions are selected and matched with target response spectrum of Indonesian code. Nonlinear analysis is performed by utilizing OpenSEES finite element software. To account the inelastic behaviour of beam and column elements, two types of plastic hinge are applied for both elements, namely finite length plastic hinges (plastic zone) and distributed hinges along the length of elements. The results show that there is a noticeable difference in seismic responses in correlation with the building model.

1. Introduction

Reinforced concrete has been popular and widely used for most of the building construction in Indonesia. Severe seismic events for the past years have increased the demand of earthquake resistant building system. The earthquake resistant building depends on its capacity to dissipate the seismic energy whilst the structure behaves nonlinearly. Therefore, nonlinear procedures of analysis are developed to assess and evaluate the nonlinear responses of the structural system as described in FEMA 356 [1]. However, the structural nonlinear responses are highly sensitive and require the accuracy of element nonlinear properties to capture the behaviour of the structural system. There are several nonlinear modeling techniques used for simulating structural nonlinear response including distributed plasticity and concentrated plasticity. Inel and Ozmen [2] evaluated the effect of plastic hinges properties in reinforced concrete building. Two proposed plastic hinge length expression were considered. The variation in displacement capacity was observed for different modelling techniques in relation to structural ductility. However, the analysis was carried based on non-linear static procedure and non-linear dynamic response was not evaluated. Botez *et al.* [3] studied the effect of plastic hinge model and distributed plasticity with non-linear progressive collapse analysis. Variation in analysis result was observed. However, the ductility in terms of column confinement was not considered.

This paper investigates the seismic nonlinear model that had been proposed and evaluate the differences in analysis result. Moreover, the requirement of structural ductility in order to keep its

Content from this work may be used under the terms of the Creative Commons Attribution 3.0 licence. Any further distribution of this work must maintain attribution to the author(s) and the title of the work, journal citation and DOI.
Published under licence by IOP Publishing Ltd

stability before collapsing is also considered in this study. The important design specification required for earthquake is the confinement in RC columns for providing ductility. Haraji [4] have studied the effect of confinement enhancement in RC column under seismic loadings. Recent study of Yuen *et al.* [5] shows the increasing ductility of reinforced concrete column by providing sufficient amount of confinement. Therefore, this study also evaluates the effect of confinement in concrete columns on seismic response. Pushover and time-history analysis are used to investigate the structure's lateral capacity and response under seismic excitation.

2. Concrete model

Many concrete stress-strain models have been proposed for the past years. Mander *et al.* [6] proposed the concrete model for both confined and unconfined with consideration of increasing ductility and strength in RC members. The concrete is modeled as unconfined concrete and confined concrete for its cover and core respectively. This study evaluates the confinement variability of concrete column in relation to seismic non-linear responses in RC building. Six-confinement configurations were selected and investigated. The core concrete parameters were developed using proposed equation by Mander *et al.* [6] The models are described in building model section.

3. Element model

The present design codes and guidelines has required the consideration of structural plastic behaviour in determining structural responses. The popular simplified approach in evaluating structure's nonlinear capacity is pushover analysis. However, nonlinear model has to be applied which consists of concentrated hinge, distributed plasticity and finite length hinge fibre-based model. This paper evaluates the variability of structural responses for variability of confinement configuration with different nonlinear fibre-based model approaches.

Fibre-based model has been widely used to simulate the nonlinearity of structural member behaviour by using finite fibre elements to capture the responses of the element. Study conducted by Chiara and Rui [7] showed that fibre model managed to simulate the propagation of nonlinearity behaviour along the cross section of the element. The distributed plasticity model predicts the nonlinear behaviour of the element, taken the nonlinear geometry and material into account, with Gauss integration points along the length of the element. The distributed plasticity fibre model is represented in figure 1.

The other modelling approach is finite length hinges approach with the principle that the plastic behavior occurs at the defined region while the rest of the element remain remains elastic. The length of plastic hinges determines the moment-rotation relationship of a particular member. As the length of plastic hinges is defined, the displacement at the tip of the column can be computed by integrating member's curvatures. Figure 2 represents the definition of plastic hinge length for cantilever member [8].

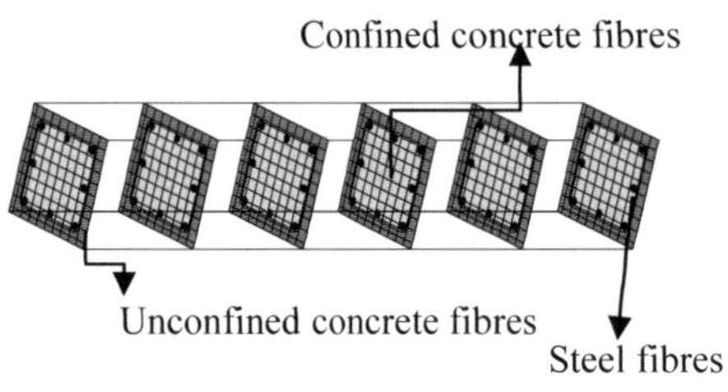

Figure 1. Fibre model for reinforced. concrete member.

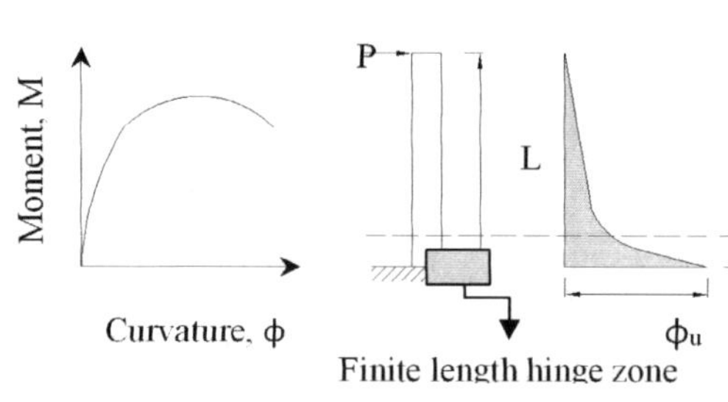

Figure 2. Plastic hinge length definition.

In this study, three proposed evaluation for determining the length of plastic hinges in concrete structure is investigated. Proposed equation by Park *et al.* [9] and Paulay and Priestley [10] are evaluated and expressed below:

$$L_p = 0.5H \tag{1}$$

$$L_p = 0.08L + 6d_b \tag{2}$$

$$L_p = 0.08L + 0.022d_b f_y \tag{3}$$

where L*p* is Length of plastic hinge, H is the height of member cross section, d*b* is the diameter of longitudinal reinforcement, and f*y* defines the yield strength of steel reinforcement.

The OpenSEES [11] finite element software is used for distributed plasticity model in this study and finite length hinge with different length of plastic hinges considered. Furthermore, the shear response in concrete column was modelled by a uniaxial hysteretic law. The section aggregator was used in OpenSEES to construct the coupling of axial/bending interaction with shear. The adopted shear-deformation law was degrading hysteretic with two hysteretic law (one degrading with ductility parameter of 0.03 and one non-degrading) in parallel. Figure 3 illustrates the adopted law with characteristic point at cracking, peak and residual strength. The shear strength, as proposed by Sezen and Mohele [12], it is expressed as follows:

$$V_{peak} = V_c(N) + V_s \tag{4}$$

where $Vc(N)$ is the concrete shear strength with contribution of axial load and V_s is the shear strength provided by ligatures. In this study, the cracking strain is computed with elastic uncracked behaviour properties. The degradation starts after the peak strength in this adopted model.

$$\gamma_{crack} = V_{crack} / GA^* \tag{5}$$

Following the study by Sezen [13], estimated the shear displacement where axial collapse occured. The peak strain is calculated as the function of steel yield stress f_y, normalised axial force v, longitudinal reinforcement ratio ρ, and shear span Lv over section height H.

$$\gamma_{peak} = \frac{f_y \rho}{5000\sqrt{v L_v / H}} - 0.0004 \tag{6}$$

Finally, the ultimate strain for the model is expressed as the ductility factor from the peak strain and shown in equation (7).

$$\gamma_{res} = \left[4 - 12V_{peak} / (A_c f_c)\right]\gamma_{peak} \tag{7}$$

4. Earthquake ground motions

Three set of ground motion records are selected and matched with target response spectrum of Banda Aceh earthquake with site class D. The records include Kobe, Loma Petra, and Managua records. Figure 4 depicts the acceleration spectrum for matched four records and target spectral acceleration.

5. Building description

Five-story reinforced concrete building is evaluated in this study. The building frame considered is interior frame and subjected to 24 kN/m dead load and 10 kN/m live load for floor and 5 kN/m on the roof designed to resist gravity load and earthquake load based on SNI-1726-2012 for Indonesian Earthquake Code with the location is in Aceh. The base support is assumed to have fixed connection for the purpose of this study. The geometry of the building is shown in figure 5.

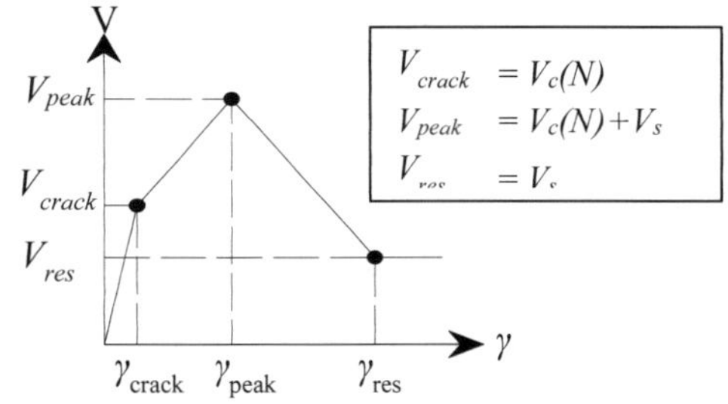

Figure 3. Monotonic shear envelope model of RC member.

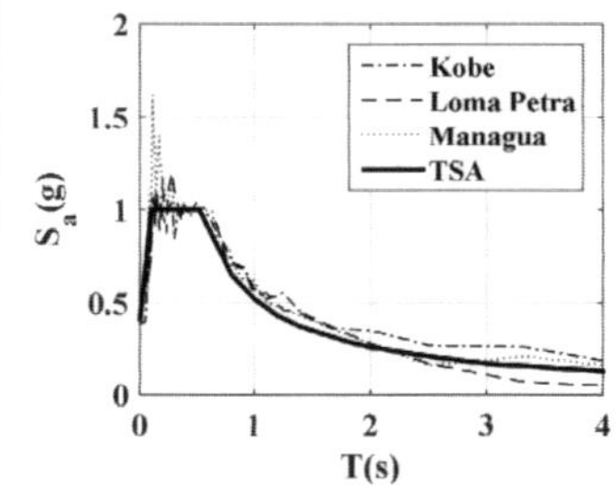

Figure 4. Response spectrum of matched records.

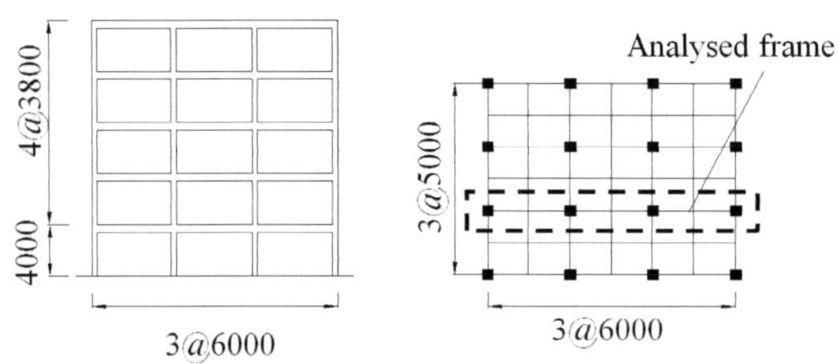

Figure 5. Building Geometry.

The dimension of column member is 500x500 mm for interior and exterior column along the height of the building and all beams are designed with 300x500 mm cross-section in addition with D10-150 shear reinforcement. The first to third story columns have 12 numbers of 22 diameter longitudinal reinforcement while for the fourth and fifth story, 8D22 is assigned. The beam reinforcement for the first and third story is typical with 5D19 and 3D19 for top and bottom reinforcement respectively. The second story beam has 6D19 for the top reinforcement instead, while for the fourth and fifth story beam, 4D19 is used for top reinforcement.

6. Analytical model

The building with 2 dimensional model is established in OpenSEES finite element software [11]. The material properties for concrete chosen is Kent-Scott-Park with degrading stiffness properties and no tensile strength. Reinforcement bar material was modelled with bilinear uniaxial material with kinematic strain hardening which is available as Steel01 in OpenSEES. The concrete material has compressive strength of 20 MPa and the steel material has yield strength of 400 MPa. Fibre model is applied for both distributed plasticity and concentrated hinges non-linear model.

The concrete fibre model comprises of concrete cover, concrete core, and reinforcement steel. The concrete cover is constructed using unconfined concrete material properties while concrete core is assigned with confined concrete properties. Flexibility based-element with 5 Gauss-Lobatto integration points is assigned for both beam and column in distributed plasticity model. Moreover, nonlinear geometry P-Δ and corotational is activated for column and beam section respectively. Furthermore, Beam and column with hinges element is assigned for finite length hinges building model. Figure 6 depicts the types of column confinement considered in this study. Core concrete uniaxial material model is computed for each confinement type and presented as follows.

IOP Conf. Series: Materials Science and Engineering **383** (2018) 012025 doi:10.1088/1757-899X/383/1/012025

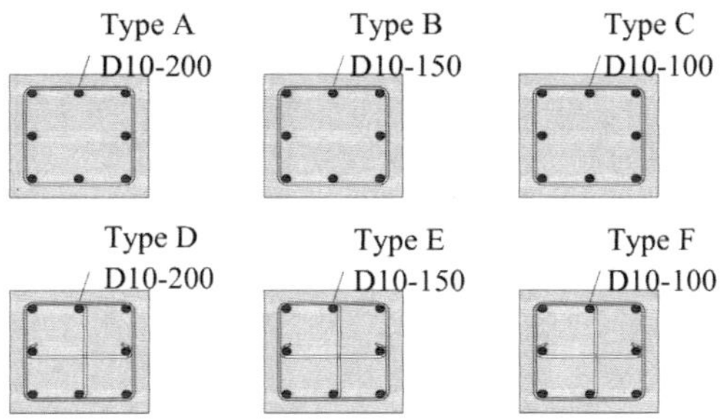

Figure 6. Column confinement model.

The building model with finite length hinges following expression as described in previous section for determining its length of plastic hinges is expressed as CH1, CH2, and CH3 respectively while distributed plasticity model is labelled as DP. The building is subjected to deadload and liveload in a factor of 1.0 and 0.3 respectively before non-linear static pushover and time-history analysis is conducted. The floor mass is computed from the gravity load applied on the structure.

7. Result and discussion

7.1. Capacity curve

Pushover analysis was performed for each building model with inverted triangle lateral load distribution corresponding to first mode period of the structure. The results of analysis for considered building models in terms of base shear versus roof displacement are presented in figure 7. The variability of confinement shows unsignificant contribution in terms of structure lateral capacity. However, the increase in ductility has been noticed fron observation of the capacity curves. The confinements with 3 legs has better ductility performance with less degradation in stiffness after peak lateral strength. Furthermore, the effect of hinge model in structure capacity curve is also compared and shown in figure 8.

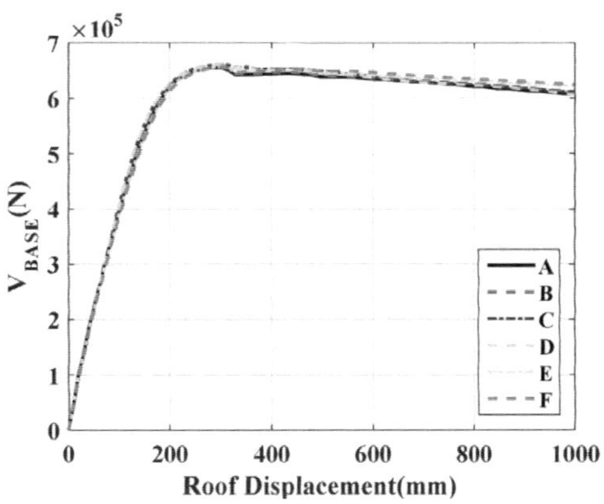

Figure 7. Capacity curve comparison for different confinement.

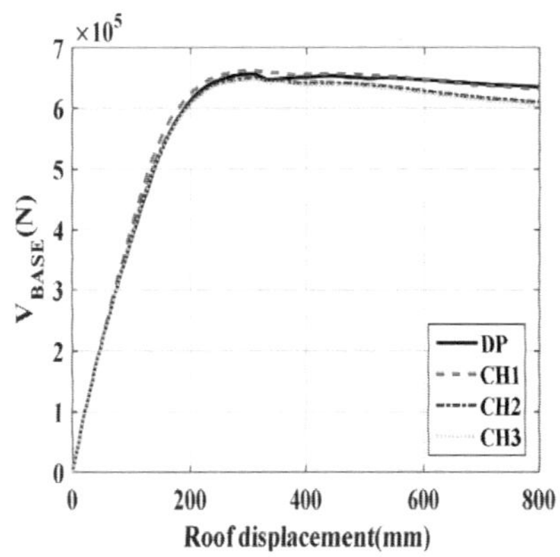

Figure 8. Capacity curve for different hinge length parameter.

7.2. Nonlinear time-history analysis

In order to study the structural behaviour considering different types of confinement and analysis model, nonlinear time-history analysis is conducted under three different excitations matched with target spectrum design as described in previous section. Figure 9 to 11 illustrated the inter-story drift profile of the building under considered ground motions and confinements.

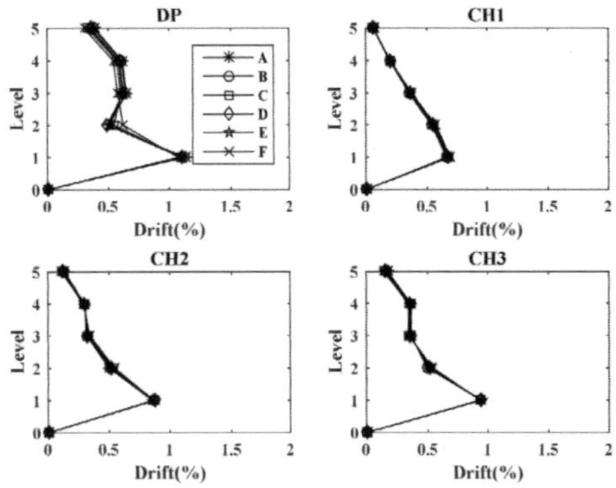

Figure 9. Inter-story drift profile of Kobe earthquake.

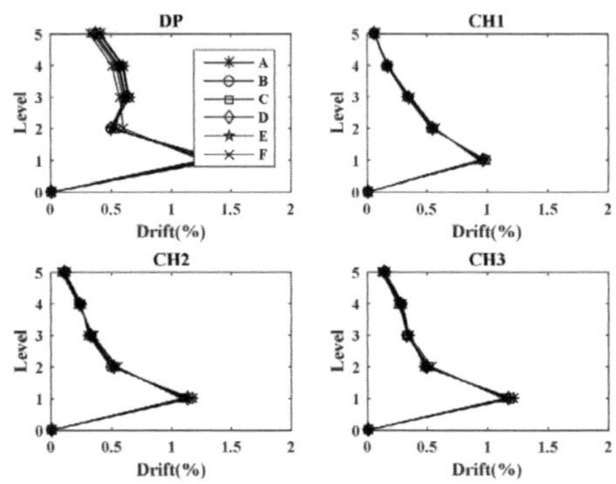

Figure 10. Inter-story drift profile of Loma Petra earthquake.

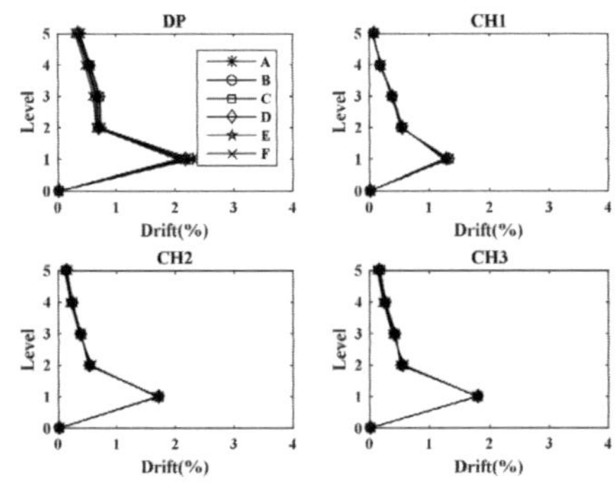

Figure 11. Inter-story drift profile of Managua earthquake.

From three ground motions considered in this study, CH1 model shows different tren of drift profile with the other model. Moreover, model CH3 has better correlation profile with DP model. The confinement variability shows no different in terms of drift profile however, slight less drift is observed in DP model by providing more confinement to the column. It may caused by higher dissipated energy as providing more confinement increases the degradation or hardening behaviour of the concrete elements. The structure has performance level between IO (Immediate-Occupancy) and LS (Life-Safety) with maximum drift in between 1% to 2%. The highest response occurs in Managua earthquake.

8. Summary and conclusion

Five-story RC building with variability of column confinement is investigated through severe nonlinear analysis model including distributed plasticity and finite length hinges with fibre model. Pushover analysis is performed to assess the structure's lateral capacity and ductility. Moreover, nonlinear time-history analysis is also conducted to simulate the building nonlinear behaviour under four ground motions which are matched with Aceh earthquake response spectrum. Based on analysis result and observation, several conclusion can be drawn as follows:

1. Ground motions matched with target response spectrum provides different responses of the structural system. As observed from the drift profile, the building is considered Immediate-Occupancy for Kobe and Loma Petra earthquake while it is life-safety for Managua earthquake.
2. The confinement spacing determines the ductility of the structure based on pushover analysis. Less degradation after peak strength is observed by providing more confinement in the column.

3. The seismic nonlinear models of finite length hinges CH2 and CH3 show good correlation with distributed plasticity model while CH1 gives inaccurate result for simulation of building nonlinear responses.

4. The inter-story drift profile for nonlinear seismic model CH2,CH3 and DP shows similar trend under three ground motions considered.

References

[1] FEMA, P. (2000). Commentary for the seismic rehabilitation of buildings. *FEMA-356, Federal Emergency Management Agency, Washington, DC.*

[2] Inel, M., & Ozmen, H. B. (2006). Effects of plastic hinge properties in nonlinear analysis of reinforced concrete buildings. *Engineering structures*, 28 (11) pp 1494-1502.

[3] Botez, M. D., Bredean, L. A., & Ioani, A. M. (2014). Plastic hinge vs. distributed plasticity in the progressive collapse analysis. *Acta Technica Napocensis: Civil Engineering & Architecture*, 57 (1) pp 24-36.

[4] Harajli, M. H., & Hantouche, E. G. (2014). Effect of active versus passive confinement on seismic response of wide RC columns with lap splices. *Journal of Structural Engineering*, 141 (9), 04014221.

[5] Yuen, T. Y., Kuang, J. S., & Ho, D. Y. (2017). Ductility design of RC columns. Part 2: extent of critical zone and confinement reinforcement. *HKIE Transactions*, 24 (1) pp 42-53

[6] Mander, J. B., Priestley, M. J., & Park, R. (1988). Theoretical stress-strain model for confined concrete. *Journal of structural engineering*, 114 (8) pp 1804-26

[7] Chiara, C., & Rui, P. (2006). Seismic response of continuous span bridges through fibre-based finite element analysis. *Earthquake Engineering and Engineering Vibration*, 5 (1), 119-31.

[8] Park, R., & Paulay, T. (1975). *Reinforced concrete structures*. John Wiley & Sons

[9] Park, R., Priestley, M. J., & Gill, W. D. (1982). Ductility of square-confined concrete columns. *Journal of the structural division*, 108 (4) pp 929-50

[10] Paulay, T., & Priestley, M. N. (1992). Seismic design of reinforced concrete and masonry buildings. John Wiley and Sons, New York, p. 767

[11] McKenna, F., Fenves, G. L., & Scott, M. H. (2000). Open system for earthquake engineering simulation. *University of California, Berkeley, CA*

[12] Sezen, H., & Moehle, J. P. (2004). Shear strength model for lightly reinforced concrete columns. *Journal of Structural Engineering*, 130 (11) pp 1692-1703

[13] Sezen, H. (2008). Shear deformation model for reinforced concrete columns. *Structural Engineering and Mechanics*, 28 (1), 39-52

Optimization of fuel consumption for municipal solid waste collection in Al Ain city, UAE

M A Maraqa[1,2*], E Z Swar Aldahab[3], M Ghanma[1] and S K Al Kaabi[4]

[1] Civil and Environmental Engineering Department, UAE University, Al Ain, UAE.
[2] Roadway, Transportation and Traffic Safety Research Center, UAE University, Al Ain, UAE.
[3] Environmental Science Mater Program, UAE University, Al Ain, UAE.
[4] Deputy General Manager, Center of Waste Management- Abu Dhabi, UAE.

*E-mail: m.maraqa@uaeu.ac.ae

Abstract. Fuel consumed in collecting municipal solid waste (MSW) accounts for a major portion of the cost of waste management. In this study, several cases were developed using the ArcGIS Network Analyst to establish optimum conditions for MSW collection in Um Gafa district in Al Ain city, UAE. A GIS model was created based on data collection and GPS tracking of collection route and bins position in the area. Results show that current waste collection at Um Gafa does not strictly follow U-turn and curb approach policies. When route optimization is applied for similar conditions, a saving of 14.3% in fuel consumption is gained. Two models were tested for optimal number and location of bins. One model was based on a 40-m service zone while the other was based on population density and land use. By adopting model 1, the number of bins was reduced by 12%, while in model 2 the number of bins was reduced by 20%. Model 2 showed higher superiority, compared to model 1, in terms of fuel consumption.

1. Introduction

Waste collection is the most fuel intensive step in waste management. It accounts for 40 to 60% of the allocated budget for municipal solid waste (MSW) management [1]. Thus, implementation of vehicle routing techniques for MSW collection is necessary for effective decision making. In general, two different approaches have been reported for optimization of MSW collection including the use of operational research [2-13] and the use of geographic information systems (GIS) [14-19].

Generally, models that are based on the operational research methodology do not provide the flexibility and responsiveness needed in real time logistic problems and thus are difficult to apply in practice. Moreover, most of these models do not account for heterogeneous vehicle fleets or maximum loading capacity of trucks. However, the GIS approach is capable of effective handling, display and manipulation of geographic and spatial information as demonstrated by several investigators [14, 16, 19].

The United Arab Emirates (UAE) is confronted with enormous solid waste management challenges that could strain municipal financial resources and the handling of the ever increasing waste volumes. Nonetheless, the government is working hard to reduce the waste generation rate, and started privatizing solid waste management in order to meet the waste collection and disposal demand.

Content from this work may be used under the terms of the Creative Commons Attribution 3.0 licence. Any further distribution of this work must maintain attribution to the author(s) and the title of the work, journal citation and DOI.
Published under licence by IOP Publishing Ltd

However, limited effort has been directed towards assessing the cost of waste services and increasing their cost effectiveness. Particularly, no work has been done to optimize fuel consumption during MSW collection. Thus, the main objective of this study was to develop a GIS-based model to calculate fuel consumption of vehicles that collect MSW. The model was then used to explore optimum conditions for waste collection in Um Gafa area in the city of Al Ain. Results of this study can be used to improve the efficiency of the waste management system in Al Ain City, thereby reducing the cost of waste collection which could ultimately results in environmental benefits.

2. Methodology

The approach followed in this work consisted mainly of three main steps. First, develop a GIS-based model to explore different scenarios to arrive at the optimal route for waste collection in the study area. Second, calculate fuel consumption and emitted greenhouse gases under the current conditions and for the explored scenarios without altering waste bin numbers or locations. Third, investigate the adequacy of the number and positions of existing collection bins and conduct route optimization of the proposed bins location. Before proceeding with the details of the methodology, a brief description of the study area is introduced.

2.1. Site description

Al Ain City covers an area of approximately 13,100 km^2, with a population of 650,221 as of 2015 [20]. The topography of Al Ain is nearly flat with the exception of the area of Hafeet Mountain. Management of MSW in the city is currently under the control of the Center of Waste Management – Abu Dhabi (CWM-AD). From a waste collection perspective, the city is divided into the north and south zones. Collection of waste from each zone is carried out by two private companies (MBM Dalla and Lavajet). Collected MSW is sent to a sorting station at Seeh Al Hemmah, nearby the compost plant. The station also receives wastes from the transfer stations in the peripheral townships. Organic-rich materials are diverted to the compost plant, recyclable wastes are collected and stored for further marketing, and the remaining unwanted materials are transported to a central single-lined landfill located about 18 km west of the transfer station.

Studies by the CWM-AD [21] estimated that Al Ain City produced between 1.4 to 1.7 kg/cap/day of solid waste between 2011 and 2015. The total amount of waste produced in 2015 in the city was about 2.5 million tons. Non-hazardous waste formed 99.6% of the total produced waste. Out of the total non-hazardous waste generated, MSW totaled 327,627 tons, 59% of which was disposed in a sanitary landfill while 17% was composted, 16% was recycled, and 8% was disposed in some dumpsites.

To carry on this study, an area within Al Ain City called Um Gafa was selected. Um Gafa is located in the south zone of Al Ain (figure 1) and covers an area of approximately 44.6 km^2, with a population of 6,772 in 2015 [20]. Um Gafa area includes one main road with a median that is central to the overall district. The main road has a dual 2-lane, each 3.7-m wide with a posted speed of 80 km/hr. Other roads are local streets with single lane (two-way) and a posted speed of 40 km/hr. Most of Um Gafa streets are not congested with traffic. In addition, the area includes 6 roundabouts and has no traffic signals. The area contains 617 large houses and 32 buildings, each consisting of 4 apartments. All the houses are adjacent to each other. Twenty-one of the apartment buildings are adjacent to each other, while the 11 others are isolated. The area also contains 21 commercial units, 4 schools, one garden and 23 farms. It should be mentioned that Um Gafa area has been selected as a case study as it is one of largest districts in Al Ain city with a high number of bins. During data collection, it has been found that Um Gafa has the least detected map errors among other areas in Al Ain.

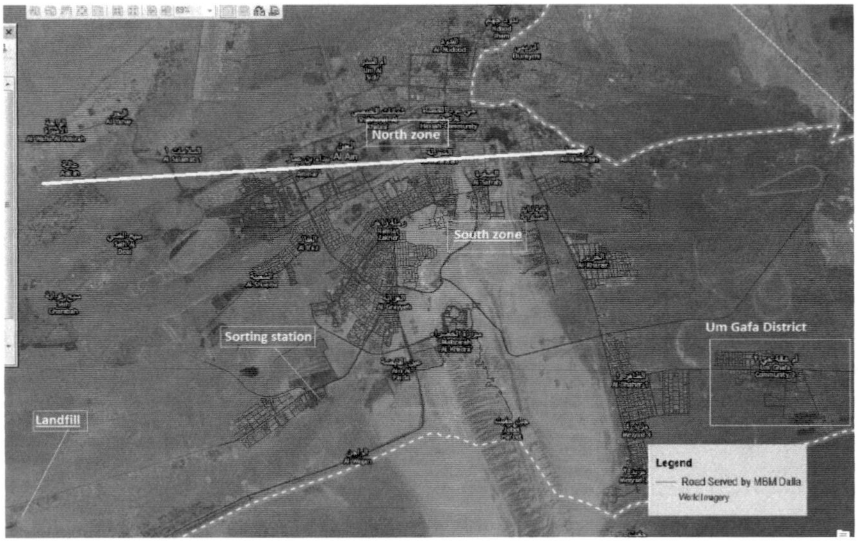

Figure 1. Location of Um Gafa area and waste management facilities in Al Ain City.

Collection of MSW in the Um Gafa is carried out by MBM Dalla on behalf of CWM-AD. MSW collection service in Um Gafa follows a communal collection system, by which waste is carried out manually and/or mechanically. The total number of waste bins in the area is 661 each of 1.3 m^3. The existing bins are not evenly distributed within the area and no previous study has been done to optimize the location of the bins. The amount of waste collected during a trip is around 16.8 tons per day. The Um Gafa area is swept by an 18-m^3 loaded compaction truck operated using diesel fuel. The collection is made by a team of 3 persons; the driver and two collectors in the truck. Besides, laborers carrying hand cart ensure the collection of waste from hardly accessible places. The waste is then transferred to the municipal sorting station directly.

2.2. Data collection and model development

In this study, a GIS-based model using ArcGIS Network Analyst tool was developed in order to establish optimum conditions for MSW collection in Um Gafa area, with an objective of minimizing fuel consumption. The GIS model was created based on data collection and GPS tracking of collection route and bins position. The ArcGIS Network Analyst tool provides network-based spatial analysis including routing, travel directions, closest facility, and service area analysis. It enables users to dynamically model realistic network conditions, including turn restrictions, speed limits, height restrictions, and traffic conditions [22]. The tool uses the Dijkstra's Algorithm [22] in order to solve the routing problem.

In developing the GIS-based model, spatial data were created from existing maps by manual digitizing. For this purpose, a shape file of Al Ain map was collected from Al Ain Town Planning Department including attributes data (area boundary, name and type of roads and their width and traffic volume details). Errors detected in the collected map (i.e., duplicate objects, undershoots and zero length objects) have been corrected manually. U-turn and curb approach polices have been collected from Al Ain Transportation Police Department. The bins locations have been collected manually using GARMIN eTrexH GPS. The current run route has been tracked by going on trip with waste vehicle's staff during their duty. The average speeds, acceleration, deceleration of the vehicle, and the average service time of the bins have been collected during the trip. The information was integrated to form a database of the case study. Data that were added manually to the generated maps include one-way and two-way roads, U-turn polices, curb approach polices, driving time, and service time of the bins.

IOP Conf. Series: Materials Science and Engineering **383** (2018) 012026 doi:10.1088/1757-899X/383/1/012026

Different layers of drawing were created including route analysis layer which contains information related to curb approach and override parameters. Each route is composed of three feature layers: stops, barriers, and routes. The stops feature layer stores the network locations that are used as stops in route analysis. Barriers, on the other hand, are used in route analysis to denote points where a route can't traverse. The route feature layer stores the resultant route of a route analysis.

2.3. Explored cases

The developed GIS model was used to investigate several case scenarios with the objective of optimizing travel distance. Fuel consumption and emitted green-house gases (GHGs) under the current conditions and for different explored scenarios were then determined using SIDRA TRIP software [23]. Due to lack of microscopic GPS data of the study area, the quick scenario analysis incorporated in SIDRA TRIP was used. Vehicle profile for heavy vehicles reported by Akcelik and Associates [24] was used. This was done by specifying the vehicle type, travel distance, travel time, number of stops, stop duration, road grade (slope of the road), average speed, average acceleration and average deceleration.

During the tracking of the current waste collection route in Um Gafa area, some violations of traffic rules have been noticed such as U-turn and curb approach policies. According to Abu Dhabi Traffic Engineering Department, the U-turn policy dictates that waste collection vehicles are allowed to make U-turns only at junctions (junctions with three or more adjacent edges) and dead ends (junctions with exactly one adjacent edge). The curb approach policy dictates that the waste collection vehicles are allowed to collect bins only on the right-hand side of the vehicle, U-turns at the bins location is prohibited. Thus, two cases have been investigated without re-locating the waste bins or changing their numbers. In Case 1, route optimization was done with the implementation of traffic rules. In Case 2, route optimization without implementing U-turn and curb approach policy. Case 2 could be used to make a comparison with the current situation, where U-turn and curb approach policies are not adhered to.

Another two cases (Case 3 and 4) were explored with the objective of finding the adequate number and optimal location of bins that should be used in the area. In Case 3, route optimization was done with bins relocation based on a 40-m service zone. In Case 4, route optimization was done with bins re-location based on population density and land use, without violating the 40-m service zone. For Case 4, the land use in the district was categorized as residential, commercial, farms and public uses. Then, the amount of waste generated from each land use was estimated and the needed collection bins were determined. The bins where then allocated manually with reference to waste generation, population density and road network (intersections). For Case 3 and 4, route optimization was carried out with the implementation of the U-turn and curb approach policies.

3. Results and discussion

3.1. Current situation

Figure 2 illustrates the actual tracking of the journey made by waste collection truck in Um Gafa area. The amount of waste collected during a trip was around 16.8 tons per day. The distance traveled by the truck was 85.7 km and lasted 9.1 hours, including 2.1 hours of driving, 6.4 hours to load/unload bins, and 30 min break. The average time for collecting one bin as obtained from field was 36 sec, while the average speed of the truck was 40 km/hr. The fuel consumption was 32.8 L. The estimated emission of CO_2, CO, hydrocarbons (HC), and NOx by SIDRA TRIP were 44200, 425, 68, and 17 g, respectively.

3.2. Optimized cases without bins re-location

Table 1 summarizes the results of optimization of Case 1 and 2. The table also lists the parameter values of the current situation. Results showed that Case 1 and 2 lead to some saving in fuel consumption compared to that under current conditions because of reduced collection time and reduced distance. Consequently, emitted pollutants were reduced.

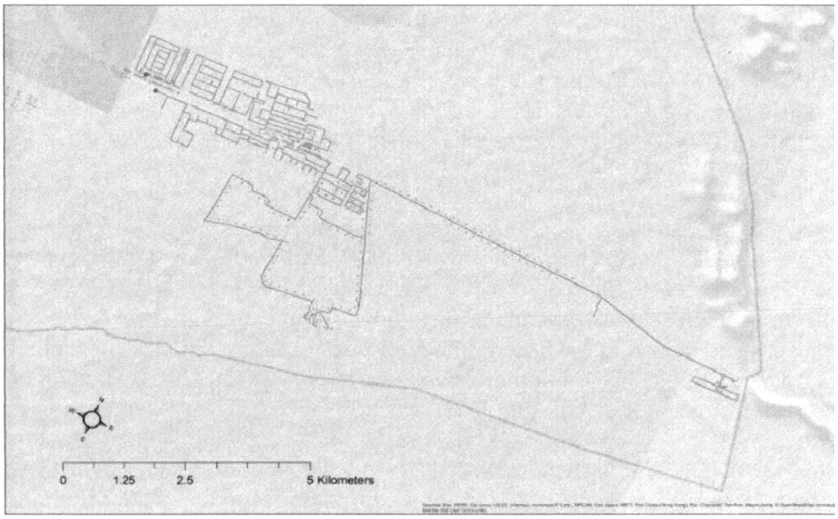

Figure 2. Current waste collection route in Um Gafa area.

The results of Case 2 (which was based on route optimization without implementing traffic rules) could be compared with those obtained for the current situation. For Case 2, the travel distance was reduced to 73.6 km (compared to 85.7 km under the current conditions). This resulted in fuel reduction of about 14.3% and a reduction of CO_2 emission by 7.2%. Case 1, however, should be the one adopted in case bins were not re-located because it involves adherence to traffic rules. Compared to the current situation, Case 1 results in fuel reduction of 5% and in CO_2 reduction of 3.6%.

Table 1. Results of the optimized cases without bins relocation.

Parameter	Current situation	Case 1	Case 2
Travel distance (km)	85.7	80.3	73.6
Travel time (hr)	9.1	8.9	8.5
Fuel consumption (L)	32.8	31.2	28.1
CO emissions (g)	425	418	410
CO_2 emissions (g)	44200	42600	41000
HC emissions (g)	68	67.2	65.6
NOx emissions (g)	17.9	17.4	16.1

3.3. Optimized cases with bins re-location

Currently, there are 661 collection bins in the study area, each with a capacity of 1.3 m^3. The bulk density of the waste in the bin is about 120 kg/m^3 and the daily quantity of waste collected in the area is 16.8 ton. Waste in the area is collected once a day. The total capacity of the bins is 859.3 m^3 which could accommodate 103 ton. Thus, the bins on the average are 16.5% full. Observations made during the site visit revealed that some bins were more than 50% full, while others were less than 10% full and some were even empty. According to Ahmed [25], the average filling of bins should range from 30% to 60%. Assuming a filling rate of 45%, the minimum number of bins that are required in the area at the present time can be calculated as:

$$N = \frac{W}{\rho_w \times S \times Fl \times CF} \quad (1)$$

where, N is the number of collection bins, W is the total quantity of waste generated per day (kg), ρ_w is the density of waste (kg/m^3), S is the size of bins (m^3), Fl is the average filling rate of the bin, and CF

is the collection frequency. Thus, the minimum number of bins in Um Gafa area would be 240, which constitutes 36% of what is provided in the area. Hence, there is a necessity to explore the possibility of reducing the number of bins on the road subject to the condition that the preferable walking distance of people to drop their MSW into the collection bin does not exceed 40 m [25]. Two models were developed for optimum waste bins location and number. In model 1, bins were located according to the requirement of a 40-m service zone (Case 3). In model 2, the number and size of bins were estimated based on population density and land use (Case 4).

In Case 3, collection bins were allocated manually to find the optimum number of bins assuming a 40-m service zone. Meanwhile, the bin size was changed from 1.3 to 3 m^3 near commercial areas and farms. By doing so, smaller bins serving commercial centers and farms were replaced by fewer, larger ones. This would positively influence routing optimization by reducing the number of stops. We found that the optimum number of bins was 572. It should be mentioned that priority to new bin locations was given to existing bins and cross-roads in order to facilitate social acceptance. Also every two bins of size 1.3 m^3 were replaced by one bin of size 3 m^3 near commercial areas and farms, noticing that, the average time required for collecting a bin of 3 m^3 is 44 sec [26]. After relocating the bins, optimization of vehicle routing was performed based on distance with the implementation of traffic rules. The results in table 2 show that the travel distance was shortened by 4% compared to Case 1 in table 1. The improvement was more emphatic in terms of the travel time, which was 8.1 hours including break time. Reductions in emissions were 7.4% for CO_2, 9.2% for CO, 9.5% for HC, and 12% for NOx.

Table 2. Results of the optimized cases with bins relocation.

Parameter	Case 3	Case 4
Travel distance (km)	80.1	78.8
Travel time (hr)	8.1	6.7
Fuel consumption (L)	29.6	27.1
CO emissions (g)	380	311
CO_2 emissions (g)	39500	38054
HC emissions (g)	60.8	57.4
NOx emissions (g)	15.2	13.8

In Case 4, optimization of the bins location was based on land use and waste generation. To do so, the generation rate per person in houses and apartments was assumed to be 2.33 and 1.85 kg/cap/d, respectively [27]. The number of residents in the house was taken as 10 and those in the apartment ranged between 4-5 persons with an average of 4.5 [20]. No data were available for waste generation rates from commercial, school and agricultural units in the area. As such, waste generation rates of commercial and school units were based on information provided by CalRecycle [28], those for gardens were based on information provided by NY CEQR [29], while those for agricultural units were based on a conducted site survey during this study. Meanwhile, the average number of employees in the existing commercial units was estimated to be 5 based on a site survey, while the area size of the garden and the farms was measured by ArcGIS. The number of students in the school (i.e., 900) represents the average among the 4 different schools in the area. As for the garden, the number of visitors (based on the sold tickets) fluctuated from 80-100 during a weekday but reached 350 during weekends. Thus, the assumption was made that there were 400 visitors of the garden each day. For the farms, the daily waste generation rate was taken as 40% based on observations made on site. The number and size of bins assigned for each land use were then determined based on (1) a filling rate of the bin that does not exceed 45% with a bulk density of 120 kg/m^3 and (2) a service zone of each bin that does not exceed 40 m. A summary of the results is listed in table 3 and the details of the entries are described below.

Based on an average filling rate of 45%, 205 bins of 1.3 m^3 were needed to serve the houses. However, based on a 40-m service zone, 392 bins were needed with each bin serving two adjacent houses. Thus, 392 bins (1.3 m^3 each) were used for houses. As for the apartments, 11 bins (1.3 m^3 each) were needed to serve the 11 isolated apartment buildings and 12 were needed to serve the 21 adjacent apartment buildings whereby two buildings were served by one bin. For commercial units, 3 bins (3 m^3 each) were needed to serve the 6 adjacent commercial units and 15 bins (1.3 m^3 each) were needed to serve the 15 isolated commercial units. All 4 schools are isolated from each other, so one bin (4 m^3) was needed to serve each school. Meanwhile, one bin (4 m^3 size) was needed to serve the garden. For the farm units, 8 bins (3 m^3 each) were needed to serve the 16 adjacent farm units and 7 bins (1.3 m^3 each) were needed to serve the 7 isolated ones.

Table 3. Generation rate at Um Gafa area and required number of waste bins.

Source	Number of units	Unit of measurement	Daily generation rate per unit (kg/d)	Daily quantity (kg)	Number of bins	Size of bins (m^3)
Large houses	617	kg/person/d	2.33×10	14,376	392	1.3
Apartments	128	kg/person/d	1.85×4.5	1,066	23	1.3
Commercial units	21	kg/employee/d	11.3×5	1,190	15 3	1.3 3
Schools	4	kg/student/d	0.5×900	1,800	4	4
Gardens	1	kg/visitor/d	0.45×400	180	1	4
Farms	23	kg/unit/d	0.4×1.3×120	1,435	7 8	1.3 3

It should be mentioned that the total estimated waste generation based on the entries in table 3 is 20,047 kg, which exceeds the 16,800 kg collected daily by 20%. This is due to some conservative assumptions made in the generation rates of the different units. If all the bins used were of 1.3 m^3 size, then the number of bins needed, if distributed based on population density and waste generation, would be 485. This number is about 20% less than those actually present on site. As indicated before, selection of other bin sizes (such as 3 m^3 or 4 m^3 bins), was made, whenever possible, in order to reduce the collection time.

Collection bins were allocated manually with reference to land use, population density and road network (intersections), but without referencing to the existing bin locations. Route optimization based on distance was then conducted as shown in figure 3. Results of the optimization (table 2) showed that the travel distance was 78.8 km shorter than that of Case 1 (80.3 km). Furthermore, a major reduction in working hours was achieved (6.7 hours including break time) as compared to 8.9 hours of Case 1. Fuel consumption was reduced to 27.1 L as opposed to 31.2 L of Case 1. Reductions in gaseous emissions were 10.6% for CO_2, 25% for CO, 14% for HC and 20% for NOx.

4. Conclusion

In this study, an optimization was developed using ArcGIS in order to improve the efficiency of collection and transportation of MSW in Um Gafa district in Al Ain City. Results show that with the use of route optimization a saving, relative to the current practice, of 14.3% in fuel consumption was gained. In addition, emitted gases were reduced by 7.2% for CO_2, 3.5% for CO, 3.5% for HC, 10.1% for NOx. Two models were developed for optimum waste bins distribution and number. In model 1, bins were located according to the requirement of a 40-m service zone. With this approach, 12% of the bins were decreased relative to the present ones. In model 2, the number and size of bins were estimated based on population density and land use. This has resulted in 20% less bins than those actually present on site. Results of the optimal route for waste collection using the two models demonstrate that the proposed models are significantly efficient in terms of travel distance and collection time, with consequent fuel consumption and gas emission savings. Furthermore, model 2 was found to be more efficient than model 1.

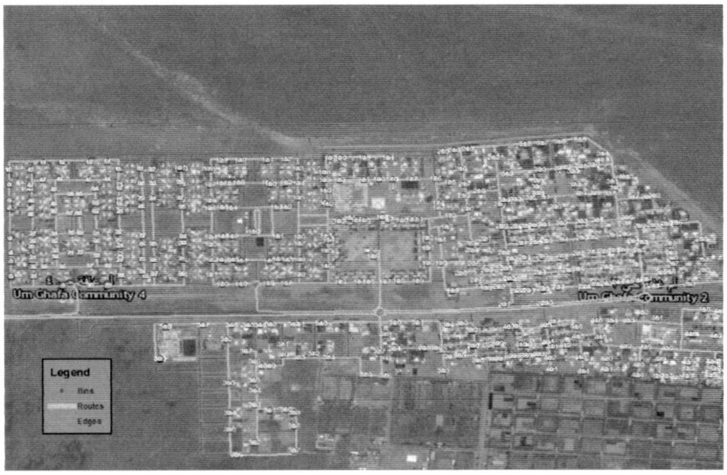

Figure 3. Optimal waste collection vehicle trip with bins located based on land use and population density.

Acknowledgement
This project was funded by the UAE University under grant 21N136. The authors are grateful to the CWM-AD for providing necessary information and facilitating the collection of field data.

References

[1] Bueno B, 2011. Networks and vehicle routing for municipal waste collection. *Networks* **4**, 65–94.
[2] Li J Q, Borenstein D and Mirchandani P B, 2008. Truck scheduling for solid waste collection in the City of Porto Alegre, Brazil. *Omega* 36, 1133–49.
[3] Solomon M M, 1987. Algorithms for the vehicle routing and scheduling problems with time window constraints, *Oper. Res.* 35, 254–65.
[4] Chang N-B and Wei Y L, 2002. Comparative study between heuristics algorithm and optimization technique for vehicle routing and scheduling in the solid waste management system. *Civ. Eng. Environ. Syst.* 19, 41-65.
[5] Teixeira J, Antunes A P and de Sousa J P, 2004. Recyclable waste collection planning-a case study. *Eur. J. Oper. Res.* 158, 543-54.
[6] Nuortio T, Kytojoki J, Niska H and Braysy O, 2006. Improved route planning and scheduling of waste collection and transport. *Expert Syst. Appl.* 30, 223–32.
[7] Vujic G, Gonzalez-Roof A, Stanisavljevic N, Ragossnig A M, 2015. Municipal solid waste development phases: Evidence from EU27. *Waste Manage. Res.* 33, 1112-20.
[8] Badran M F and El-Haggar S M, 2006. Optimization of municipal solid waste management in Port Said-Egypt. *Waste Manage.* 26, 534–45.
[9] Kim B-I, Kim S and Sahoo S, 2006. Waste collection vehicle routing problem with time windows, *Comput. Oper. Res.* 33, 3624–42.
[10] Benjamin A M and Beasley J E, 2010. Metaheuristics for the waste collection vehicle routing problem with time windows, driver rest period and multiple disposal facilities. *Comput. Oper. Res.* 37, 2270-80.
[11] Johansson O M, 2006. The effect of dynamic scheduling and routing in a solid waste

management system. *Waste Manage.* 26, 875- 85.

[12] Sonesson U, 2000, Modelling of waste collection – a general approach to calculate fuel consumption and time. *Waste Manage. Res.* 18, 115-23.

[13] Angelelli E and Speranza M G, 2002. The application of a vehicle routing model to a waste-collection problem: Two case studies. *The Journal of the Operational Research Society* 53, 944-52.

[14] Ghose M K, Dikshit A K and Sharma S K, 2006. A GIS based transportation model for solid waste disposal – A case study on Asansol municipality, *Waste Manage.* 26, 1287–93.

[15] Ericsson E, Larsson H and Brundell-Freij K, 2006. Optimizing route choice for lowest fuel consumption – Potential effects of a new driver support tool. *Transport. Res. C-Emer.* 14, 369–83.

[16] Chalkias C and Lasaridi K, 2009. A GIS based model for the optimisation of municipal solid waste collection: The case study of Nikea, Athens, Greece. *WSEAS Transactions on Environment and Development* 5, 640-50.

[17] Kashid S D, Nalawade D B, Nagne A D, Dhumal R K, Kale K V, 2015. Solid waste and waste bin management by using RS & GIS: A review, *International Journal of Scientific and Engineering Research* 6, 554-57.

[18] Tavares G, Zsigraiova Z, Semiao V and Carvalho M G, 2009. Optimisation of MSW collection routes for minimum fuel consumption using 3D GIS modelling, *Waste Manage.* 29, 1176–85.

[19] Nithya R, Velumani A and Senthil Kumar S R R, 2012. Optimal location and proximity distance of municipal solid waste collection bin using GIS: A case study of Coimbatore city. *WSEAS Transactions on Environment and Development* 8, 107-19.

[20] SCAD, 20015. Statistics Center- Abu Dhabi. https://www.scad.ae/ Accessed June 10, 2015.

[21] CWM-AD, 2015. Personal communication with Salem Al Kaabi, Deputy General Manager, Centre of Waste Management - Abu Dhabi.

[22] Karadimas N V, Doukas N, Kolokathi M and Defteraiou G, 2008. Routing optimization heuristics algorithms for urban solid waste transportation management, *WSEAS Transactions on Computers* 7, 2022-31.

[23] Ntziachristos L and Samaras Z, 2000. *COPERT III computer programme to calculate emissions from road transport- Methodology and emission factors* (version 2.1), Technical report No 49, European Environment Agency, Copenhagen.

[24] Akcelik and Associates, 2007. *SIDRA TRIP user guide*, version 1.0. Akcelik and Associates Pty Ltd, Melbourne, Australia.

[25] Ahmed S M, 2006. Using GIS in solid waste management planning: A case study for Aurangabad, India, MSc. Thesis, Linkopings University, Linkoping, Sweden.

[26] Katkar A A, 2012. Improvement of solid waste collection by using optimization technique. *International Journal of Multidisciplinary Research* 2, 298-307.

[27] Abu Qdais H A, Hamoda M F and Newham J, 1997. Analysis of residential waste at generation sites. *Waste Manage. Res.* 15, 395–406.

[28] CalRecycle, 2006. Waste characterization, Residential developments: Estimated solid waste generation rates, http://www.calrecycle.ca.gov/ Accessed March 1, 2017.

[29] NY CEQR, 2011. Phased redevelopment of Governors Island, Final generic environmental impact assessment, Chapter 13: Solid waste and sanitation services, CEQR No. 11DME007M. http://www.nyc.gov/html/oec/ Accessed March 20, 2017.

IOP Conf. Series: Materials Science and Engineering **383** (2018) 012027 doi:10.1088/1757-899X/383/1/012027

Laboratory testing of different window design cases for noise transmission

M A Maraqa[1,2*], Y S Hawas[1,2], Md D Alam[2], J El Zarif[3] and K N Aljunadi[3]

[1]Civil and Environmental Engineering Department, UAE University, Al Ain, UAE
[2]Roadway, Transportation and Traffic Safety Research Center, UAE University, Al Ain, UAE
[3]Infrastructure and Municipal Assets Sector, Abu Dhabi City Municipality, Abu Dhabi, UAE

*E-mail:m.maraqa@uaeu.ac.ae

Abstract. We investigated the effect of different window designs employed in Abu Dhabi Emirate- UAE on the level of noise transmission. For that, 21 window cases were tested in the laboratory. Several factors were considered, including window glass thickness, glazing, lamination, opening style, area, frame type, frame style, and use of shutters. Results showed that sound transmission loss is improved significantly with the use of shutters and the use of hinged, not sliding, windows. The instalment of glass thicker than 6 mm did not cause any improvement. As for the double glazed windows, no noticeable improvement was detected with an air gap of 12 mm or less, but increasing the air gap to 20 mm causes a moderate improvement. Slight improvement was found when a lamination layer or uPVC frame (instead of aluminum) was used. It was further found that box umbrella and curtain wall frames did not seal as well as half umbrella frames, and thus resulted in a lower acoustic performance of the windows.

1. Introduction

Noise is a threat to human health and well-being. Prolonged exposure to high levels of noise causes impaired hearing, and increases stress and blood pressure, causing gastric problems as well as headaches, migraines, and anxiety or depression [1]. Although noise problems have been well recognized in many countries, the problem continues to grow, accompanied by an increasing number of complaints from affected individuals. From a global perspective, the growth in urban noise pollution is considered unsustainable, because not only it has adverse effects on health but it also affects future generations by degrading residential, social and learning environments [2]. With the realization that noise is a global issue [3, 4], precautionary actions are needed for noise protection in any environmental planning situation.

Windows play a significant role in achieving a good acoustic environment. In fact, windows provide an acoustic weakness to the facade of a building as noise is transferred more easily through glazing than through external walls [5]. Thus, improving the sound insulation of windows helps reduce the ingress of external noise. Several factors could affect noise transmission through windows, such as glass thickness, sealant type, type of glazing (single, double or triple), air gap thickness in multiple glazing, frame type, frame material, external shutters, use of secondary windows, window area, opening style, lamination, etc. Therefore, particular attention to the construction detail of windows is

Content from this work may be used under the terms of the Creative Commons Attribution 3.0 licence. Any further distribution of this work must maintain attribution to the author(s) and the title of the work, journal citation and DOI.
Published under licence by IOP Publishing Ltd

needed to effectively lower the level of external noise breach into a building. Such details should be incorporated into building codes to assure optimal window design.

The objective of this study was to assess sound transmission loss of different window designs currently used in the Emirate of Abu Dhabi, UAE. The motivation to carry on the study was driven by the desire of Abu Dhabi City Municipality (the funding agency) to develop specifications to control noise propagation through windows. Findings of this study are considered an essential component towards meeting such a goal. Once developed, the specification could be incorporated in the Abu Dhabi International Building Code [6] to guide and assist designers to achieve the required drop in the noise level between the two sides of the window.

2. Methodology

2.1. Tested cases

The used window sizes in Abu Dhabi Emirate vary widely. Some windows could be as small as 1 m^2, while others could be as large as 9 m^2 or larger, with the majority of windows fall in the range between 2 and 4.5 m^2. These figures are approximate and are based on a visual survey of windows on Al Salam Street in Abu Dhabi City. Furthermore, variation occurs in window configurations, but the most commonly used ones are rectangular. Based on that, three window sizes were selected for laboratory testing for noise transmission loss. All tested windows had a fixed part on the top (60 cm high). The windows could be sliding or hinged. The hinged windows had a similar configuration as that of the sliding windows with the exception that the sliding panes were fixed from the top and open to the outside. For all windows, the glass was float, except one case where tempered glass was used.

Given that the window glass thickness used in Abu Dhabi Emirate varies from 6 to 19 mm and the air gap thickness in double glazing ranges from 12 to 20 mm, many different alternatives could be envisioned. Testing for all possible alternatives would be costly and time-consuming. As a result, the acoustic performance of only some glazing systems and window design alternatives was obtained. In total, 21 window cases were tested for noise transmission loss in the laboratory. These cases are listed in table 1. Factors that were considered include window size, glass thickness, glazing type (single versus double), air gap thickness between glazing, use of external shutters, type of window (sliding vs hinged), type of frame (half umbrella, full umbrella, or curtain wall), frame material (aluminum vs uPVC), glass type (float vs tempered), and glass lamination.

2.2. Window fabrication and installation

Window fabrication was carried out by Zoltic Aluminum and Glass Company as per the specifications listed in table 1. Window frames and glass panes were transported to the laboratory for installation. For each case, the frame was fixed first in the provided opening and the sides were sealed with silicone (Dow Corning 813c). The bottom part of the frame was installed near the floor for all cases. Glass panes were then fixed to the frame and were sealed with rubber (Premium Quality Seal, UAE). For each case, the window panes were checked to make sure they are completely closed before testing. For all cases (except case 18), float, transparent glass was used (Emirates Float Glass LLC, UAE). For case 18, tempered glass was used (City Land Glass Factory, UAE). Laminated glass of case 4 was obtained from Emirates Float Glass LLC. For the tested double glazed cases, aluminum spacer bars were used (Gulf Extrusions Company). Wood shutters used in case 11 and 14 cover the whole window length. The shutter is made of solid timber, with pieces (5 cm x 1 cm thick) that overlay each other. Aluminum shutters used in case 12 and 15 were made of hollow aluminum pieces, each 11 cm wide. The shutters have been entirely closed during testing and were placed at the outer edge of the opening of the window holder (i.e., adjacent to the source room side), with a 5-cm distance from the window frame.

2.3. Window testing

Laboratory testing of noise transmission through the various window cases was carried out at Al Futtaim Exova in Dubai, UAE. The tests were intended to determine the acoustic performance of the different window samples. The test methods were conducted in accordance with the ASTM methods. At the laboratory, the test chamber has been constructed according to specifications detailed in ISO 140-1:1998. The acoustic test facility comprises an adjacent pair of acoustically sealed semi-reverberant chambers. Both chambers have been installed on separate, isolated acoustic floor systems to limit sound and vibrational energy transference (flanking) between the chambers and from the immediate surroundings. The walls of the test facility are formed from sound absorbent acoustic paneling that has been specially designed to provide a high sound transmission loss capability while producing the desired reverberant levels for the semi-reverberant acoustic test chamber. The test specimen is mounted into the opening between the test chambers, provided by the separate, isolated sample holder. The dimensions of the noise source room are 7.08 m long, 5.753 m wide and 4.425 m high. The receiving room is 7.36 m long, 5.98 m wide and 4.6 m high. As for the opening of the testing chamber, the dimensions are 3.7 m long, 0.35 m wide and 2.7 m high.

Table 1. List of tested cases [a].

Case	L (m)	H (m)	Glazing	Frame style	Opening style	Shutter type	Frame type
1	1.8	1.8	Single 6mm	HU	Hinge	No	LA
2	1.8	1.8	Single 6mm	HU	Sliding	No	LA
3	1.8	1.8	Single 8mm	FU	Sliding	No	MA
4	1.8	1.8	Single 12.76 (6-0.76-6) laminated	HU	Sliding	No	LA
5	1.8	1.8	Single 12mm	FU	Sliding	No	MA
6	1.8	1.8	Double 6mm-12mm-6mm	FU	Sliding	No	MA
7	1.8	1.8	Double 6mm-12mm-12mm	CW	Hinge	No	HA
8	1.8	1.8	Double 6mm-20mm-6mm	CW	Hinge	No	HA
9	1.8	1.8	Double 12mm-12mm-12mm	CW	Hinge	No	HA
10	1.8	1.8	Single 6mm	HU	Hinge	No	uPVC
11	1.8	1.8	Single 6mm	HU	Sliding	Wood	LA
12	1.8	1.8	Single 6mm	HU	Sliding	Aluminum	LA
13	1.8	1.8	Double 6mm-12mm-6mm	CW	Hinge	No	HA
14	1.8	1.8	Double 6 mm-12mm-6mm	FU	Sliding	Wood	MA
15	1.8	1.8	Double 6 mm-12mm-6mm	FU	Sliding	Aluminum	MA
16	2.4	1.8	Single 6mm	HU	Sliding	No	LA
17	1.2	1.8	Single 6mm	HU	Sliding	No	LA
18	1.8	1.8	Single 6mm (tempered)	HU	Hinge	No	LA
19	1.8	1.8	Double 6mm-12mm-6mm	HU	Hinge	No	LA
20	1.8	1.8	Single 8mm	HU	Sliding	No	LA
21	1.8	1.8	Single 12mm	HU	Sliding	No	LA

[a] HU is for half umbrella, FU is for full umbrella, CW is for curtain wall, LA is for light-weight aluminum, MA is for medium-weight aluminum, HA is for heavy-weight aluminum, and uPVC is for unplasticized polyvinyl chloride.

The sample holder at Exova's laboratory comprises of 350 mm wide steel channel which is directly connected to a concrete slab which is isolated from the ground by a high-density mineral wool matt. The infill lining of the steel channel section is high-density hardwood. Both the source and receiving rooms are separate constructions, isolated by high-density mineral wool blankets from the concrete slab. Reduction of the size of the opening to accommodate the tested window cases was carried out using four layers of gypsum board sheets (2 cm thick each, filled with fiber glass).

The source and receiving rooms are covered by a secondary external enclosure which is only connected at doorways by heavy density flexible loaded vinyl sheeting. Similarly, the gaps between the sample holder frame and the source and receiving rooms are sealed by flexible mastic. The entry to both the source and receiving rooms is protected by two sets of isolated doorways. Calibration of the acoustic laboratory of Exova was conducted by PKA Acoustic Consulting [7].

The instruments used for measurement of sound transmission were all made by Bruel & Kjaer and include a sound level meter (Model 2250), a pre-amplifier (Model ZC-0032), a microphone (Model 4189), a calibrator (Model 4231), an omnidirectional loudspeaker (Model 4292), a loudspeaker amplifier (Model 2716), and a pink noise source (Model 2250). Measuring instruments and related test equipment were all calibrated.

Sound transmission testing was carried out as per the ASTM E90-09 [8] requirements. The test was performed to ascertain the airborne sound insulation capabilities of the sample windows under laboratory conditions, from the source room to the receiving room. The sound transmission test sample was installed and sealed in the available aperture test opening between the acoustic rooms. To restrict the passage of any flanking noise between the chambers, all instances of bridging between the acoustic chambers were investigated and cleared following the installation of the sample. The reverberant levels of the chamber were retested for each new test arrangement or set up.

During testing, the sound generator was set up in the corner of the sound source chamber at two different locations. The sound source was generated in the form of an impulse sound energy, which is a signal produced internally from the sound level meter and fed into the loudspeaker. The level of the impulse signal was sufficiently above the background ambient noise level to permit evaluation of the decay curves (at least 30 dB above ambient).

Sound levels in the 1/3 octave bands from 100 to 5,000 Hz were measured, in both the source and receiving rooms using a rotating microphone boom. The rotating microphone was positioned at two points in the source room and two points in the receiving room. The arm of the rotating boom holding the microphone was 1.2 m and was held at an elevation of 1.4 m above the ground. The arm was set to rotate one full revolution in 64 seconds with an angle of inclination within 10 degrees from a horizontal plane. The microphone was set at a minimum distance of 1.5 m from the source to ensure that measurements were not taken in the direct field. Also, in all measurements, the microphone was moved to a minimum distance of 1 m from walls and extended objects. The test for each sample was taken at six microphone receiver positions, with at least 1.5 m between each position. For each microphone position, a minimum of 15-second recording period was done. The distance between room boundaries, diffusers, and the microphone was at least 1 m. Figure 1 shows the setup for one of the tested cases.

Figure 1. Testing setup of case 5.

Background noise measurement was also measured in the receiving room. The average background noise level in the receiving room ranged between 7.07 to 13.66 dB, with generally higher background noise occurred at the border of the 1/3 octave bands as compared to values at intermediate frequencies. The background noise level in the receiving room, even at low frequency, was considered negligible and hardly detected by human ears. Therefore, correction for background noise was not necessary since the background level was more than 10 dB below the combination of signal and background [8].

Reverberation time was measured according to the ASTM method C423-09a [9] by setting up the sound generator in the corner of the receiving room at two different locations, while the rotating microphone was positioned at two points. In all cases, initiation of the sound generator was controlled by a wireless transmission meter (AKG, Model WMS450). Variation in the reverberation time was small among the tested frequencies, albeit there was a drop in the reverberation time with the increase in frequency. The overall average reverberation time for the tested cases ranged between 2.30 and 2.39 sec.

2.4. Calculation of sound reduction

The results for each tested case were reported as the sound transmission class (STC) according to the ASTM method E413-10 [10]. STC is a single-number rating used to compare the sound transmission loss of a material with the subjective notion of sound insulation from sources. The calculation of the sound transmission loss follows the correction of the receiver side readings for the ambient and reverberation time factors. Sound transmission loss (STL) is defined as:

$$STL = L_1 - L_2 + 10log\left(\frac{S}{A}\right) \qquad (1)$$

where L_1 and L_2 are the sound levels in the source and receiving rooms (dB), respectively, S is the area of the of test specimen that is exposed in the receiving room (m^2), and A is the sound absorption of the receiving room with the test specimen in place (m^2). The sound absorption of the receiving room is given by the Sabine equation:

$$A = 0.921\frac{Vd}{c} \qquad (2)$$

where V is the volume of the reverberation room (202 m^3 in our case), c is the speed of sound (340 m/s), and d is the sound decay rate (dB/s). The sound decay rate is estimated as:

$$d = \frac{60}{T_2} \qquad (3)$$

where T_2 is the reverberation time (i.e., the time that corresponds to a drop of 60 dB in the receiving room). STC is derived by fitting a reference rating curve to the STL values measured for the 16 contiguous 1/3 octave frequency bands with nominal mid-band frequencies of 125 Hz to 4,000 Hz inclusive. A reference rating curve is fitted to the corresponding 16 measured STL values. Comparison with the reference curve results in a set of positive and negative deviations for each frequency band. The reference curve is increased or decreased in 1 dB intervals to achieve the maximum sum of deficiencies (STL values less than the reference rating curve), less than or equal to 32 dB, and no single deficiency is greater than 8 dB. Then, the STC value is determined as the numerical value of the reference contour at 500 Hz.

3. Results and discussion

STL curves were generated for all tested cases. An example of an STL curve is shown in figure 2. The curve for each case was generated by subtracting the average sound level at a given frequency in the receiving room from the mean sound level at the same frequency in the source room using the collected raw data. The STC value for each case was then determined as listed in table 2. The effect of changes in window design on the STC value is presented below. Factors that were considered are glass thickness, glass lamination, glazing, air gap between glass panes, frame material, frame type, window area, use of shutters, and glass type.

The impact of glass thickness was evaluated by considering case 2, 20 and 21 are considered. These cases were identical except for the glass thickness (6 mm for case 2, 8 mm for case 20, and 12 mm for case 21). All these cases resulted in equal STC values of 23 dB, indicating no impact on sound transmission for the tested range of glass thickness. However, from the STL values for these cases (not shown here), there appeared to be a small improvement in sound transmission loss in selecting a thicker glass. The effect of glass thickness on sound transmission could also be evaluated by considering case 3 and 5. These two cases were similar with the exception of glass thickness (8 mm for case 3 and 12 mm for case 5). The STC value for case 3 was 21 dB and that for case 5 was 20 dB. Similar results were also obtained in the experiments of Tadeu and Mateus [11] for the glass thickness of 4 and 8 mm. These results, once again, indicate no impact on sound transmission due to changes in glass thickness, and thus the selection of a thicker glass for sound reduction would not be justified.

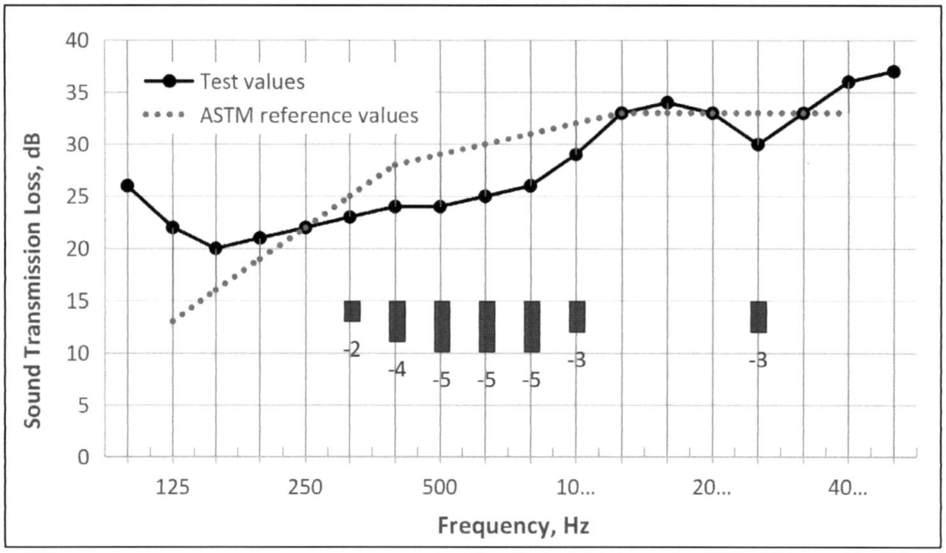

Figure 2. STL at the 1/3 octave bands of case 1.

The effect of glass lamination could be evaluated by considering case 4 and 21. The two cases were similar with the exception of lamination (case 4 was laminated by a 0.76 polyvinyl butyral (PVB) layer sandwiched between two 6-mm glass sheets while in case 21 single glazing of 12 mm glass panes was used). The STC value for the case with laminated glass was 24 dB, while that for the non-laminated glass was 23 dB. Thus, the use of a 0.76 mm lamination resulted in only 1 dB reduction in sound transmission as compared to a non-laminated case. It has been reported that laminated glass has an advantage of addressing specific noise reduction efficiency [12]. The plastic interlayers change the frequency response of the composite sheet in comparison with the same weight of ordinary glass and also absorb some of the sound energy [13]. Although not investigated in this study, it is possible that a thicker laminated layer could lead to higher STC values.

The effect of glazing could be evaluated by considering case 5 and 6. Both cases are similar with the exception that case 5 included a single glazed window using 12 mm glass, while the window in case 6 was double glazed using 6 mm glass panes with an air gap of 12 mm in between. For both cases, the STC value was the same (20 dB), indicating that a slight air gap would not improve the performance of windows from a sound reduction point-of-view. The low STC value of the two cases as compared to those of the other tested cases could be mainly attributed to the use of a box umbrella frame, which did not provide a good seal with the glass panes as will be discussed later.

The impact of the air gap thickness between glass panes in double glazed windows could be evaluated by considering case 8 and case 13. In both cases, double glazed curtain wall windows of 6 mm glass were used. The difference between the two cases was in the thickness of the air gap. In case 8 an air gap of 20 mm was used, while in case 13 the air gap was 12 mm. The STC value for case 8 was 29 dB while that for case 13 was 25 dB. As noted by Pilkington [14], a standard double glazed unit does not reduce sound transmission much more than a monolithic glass. What matters is the thickness of the air space between glass panes, but only for really wide cavities. Tadeu and Mateus [11] found that the sound reduction index improves, if the air space between glass panes is close to or greater than 50 mm. However, beyond a spacing of about 200 mm, it becomes uneconomical to install such windows because the incremental acoustic improvement is small.

Table 2. STC values of the tested cases.

Case	STC (dB)	Case	STC (dB)	Case	STC (dB)
1	29	8	29	15	32
2	23	9	26	16	23
3	21	10	31	17	23
4	24	11	32	18	29
5	20	12	34	19	29
6	20	13	25	20	23
7	25	14	29	21	23

The effect of changing the glass thickness in double glazed windows could be evaluated by considering case 7, 9 and 13. In the three cases, hinged, double glazed curtain wall windows were used. The difference between these cases was that case 7 consisted of 6 mm and 12 mm glass panes with 12 mm air gap, while case 9 consisted of two 12 mm glass panes with 12 mm air gap, and case 13 was made of 6 mm glass sheets with 12 mm air gap. The STC value for case 7, 9 and 13 were 25, 26, and 25 dB, respectively. Apparently, changes in the glass thickness within the same window do not cause a noticeable improvement in the STC value.

The two types of frame materials used in this study were aluminum and uPVC. The impact of frame material could be evaluated by considering case 1 and 10. Both cases were similar except in the the material of the frame (the frame in case 1 was made of aluminum while that in case 10 was made of uPVC). For case 1, an STC value of 29 dB was obtained, while for case 10 an STC value of 31 dB was obtained. Thus, uPVC frames result in 2 dB increase in the STC value compared to aluminum frames. The improvement in the performance of the uPVC frame was mainly due to a higher sound transmission loss at low and intermediate frequencies (not shown here).

Evaluation of the effect of frame type on sound transmission through windows could be made by considering case 3 and 20. Both cases were similar except that in case 3 a full (box) umbrella frame was used, while in case 20 a half umbrella frame was used. Results revealed that a half umbrella frame results in a better performance regarding STL, with the STC value for case 20 being a 2-dB higher than that of case 3. A similar conclusion would be obtained if one compared the STC values of case 5 and 21. The difference between these cases was in the type of umbrella frame used (in case 5 a box umbrella frame was used while in case 21 half umbrella frame was used). The STC value for case 21 was 3 dB higher than that of case 5. In conclusion, half umbrella frames result in a 2-3 dB increase in the STC value compared to full umbrella frames.

Additional comparison of the effect of frame type could be made considering case 13 and 19. The difference between these cases was that in case 13 a curtain wall frame was used while in case 19 a half umbrella frame was used. The STC for case 13 was 25 dB while that for case 19 was 29 dB. Thus, a half umbrella hinged frame would result in a 5 dB increase in the STC value compared to similar windows made of curtain walls. In general, the effect of frame type depends on the sealing mechanism (air leaks/tightness). Tadeu and Mateus [11] studied the sealing effect on window considering the sound reduction index. They found that the sound reduction index is 18 dB for a window with a small gap of 2 mm, and it is 20 dB for a window with poorly sealed frame. However, the same well-sealed window has a sound reduction index of 29 dB which is almost similar to the same window without a frame. Moreover, Park and Kim [15] found that windows with the highest performance grade airtightness had better sound insulation that those with poor or unmeasurable airtightness.

The effect of shutters could be evaluated by considering case 2, 11 and 12. These cases were identical had the types and presence of shutters varied (case 2 was tested without a shutter, while case 11 and 12 were tested with wood and aluminum shutters, respectively). The STC values for case 11 and 12 were 9-11 dB higher than that of case 2. Comparison between cases 11 and 12 showed that windows with aluminum shutters had a slightly better sound insulation (about 2 dB) than those with wood shutters. A similar conclusion would be reached if one compared the STC values for case 6, 14, and 15. The difference between these cases is that case 6 was tested without a shutter, while case 14 and 15 were tested with wood and aluminum shutters, respectively. The STC values for case 14 and 15 were 9-12 dB higher than that of case 6, with the windows with aluminum shutters showing a slightly better sound insulation (about 2 dB) than that of the wood ones. A possible reason for the slightly better performance of aluminum shutters as compared to wood shutters that are used in this study is that in the former hollow aluminum was used while in the latter solid wood was used. The air gap in the aluminum shutter could have resulted in an improvement over solid wood shutters. Another possible reason could be due to the presence of more openings in the wood shutter than in the aluminum shutter due to variations in the width of the pieces making the two shutters (the width of wood pieces was 5 cm and of aluminum pieces was 11 cm).

Evaluation of the effect of window area on the sound transmission loss revealed that the STC value is independent of the window area because the results are corrected for window size in obtaining the STC value. This is evident by comparing the STC values for case 2, 16 and 17. These windows differed only in their areas, but the STC values were the same for all of them.

The two types of window opening styles that were tested in this study were sliding and hinged. The effect of window opening style could be demonstrated by comparing the results of case 1 (sliding window) and case 2 (hinged window). The STC value for case 1 was 29 dB while that of case 2 was 23 dB. Thus, the use of hinged windows results in a significant increase of about 6 dB in the STC value as compared to sliding windows, assuming other factors remain the same. A similar conclusion could be reached if the assessment is made by considering case 6 and 19. These cases differ in two aspects: (1) in case 6 a sliding window was used while in case 19 a hinged window was used, and (2) in case 6 a box umbrella frame was used while for case 19 a half umbrella frame was used. The STC value for case 6 was 20 dB and that for case 19 was 29 dB. Given that half umbrella frames gave STC values of 2-3 dB higher than those of box umbrella frames (as explained previously), hinged windows would then have STC values that are 6 to 7 dB higher than those of sliding windows. The outperformance of hinged windows compared to sliding windows, as noticed here, is consistent with the findings reported in the literature. Milgard [16], for example, reported that hinge style windows provide a tighter seal against the frame as compared to sliding windows, therefore reducing noise transmission.

Two types of glass were tested in this study; float and tempered. The effect of glass type could be demonstrated by considering case 1 (float glass) and case 18 (tempered glass). The STC values for both cases were 29 dB. Thus, the use of tempered glass over float one does not add value to sound transmission loss.

4. Conclusion

Based on laboratory testing of 21 window cases, it was concluded that sound transmission loss through windows is improved significantly with the use of shutters (when completely closed) and the use of hinged, not sliding, windows. However, using glass thicker than 6 mm or using tempered glass instead of float glass does not cause any improvement. Meanwhile, use of double glazed windows with an air gap of 12 mm or less does not result in any tangible effect. However, increasing the air gap from 12 to 20 mm causes a moderate improvement. The use 0.76 mm PVB lamination layer and the use of uPVC frames (instead of aluminum frames) resulted in a slight improvement. It was also concluded that the current design of box umbrella and curtain wall frames do not seal well as compared to those of half umbrella frames, and thus result in lower STC values.

Acknowledgement
This project was undertaken as part of a scientific collaboration between Abu Dhabi City Municipality (ADM) and the Roadway Transportation and Traffic Safety Research Center (RTTSRC) at the UAE University through project number ADM 282/2015. Particular thanks go to Sami A. Alhashmi, Husain M. Alsaeedi, and Shamsa S. Almuharrami at ADM.

References
[1] Goines L and Hagler L, 2007. Noise pollution: A modern plague, *Southern Medical Journal* 100, pp. 287–94
[2] Berglund B, 1998. Community noise in a public health perspective, In V C Goodwin and D C Stevenson (eds.) Inter Noise 98. Sound and Silence: Setting the Balance, Vol. **1**, 19–24, New Zealand Acoustical Society, Auckland, New Zealand
[3] Lang W W, 1999. Is noise policy a global issue, or is it a local issue? In: J Cuschieri, S Glegg and Y Yong (eds.) Internoise 99 – The 1999 International Congress on Noise Control Engineering, 6-8 December 1999, Fort Lauderdale, Florida, USA, pp. 1939–43
[4] Sandberg U, 1999. Abatement of traffic, vehicle and tire/road noise – the global perspective. In: J Cuschieri, S Glegg and Y Yong (eds.) Internoise 99 – The 1999 International Congress on Noise Control Engineering, 6-8 December 1999, Fort Lauderdale, Florida, USA, pp. 37–42
[5] Adelaide City Council, 2014. Noise Technical Fact Sheet 3 - Sound insulation of windows, www.adelaidecitycouncil.com/city-living/home-propertymanagement/noise/resources/, Accessed May 15, 2014
[6] ADIBC, 2013. Abu Dhabi International Building Code. Department of Municipal Affairs, Abu Dhabi, UAE. ISBN: 978-1609835712
[7] PKA Acoustic Consulting, 2010. Report on acoustic calibration, Exova Acoustic Laboratory, Dubai, United Arab Emirates, Project 209069, Artarmon, NSW
[8] ASTM E90-09, 2009. Standard test method for laboratory measurement of airborne sound transmission loss of building partitions and elements, ASTM International, West Conshohocken, PA, http://www.astm.org/Standards/E90.htm
[9] ASTM C423-09a, 2009. Standard test method for sound absorption and sound absorption coefficients by the reverberation room method, ASTM International, West Conshohocken, PA, www.astm.org/Standards/C423.htm
[10] ASTM E413-10, 2010. Classification for rating sound insulation, ASTM International, West Conshohocken, PA, www.astm.org/ Standards/C413.htm
[11] Tadeu A J B and Mateus D M R, 2001. Sound transmission through single, double and triple glazing. Experimental evaluation. *Applied Acoustics* 62, pp. 307–25
[12] Gervais R, 2011. Home Recording Studio: Build it Like the Pros. 2nd edition, Thomson Course Technology, Boston, Mass. ISBN: 978-1435457171
[13] Lyons A, 2010. Materials for Architects and Builders, 4th ed., Elsevier, Boston. ISBN: 978-1856175197
[14] Pilkington, 2007. Technical Bulletin - Noise control with glass, http://www.pilkington.com/europe/uk+and+ireland/english/products/bp/ downloads/byproduct/noisecontrol/default.htm, Accessed March 1, 2016
[15] Park H K and Kim H, 2015. Acoustic insulation performance of improved airtight windows. *Construction and Building Materials* 93, pp. 542–50
[16] Milgard, 2016. http://www.milgard.com/professionals/technical-resources, Accessed March 1, 2016

IOP Conf. Series: Materials Science and Engineering **383** (2018) 012028 doi:10.1088/1757-899X/383/1/012028

Study on forming and laying process of high strength pipeline steel

Lian Yeda*, Ma Qiang, Zhang Bing, Zhang Yuqiang and Li Hao

School of Mechanics and Civil Engineering & Architecture, Northwestern Polytecnical University, 710072, Xian, China

*Email: lianyeda@qq.com

Abstract. In order to analyze the simulation results of UOE molding and solve the buckling problem of high strength pipeline steel in the process of ocean laying, taking X80 pipeline steel sample as the object of study, a better mechanical model of kinematic hardening was adopted. With the aid of ABAQUS finite element analysis software, the finite element analysis of UOE simulation results ,the loading of UOE simulation results and buckling in ocean laying are carried out. The results show that: Through the extraction of the load at the end of the welded tube and the strain at the buckling position, draw it into curves and observe the position of the strain corresponding to the inflection point of the curve. The critical strain of welded pipe subjected to buckling under different working conditions is analyzed, so it is found that the critical strain increases with the increase of internal pressure and the critical strain becomes smaller with the increase of the external pressure. The analysis method and results can also provide reference for the actual production, laying and installation of oil and gas pipelines.

1. Introduction

High-strength pipeline steel is one of the popular research fields in the pipeline engineering industry at home and abroad, which is characterized by high strength and high toughness.The UOE molding process in the pipeline steel forming method is a commonly used molding method in the world today. In the process of laying the pipeline in the ocean, the pipeline is prone to buckling for environmental factors, so the problem of buckling failure of pipeline steel is one of the problems to be solved in the pipeline laying process.

Xing et al. [1] studied the buckling analysis of the submarine pipeline under axial load based on the Euler-Bernoulli theory, and obtained the failure criterion in this buckling condition.Wang Xiaoxiu et al. [2] studied the three-dimensional finite element simulation of UOE pipe forming based on elasto-plastic finite element method, verified the accuracy of the model, and studied the variation of sheet size in UOE forming process. Edited et al. [3] simulated the UOE molding of high-strength pipeline steel to verify the correctness of the mechanical model selection. At the same time, the load applied to the mold in each molding step is also analyzed, and it is verified that the load applied by the process machine in the molding process is within the planned range. Liang Zhenting [4] studied the various pipelining methods in the ocean, analyzed the buckling problem in marine laying and proposed the solution to the buckling problem.

In this paper, X80 pipeline steel is selected as the sample, and the ABAQUS finite element analysis software is used to establish the kinetic hardening mechanical model which is most consistent with

Content from this work may be used under the terms of the Creative Commons Attribution 3.0 licence. Any further distribution of this work must maintain attribution to the author(s) and the title of the work, journal citation and DOI.
Published under licence by IOP Publishing Ltd

actual situation. It is applied to the finite element simulation of buckling analysis of UOE forming and marine pipe laying.

2. Establishment of the kinetic hardening model of Pipeline Steel

In recent years, the research on the kinetic hardening model of steel under large deformation conditions has been studied in the literature [5]. Rudolph et al. [6] used ABAQUS finite element analysis software to analyze the constitutive model of metallic materials.In this paper, the unidirectional tensile test and tensile-compression test of the round bar intercepted in X80 sheet are carried out, and the basic mechanical parameters of the material are obtained.The ABAQUS finite element analysis software was used to establish the same simulation scheme as the experiment. The kinematic hardening model of ABAQUS material library and the kinematic hardening model defined by UMAT are used to calculate and compare the stress-strain curves and the experimental curves respectively.The results show that the effect of the kinematic hardening model defined by the UMAT subroutine is in accordance with the experimental results.

3. UOE finite element modeling simulation

U molding process is to use the vertical die to load the sheet until the ends of the horizontal angle of about 90 ° to stop loading, then use both sides of the roll to load the plate horizontally to the sheet metal contact, and then unload to make the sheet rebound into "U"; O molding process is from the U mold out of the sheet into the fixed O mold lower part and the upper part of the O die is loaded downward to move the ends of the sheet to both sides. When the O die is loaded down to the maximum displacement, the sheet is essentially O-type, and the upper part of the O-die is withdrawn along the loaded reverse path; After leaving the O molding die, the sheet contacts the expanding die, the eight molds in the middle of the sheet load to the sheet metal along the circular radial outward to complete the final expansion process of UOE molding.

3.1. Analysis of UOE simulation results

In the process of bending edges,the mold exerts a load on the sheet, the stress distribution of the sheet is relatively concentrated,as shown in figure 1 (a).The stress value produced by the outer layer of the sheet shown in figure 1 (b) is larger, the stress value produced by the intermediate layer is smaller and the stress range is spread. As shown in figure 1 (c), the degree of loading of the sheet is gradually increased, and the degree of bending produced by the sheet is also increased. The change of the equivalent stress is larger and the equivalent stress concentration also spreads gradually to the edge of the sheet metal.When the crimping process is substantially completed, the position of the maximum sheet stress is located near the middle layer, as shown in figure 1 (d).

In the bending process of U forming,when the die moves down vertically, both sides of the sheet are bent. At this time, the maximum stress is located at the joint of the plate and the die. As the die is loaded gradually, the stress changes gradually to both ends, which makes the stress on the outer surface of the sheet larger and the stress on the inner surface smaller, as shown in figure 2 (a). Figure 2 (b) shows the die exerts a horizontal load on the sheet, so that the sheet is drawn closer to the interior. The maximum position of the sheet metal is derived from the edge of the sheet, and the stress in the other positions of the sheet is not significantly changed. In this molding process, the sheet is made "U" by pressing down the mold, and then the horizontal roller mold continues to load the sheet, The plastic deformation of the sheet is completed.Finally, the horizontal loading die is unloaded to make the plate rebound, from figure 2 (c) can be seen in the middle part of the sheet stress is greater than the external surface stress.

As can be seen from figure 3 (a), the sheet is deformed at the bottom of the forming process in the O molding process, the inner layer is compressed to produce compressive stress, and the outer layer is stretched to produce tensile stress. When the applied load is larger, the pressure of the sheet is also increased, the two ends of the sheet and the central part began to protrude outward, gradually forming as shown in figure 3 (b). As can be seen from the figure 3 (b), the stress of the sheet is gradually increased, and spread to the center and both ends of the sheet and the final shape is O type. As can be seen from figure 3 (c), the sheet is basically compressed and molded, the compression position is the greater stress,and the stress on the outer surface is greater than the inner surface.The upper end of the mold moves upwards, the O molding of the sheet material is unloaded, and the sheet material generates local rebound, as shown in figure 3 (d).

The expansion process is shown in figure 4. As shown by figure 4 (a), due to the impact of incomplete contact with the mold, the plastic deformation first appears near the edge of the sheet, and because the curvature is small, the position of the maximum stress is also located, as shown in figure 4 (b).With the expansion of the volume gradually increased, local plastic deformation gradually extended to the inner wall of the sheet. The sheet material gradually enters the plastic stage, from inside to outside, from the center to the sides, where the internal stress state of the sheet is mostly tensile stress, as shown in figure 4 (c).When the mold is unloaded, the sheet is spring back. As shown in figure 4 (d), after the mold is completely unloaded, the residual stress will be greatly reduced before the expansion due to the expansion process. It can be seen from the figure 4 (d) that the residual stress is mainly distributed in the area where the sheet is subjected to the maximum diameter.

(a) (b) (c) (d)

Figure 1. Bending edge molding process.

(a) (b) (c)

Figure 2. U molding local simulation.

(a) (b) (c) (d)

Figure 3. O molding local simulation.

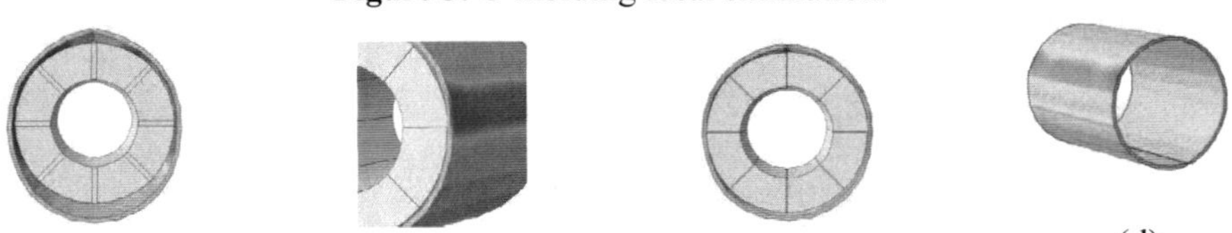

(a) (b) (c) (d)

Figure 4. Local simulation of expanding stage.

3.2. Load Analysis of UOE simulation results

In order to visually understand the stress changes in the molding process,the sheet on the mold by the loading capacity of a large number of units were removed alone. The main method of selecting the points is to divide the sheet into two parts and half to be the object of study, and then select the three

IOP Conf. Series: Materials Science and Engineering **383** (2018) 012028 doi:10.1088/1757-899X/383/1/012028

units in the upper and lower layers respectively, and select the position as shown in figure 5. The stress change of the unit in the whole process is extracted, and the stress forming step curve is plotted, as shown in figure 6.

As can be seen from figure 6, the forming process of maximum load is the crimping, O forming and expanding, respectively. In other molding processes, the load applied to the sheet is small, and the stress curve can also be used to determine the UOE sheet strength.

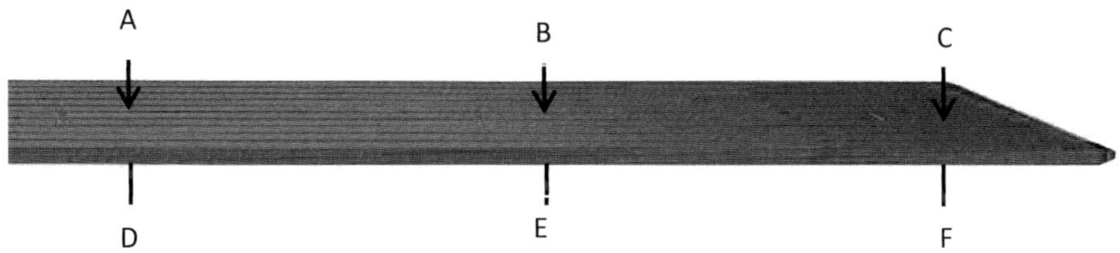

Figure 5. Unit location selection.

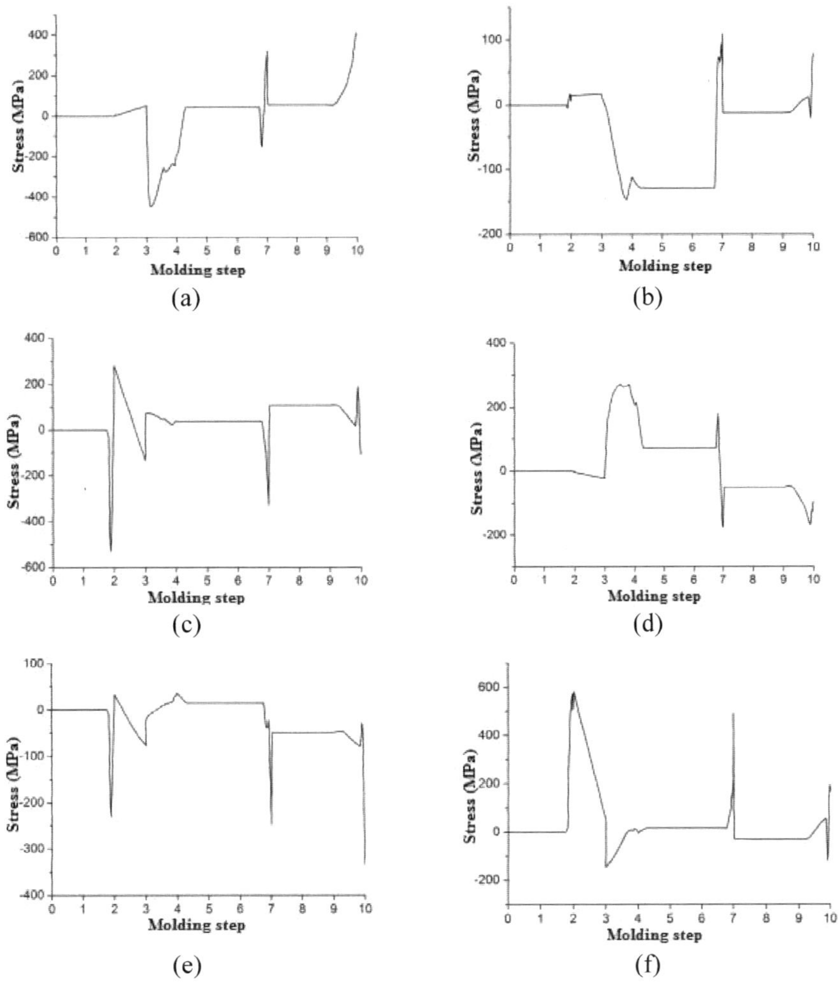

Figure 6. Stress molding step diagram.

4. Finite element simulation analysis of buckling in marine laying

YAO Baoheng et al. [7] established the vibration equation in the longitudinal direction for the submarine pipeline, and solved the influence of the wave frequency on the stress distribution and amplitude of the pipeline in different marine environments by Galerkin truncation method. Fabian [8] studied the application of joint load to the pipeline in the elastic deformation range to cause buckling, the results were analyzed and studied. As shown in figure 7, the kinetic hardening model defined by UMAT is loaded without the application of internal pressure until buckling occurs and the buckling position is in the middle of the pipe compression surface.Under the ideal condition, the finite element simulation is carried out.Compared with a large number of tests, it can be seen that the buckling location occurs in the middle of the model under the kinetic hardening model, which is more in line with the actual situation. At the same time, the data of the cloud image are extracted, and the relationship curve between load and strain is drawn, and then compared with the experimental results of Fathi [9].

When the critical strain of the simulation results is 1.41%, the pipe is buckled and the load on the pipe is 1034 kN. And the critical strain is 1.63% when the pipe is buckled on the Fathi test curve, the load of the pipeline is 1051 kN. By comparison, it is found that the results of the two are basically consistent, and the accuracy of the model is verified as shown in figure 9.

In the process of oil and gas transportation, the internal pressure is required to be applied to the pipeline. Therefore, when the finite element simulation is carried out, the relationship between the critical stress and the critical strain is observed by applying 8Mpa, 10Mpa and 12Mpa pressure respectively. Figure 8 shows the buckling simulation cloud diagram with internal pressure applied to the pipe and figure 10 is the strain-load curve under three internal pressures.

As shown in table 1, the critical strain of the buckling increases with the increase of the internal pressure of the pipeline within a certain range, and the critical load decreases.

Table 1. Analysis of pipe buckling results.

Pipe internal pressure (MPa)	Pipe external pressure (MPa)	Critical strain (%)	Critical load (kN)
8	0	1.07	925
10	0	1.13	868
12	0	1.21	798

When the pipeline has not yet run, there is no internal pressure,so in the laying process of marine pipelines, the pipeline is mainly affected by bending moment, axial force and the role of external pressure. Figure 11 shows the buckling simulation of the pipe when subjected to external pressure. Figure 13 shows the strain-load curves under three external pressures.

As shown in table 2, the critical strain of the buckling decreases with the increase of the external pressure of the pipeline within a certain range, and the critical load decreases.

Table 2. Analysis of pipe buckling results.

Pipe internal pressure (MPa)	Pipe external pressure (MPa)	Critical strain (%)	Critical load (kN)
0	0.5	1.23	1030
0	1	1.18	1017
0	2	0.98	991

When the pipeline in the submarine officially operates, not only the internal internal pressure, the sea also caused external pressure on the pipeline, the pipeline is mainly subject to bending moment, axial force, internal pressure and the role of sea pressure. Figure 12 shows the buckling simulation of the pipe when subjected to internal and external pressure. Figure 14 shows the strain-load curves under three external pressures.

Table 3. Analysis of pipe buckling results.

Pipe internal pressure (MPa)	Pipe external pressure (MPa)	Critical strain (%)	Critical load (kN)
8	2	1.07	972
10	2	1.08	925
12	2	1.11	871

As shown in table 3, the external pressure does not change and the internal pressure increases within a certain range, and the critical strain is increased and the critical load is reduced when the pipe is buckled.

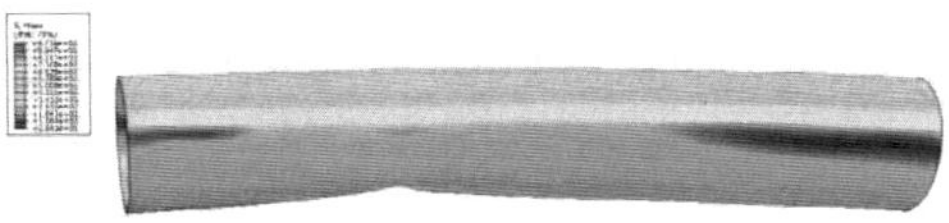

Figure 7. Buckling simulation of noninner pressure welded pipe.

Figure 8. Buckling simulation under the condition of internal pressure.

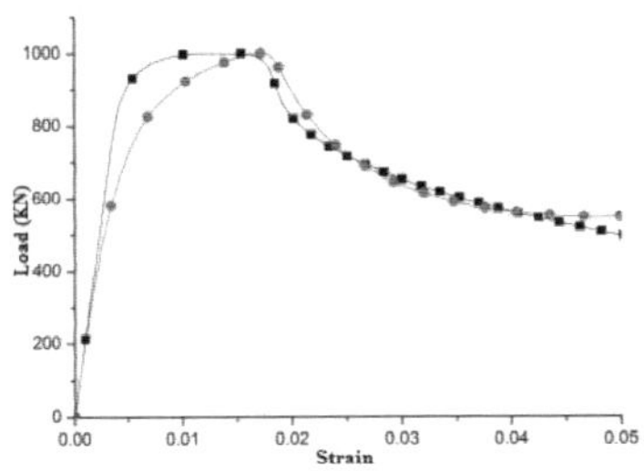

Figure 9. Load strain curve without internal and external pressure.

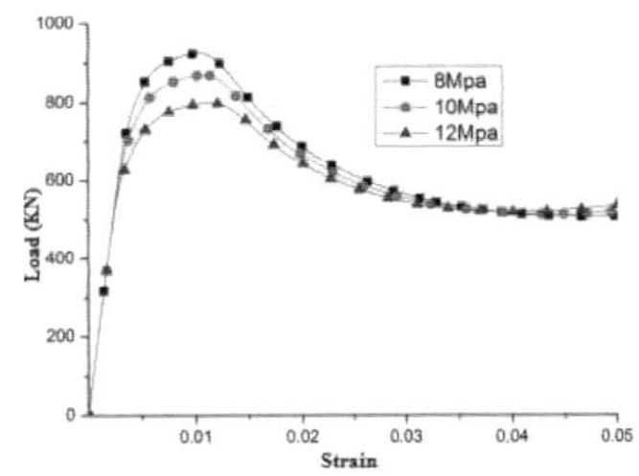

Figure 10. Load strain curves under three internal pressure conditions.

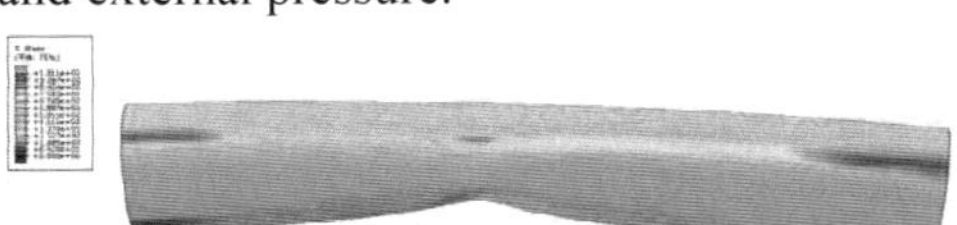

Figure 11. Buckling simulation under external pressure.

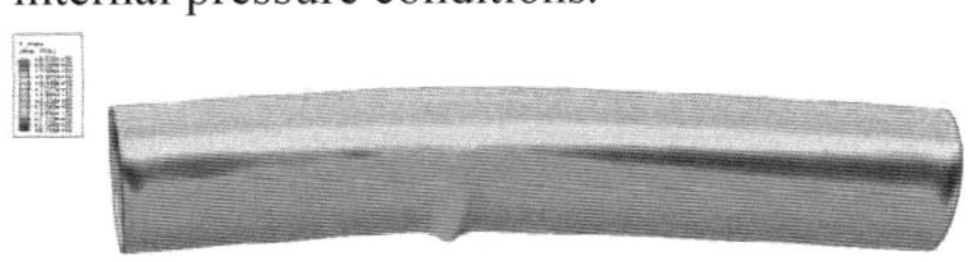

Figure 12. Buckling simulation under the condition of internal and external pressure.

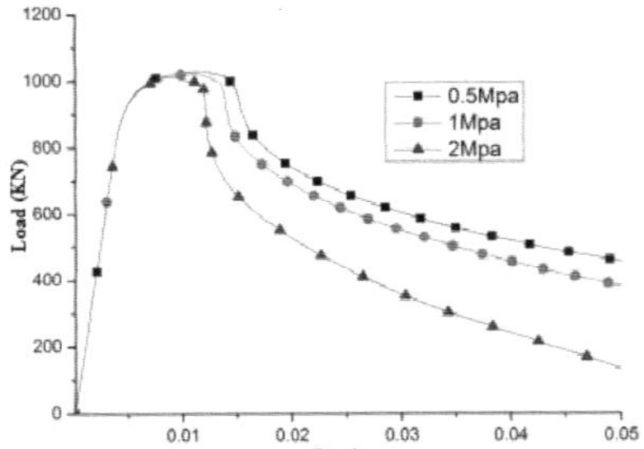

Figure 13. Load strain curves under three external pressures.

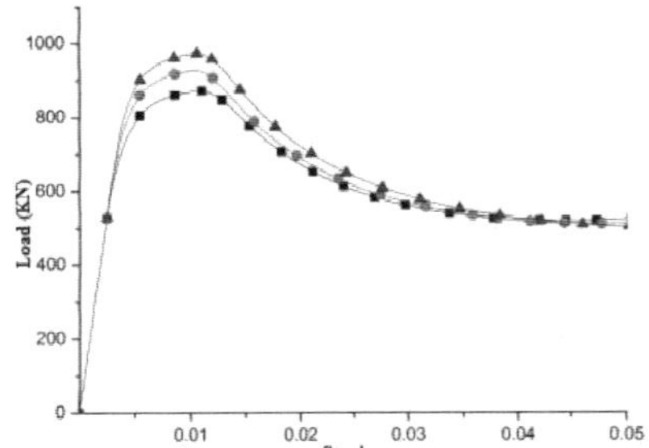

Figure 14. Load strain curves of three kinds of internal and external pressure.

5. Conclusion

Based on the basic theory of elasto-plasticity and the X80 round bar test, this paper establishes the kinetic hardening mechanical model which is the most consistent with the actual situation. For the UOE forming process and the buckling problem in the pipeline laying, ABAQUS finite element software is used to simulate the simulation. As follows:

(1) Establish the same simulation scheme as the test. The kinematic hardening model of the ABAQUS material library and the kinematic hardening model defined by UMAT are calculated respectively. The data obtained from the two methods are compared with the experimental data. It can be seen that the effect of the kinematic hardening model defined by the UMAT subroutine is consistent with the experimental results.

(2) A detailed stress-strain analysis is performed for each step of UOE molding. The three-dimensional model of ABAQUS can be used to understand the deformation of the sheet during UOE molding. In the simulation process, some important positions on the sheet and the stress, strain and load of the mold are output, so that the weak trend of the structure and the stress change inside the plate can be understood in advance.

(3) By using ABAQUS finite element simulation, a simplified pipeline model is established and loaded. For different working conditions, according to the load and strain data to draw the curve, the critical load and critical strain are obtained when the pipeline is buckled. The results show that if the internal pressure increases, the critical strain will increase; if the external pressure increases, the critical strain will decrease. The critical load will be reduced regardless of the increase of internal and external pressure.

Acknowledgement

This work is supported by National Natural Science Foundation of China Under Grant No. 51308458.

References

[1] Xing Jingzhong, Liu Chuntu and Xu Yongjun. Buckling analysis of buried spanning submarine pipeline[J]. *Engineering Mechanics*, 2006, 23 (2) pp 173–76(in Chinese)

[2] Wang Xiaoxiu, Zou Tianxia and Li Xinwen, et al. Simulation and analysis of UOE pipe production processing by three-dimensional finite element method[J]. *Baosteel Technology*, 2014(1) pp 40–46(in Chinese)

[3] Edited, By T, Cui-Yan L I. Numerical Simulation Analysis on High Strength UOE Steel Pipes[J]. *Welded Pipe & Tube*, 2010

[4] Liang Zhenting. Analysis on mechanical performance for deepwater pipe-laying[D]. Hangzhou: Zhejiang University, 2008(in Chinese)

[5] Wallin M, Ristinmaa M and Ottosen N S. Kinematic hardening in large strain plasticity[J]. *European Journal of Mechanics - A/Solids*, 2003, 22 (3) pp 341–56

[6] Rudolph J, Gilman T, Weitze B, et al. Using Nonlinear Kinematic Hardening Material Models for Elastic-Plastic Ratcheting Analysis[C]// ASME 2015 Pressure Vessels and Piping Conference. American Society of Mechanical Engineers, 2016

[7] Yao Baoheng, Ren Ping and Li Zhigang, et al. Dynamic analysis of pipeline laying in deep ocean based on Galerkin truncation method[J]. *Journal of Vibration and Shock,* 2008, 27 (10) pp 43–45(in Chinese)

[8] O, Fabian. Collapse of cylindrical elastic tubes under combined bending pressure and axial loads[J]. *International Journal of Solids and Struetures*. 1977(13) pp 1257–70

[9] Fathi A, Cheng J J R, Adeeb S, et al. Critical Buckling Strain in High Strength Steel Pipes Using Isotropic-Kinematic Hardening[C]//International Pipeline Conference. 2010 pp 39–47

Feasibility study on further utilization of timber in China

Karol Sikora*, Jianli Hao, Isaac Galobardes, Weiqi Xing, Subo Wei and Zitong Chen

Xi'an Jiaotong-Liverpool University, Suzhou, P. R. China

*E-mail: Karol.Sikora@xjtlu.edu.cn

Abstract. The use of timber in building industry has expanded in many countries over last 20 years due mainly to its lightweight, little adverse impact to the environment, and strong anti-seismic properties. Despite the fast development of the construction industry in China in recent years, timber products utilization is still very limited. Therefore, the purpose of this study was to investigate the feasibility to further utilize timber based products in the building industry in China. A review of engineered wood products utilization in China, and oversees was conducted first. The general properties of timber products were followed. Based on this, a survey focusing on the current situation and suitability of localised grown timber products was carried out to study the feasibility. This paper concludes that timber based products have a great future in China. It recommends that it is imperative to further research how to promote timber's utilization in China to be in line with the sustainable development.

1. Introduction

Timber is a traditional material used for construction in ancient China. Wood as a material is natural, created during a tree growth, and mature trees of whatever type can be source of structural timber [1]. In general timber can be characterized by high strength, high durability and good finished appearance.

Timber resource distribution in China is uneven. There are three main forest regions: northeastern China and Inner Mongolia (28% of forest area), Sichuan and Yunnan provinces in southwest (19%), and 10 other provinces in southern China (36%), which total forest areas occupy 83% [2], see figure 1. The dominant timber species are fir, oak, pine, Yunnan poplar, birch, spruce and eucalyptus [3]. Compared with the northeast traditional forests that Coniferous trees like Pinus koraiensis, Larix, picea and red spruce, coastal region (southeast forest region) is an important forest region which produce the fast-growth trees and industrial material forests, such as cedar, eucalyptus and horsetail pine.

Figure 1. China's three main forest area location (10 million ha) [2].

Content from this work may be used under the terms of the Creative Commons Attribution 3.0 licence. Any further distribution of this work must maintain attribution to the author(s) and the title of the work, journal citation and DOI.
Published under licence by IOP Publishing Ltd

2. Current utilization of timber in China and other parts of the World

Based on the National Forest Investigation in 2014, usable forest is about 62.9308 million hectares. Compared with the seventh National Forest Investigation in 2008, it has increased by 2.8564 million hectares. However, China's average volume of standing timber per hectare is 73.13 m^2, which is just 69% of the world average. Having in mind the deforestation in the 20th century, the situation of forestry in China is described as having inadequate forest resources and declining in forest quality [3]. Besides, insufficient forest resources, the limitation of feeling trees adopted by the NFPP (Natural Forest Protect Program) also widen the gap of China's timber supply and demand [2]. During the Great Leap Forward (an economic campaign of P. R. China from 1958 to 1961), the number of forest resources began to decline and until the late 1970s the forest resources nearly depleted. Since 1980s, there is no development in timber construction field in China [4]. From then on, due to the shortage of timber resource, government set strict rules to limit the use of timber and appealed to use steel and concrete. Thus, the timber construction decreased rapidly and the relevant studies were stopped [5]. Nowadays, supply shortage of the wood resources has become the bottleneck of the further development of the national wood industry. The production of timber plantation cannot meet the increasing demand of wood consumption, so import is considered to be the primary solution to this issue.

When timber construction development stagnated in China, it progressed rapidly in other countries [4]. Take Canada as an example, timber industry has become one of the pillar industries; the standards and technology have developed well [6]. In China, merely 1% of new residences are built with timber; only 26% of wood based panels are used as construction materials and most are finishing material. In comparison, timber utilization rate in construction among Canada, USA and Japan are over 40% [5].

In USA vast majority of the low-storey apartments are timber or half-timbered structures [7]. 1.5 million residences are built in America annually and 70% are timber [8]. And in Japan, timber residences count for about 45% of the total residences, and apartments that under 4-storey are almost all in timber construction [7]. In addition, multi-storey and large-span timber constructions have become two focuses since the development of cross-laminated timber (CLT), Glulam and structural composite lumber (SCL). Many countries, such as Japan, Canada, USA and European ones, have set standards relevant to multi-storey and large-span timber construction to guide the design and construction [9, 10].

3. Physical properties of timber

Timber is a Visco-elastic solid material affected by instantaneous elastic and creep deformation, and its elasticity is primarily related to water content, density and angle between loading direction and grain. Due to the hygroscopic nature of timber, creep is accelerated by moisture variations, resulting in the mechano-sorptive effect [11]. Timber is anisotropic, making the properties have much higher stiffness and strength parallel to grain than across the grain. In addition, mechanical properties of timber vary with moisture content, density and type of loading.

Timber is the only primary material that comes from a renewable resource. Timber is reusable, recyclable and biodegradable and the growth cycle is repeatable. Compared with other commonly used construction materials, timber possesses the characteristics of low thermal conductivity, high strength to weight ratio resulting in less labour and machinery when using in construction shown in table 1. Besides, manufacturing of timber products needs much less energy than other materials [1].

Table 1. Strength/density ratios for some structural materials [1].

MATERIAL	DENSITY (kg/m^3)	STRENGTH (MPa)	STRENGTH/DENSITY [10^{-3} $MPa.m^3/kg$]
Structural steel	7800	400-1000	50-130
Concrete (compression)	2400	30-120	13-50
Clear softwood (tension)	400-600	40-200	100-300

Clear softwood (compression)	400-600	30-90	70-150
Structural timber	400-600	15-40	30-80

As for the disadvantages of the timber, its moisture content and anisotropic nature can influence the strength of timber. The dry shrinkage and wet expansion characteristics will change the size of the timber, and the difference of shrinkage and swelling in each direction can cause the cracking and warping due to its anisotropy. Additionally, according to tests on large numbers on small clear specimens, there is a significant variation in strength of similar specimens [12]. Hence, the large variability of timber is another disadvantage that needs to be considered. Moreover, timber has high tendency to deteriorate, from bacteria, fungi, insect, chemicals, mechanical, sunlight, water and fire. To mitigate these processes, some measures have to be adopted such as seasoning process, so that to prevent the timber from possible decaying and make it suitable for use. Beyond that, timber itself has natural defects due to abnormal growth; shakes, rind gall, and wane-edges. All of them can weaken the applicability of timber [1].

4. Challenges and potentials with new developments

4.1. Engineered wood products

In the traditional Chinese buildings, logs and sawn timber are normally used for roof, floor, and wall framing. Recently, engineered wood products are produced, by processing the natural material, to maximize the properties. It results in improvements of technologies and new products of better parameters than of natural product allowing utilization of species that were not considered before [3].

Engineered wood products consist of broad class of structural wood products that are commonly used as structural materials in construction. These products are made from veneers, strands and sliced wood, which are grouped into a specific structure and then bonded together with adhesives to make panels, timber-like or shaped structures by high-pressure and/or high-temperature. There are several typical engineered wood products, like veneer-based material such as plywood, composite material such as oriented strand board (OSB) and laminates such as cross laminated timber (CLT) [12].

4.2. Challenges from Chinese perspective

A few reasons explaining why timber is not widely utilized in constructions in China have been identified. First of all, there is no proper policy to support timber constructions in China. According to Design of Timber Structures GB 50005-2003 [13] the number of stories is confined to only three, and the maximum length as well as area of building are also limited. Chinese designers have been familiar with concrete and masonry construction rather than timber constructions. Thus, if developers want to build using timber, it will cost much more than with traditional materials, because they need to employ designers with high expertise, mainly from abroad. Moreover, complex technology and trained employees are required for production, construction, and maintenance, which increase costs of these further steps. Currently the technology of timber and engineered-wood is not suitable for skyscrapers. Until now, the highest wooden building using CLT in the world, is 53 meters tall; Brock Commons which located in Vancouver at the University of British Columbia [14]. And China has the significant acceleration of urbanization, and there is huge demand for very high buildings. Lastly, common perception on timber, as being not a solid material and there are doubts about its fire resistance, especially in comparison to reinforced concrete, is also a barrier [10].

4.3. Potentials to be used in China

First of all, concrete as a main resource of construction in China, has caused a severe problem to the environment. To mitigate the problem of vast energy consumption and severe environmental impact, timber can be used as construction material since it is renewable, durable, energy-efficient and cost-efficient, acting as substitutes of concrete to some extent. Concrete buildings in China are more than 99% of the total amount [5], therefore the potential for timber is great enough, even due to possibility of changing this monotony in using mainly concrete. Furthermore, since natural forest recovery, artificial forest, timber resources are abundant enough to resume the utilization in China.

New technologies within engineered-wood products have been researched and developed recently in China and other countries. Products such as glulam, OSB and CLT rely mainly on import, so there is a gap for local producers. With the more advanced technology, engineered-wood products are more likely to be competitive, in cost, suitability and sustainability, than other traditional materials. In addition, a series of standards related to timber buildings are under development and review in China. In 2015, the polity was introduced that 30% of prefabricated materials should utilize wood in Chinese construction industry by 2025 [15]. Furthermore, the height limit is being to be increased from three to four for timber buildings soon in China. With these specific construction standards, timber is more suitable for the particular areas of China, especially those urban areas where low-rise house is allowed only and suburb areas with lower land price and beautiful scenery.

There is an increasing demand for the villa-type wooden structure houses in metropolitans such as Shanghai, Beijing and Shenzhen, and timber structures. Furthermore, because of living habits, foreigners who work in China are another potential group that inclines to timber-framed building. Another possible potential for timber is related to active seismic zones in China. Light weight and ductility of timber allows it to absorb seismic energy without collapse. This has been proved only timber-framed structures were intact in Wenchuan earthquake in 2008, while other buildings were damaged badly [16].

5. Survey results and analysis

To investigate the current situation and suitability of Chinese-grown timber for the manufacture of laminated panel a survey of the construction industry was made. A questionnaire was designed (in Chinese and English language) consisting of over 20 questions related to current usage of different materials for construction projects, and predictions for future, based on the current trends. Feedback was obtained from carefully selected 9 industry members representing consulting and architecture companies who are the main decision-makers on the type of the construction material used for the projects.

From feedback, participants were mainly from private companies (7 participants) with less than 300 employees and the other two participants represented public company and public institution. Their field of business covers consulting (5 of 9), architecture (3 of 9) and research (1 of 9). Although area served are national (77.8%) or worldwide (22.2%), their companies are concentrated on Shanghai and Suzhou, Yangtze Delta Region. As presented in figure 2, almost all participants (8 of 9) mentioned that they used solid wood before, and some used Glulam (4 participants), laminated-veneer lumber (3), I-joist (2) and OSB (2) on their projects. Also recently developed composites: CLT (1 participant) and bamboo structural elements (2 participants) were used in their projects in China.

The three most important characteristics that participants take into consideration when selecting materials are strength (66.7%), durability (55.6%) and fire resistance (55.6%), see figure 3. However, material cost (44.4%) and construction cost (44.4%) are also important for these companies. In terms of the familiarity with engineered-wood products, more than half (5 participants) only heard about them but have never used; two participants had limited contact with engineered-wood products and know the functions and three are familiar with engineered-wood products well.

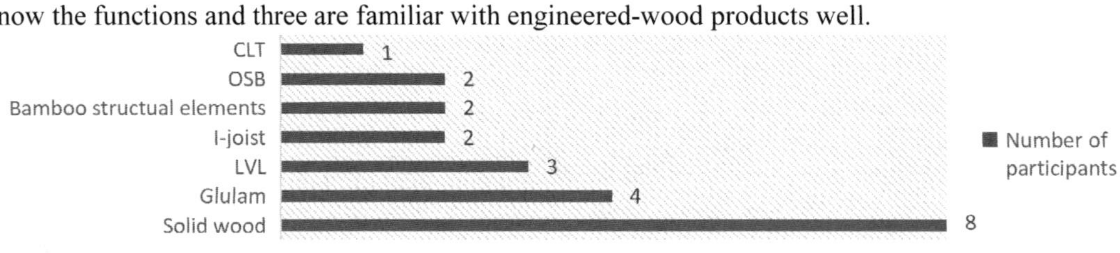

Figure 2. Participants' response on 'Wood-based products that used before'.

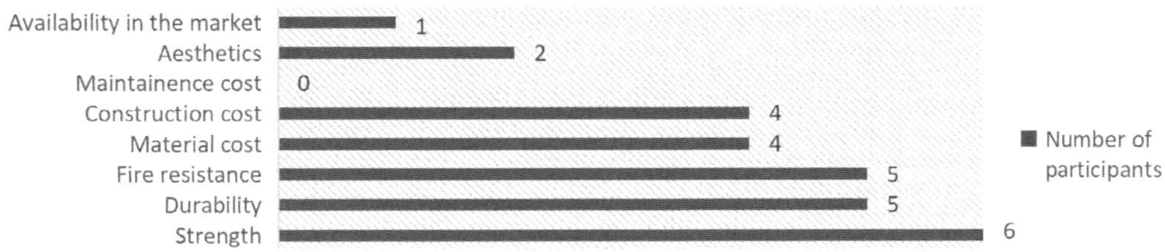

Figure 3. Participants' response on: 'The most important characteristics of material in construction'.

Based on results, CLT is seldom used, which is consistent with the current situation in China. Moreover, there is no specific answer about which type of construction has the highest potential for using CLT. With regard to the possibility of using CLT in the future, three participants think CLT will not be used and five consider the possibility is less than 50% in the next two years, while one people expects it is high possible (larger than 50%) to use CLT in the next projects. From the perspective of the companies, availability in the market and material costs are the biggest barriers that limit the use of CLT, which was pointed out by seven participants. Other factors include technology (2 of 9) and strength performance (2 of 9). In addition, both literature review and questionnaire results have highlighted lack of knowledge of the material has limited its usage.

The participants of the survey were asked to rank the materials: CLT, engineered bamboo, steel, concrete, masonry and timber-frame from the different aspects. According to overall tendency of the feedback, the characteristics of these construction materials can be concluded generally in a way as shown in table 2. Timber products were considered as well performing in the aspects of aesthetics, maintenance cost and environmental friendliness. It was found that engineered-wood products can be cost-efficient. In addition, its durability and strength performance can be as good as concrete and steel; some are even better with the special manufacturing. And for fire resistance, although fire will ignite timber, it takes time to burn the structure to collapse.

Table 2. The characteristics ranking of construction materials.

CHARACTERISTIC	BEST	WORST
Acoustic performance	Concrete	Steel
Aesthetics	Engineered bamboo	Steel, Masonry
Availability in the market	Concrete	CLT
Material cost (low to high)	Concrete	Steel
Construction cost (low to high)	Masonry	Steel
Maintenance cost (low to high)	CLT	Steel
Durability	Concrete	Timber-frame
Strength performance	Steel	Timber-frame
Fire performance	Concrete	Timber-frame
Environmental friendliness	Timber-frame	Concrete

6. Conclusions

The results presented in this study clearly demonstrates that engineering timber could be further used in construction industry in China due to its advantages. It should be noted, however, there are many obstacles resulted from engineering timber technology development and lack of relevant legislations, design and construction codes. To improve the situation, it is suggested that more research effort from the main stakeholders, such as government and universities, in the construction industry in China should be made to address different aspects of timber's further utilization. Moreover, the industry survey revealed that bamboo can have great potential in China due to the availability and properties. It is recommended that comprehensive studies on utilization of bamboo in constructions should be performed.

Acknowledgments
The authors would like to thank Xi'an Jiaotong-Liverpool University support for funding the Summer Undergraduate Research Fellowship (SURF) to undertake the project.

References

[1] Thelandersson S, Larsen HJ. *Timber Engineering.* John Wiley & Sons Ltd, 2003
[2] Yang HQ, Nie Y, Ji CY. Study on China's timber resource shortage and import structure: Natural forest protection program outlook, 1998 to 2008. *Journal of Forest Production.* 2010: 60(5): pp 408-414
[3] Zhang CC, Gao XL. Analysis on timber supply capacity of forest resources in China. *Forest Resources Management.* 2014: 6(7): pp 51-55
[4] Zhou HB, Fei BH, Ren HQ. Evolutions of Research on Wood Frame Housing around the World. *China Wood Industry* 2006: 20(4): pp 1-4
[5] Fei BH. The Analysis of Timber- framed Building Development in China. Paper presented at: *International timber construction industry exchange conference, 2010, China*
[6] Wang CY, Wang X. Reflection of the Development of Timberwork Building in China. *Lín Chǎn Gōng Yè. 2013*: 40(5) pp 14-25
[7] Lv JG. The development analysis of timber structural construction at home and abroad. *Architecture and Decoration.* 2012: 2 pp 37-38
[8] Wang XJ, Wng T, Jiang SL. Present Situation of the Production of Wooden Houses Abroad. *VIBP.* 2005: 8(5) pp 5-8
[9] Xiong HB, Ou YL, Wu Y. State-of-the-art Research of Tall Wood Buildings. *Journal of Tongji University.* 2016: 44(9): pp 1297-1306
[10] Riala M, Hola L. Multi-storey timber construction and bioeconomy – barriers and opportunities. *SCAND J FOREST RES.* 2014: 9(4) pp 367-77
[11] O'Ceallaigh C., Sikora K., McPolin D., Harte A. M., Viscoelastic creep in reinforced glulam, World Conference on Timber Engineering, Vienna, Austria, Eds.: J. Eberhardsteiner, W. Winter, A. Fadai, M. Pöll, Publisher: Vienna University of Technology, Austria, 2016
[12] Harte A., McPolin D., Sikora K., O'Neill C., O'Ceallaigh C., Irish Timber – Characterisation, Potential and Innovation, Civil Engineering Research in Ireland, Belfast, UK, 2014
[13] Glulam LTD. *What is Glulam.* [Web Page]. Available from: http://glulambeams.co.uk/about-glulam/what-is-glulam [cited 2017 August 16].
[14] The Highest Wood Frame Housing in the World: Brook Commons Project. *Construction Technology.* 2016: (5): pp 34-35
[15] Yu L. Department of housing construction: Prefabricated construction will be more than 50% of new constructions in 2025. *Brick and Tile.* 2016: 6: pp 41
[16] Wang CY, Wang X. Reflection of the development of timber building in China. *China Forest Products Industry.* 2013: 40: pp 14-25

IOP Conf. Series: Materials Science and Engineering **383** (2018) 012030 doi:10.1088/1757-899X/383/1/012030

Evaluation of RC building strengthened with column jacketing method with consideration of soft-story

A H Sinaga[1], H Wijaya[2*] and D R Teruna[3]

[1]Department of Civil Engineering, University of Cendrawasih, Jayapura, Indonesia
[2]Department of Civil and Construction Engineering, Swinburne University of Technology, Melbourne, Australia
[3]Department of Civil Engineering, University of Sumatera Utara, Jln. DR. Mansyur Kampus USU Medan 20155, Indonesia

*E-mail: 101141353@student.swin.edu.au

Abstract. The assessment and strengthening method of existing reinforced concrete buildings in terms of seismic responses have been highly important for the past decades especially buildings with soft-story mechanism which are found to be more vulnerable. Several strengthening techniques have been proposed including Fibre Reinforced Polymer and steel jacket of RC column elements. This paper aims to evaluate the effectiveness of two strengthening techniques with variability of soft-story under severe earthquake excitations. Five-story RC building is investigated and non-linear dynamic analysis is performed to capture the inelastic responses of the structure. Four sets of ground motion are selected and matched with target response spectrum for Aceh earthquake. Seismic responses of the considered models and strengthening techniques are compared.

1. Introduction

Recent reported earthquake damage in Indonesia [1] has shown the higher percentage of structural failure associated with soft-story mechanism especially for reinforced concrete building. The soft-story mechanism occurs when the stiffness of one story is much less than the adjacent story stiffness. The structure was designed with inadequate seismic vulnerability assessment. Rajeev [2] reported the effect of soft-story parameters in seismic fragilities of the structure. The structure with higher soft-story level was found to be more vulnerable in moderate to high intensity earthquake. Therefore, strengthening technique to overcome the challenge in soft-story building have been proposed and developed for the past decades. The most common retrofitting techniques are FRP and steel jacketing which have the efficiency in terms of workability and implementation. Evaluation of each strengthening method is critical to obtain the most efficient technique in retrofitting of the building.

The applications of FRP and steel jacketing method and finite element modelling techniques for reinforced concrete column have been reported [3,4]. Recently, Seifi et al. [4] proposed the physical model for FRP confined concrete column calibrated with the experimental data. Moreover, steel jacketed concrete column model was also proposed Campione et al. [5]. Recent Study conducted by Futardo et al. [6] reported the efficiency of strengthening technique in soft-story building. The study shows the method of column jacketing is effective in reducing seismic responses in soft-story building. However, the application of the strengthened column model in building with variability of soft-story

Content from this work may be used under the terms of the Creative Commons Attribution 3.0 licence. Any further distribution of this work must maintain attribution to the author(s) and the title of the work, journal citation and DOI.
Published under licence by IOP Publishing Ltd

level has yet been evaluated. The premise of this study is to illustrate the efficiency of retrofitting techniques in reinforced concrete building with different soft-story parameters.

2. Building model

Five-story reinforced concrete building is employed to illustrate the effect of strengthening technique in different soft-story effect. Figure 1 depicts the building geometry and main element detailing. The building is designed to resist 20 kN/m gravity load and seismic load based on based on SNI-1726-2012 for Indonesian Earthquake Code with the location is in Aceh. The concrete has compressive strength of 25 MPa and the yield strength of the reinforcement steel is 420 MPa.

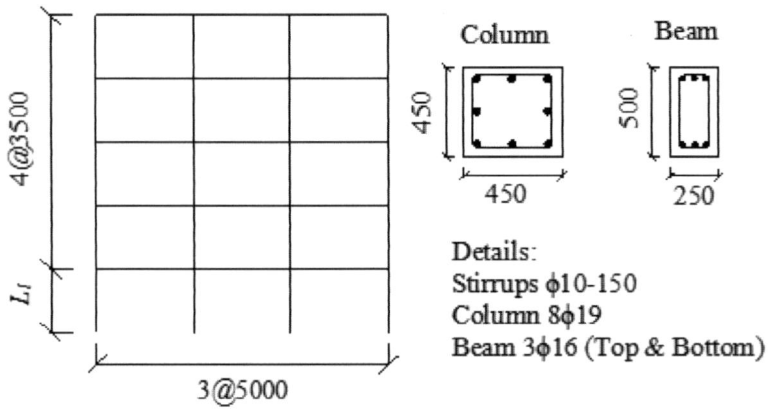

Figure 1. Building details.

3. Impact of soft-story

The definition of soft-story is that the stiffness of lateral resisting force in any story is less than 70% of the stiffness of the adjacent story which induces the localized concentration of drift. The soft-story is quantified by story lateral stiffness, therefore the parameter of soft-story used in this paper is the relative height between the first-story column and the adjacent floors column height.

$$SS = \frac{K_2}{K_1} = \left(\frac{L_2}{L_1}\right)^3 \tag{1}$$

where K_1 and K_2 are the lateral stiffness of first and second story while L_1 and L_2 defines the column height of the first and second story respectively. In this study, the parameter *SS*(Soft-Story) is chosen as 40%, 60%, 80% and 100% to illustrate the sensitivity of the parameter to the structural responses. The story height for each parameter of *SS* is summarized in table 1.

Table 1. First story height for different SS.

SS	%	First story column height (mm)
1	100	3500
0.8	80	3770
0.6	60	4150
0.4	40	4750

4. Element model

The building model is constructed in OpenSees finite element software [7]. *Concrete01* material is applied to model the confined and unconfined concrete properties based on Mander et al. [8] respectively. Moreover, *Steel02* material based on Giuffre-Menegotto-Pinto is used to model the reinforcement bar. The beam and column elements are constructed with flexibility-based fibre-section element with 5 Gauss-Lobatto integration points. Section aggregator is used to model the shear-axial and flexural response. The method of including shear can be found in [9]. Furthermore, P-Delta effect and Corotational transformation is included to compute the second order effect for columns and beams respectively. Stiffness proportional damping is used and calibrated to be 5% for the first elastic mode. For this study, the base support is assumed to be fixed.

5. Strengthening technique and modelling

In this study, the soft-story column is located at the first-story where the failure is most-likely to occur. Therefore, only the first-floor column is strengthened to observe the effect of the retrofitting method.

5.1. FRP jacketing

The base-column is strengthened with 3 layers of MasterBrace FIB 450/50 Carbon Fibre Sheet with fabric thickness of 0.255 mm, fabric width of 500 mm, tensile strength of 4900 MPa and 230 GPa tensile elastic modulus with near surface-mounted (NSM) system of 13 mm diameter reinforcement bar. The *ConfinedConcrete01* material available in OpenSees [7] is utilized to assign the effect of FRP confinement in RC column and applied in fibre-section element. The application of this method can be found in [4]. The experimental conducted in [4] showed the failure mode of FRP jacketed column was due to the buckling of the bar. Therefore, the *ReinforcingSteel* material is assigned to model the reinforcement bar to incorporate the buckling effect based on Dhakal and Maekawa model and rupture behavior.

5.2. Steel jacketing

The steel jacketing technique used in this study is based on the work of Rosario Montouri and Vincenzo Piluso [10]. The column is strengthened with angles and battens to increase the stiffness and confinement. Angle size of 150.150.12 is used with batten width and thickness of 15mm and 3mm respectively with 250 MPa yield strength applied 0.5 m from the column end. Figure 2 shows the strengthened column in this study. The physical model approach used in this study was proposed by Campione et al. [5]. The angles properties were modified from the work of by defining the interface and contact between the steel angles and reinforced concrete column. The maximum load can be resisted by the steel angles is the function of interface stress along the direction of the contact surface as shown in equation (2).

$$P_a^* = 2.n_a.n_{la}.l_0\left(c_0 + \mu.f_{le,\max}\right) \tag{2}$$

where n_a is the number of steel angles, n_{la} is the width of each side of the angles, l_0 is the total length of the angles, c_0 is the cohesive strength, μ is the friction coefficient [5] and $f_{le,max}$ is the maximum confinement pressure determined based on [9]. The yield strength of the angles is then defined based on the force over total-area relationship as defined in equation (3).

$$f_{y(angle)} = P_a^* / A_{(angle)} \tag{3}$$

$$E_{(angle)} = f_{y(angle)} / \varepsilon_{0(angle)} \tag{4}$$

The steel-jacketed concrete column is modelled by utilizing fibre-section element with the angles with the angle is modelled with *Steel02* material with properties as described previously.

IOP Conf. Series: Materials Science and Engineering **383** (2018) 012030 doi:10.1088/1757-899X/383/1/012030

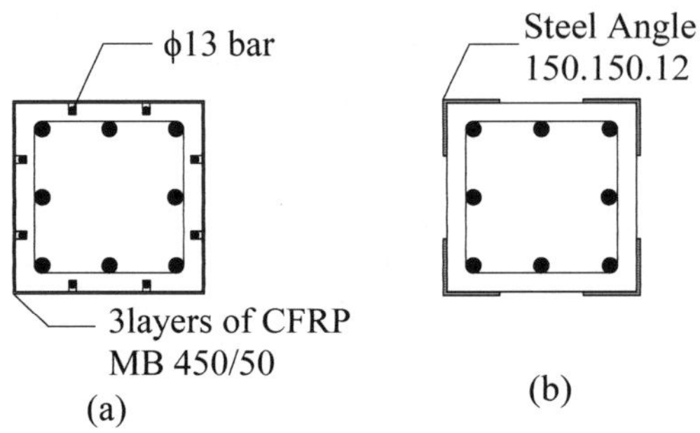

Figure 2. Strengthening model (a) FRP jacketing with NSM (b) Steel jacketing.

6. Earthquake ground motions

Four set of ground motion records are selected and matched with target response spectrum of Banda Aceh earthquake with site class D. The records include Kobe, Loma Petra, Tabas and Managua records. Figure 3 depicts the acceleration spectrum for matched four records and target spectral acceleration.

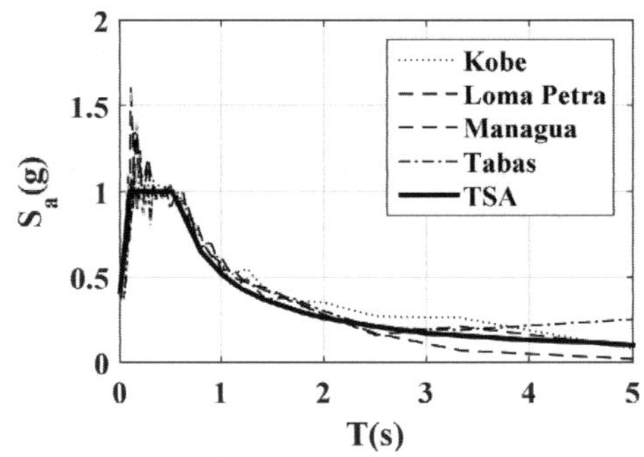

Figure 3. Matched and target response spectrum.

7. Result and discussion

Inter-story drift of each building models considering different strengthening method and soft-story variability is compared. Inter-story drift represents the structural damage as described in FEMA356 [11] which is 1.0% for Immediate Occupancy, 2.0% for Life-Safety and 4.0% for Collapse Prevention. Figure 4 to 7 show the maximum inter-story drift under four different ground motions as described in previous section.

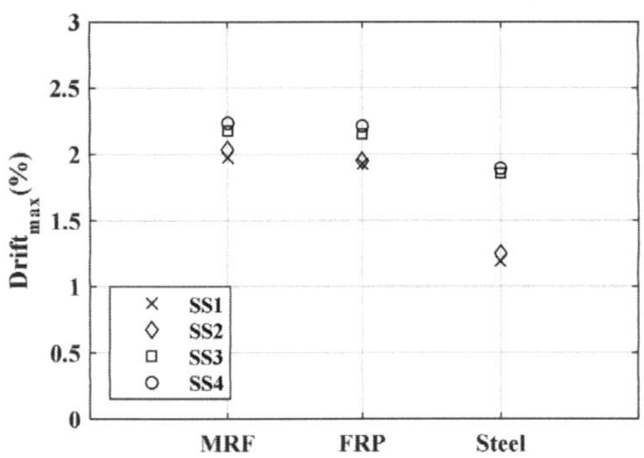

Figure 4. Maximum drift for Kobe.

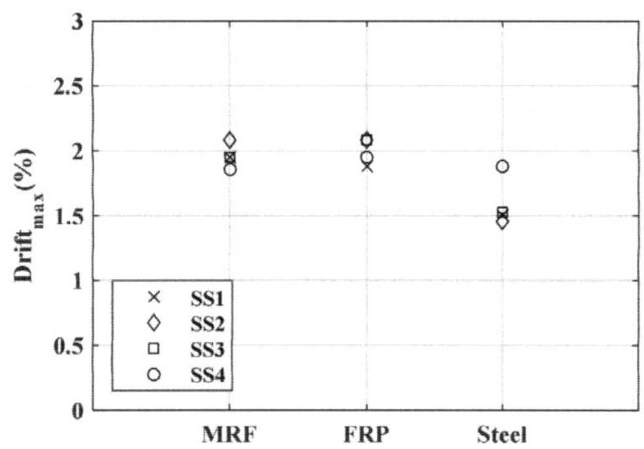

Figure 5. Maximum drift for Loma Petra.

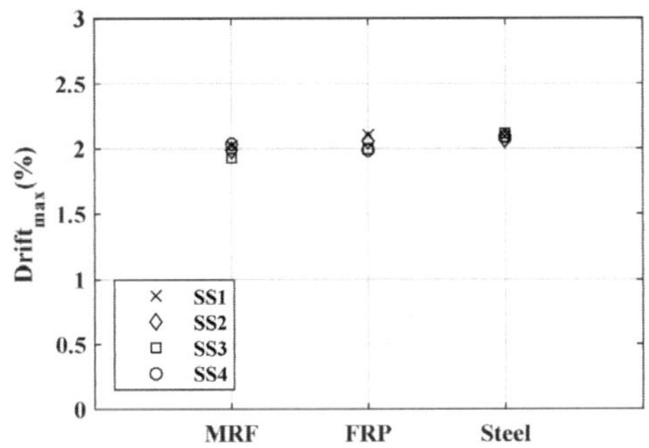

Figure 6. Maximum drift for Managua.

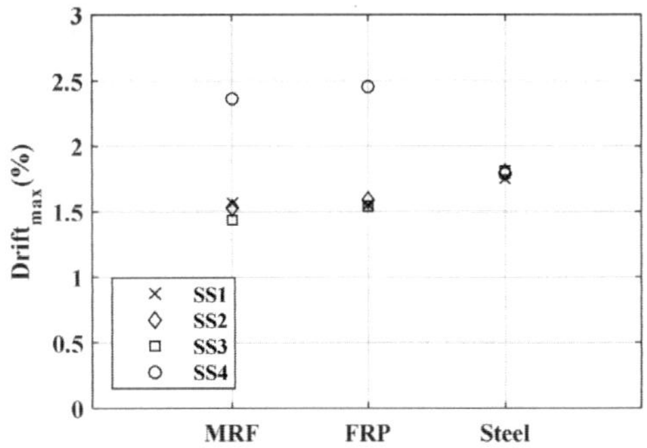

Figure 7. Maximum drift for Tabas.

The result in terms of maximum story drift from four set of ground motions considered is compared. For building with soft-story parameters of 1.0 and 0.8, both FRP and steel jacketing provides better responses than unstrengthened building for Kobe and Loma Petra records. It is noticed that in this case, the strengthening method with FRP and near surface mounted system is ineffective in increasing structural performance due to buckling of NSM bar. For Kobe and Loma Petra earthquake, the result shows that steel jacketing technique is effective in reducing the seismic response for *SS1* and *SS2*. However, the effect of steel jacketing method is found inefficient when the soft-story parameter increases to *SS3*. Furthermore, to observe the effect of strengthening methods in structure lateral capacity, pushover analysis is performed with inverted triangle lateral force distribution. Figure 8 to 10 depicts the capacity curve of each base column model and soft-story parameter.

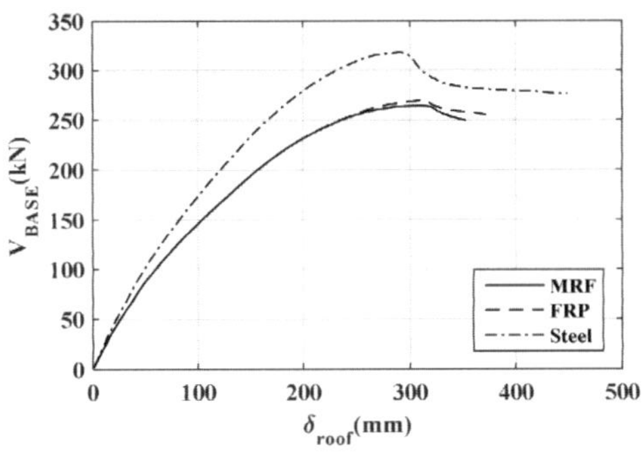

Figure 8. Capacity curve for *SS1*.

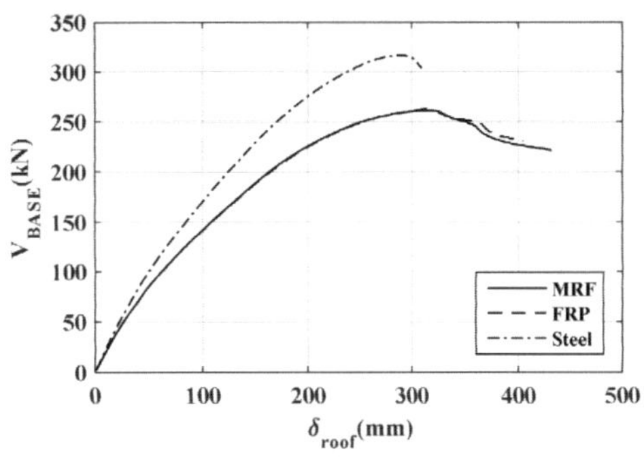

Figure 9. Capacity curve for *SS2*.

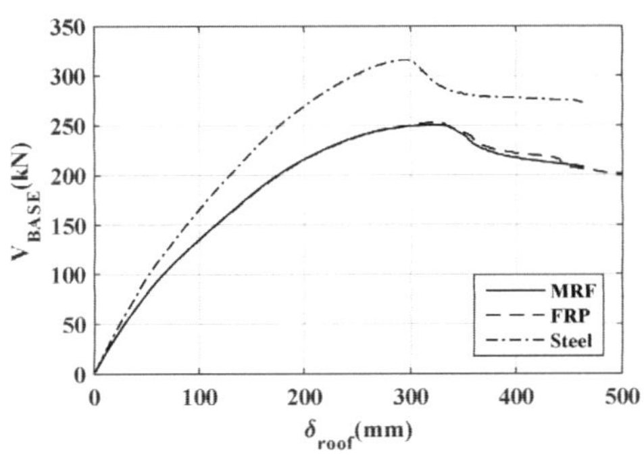

Figure 10. Capacity curve for *SS3*.

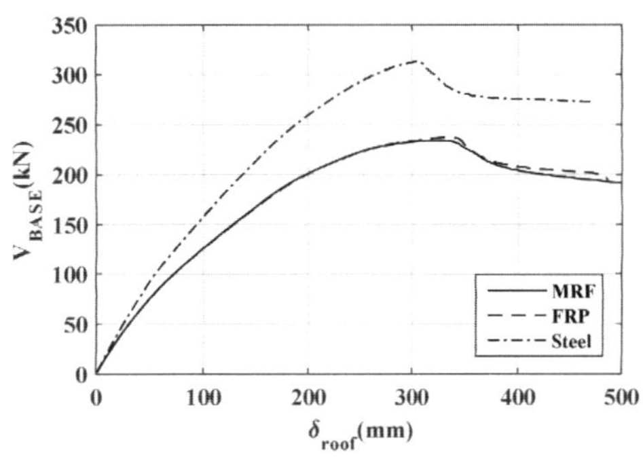

Figure 11. Capacity curve for *SS4*.

8. Conclusion

Five-story reinforced concrete building with variability of first story column height to represent the soft-story is evaluated under two different jacketing retrofitting technique including FRP jacket with near surface mounted system and steel jacket with angles and battens. Pushover and nonlinear dynamic analysis are performed to assess the nonlinear responses of the structure. Based on the analysis results, several observations are made as follows:

1. The soft-story affect the lateral capacity of the reinforced concrete building. It is observed that the increasing height of the first story results in decreasing peak value of base shear in pushover curve for all building models.
2. Base column strengthened with FRP and near mounted surface system is noticed not efficient in strengthening the structure. More numbers of NSM bars and layers of CFRP may be required for the strengthening method to be applicable. Furthermore, providing the FRP jacket not only at the base column may also be considered.
3. Steel jacketing techniques shows efficiency in increasing structure lateral stiffness and decreasing the nonlinear response under earthquake excitation.
4. Effectiveness of strengthened column varies under different ground motions.

Further study shall be made incorporating more parameters involved in comparing the strengthening method for soft-story building including different location of strengthened elements, concrete jacketing technique, structural vulnerability and different numerical model for strengthened elements.

References

[1] Ghobarah A, Saatcioglu M, & Nistor I (2006) The impact of the 26 December 2004 earthquake and tsunami on structures and infrastructure Engineering structures, 28 (2), pp 312-26
[2] Rajeev P, & Tesfamariam S (2012) Seismic fragilities for reinforced concrete buildings with consideration of irregularities Structural Safety, 39, pp 1-13
[3] Garcia R, Hajirasouliha I, & Pilakoutas K (2010) Seismic behaviour of deficient RC frames strengthened with CFRP composites *Engineering Structures*, 32 (10), 3075-85
[4] Seifi A, Hosseini A, Marefat M S, & Zareian M S (2017) Improving seismic performance of old-type RC frames using NSM technique and FRP jackets *Engineering Structures*, 147, pp 705-23
[5] Campione G, Cavaleri L, Di Trapani F, Ferrotto M F, Macaluso G, & Papia M (2015) Modelling steel jacketed RC columns: Remarks by experimental-numerical comparisons *In OpenSees Days, 2nd Italian Conference* pp. 10-11
[6] Furtado A, Rodrigues H, Varum H, & Costa A (2017) Evaluation of different strengthening techniques' efficiency for a soft storey building *European Journal of Environmental and Civil Engineering*, 21 (4), pp 371-88
[7] McKenna F, Fenves G L, & Scott M H (2000) Open system for earthquake engineering simulation *University of California, Berkeley, CA*
[8] Mander J B, Priestley M J, & Park R (1988) Theoretical stress-strain model for confined concrete *Journal of structural engineering*, 114 (8), pp 1804-26
[9] Franchin P, A O L O, & Pinto P E (2007, April) Transitability of mainshock-damaged bridges *In Proc. 1st joint US Italy workshop on seismic design of bridges* Pavia, Italy: IUSSPress
[10] Montuori R, & Piluso V (2009) Reinforced concrete columns strengthened with angles and battens subjected to eccentric load *Engineering Structures*, 31 (2), pp 539-50
[11] FEMA 356 (2000) Prestandard and commentary for the seismic rehabilitation of buildings

Evaluation on R&D projects in Korea to predict and reduce natural disaster

Lee Kangguk[1,*], Wonhwa Hong[2] and Kim Su-Young[3]

[1]Disaster and Safety Management Institute, Kyungpook National University, Korea
[2]School of Architectural, Civil, Environmental and Energy Engineering Kyungpook National University, Korea
[3]Disaster and Safety Management Institute, Kyungpook National University, Korea

*Email: ggyi@naver.com

Abstract. This study is intended to evaluate and analyze R&D projects for natural disaster prediction and reduction that have been conducted from 2003 up to 2016 in Korea and use it as foundational material for future R&D projects for disaster prevention technology in Korea. In consideration of essentiality about the characteristics of Korea's natural disaster prediction and reduction R&D projects and achievement of project goals, scientific, technological, and social performances are arising. Concerning "technical performance", in 2016, there were 4 patents registered receiving over 80 points (out of 100) (50%), so in the future, technology level is expected to rise greatly. Moreover, a lot of research results are being applied practically to the affair of Ministry of Public Safety and Security.

1. Introduction

In Korea, natural disaster prediction and reduction R&D projects have been carried out in order to establish and promote scientific technology promotion policy in the area of disaster and safety management and encourage R&D associated with disaster prevention technology and disaster prevention industry to protect people's life, properties, and major facilities [1]. This study is intended to evaluate and analyze R&D projects for natural disaster prediction and reduction that have been conducted from 2003 up to 2016 in Korea and use it as foundational material for future R&D projects for disaster prevention technology in Korea.

2. The contents of R&D projects for natural disaster prediction and reduction in Korea & the management system

2.1. The promotional process and contents of the projects

Until 2003, development projects for natural disaster prevention technology in Korea had been transferred to national R&D projects, and according to 'pan-ministry flood prevention measures', the area of flood disaster came to be separated as projects to develop flood disaster prevention technology from 2004. Since then, revisions have been repeatedly made according to the reorganization of Korea's government agencies and directions of government policy [2]. It is manly about devising mid- and long-term plans for the 10-year R&D project for natural disaster reduction and building up disaster prevention infrastructure to realize a safe nation with no concern for disasters like flood or landslides [3]. According to the designation of sunset industry in 2016, continuous tasks were completed, and

Content from this work may be used under the terms of the Creative Commons Attribution 3.0 licence. Any further distribution of this work must maintain attribution to the author(s) and the title of the work, journal citation and DOI.
Published under licence by IOP Publishing Ltd

then, the project ended. This study classifies R&D projects for natural disaster prediction and reduction that are being carried out in Korea (table 1).

Table 1. Classification on Korea's natural disaster prediction and reduction R&D projects.

Division	Contents
Evaluating the risk of natural disaster, predicting damage, & developing prevention technology	- Analyzing and evaluating disaster vulnerability and risk to prevent damages to facilities and minimize the impact and building up strategies to prevent, prepare for, and cope with disasters by developing technology to predict damage - Developing technology to build up a high-efficiency • high-performance prediction • warning system using ICT convergence technology and smart technology in order to establish a prompt and accurate prediction • warning system
Developing technology to enhance capabilities to prepare for and also cope with natural disaster	- Developing disaster management technology to cope with the increase of risk caused by climatic changes and technology to improve disaster-preventing competence to cope with disaster for less damage - Predicting volcanic disaster promptly, developing analysis technology to minimize damage and cope with it systematically, and building up an international cooperation network for collaborative coping
Technology to support victims from disasters to help them return to normal life	- Studying ways for efficient restoration about large-scale complex damaged area and establishing work continuance management technology to cope with earthquake or tsunami
Collaboratively planned multi-ministry social problem-solving projects (Disaster safety service)	- Devising a social agreement about the level of national disaster relief policies or services corresponding to the national status and developing technology to optimize disaster relief and budget operation

2.2. The project management system

In 2016, Korea's natural disaster prediction and reduction R&D projects were managed and operated by Kyungpook National University's enterprise organization for disaster safety technology development [4]. The enterprise organization is generally overseeing R&D project planning, selection, evaluation, progress management, and adjustment, performing evaluation on selecting research projects, mid-progress report sessions, annual evaluation, and final evaluation, conducting research on demands from all the industrial and academic links to discover new research projects continuously, and also managing and supporting researches conducted by research institutes. The project management system is shown as below (figure 1).

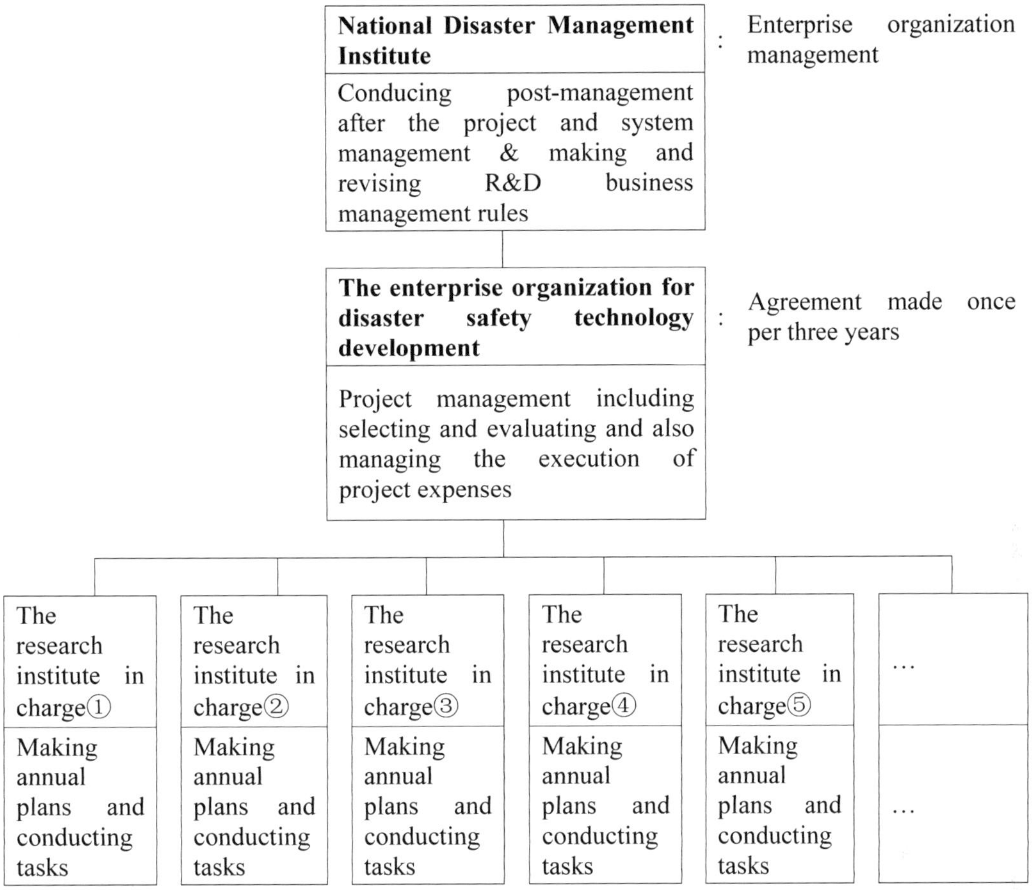

Figure 1. The management system for Korea's natural disaster prediction and reduction R&D projects.

3. The performance plans and performance achievement of Korea's natural disaster prediction and reduction R&D projects

3.1. The performance plans of Korea's natural disaster prediction and reduction R&D projects

The ultimate performance goal of Korea's natural disaster prediction and reduction R&D projects is to build up customized coping technology R&D and local disaster prevention infrastructure based on scientific technology to predict natural disaster that can occur from future climatic changes and reduce damage, realize people's safety, and establish a local disaster safety system by securing 77% of the 「Natural Disaster Monitoring • Prediction • Coping Technology」 level compared to that of advanced countries.

3.2. The performance achievement of Korea's natural disaster prediction and reduction R&D projects

Korea's natural disaster prediction and reduction R&D projects are integrated with social disaster safety technology development projects in 2017 and reorganized as disaster prediction and reduction R&D projects [5]. Therefore, in 2016, at the time of setting up performance goal indexes with the institution in charge of evaluation on government contribution specialized in investigating, analyzing, and evaluating Korea's national R&D projects in Korea Institute of Science & Technology Evaluation and Planning, only 2016 goals were established. In 2017, based on the reorganized projects, performance goal indexes will be set up again. Also, with the designation of sunset industry, budgets and tasks are reducing continuously (table 2).

Table 2. The performance goals of Korea's natural disaster prediction and reduction R&D projects in steps and the score provision of performance indexes.

Performance indexes ①	**Index weight ②**	**The target ③**	**Achieved value ④**	**Achievement rate ⑤**	**Performance index weight ⑥**	**Scores by indexes ⑦**	**The final score ⑧**
Growth engine creation	0.1	20	28	100%	1.0	3	28.9
Policy use	0.3	4	7	100%	1.0	9	
Academic performance	0.29	52.44	47.65	90.87%	1.0	7.9	
Academic performance(2) (Multi-ministry indexes)	0.01	52.44	0	0	1.0	0	
Technological value	0.3	5	5.1	100%	1.0	9	
Sum	1						

4. The performance validity and performance evaluation of Korea's natural disaster prediction and reduction R&D projects

4.1. The performance validity of Korea's natural disaster prediction and reduction R&D projects

In consideration of the characteristics of Korea's natural disaster prediction and reduction R&D projects and essentiality about the achievement of project goals, scientific, technological, and social performances are arising, and the grounds of performance are presented as follows (table 3).

Table 3. The performance validity of Korea's natural disaster prediction and reduction R&D projects.

Performance type	**Allotting of marks**	**Superiority of performance**	**Connectivity and essentiality**	**Validity of sources**
Scientific	5	• The performance of this project is constantly increasing. But the level of Korea's disaster safety technology is still low as 73%, so journals that can reach the level of advanced countries' technology have not yet produced results.	• The result of this project is practically associated with reducing natural disaster. SCI or journals published are quantitatively increasing, and superiority of their quality is also shown.	• Performance information has been entered to NTIS (National Science & Technology Information Service), and the performance has been examined by KISTEP.

Technological	5	• The result of this project is only the registration of Korean patent, and it can be applied industrially.	• The result of this project is practically associated with reducing natural disaster. The quality of patent is also higher than 'B', so its technological superiority is thought to be shown.	• Performance information has been entered to NTIS, and the performance has been examined by KISTEP.
Social	20	• The social performance of this project is constantly growing. Regarding political application corresponding to core performance, academic and social contribution is thought to be favorable.	• The study result is associated with the affairs of Ministry of Public Safety and Security (current Ministry of Public Administration and Security) and applied politically. The performance system is provided for free and is used to cope with disaster or victims practically.	• 'Superiority to cope with climatic change and 10 items selected' and DB building performance and performance to publicize for the utility of study results are thought to be valid.

4.2. The performance evaluation of Korea's natural disaster prediction and reduction R&D projects

4.2.1. Scientific performance
Regarding Korea's natural disaster prediction and reduction R&D projects, among total 233 theses, SCI thesis are total 44 and KCI (Korea Citation Index) theses are total 189, and for the last three years, the mean mrnIF (Modified Rank Normalized Impact Factor) of SCI theses is found to be 53.59, so qualitative superiority is fairly high. Also, the mrnIF of international journals of disaster risk reduction exceeds 31.32, and about scientific performance, a large number of theses are being published, so constant performance rise is expected.

4.2.2. Technological performance
Regarding the patent application and registered patent performance of Korea's natural disaster prediction and reduction R&D projects, there are recently 47 cases for the last three years, and among them, patents registered are 11 cases. This project's patent performance per 1 billion is 1.34, so it is higher than 1.23, the mean of disaster safety R&D. Concerning "technical performance" to evaluate technological power in registered patents' K-PEG (Korea - Patent Evaluation & Grading), in 2016, there were 4 patents registered receiving over 80 points (out of 100) (50%), so in the future, technology level is expected to rise greatly.

IOP Conf. Series: Materials Science and Engineering **383** (2018) 012031 doi:10.1088/1757-899X/383/1/012031

4.2.3. Social performance

Regarding constant growth in the policy use performance of Korea's natural disaster prediction and reduction R&D projects, multilateral publicity activities are being conducted including exhibitions to increase interest in disaster safety and improve recognition, societies • associations, books and videos, discussion sessions, and articles. Most of the research projects are accompanied with systems and sample products developed and delivered for free to Ministry of Public Safety and Security (current Ministry of Public Administration and Security) and research institutions for national disaster safety. Accordingly, from 2012 property and life loss from natural disaster have been continuously reducing. This is thought to be the indirect effect of this project closely associated with the affairs of Ministry of Public Safety and Security (current Ministry of Public Administration and Security).

5. Conclusion

The purpose of this study is to conduct evaluation and analysis on Korea's natural disaster prediction and reduction R&D projects conducted from 2003 up to 2016 and provide foundational material for Korea's disaster prevention technology R&D projects to be carried out afterwards. The conclusions of this study are as written below:

1. Korea's natural disaster prediction and reduction R&D projects are being carried out in the aspects of developing technology to evaluate the risk of natural disaster and predict and prevent damage, developing technology to enhance competence to prepare for and cope with natural disaster, developing technology to support overcoming from disaster to help return to normal life, and carrying out social problem-solving pan-ministry collaborative planning projects (disaster safety service).

2. In consideration of essentiality about the characteristics of Korea's natural disaster prediction and reduction R&D projects and achievement of project goals, scientific, technological, and social performances are arising.

3. Regarding the scientific performance of Korea's natural disaster prediction and reduction R&D projects, a large number of excellent theses are being presented, so the performance is expected to rise constantly. Concerning "technical performance", in 2016, there were 4 patents registered receiving over 80 points (out of 100) (50%), so in the future, technology level is expected to rise greatly. Moreover, a lot of research results are being applied practically to the affair of Ministry of Public Safety and Security (current Ministry of Public Administration and Security).

Acknowledgement

This research was supported by a grant through the Disaster and Safety Management Institute funded by Ministry of Public Safety and Security of Korean government.

References

[1] National Institute For Disaster Prevention, A Study for the Establishment of Overall Plan on the Disaster and Safety Management Technology, 2007

[2] KNU Industry-Academic Cooperation Foundation, Disaster Safety Technology Development Project's Planning Research for Development of performance indicators, 2016

[3] Government R&D Project budget allocation adjustments(plan). National Science & Technology Council , 2015

[4] National Research and Development Project Evaluation Report, KISTEP, 2012

[5] National Disaster Management Institute, Activiation and general adjustment strengthening plan of disaster safety R&D, 2014

Program spatial movement of high-speed vehicles

Viktor Volodymyrovych Kravets[1], V Sakhno[2], K Bas[1,2,*] and Vl Kravets[3]

[1]Department of Automobiles and Automobile Economy, State Higher Education Institution "National Mining University", 19 Dmytro Yavornytsky av., 49600, Dnipro, Ukraine.
[2]Department of Motor Vehicles, National Transport University, 1 Suvorov str., 01010, Kyiv, Ukraine.
[3]Department "Life safety" of the National University of Railway Transport named after Academician V. Lazaryan, 2 Lazaryan str., 49010, Dnipro, Ukraine.

*E-mail: k.m.bas.69@gmail.com

Abstract. Program spatial movement of a high-speed vehicle has been synthesized with the help of hodograph in terms of spiral paths. A mathematical model of a primary road in the form of a line over ruled bearing surface in the neighbourhood of the program trajectory has been developed. Vector equations of the elements of ruled surface and bearing surfaces oriented orthogonally to inertia forces have been set up. The inertia forces effecting a mass of a high-speed vehicle in the context of non-uniform motion along spiral program path has been demonstrated in a vector form.

1. Introduction

The task of a vehicle is to transport certain freight from one specified spatial point to another while taking into consideration additional conditions. In terms of ground transport, the boundary points and motion route are on the Earth's surface. As for the railway transport, motion route is determined by a rigid route being a single nominal calculation program trajectory. In the context of motor transport, motion route determines the neighbourhood of the admissible trajectories [1, 2]. Neighbourhood of program trajectory in the shape of a line on the bearing surface forms a highway. The development of mathematical model of the bearing surface of a high way as a constituent part of calculation experiment technique providing dynamic design of a "vehicle – road" system is proposed [3-7].

2. Statement of the problem

It is assumed that at initial time and final time (boundary conditions), phase state of a vehicle has been specified. Time for the vehicle transport from one spatial point to another as well as a shape of the spatial trajectory compatible with probable motion of the vehicle is among the heuristically specified additional conditions. In this context, program motion of the vehicle is considered as the motion of a material point in terms of the required mode (i.e. accelerated, decelerated, and uniform) along the selected spatial curvilinear trajectory. The spatial curvilinear trajectory is represented with the help of vector-radius of a material point depending on time as a parameter that is program notion of the vehicle is determined by means of a hodograph of the radius-vector. The hodograph of program motion is taken as a bearing surface. The bearing surface is found in the class of ruled surfaces. It is required to develop mathematical model of a bearing surface corresponding to program spatial transport of the high-speed vehicle.

Content from this work may be used under the terms of the Creative Commons Attribution 3.0 licence. Any further distribution of this work must maintain attribution to the author(s) and the title of the work, journal citation and DOI.
Published under licence by IOP Publishing Ltd

3. Hodograph of the program motion

It is proposed to form a variety of program spatial motions of vehicles in the class of spiral lines with the help of following hodograph [8, 9]:

$$\bar{r}(t)=\left\|\rho_0\rho_1\rho_2\rho_3\right\|\left\|\begin{matrix}1\\t\\t^2\\t^3\end{matrix}\right\|\left(\bar{i}\cos\omega t+\bar{j}\sin\omega t\right)+\bar{k}\left\|h_0h_1h_2h_3\right\|\left\|\begin{matrix}1\\t\\t^2\\t^3\end{matrix}\right\|, \tag{1}$$

where: $\rho_i h_i\,(i=0,1,2,3)$ are running parameters being determined according to the specified boundary conditions; t is time (independent parameter); ω is average angular turning velocity, i.e. $\omega=\frac{\varphi_0}{t_0}$; φ_0 is full turning angle; t_0 is the required spatial transport time; $\bar{i},\bar{j},\bar{k}$ is the basis of ground Cartesian reference system.

Method of the program motion hodograph development within the considered spiral trajectory class is to determine the introduced running parameters according to the required boundary conditions. Additional conditions also include such limitations on technical applicability as the geometry of the trajectory shape as well as the velocity motion mode along the trajectory.

4. Natural trihedral of the program motion hodograph

Natural trihedral of the program motion hodograph is movable orthogonal coordinate system connected with a vehicle being considered as a material point of mass m. In this context, basis of the natural trihedral is convenient to be determined according to the hodograph (1) of the program motion in the form of:

$$\begin{aligned}\bar{\tau}&=\frac{1}{\left(\dot{\bar{r}}\cdot\dot{\bar{r}}\right)^{\frac{1}{2}}}\dot{\bar{r}}\\ \bar{n}&=-\frac{1}{\left(\dot{\bar{r}}\cdot\dot{\bar{r}}\right)^{\frac{1}{2}}\left|\dot{\bar{r}}\times\ddot{\bar{r}}\right|}\dot{\bar{r}}\times\left(\dot{\bar{r}}\times\ddot{\bar{r}}\right)\\ \bar{b}&=-\frac{1}{\left(\dot{\bar{r}}\cdot\dot{\bar{r}}\right)^{\frac{1}{2}}\left|\dot{\bar{r}}\times\ddot{\bar{r}}\right|}\dot{\bar{r}}\times\left[\dot{\bar{r}}\times\left(\dot{\bar{r}}\times\ddot{\bar{r}}\right)\right]\end{aligned} \tag{2}$$

where: $\bar{\tau}$ is a unit vector of a tangent; $\bar{n}$ is a unit vector of a principal normal; $\bar{b}$ is a unit vector of a binormal; $\dot{\bar{r}}$ and $\ddot{\bar{r}}$ are 1st and 2nd time derivatives from $\bar{r}(t)$.

Represent the equation of a normal plane passing through the current point $\bar{r}_M(t)$ of hodograph, in a vector form:

$$\left[\bar{\rho}-\bar{r}_M(t)\right]\cdot\bar{\tau}=0, \tag{3}$$

where $\bar{\rho}$ is a radius-vector of a random point of the normal plane within the earth coordinate system.

5. Bearing surface

Bearing surface is being formed as a ruled surface (Shukhov surface) which guiding line is determined with the help of program motion hodograph and rectilinear generators are within the cross of normal and bearing surfaces corresponding to the current hodograph point. Vector equation of bearing surfaces of the hodograph is:

$$\left[\bar{\rho}-\bar{r}_M(t)\right]\cdot\bar{g}_n=0, \tag{4}$$

where $\bar{g}_n$ is a unit vector of a projection of the resulting inertia forces on a normal plane. Then, the generators of the bearing surface in space can be specified as the cross of two planes (3), (4):

$$\begin{cases} \left[\bar{\rho}-\bar{r}_M(t)\right]\cdot\bar{\tau}=0 \\ \left[\bar{\rho}-\bar{r}_M(t)\right]\cdot\bar{g}_n=0 \end{cases}, \tag{5}$$

or with the help of the equation of a straight line lying within the normal plane (3) and passing through the hodograph point $\bar{r}_M(t)$ (1) in the direction being perpendicular to the projection of resulting inertia forces on a normal plane, i.e.

$$\left[\bar{\rho}-\bar{r}_M(t)\right]\times\bar{s}=0, \tag{6}$$

where $\bar{s}$ is a unit guiding vector of the forming bearing surface. In this context,

$$\bar{s}=\bar{\tau}\times\bar{g}_n. \tag{7}$$

6. Inertia forces

Inertia forces [10, 11] include the Earth gravity force as a vector directed to a center of a spherical surface:

$$\bar{G}_0=-\gamma\frac{M_0 m}{R_0^2}\cdot\frac{\bar{R}_0}{R_0}. \tag{8}$$

Acceleration

$$\bar{g}_0=\gamma\frac{M_0}{R_0^2}\cdot\frac{\bar{R}_0}{R_0} \tag{9}$$

Corresponds to the force (8), where: M_0 and R_0 are mass of the Earth and its radius; γ is gravitational constant; $\bar{R}_0$ is a radius-vector of a vehicle at the spherical earth's surface; the radius-vector starts in the center of the Earth. Then, (8) with allowance for (9) can be represented in the form:

$$\bar{G}_0=-\bar{g}_0 m. \tag{10}$$

Inertia force, depending upon the Earth rotation (centrifugal force) is determined by means of the active formula:

$$\bar{G}_C=-\bar{\Omega}_0\times\left(\bar{\Omega}_0\times\bar{R}_0\right)m. \tag{11}$$

Acceleration

$$\bar{g}_C=\bar{\Omega}_0\times\left(\bar{\Omega}_0\times\bar{R}_0\right). \tag{12}$$

Corresponds to the force (11), where $\bar{\Omega}_0$ is a vector of angular velocity of the Earth rotation. Then, (11) with allowance for (12) can be represented in the form:

$$\bar{G}_C=-\bar{g}_C m. \tag{13}$$

Inertia force depending upon the Earth rotation and high-speed motion of the vehicle (Coriolis force) is determined by means of the vector formula:

$$\bar{G}_{Cor}=-2\left(\bar{\Omega}_0\times\dot{\bar{r}}_M\right)m. \tag{14}$$

Coriolis acceleration

$$\bar{g}_{Cor}=2\left(\bar{\Omega}_0\times\dot{\bar{r}}_M\right). \tag{15}$$

Corresponds to the force (14). Then, (14) with allowance for (15) can be represented in the form

$$\bar{G}_{Cor}=-\bar{g}_{Cor}m. \tag{16}$$

Inertia force depending upon high-speed motion of the vehicle along the spherical earth's surface (centrifugal inertia force) is:

$$\bar{N}_C=\frac{\left(\dot{\bar{r}}_M\cdot\dot{\bar{r}}_M\right)}{\left(\bar{R}_0\cdot\bar{R}_0\right)}\bar{R}_0 m. \tag{17}$$

Centripetal acceleration

IOP Conf. Series: Materials Science and Engineering **383** (2018) 012032 doi:10.1088/1757-899X/383/1/012032

$$\bar{g}_N = -\frac{\left(\dot{\bar{r}}_M \cdot \dot{\bar{r}}_M\right)}{\left(\bar{R}_0 \cdot \bar{R}_0\right)} \bar{R}_0 \tag{18}$$

Corresponds to the force (17). Then, (17) with allowance for (18) can be represented in the form

$$\bar{N}_C = -\bar{g}_N m . \tag{19}$$

Inertia force depends upon the non-uniform transport motion along the curvilinear program trajectory. Tangential inertia force is:

$$\bar{N}_\tau = -\frac{\left(\dot{\bar{r}}_M \cdot \ddot{\bar{r}}_M\right)}{\left(\dot{\bar{r}}_M \cdot \dot{\bar{r}}_M\right)} \dot{\bar{r}} m . \tag{20}$$

Tangential acceleration

$$\bar{w}_\tau = \frac{\left(\dot{\bar{r}}_M \cdot \ddot{\bar{r}}_M\right)}{\left(\dot{\bar{r}}_M \cdot \dot{\bar{r}}_M\right)} \dot{\bar{r}}_M \tag{21}$$

Corresponds to the force (20). Then, (20) with allowance for (21) can be represented in the form

$$\bar{N}_\tau = -\bar{w}_\tau m . \tag{22}$$

Normal inertia force is:

$$\bar{N}_n = \frac{\left[\dot{\bar{r}}_M \times \left(\dot{\bar{r}}_M \times \ddot{\bar{r}}_M\right)\right]}{\left(\dot{\bar{r}}_M \cdot \dot{\bar{r}}_M\right)} m . \tag{23}$$

Normal acceleration

$$\bar{w}_n = -\frac{\left[\dot{\bar{r}}_M \times \left(\dot{\bar{r}}_M \times \ddot{\bar{r}}_M\right)\right]}{\left(\dot{\bar{r}}_M \cdot \dot{\bar{r}}_M\right)} . \tag{24}$$

Corresponds to the force (23). Then, (23) with allowance for (24) can be represented in the form

$$\bar{N}_n = -\bar{w}_n m . \tag{25}$$

Resulting inertia force is being determined as a vector sum (10), (13), (16), (19), (22) and (25)

$$\bar{I} = \bar{G}_0 + \bar{G}_C + \bar{G}_{Cor} + \bar{N}_C + \bar{N}_\tau + \bar{N}_n \tag{26}$$

or within projections of the natural trihedral axis (2):

$$\bar{I} = \bar{I}_\tau + \bar{I}_n + \bar{I}_b . \tag{27}$$

In this context, $\left(\bar{I}_n + \bar{I}_b\right)$ determines the projection of the resulting inertia forces on a normal plane (3).

Then, a unit normal vector of the bearing surface (4) is:

$$\bar{g}_n = \frac{\bar{I}_n + \bar{I}_b}{\left|\bar{I}_n + \bar{I}_b\right|} . \tag{28}$$

It should also be noted that a unit vector of the direction of local vertical line of the earth coordinate system is determined in the form of:

$$\bar{k} = -\frac{\bar{G}_o + \bar{G}_C}{\left|\bar{G}_o + \bar{G}_C\right|} . \tag{29}$$

7. Summary

It is proposed to consider program spatial motion of a high-speed vehicle with the help of a hodograph in the class of spiral trajectories. Mathematical model of a primary road is built in the form of a line at a ruled bearing surface in the neighbourhood of a program trajectory. Spiral program trajectory is a guideline of a ruled bearing surface; rectilinear generators in each point of a hodograph are found at the cross of normal and bearing surfaces. Equations of normal and bearing surfaces of a program hodograph are demonstrated in a vector form. Vector equation of generators oriented orthogonally to the resulting inertia forces is set up. Components of inertia forces including gravitation force are represented in a vector form; the forces effect the mass of a high-speed vehicle in terms of non-uniform motion along the curvilinear program trajectory at the spherical earth's surface.

References

[1] Zhang J, Zhou S and Huang C 2010 A new method for the fault diagnosis of the train wheelset based on characteristic spectrum analysis. *Proceedings of the 29th Chinese Control Conference, CCC 10* pp. 3988–92

[2] Ziborov K A, Protsiv V V and Fedoriachenko S O 2013 On formation of kinematical and dynamical parameters of output elements of the mine vehicles in transient *Scientific Messenger of the National Mining University* 4 pp 64–68

[3] Bass K , Plakhotnik V and Krivda V 2013 Influence of the road profile and dump truck parameters on the exploitation. *Scientific Messenger of the National Mining University* 4 pp 59–63.

[4] Bass K, Kuvayev S, Plakhotnik V and Krivda V 2014 Planar and spatial mathematical motion simulation of open pit mining vehicles. *Scientific Messenger of the National Mining University* 3 pp 60-65.

[5] Sakhno V, Poliakov V, Timrov O and Kravchenko O 2016 Lorry convoy stability taking into account the skew of semitrailer axes *Transport Problems* 11 pp 70–76.

[6] Kravets V, Bass K, Kravets T and Tokar L 2015 Dynamic design of ground transport with the help of computation experiment *MMSE Jornal* 1 pp 105–11

[7] Kravets V.V, Bass K.M, Kravets T.V, Zubariev M.S and Tokar L.A 2016 Kinetostatics of wheel vehicle in the category of spiral-screw routes, *MMSE Journal 5*, pp 87-99

[8] Басс К М, Кравец В В, Кравец Т В 2012 Математическая модель дорожной поверхности скоростной автомагистрали в развязках и поворотах. *Міжвузівський збірник "Наукові нотатки" 36* (Луцьк) pp 23-25.

[9] Kravets V, Kravets T, Bass K and Tokar L 2014 Mathematical model of path and hodograph of surface transport *Transport problems 1* (Katowice) pp 830–41.

[10] Martynyuk A.A, Lobas L.G, Nikitina N.V 1981 *Dynamics and sustainability of transport vehicle wheelset movement* (Kiev: Tekhnika) p 223

[11] Kravets V, Kravets T 2008 Evaluation of centrifugal, coriolis and giroscopic foces on a railroad vehicle moving at high speed *Int. Appl. Mech.* 44 pp 101-09.

Stochastic optimization of reinforced concrete elements' design

J Venclovský*, **P Štěpánek and I Laníková**

Institute of Concrete and Masonry Structures, Faculty of Civil Engineering, Brno University of Technology, Veveří 331/95, 602 00 Brno, Czech Republic

*E-mail: venclovsky.j@fce.vutbr.cz

Abstract. Main goal of the paper is to present an algorithm for stochastic optimization of design of steel-reinforced concrete element's cross section. Firstly, the deterministic problem is introduced and described, followed by the description of uncertainties involved in the process and stochastic reformulation of the problem. Afterwards, the algorithm itself is introduced. This algorithm is based on internal cycle of deterministic optimization using reduced gradient method and external cycle of stochastic optimization using regression analysis. The probability is assessed on orthogonal grid via modified bisection method. The paper concludes with presentation of the performed calculations and their results.

1. Introduction

It is possible to see various applications of mathematical optimization in civil engineering (structural design, reconstruction of transportation networks etc.) Initially, deterministic approaches have been introduced to solve these issues (see [1]). But despite their complexity, these approaches are insufficient to comprehend the probabilistic nature of mentioned problems and thus provide only suboptimal solutions. Hence the effort prevails to reconsider these deterministic approaches and deal with uncertainties involved in said issues in less straightforward way (see [2, 3]).

Paper focuses on optimization of the structural design. There are various approaches to the optimization. The difference between deterministic and stochastic optimization was already mentioned. This paper is aimed on stochastic optimization. Further differences derive from the subject of optimization. In this case, it is the shape of the cross section.

It is also possible to distinguish various optimization algorithms. There are analytic algorithms (e.g. [4]) and simulation algorithms (e.g. [5]). Analytic algorithms are based on probabilistic analysis of the problem and it is known, that they cannot be used to achieve a high precision. Simulation algorithms are based on discretization of probability into particular scenarios followed by simulation using these scenarios (every particular scenario is actually deterministic). Since the precision that needs to be achieved can be very high (even 10^{-6}) it seems that the only possibility to achieve such precision is to use simulation and heuristic algorithm simultaneously.

This paper aims to introduce an optimization algorithm, which works in two cycles. The internal cycle is deterministic and based on reduced gradient method. The external cycle deals with probability and as mentioned before it is heuristic, based on simulation and regression analysis. It works in the space of stochastic variables rather than the space of the design variables. The algorithm is aimed to reach 99.99 % probability that the structural element satisfies ultimate and serviceability limit states.

Content from this work may be used under the terms of the Creative Commons Attribution 3.0 licence. Any further distribution of this work must maintain attribution to the author(s) and the title of the work, journal citation and DOI.
Published under licence by IOP Publishing Ltd

IOP Conf. Series: Materials Science and Engineering **383** (2018) 012033 doi:10.1088/1757-899X/383/1/012033

2. Deterministic approach

It is necessary to begin by introducing the problem in its deterministic form. In deterministic approach, all probability inputs are replaced with their design values. Furthermore, the principle of safety coefficients is applied. Simplified definition of the optimization task therefore is:

$$\min_{x} f(\boldsymbol{x}), \text{ subject to:} \tag{1}$$

$$L(\boldsymbol{x}) \leq R(\boldsymbol{x}), \tag{2}$$

$$w(\boldsymbol{x}) \leq w_{\max}, \tag{3}$$

$$\boldsymbol{x} \in \boldsymbol{C}. \tag{4}$$

All conditions and variables are described in more detail in the following text. It is necessary to mention, that besides the vector of design variables $\boldsymbol{x}$, there are another inputs into the task (1-4), namely prescribed loads and material characteristics.

2.1. Vector of design variables $\boldsymbol{x}$.

This paper restricts itself to one-dimensional construction elements – beams. The length of the beam is obviously set and cannot change. Vector $\boldsymbol{x}$ in task (1-4) therefore contains variables, which describe shape of the beam's cross section, including the area of reinforcing steel (see figure 1 as an example).

2.2. Objective function

The objective function $f(\boldsymbol{x})$ in (1) can represent anything from simple cost of materials to multi-criteria function describing environmental aspects through construction, maintenance and disposal of the structure. To maintain simplicity, the objective function, which calculates the cost of materials, is used in this paper:

$$f(\boldsymbol{x}) = V_c(\boldsymbol{x}) \times C_{cv} + W_s(\boldsymbol{x}) \times C_{sw}, \tag{5}$$

where V_c is volume of used concrete, C_{cv} is cost of concrete per unit of volume, W_s is weight of used steel and C_{sw} is cost of steel per unit of weight.

2.3. Ultimate limit state

The condition (2) represents ULS in task (1-4) – $L(\boldsymbol{x})$ is the effect of load and $R(\boldsymbol{x})$ is the structural resistance. This description of ULS is of course very simplified. In fact, ULS consists of many equations and a condition limiting the values of strain of concrete and steel.

In order to evaluate ULS it is necessary to calculate the vector of deformations first. This is achieved using matrix calculations, which are in principle based on solving the following differential equation via the finite element method:

$$EIw^{(4)} = q, \tag{6}$$

where E is the Young modulus of elasticity, I is the moment if inertia, w is the deflection and q is the prescribed uniformly distributed load over unit of length. Solving this equation via matrix calculations based on FEM gives:

$$\boldsymbol{KU} = \boldsymbol{F}, \tag{7}$$

where $\boldsymbol{K}$ is the stiffness matrix, $\boldsymbol{U}$ is the vector of deformations and $\boldsymbol{F}$ is the vector of load.

Using this equation, the internal forces are calculated. From them, the so-called strain parameters (see [5]) and subsequently the strains of concrete and steel in decisive points are computed and they are compared to their limiting values.

2.4. Serviceability limit state

As it is obvious from condition (3), the SLS in this paper is restricted to limiting deflection. The condition (3) is again very simplified and actually represents large number of equations. The evaluation of SLS is similar to evaluation of ULS with the difference, that the possible formation of cracks must be taken into account. These cracks then affect the bending stiffness B_i:

$$B_i = EI_i , \tag{8}$$

which affects the moment of inertia I_i – one of the inputs into the differential equation and therefore the matrix calculation. Whether the cracks form is determined from the vector of internal forces. It is therefore necessary to find the state of equilibrium (of variables' values in the mentioned equations, see [5]).

2.5. Design variables' boundaries

Condition (4) define boundaries of the design variables' values. It also represents conditions following mutual relations among these variables – e.g. maximal and minimal area of steel reinforcement in relation to area of the concrete cross section.

2.6. Optimization algorithm

Optimization itself uses the reduced gradient method, specifically the solver CONOPT from the optimization software GAMS. This sets some rules for the performed calculations. All relations (equations) and their first derivatives must be continuous. In some cases (stress-strain diagrams, mentioned stiffness in SLS), this is achieved using Hermit interpolation (see [5] again).

3. Stochastic approach

The designing process involves a lot of uncertainties. Among these are:

- randomness of physical quantities used in the design (as its natural characteristic)
- statistical uncertainties during the description of a quantity caused by a lack of data,
- model uncertainties caused by inaccuracies in the calculation model in comparison with real structural behavior,
- uncertainties caused by inaccuracies of the limit states definitions,
- human element deficiencies within the design procedure and execution and usage of the structure.

In the deterministic approach all these uncertainties are included in the calculation by using design values and/or via using various safety coefficients. It is obvious, that for desired probability of structure failure, the stochastic approach provides more precise and improved solution.

By embodying uncertainties, the original task (1-4) is reformulated to its stochastic form:

$$\min_{\boldsymbol{x}} M(f(\boldsymbol{x},\xi)) \text{, subject to:} \tag{9}$$

$$P(L(\boldsymbol{x},\xi) \le R(\boldsymbol{x},\xi)) \le \alpha , \tag{10}$$

$$P(w(\boldsymbol{x},\xi) \le w_{\max}) \le \beta , \tag{11}$$

$$\boldsymbol{x} \in \boldsymbol{C}, \tag{12}$$

where M is mean, ξ represents uncertainty, α is a desired probability that the ULS condition holds and β is a desired probability that the SLS condition holds.

4. Example

Proposed solution is presented directly on a simple example.

The task is to design 5 m long cantilever beam loaded with uniformly distributed load q and normal force N as shown on figure 1.

Cross section of the beam is described by 8 variables b_1, b_2, b_3, h_1, h_2, h_3, A_{s1}, A_{s3} – 6 of them describe the shape of cross section, while 2 describe the area of reinforcing steel (see figure 1).

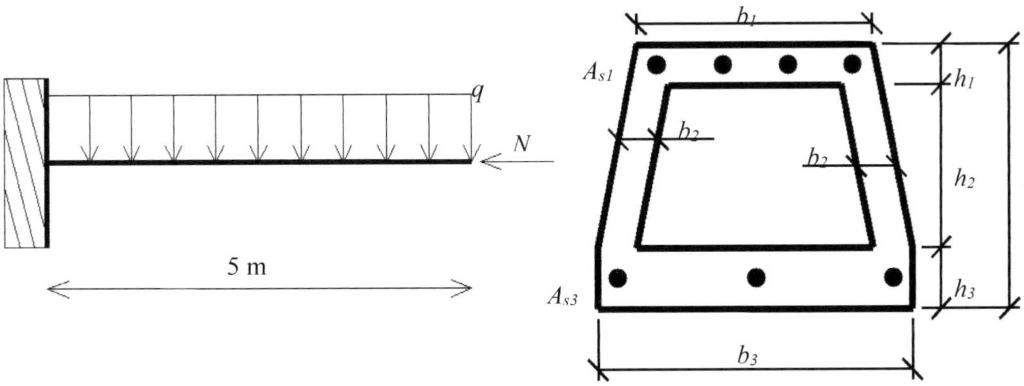

Figure 1. Scheme of the cantilever, $\boldsymbol{x} = (b_1, b_2, b_3, h_1, h_2, h_3, A_{s1}, A_{s3})$.

In the FEM calculation, the cantilever is divided into 10 elements e_1, …, e_{10}, each 0.5 m in length.

The distributed load q [kN/m] and normal force N [kN] are random variables with gamma distribution. Probability distributions of q and N are:

$$q \sim \Gamma(1500; 0.015),\ N \sim \Gamma(148; 0.05). \tag{13}$$

In order to better understand the distributions, the 0.05 quantile of distribution of q is 21.6 kN/m, 0.5 quantile is 22.5 kN/m and 0.95 quantile is 23.5 kN/m. Analogically, 0.05 quantile of distribution of N is 6.4 kN, 0.5 quantile is 7.4 kN and 0.95 quantile is 8.4 kN. All these values are approximate.

To maintain simplicity, the probabilities α and β from conditions (10) and (11) are replaced with only one probability γ. The mentioned conditions are thus replaced with following condition:

$$P(L(\boldsymbol{x},\xi) \leq R(\boldsymbol{x},\xi) \wedge w(\boldsymbol{x},\xi) \leq w_{\max}) \leq \gamma\,. \tag{14}$$

The value γ is set to 99.99 %.

There are some other restrictions regarding design variables. The overall height of cross section is limited to 350 mm. Each of the heights h_1, h_2, h_3 has to be minimally 100 mm. The width b_2 is at least 100 mm, b_1 and b_3 at least 200 mm but no more than 900 mm. For widths b_1, b_3 the following condition needs to be fulfilled:

$$b_k \geq 2\times 50 + A_{sk}/(\pi \times 20^2/4)\times(50+20)\,. \tag{15}$$

This relation represents, that the given width has to accommodate required area of reinforcing steel; reinforcing steel is composed of bars 20 mm in diameter, minimum distance between the surface of any steel bar and the concrete surface and between any two steel bars is 50 mm.

The considered costs are 2500 CZK per 1 m^3 of concrete and 30 CZK per 1 kg of steel.

5. Solution algorithm

The algorithm uses regression analysis to iterate towards the suitable solution of the task. In this chapter, the algorithm is described, including the method of probability assessment.

5.1. Probability assessment

The probability is assessed on an orthogonal mesh via method, that can be described as modified bisection method. The principle of this method is to find the boundary between scenarios, which satisfies deterministic conditions of the task, and scenarios, which do not.

Orthogonal mesh in its maximum size (bisection divides maximum mesh into smaller ones) consists of 514×514 points. The mesh layout is thicker toward the expected location of the mentioned boundary.

IOP Conf. Series: Materials Science and Engineering **383** (2018) 012033 doi:10.1088/1757-899X/383/1/012033

5.2. Calculation initialization

The calculation is initialized by selecting given number of scenarios of random variables. For each scenario the deterministic optimization is performed and probability, that thus acquired solution holds ULS and SLS conditions, is assessed (assessment of the left hand side of condition (14)). In this paper, the example defined in chapter 4 is initialized by 16 scenarios, which are made of combinations of values of q and N as follows:

$$q \in \{20.25; 21.75; 23.25; 24.75\}, \tag{16}$$

$$N \in \{5.25; 6.75; 8.25; 9.75\}. \tag{17}$$

5.3. Regression analysis

Acquired points are fitted with a polynomial function via least squares method – for probability it is the 3rd degree polynomial, for objective function it is the 1st degree polynomial. Degrees of polynomials were chosen appropriately in regard to values of probability and objective function observed on a greater sample than just the mentioned 16 scenarios.

Regression analysis is weighted. The error in each point i is multiplied by $1/|p_i-\gamma|$, where p_i is the assessed left hand side of condition (14) for point i. This way, the regression resembles interpolation while near the solution.

5.4. Iteration

After initialization, the iterative process follows. According to regression analysis, the point that satisfies condition (14) and at the same time has the minimal value of objective function is selected as a possible solution. For this scenario the deterministic optimization is performed and probability that thus acquired solution satisfies deterministic conditions is assessed.

If the resulting probability is greater and also sufficiently close to value γ, the algorithm ends and the current deterministic solution is the solution of the whole task.

In the opposite case, this newly assessed point is added as another point to the regression analysis. Regression, new point selection, deterministic optimization and probability assessment are all performed again. This iterative procedure continues, until the assessed probability is higher and also sufficiently close to value γ.

5.5. Heuristic algorithm

Proposed algorithm actually does not solve minimization task (9) but the task:

$$\min_{\xi} \min_{\boldsymbol{x}} f(\boldsymbol{x},\xi)\,. \tag{18}$$

Therefore, the algorithm is heuristic and provides only suboptimal solution.

6. Calculation results

Table 1 contains the values of design variables, which were found as a solution to the example from chapter 4.

Table 1. Values of design variables b_1, b_2, b_3, h_1, h_2, h_3 [mm] and A_{s1}, A_{s2} [mm^2] in finite elements e_1, …, e_{10} as a solution to example defined in chapter 4.

	e_1	e_2	e_3	e_4	e_5	e_6	e_7	e_8	e_9	e_{10}
b_1	900.0	781.9	655.5	546.1	445.1	353.0	900.0	470.1	200.0	200.0
b_2	100.0	100.0	100.0	100.0	100.0	100.0	100.0	100.0	100.0	100.0
b_3	900.0	900.0	899.6	712.5	542.3	392.1	402.0	200.0	200.0	200.0
h_1	100.0	100.0	100.0	100.0	100.0	100.0	100.0	100.0	100.0	100.0
h_2	150.0	150.0	150.0	150.0	150.0	150.0	150.0	150.0	100.0	100.0
h_3	100.0	100.0	100.0	100.0	100.0	100.0	100.0	100.0	100.0	100.0
A_{s1}	3590.4	3060.2	2493.0	2002.2	1548.9	1135.6	360.7	196.4	92.8	60.0
A_{s2}	210.0	195.8	180.6	152.5	126.8	103.7	170.2	102.4	60.0	60.0

The probability, that solution satisfies ULS and SLS was assessed as 99.99017 %. This probability was subsequently tested by the Monte Carlo simulation. From 10^5 scenarios, 9 failed the deterministic conditions which is in accordance with previous probability assessment.

The solution was found for the scenario $q = 24.6767$ kN/m, $N = 4.7635$ kN. Its objective function value is 3575.08 CZK.

Evidently, the values from table 1 are not suitable for an actual design purposes. They serve rather as a guide to arrange the cantilever's cross section over its length.

Table 1 contains one peculiar value (b_1 in e_7). This is caused by the deterministic optimization algorithm – the GRG method. Since the only function of width b_1 is to accommodate reinforcement area A_{s1}, this value can obviously be lowered. Nevertheless, the optimization algorithm kept the current value, because the minimum (defined as a reduced gradient sufficiently close to zero) was already reached.

7. Conclusions

The algorithm to stochastically optimize design of reinforced concrete structural elements was developed. This algorithm was successfully tested on a simple example. Probability assessment of proposed algorithm was successfully tested by Monte Carlo method. The future work should focus on reaching higher precision of probability of failure as well as on using the algorithm in issues with more random variables.

Acknowledgment

This contribution was prepared within a scientific-research work of the project FAST/FSI-J-17-4753 "Advanced methods of design optimization of structural elements", supported by the Brno University of Technology, Faculty of Civil Engineering and Faculty of Mechanical Engineering.

References

[1] Chakrabarty B K 1992 A model for optimal design of reinforced concrete beam *Journal of Structural Engineering* 108 3238–42

[2] Frangopol D M, Kongh J S and Ghareibeh E S 2001 Reliability-based life-cycle management of highway bridges *Journal of Computing in Civil Engineering* 15 27–34

[3] Ziemba W T and Wallace S W 2004 Applications of Stochastic Programming *SIAM conference*

[4] Žampachová E 2009 *Approximations in stochastic optimization and their applications* (Ph.D. thesis, Brno University of Technology)

[5] Plšek J 2011 *Design optimisation of concrete structures* (Ph.D. thesis, Brno University of Technology)

Optimization of energy pile conductance using finite element and fractional factorial design of experiment

Khaled Ahmed[1,*], Mohammed Al-Khawaja[1] and Muhannad Suleiman[2]

[1] Mechanical & Industrial Engineering Department, Qatar University, Doha 2713, QATAR
[2] Civil and Environmental Engineering, Lehigh University, PA 18015, USA

*E-mail: kh.ahmed@qu.edu.qa

Abstract. In Ground Source Heat Pumps (GSHP), Energy Piles pose as heat exchangers that transfer the heat from the buildings to the shallow ground lower temperature in order to decrease the energy consumption whilst cooling the buildings. These piles are mainly designed for highest possible thermal conductance. In this paper, nine factors influencing the thermal conductance of the energy pile are defined and statistically evaluated. These nine factors are; number of tubes, pile diameter, tube diameter, tube thickness, tube location, pile conductivity, tube conductivity, soil conductivity, and water flow rate. The thermal conductance of the energy pile is calculated using finite element model. The significance of these factors is evaluated using fractional factorial uniform design of experiment. The results show significance increase in the pile thermal conductance with the increase of the tube diameter, number of tubes, water flow rate, and tube and pile thermal conductivities. Furthermore, the tubes location near the pile outer surface show significant increase in the pile thermal conductance. On the other hand, decreasing pile diameter slightly increases the pile thermal conductance. Nevertheless, the soil thermal conductivity has shown insignificant effects on the pile thermal conductance.

1. Introduction

Ground Source Heat Pumps (GSHP) are a geothermal free form of energy utilizes the constant temperature of the shallow ground all year round to reduce energy consumption in cooling buildings through its energy piles [1]. The significant advantage of using energy piles over boreholes systems is that they require no additional structural or hydraulic measures because they are installed within elements that are already needed for structure [2]. Energy piles utilize renewable geothermal energy for buildings heating and cooling purposes and need proper design and sizing in order to end up with high plant efficiency [3].

Research on the controlling factors of the energy piles have shown that maximizing the pile surface, maximizing the concrete thermal conductivity, and maximizing the number of water tubes will increase the heat exchanging through the energy pile [4]. These reports neither include all elements affecting the thermal conductance of energy pile nor their interconnection effects. The analytical formulae proposed by [5] have shown nine factors affecting the energy pile steady state thermal conductance. These nine factors are; number of tubes, pile diameter, tube diameter, tube thickness, tube location, water flow rate, and the thermal conductivities of the pile, tube and soil. Investigating these factors together require unreachable number of experiments to evaluate the interrelation between

Content from this work may be used under the terms of the Creative Commons Attribution 3.0 licence. Any further distribution of this work must maintain attribution to the author(s) and the title of the work, journal citation and DOI.
Published under licence by IOP Publishing Ltd

these controlling factors. Statistical design of experiments methods like uniform fractional factorial design can solve this problem.

Based on either the finite-element method or the finite-volume method, various numerical approaches for full discretization of the Ground source heat exchangers like boreholes or energy piles have been formed. These models are employed to solve the heat-exchanging problem to optimize the heat exchanger geometry [6-9]. These models require extensive CPU time for being 3D models or for solving the transient effects. In order to decrease the time of calculations, the current analysis will be restricted to 2D steady state model.

The objective of the current work is to build a 2D steady state finite element thermal model to predict the energy pile thermal conductance at different combinations of the controlling factors. Then statistical regression method will be used to define a correlation between these controlling factors based on the significance of each of these factors on the energy pile steady state thermal conductance whilst changing other factors using uniform fractional factorial design of experiment method.

2. Energy pile factors and FE model

The current work considers the energy pile cross sectional in plane factors assuming same behaviors and relations through the pile height. Figure 1 shows schematic configurations of the studied energy pile in the current work.

The energy pile is symmetrical with repeated pattern. Half of the repeated sector can represent the whole energy pile cross section as shown in figure 1. The angle "θ" depends on the number of U-tubes in the pile and equal ($90/n$) where 'n' is the number of U-tubes. Lines "O-A" and "O-B" are symmetry lines. The energy pile model considers the variation of five geometrical factors; number of U-tubes (n), pile diameter (d_p), tube inner diameter (d_i), tube thickness (t), and tubes spacing (S). In addition, it considers the variation of thermal conductivities of the pile (K_p), the HDPE tube (K_t) and the sand (K_s). The study will consider one operational factor, which is the heat transfer coefficient of the circulating water (H). The sand width will not affect the steady state conductance of the energy pile so that it will assumed with constant value in the current study (L_s=1.0 m). The energy pile studied factors throughout the current study are listed in table 1.

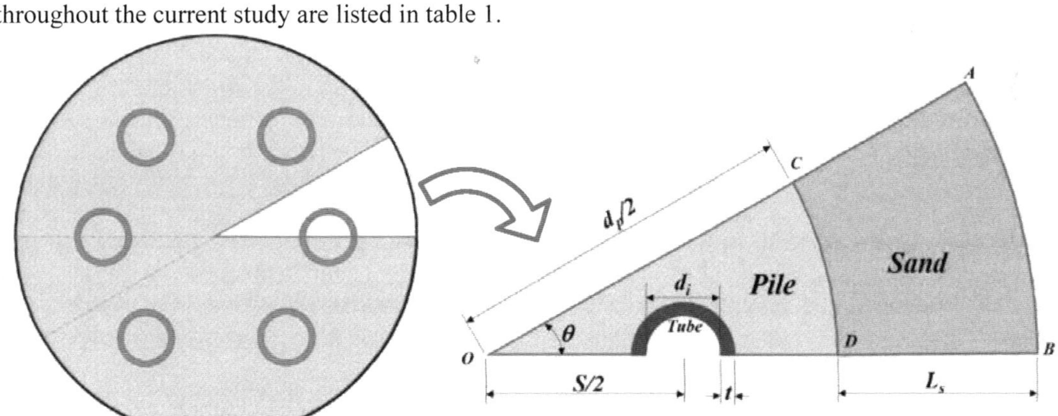

Figure 1. Energy pile geometrical factors.

Table 1. Studied factors and their levels.

Level	n	d_p	d_i	T	S	K_p	K_s	K_t	H
-1	1	0.4	0.02	0.002	0.40 d_p	1.0	0.5	0.5	10
0	2	0.7	0.03	0.003	0.60 d_p	1.75	1	16	55
+1	3	1.0	0.04	0.004	0.80 d_p	2.50	1.5	32	100

Using Galerkin method and the divergence theorem [10], the discretized 2D finite element equation of a steady state condition, with no heat generation, takes the following form;

$$\left[\iint_A [B]^T[K][B]dA + \int_S h[N^s]^T[N^s]dS\right]\{\overline{T}\} = \int_S q^s[N^s]^T dS + \int_S hT_f[N^s]^T dS \quad (1)$$

where [B] is the temperature gradient interpolation matrix and {T} is the nodal temperature vector.

The boundary conditions of this equation take the following form with respect to figure 1;

Symmetric boundary conditions at lines "O-A" and "O-B" (Adiabatic BC).

$$\frac{\partial \mathrm{T}}{\partial \mathrm{S}} = 0 \quad (2)$$

where *S* is the normal vector at symmetric lines.

Convection heat exchange at the inner surface of the HDPE tube.

$$-\mathrm{K_s}\frac{\partial \mathrm{T}}{\partial \mathrm{S}} = \mathrm{H}(\mathrm{T_s} - \mathrm{T_f}) \quad (3)$$

where;

S is the normal vector at the inner surface of the HDPE tube.

T_s is calculated temperature at the inner surface of the HDPE tube.

T_f is the bulk temperature of the circulating water (45°C).

H is the studied heat transfer coefficient.

Specified constant temperature at the outer surface of the sand line "A-B".

$$\mathrm{T_{A-B}} = 25\ ^{\circ}\mathrm{C} \quad (4)$$

The set of linear equations represented by equation (1) and the boundary conditions represented by equations (2-4) are assembled to form the global system matrix equations to be solved to calculate the temperatures at the inner surface of the HDPE tube "T_t" and the outer surface of the energy pile at line "C-D" "T_{C-D}". Also, the heat flow at the inner tube surface and the outer surface of the sand line "A-B" are calculated for convergence checking and for thermal conductance calculations. The energy pile steady state thermal conductance is calculated by the following equation;

$$\mathrm{C_{p,\ FE}} = \frac{\mathrm{Q_{A-B}}}{\mathrm{T_t} - \mathrm{T_{C-D}}} \quad (5)$$

2.1. FE Model Validation and Mesh Density Sensitivity

The finite element model of each design point is built and solved, autonomously, using ANSYS Parametric Design Language offered by ANSYS_MAPDL. This powerful scripting language allows parameterizing the finite element model and automating all related tasks of solution and post processing.

The current finite element model is verified against an analytical model of the steady state thermal resistance, which is the reciprocal of the thermal conductance, for these energy pile configurations. The model is proposed by [5] and approximated the thermal resistance of double U-tube energy pile as follows;

$$R_p = \frac{1}{4\pi K_p}\left[\ln\left(\frac{d_p^{2n}}{2nd_oS^{2n-1}}\left(\frac{d_p^{4n}}{d_p^{4n} - S^{4n}}\right)^{\left(\frac{\mathrm{K_p} - \mathrm{K_s}}{\mathrm{K_p} + \mathrm{K_s}}\right)}\right)\right] + \frac{1}{2\mathrm{n}\pi}\left(\frac{1}{2\mathrm{K_t}}\ln\left(\frac{\mathrm{d_o}}{\mathrm{d_i}}\right) + \frac{1}{\mathrm{d_i H}}\right)_t \quad (6)$$

where;

R_p [m.K/W]: Energy pile thermal resistance

n: Number of U-tubes

$m = 2(n-1)$

d_p [m]: Pile diameter.

d_i [m]: Tube inner diameter

t [m]: Tube thickness

S [m]: Tubes spacing.

K_p [W/m.K]: Pile thermal conductivity.

K_s [W/m.K]: Soil thermal conductivity.
K_t [W/m.K]: Tube thermal conductivity.
H [W/m^2.K]: Convection heat transfer coefficient

The FE results are usually sensitive to the mesh density. A proper mesh density that gives acceptable result within applicable CPU time is important to be achieved ahead of the investigation. A model with mid factors listed in table 1 is used in the validation and the study of mesh density sensitivity. The mesh density in this model is parameterized to the number of elements through the HDPE tube thickness. The rest of elements sizes are assigned to be within the recommended aspect ratio (< 1:5). The results of the validation and mesh density sensitivity study are shown figure 2. Domain discretization with 20 elements through HDPE thickness is the most compromised pattern between predicted results and calculation time. This pattern shows deviation percentage less than 0.02% with 10 seconds of CPU time.

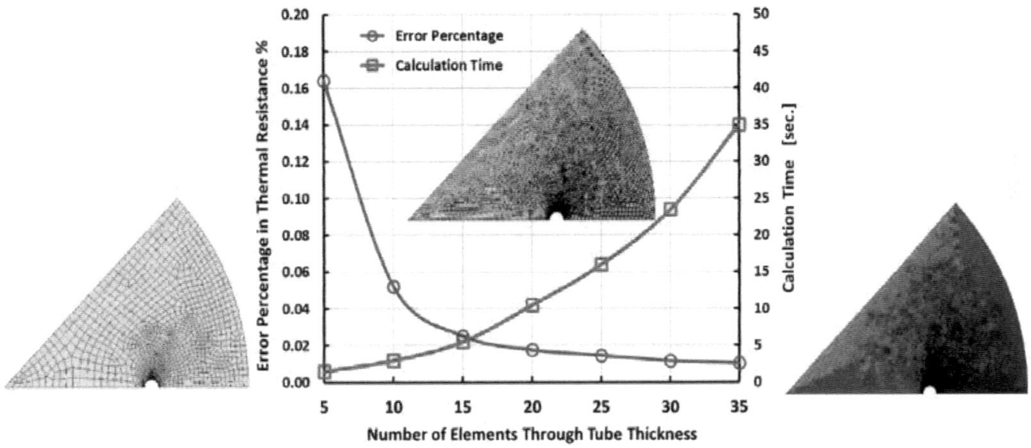

Figure 2. Deviation in the predicted energy pile thermal resistance and the CPU calculation times at different mesh densities showing the pile mesh layouts of; 5, 20, and 35 elements through tube thickness.

3. Uniform design of experiment

Design of experiment methods are widely used in factors correlations and performance optimizations of multivariable systems [11]. Fractional factorial design of experiment is optimally suitable for systems with large number of factors. The Uniform design is an efficient fractional factorial design [12]. The uniform design is one of the robust space-filling designs that is significantly important in investigating large engineering systems [13].

3.1. Design Selection

The domain of each factor of the energy pile system factors is levelled to three levels (-1, 0, +1). The related values of these levels for each factor are listed in table 1. The designs incorporated with 9 factors at three levels (3^9) can be investigated using number of simulation experimental runs as low as 9 runs and as high as 51 runs. The number of runs affect the uniformity discrepancy of the selected design. Figure 3 presents the effect of the number of runs on the discrepancy CD_2 of the design $U_n(3^9)$ [14]. The current work uses the uniform designs $U_{27}(3^9)$, $U_{36}(3^9)$ and $U_{51}(3^9)$ with 27, 36 and 51 runs respectively [14].

3.2. Signal to Noise Ratio

Measured quantities are affected by significant and insignificant factors. Significant factors produce strong signal while insignificant factors produce noise. Magnification of the signal to noise ratio emphasizes the effect of each factor on the measured data. This leads to optimizing the controlling factors for better performance. The signal to noise ratio in the current work is calculated with the Taguchi larger the better relation as follows [15];

$$S/N =- 10 \log_{10}\left(\frac{1}{m}\sum_{i=1}^{m}\frac{1}{C^2_{p,i}}\right) \tag{7}$$

where;

m: is the number of observations for each factor.

$C^2_{p,i}$: is the FE predicted thermal conductance of the energy pile at experiment number i.

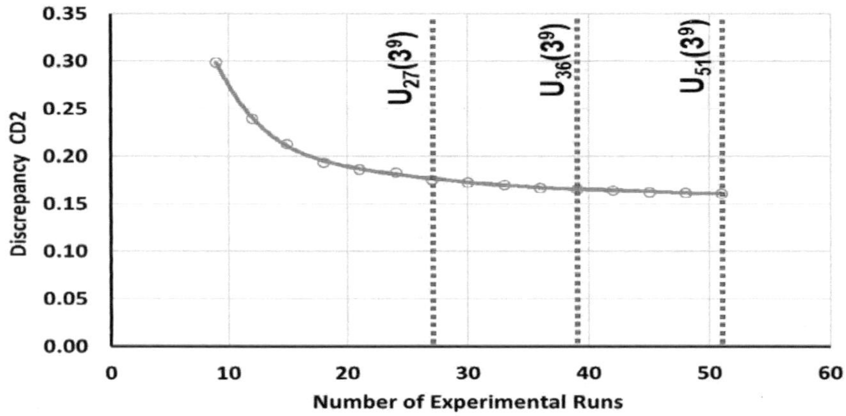

Figure 3. Effect of number of runs on the uniform design $U_n(3^9)$ discrepancy CD_2 edited based on the data published by [14].

Another method to investigate the significance for each factor is the estimation of the cubic least square regression of the observed data at all experiments. The common cubic correlation is simply expressed as follows;

$$Y = a_0 + \sum_{k=1}^{9} a_k x_k + \sum_{i=1}^{9}\sum_{j=i}^{9} b_{ij} x_i x_j + \sum_{i=1}^{9}\sum_{j=i}^{9}\sum_{k=j}^{9} c_{ijk} x_i x_j x_k \tag{8}$$

where;

Y represents a function of the equivalent thermal conductance of the energy pile including water tubes.

a_o, a_k, b_{ij} and c_{ijk} are the correlation coefficients of the cubic model. These coefficients can be calculated using stepwise least square method.

x_i represents the controlling factors as shown in table 2.

The first part is the intercept constant, the second part represents the linear weight of each factor separately, and the third part represents the cubic and the interaction between controlling factors in a pair wise manner.

Table 2. The related factors for each variable in the cubic correlation.

x_1	x_2	x_3	x_4	x_5	x_6	x_7	x_8	x_9
n	d_p	d_i	t	S	K_p	K_s	K_t	H

4. Results

4.1. Effect of factors means

The current work investigated the energy pile conductance using three designs $U_{27}(3^9)$, $U_{36}(3^9)$ and $U_{51}(3^9)$ with 27, 36 and 51 runs respectively [14]. Uniform design $U_{51}(3^9)$ is shown in table 3, and the other designs are available at the Uniform Design Tables book [14]. The significance of each factor is

IOP Conf. Series: Materials Science and Engineering **383** (2018) 012034 doi:10.1088/1757-899X/383/1/012034

measured by calculating the mean thermal conductance for each factor with all other factors. The three designs have shown consistent behaviour with the most significant factors as shown figure 4. On the other hand, the less significant factors have shown non-consistent behaviour with the three designs. The significance of the controlling factors on the mean thermal conductance is shown in figure 5. The most five significant factors with direct proportionality are the convection heat transfer coefficient "*H*", the number of tubes "*n*", the pile thermal conductivity "K_p", the distance between tubes "*S*", and the tube inner diameter "d_i", respectively in significance order. The other four factors are either less significant or with inverse proportionality.

4.2. Signal to Noise Ratio measurements

The calculated thermal conductance of the energy pile at the different runs using the uniform designs of experiment are analyzed using equation (7). The significance of the factors have been better emphasized. The three designs have shown more consistence results with the Signal to Noise ratio measure than that obtained by the means measure as shown in figure 6.

The three designs have shown that the least significant factors; are tube thickness "*t*", soil thermal conductivity "K_s", and the tube thermal conductivity "K_t". The other factors have shown highest signal to noise ratio at the highest level "+1". These factors are the convection heat transfer coefficient "*H*", the number of tubes "*n*", the pile thermal conductivity "K_p", the tube inner diameter "d_i", and the distance between tubes "*S*" respectively in significance order, as shown in figure 6. On the other hand, the pile diameter has shown the highest signal to noise ratio at level "-1". This implies that the thermal conductance of the energy pile is slightly decrease with the increase of the energy pile diameter.

4.3. Cubic Least Square model

Using the stepwise least square regression, the coefficients of the cubic equation (8) are calculated using the three uniform designs $U_{27}(3^9)$, $U_{36}(3^9)$ and $U_{51}(3^9)$. As shown in figure 7, the three designs show direct proportionality of the energy pile thermal conductance with the five most significant factors; (H, n, K_p, d_i, and S) and inverse proportionality with d_p. However, the other three factors are controversial because they have minor effects. Design $U_{27}(3^9)$ shows dependency of the pile conductance on the t and K_s, however, the other two designs show no dependency. On the other hand, design $U_{27}(3^9)$ shows no dependency on K_t, however, the other two designs show direct proportionality of the pile conductance and K_t as shown in figure 7.

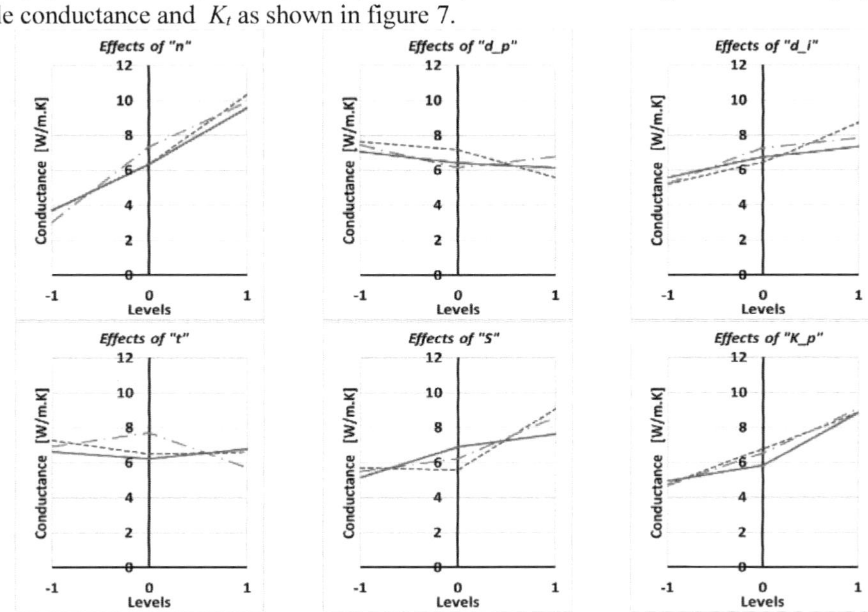

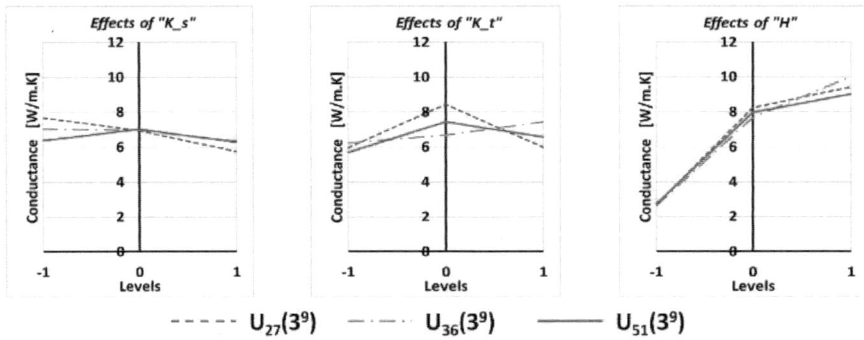

Figure 4. Effects of each factor separately on the energy pile conductance.

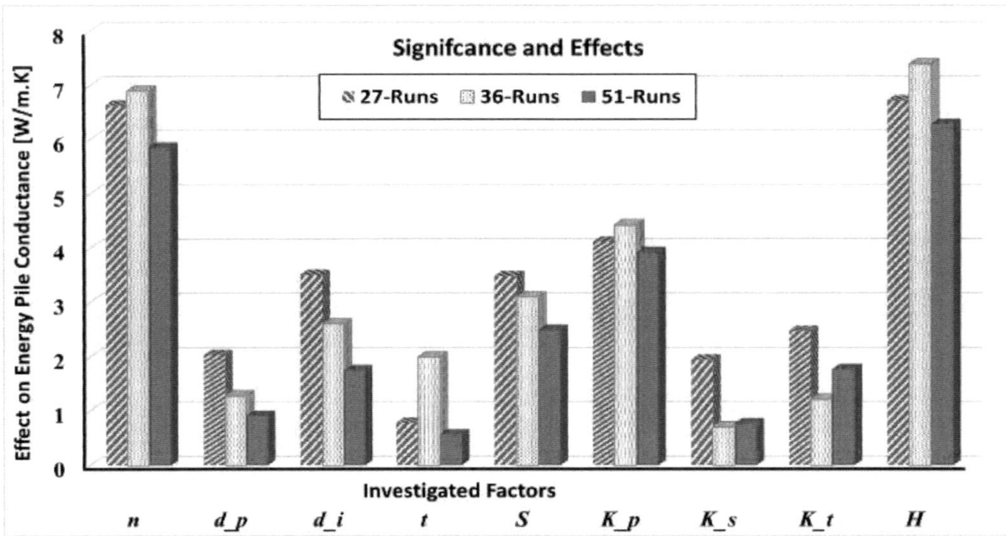

Figure 5. Effects and significance of each factor separately on the energy pile conductance.

Table 3. Uniform design $U_{51}(3^9)$ [14].

Exp.	n	d_p	d_i	t	S	K_p	K_s	K_t	H	*Pile Cond.* C_p
1	1	1	1	1	1	0	-1	-1	1	16.8501
2	0	1	-1	0	1	-1	0	1	1	6.3859
3	1	-1	1	1	-1	0	1	0	0	10.4425
4	0	-1	1	-1	0	1	-1	-1	-1	4.0793
5	-1	0	0	1	1	-1	1	0	1	4.1772
6	-1	1	0	1	0	-1	1	1	0	2.9434
7	1	0	0	-1	-1	1	1	-1	1	11.5765
8	0	0	0	0	0	0	0	0	0	8.3735
9	0	0	0	1	1	-1	-1	-1	0	6.3752
10	0	1	-1	-1	0	1	1	0	1	9.8099
11	0	1	0	1	-1	1	-1	0	-1	2.9456
12	0	1	1	0	-1	-1	1	-1	1	4.8182
13	1	1	0	1	1	1	0	1	0	16.6359
14	0	0	0	0	0	0	0	0	0	8.3735
15	-1	0	0	1	-1	0	0	-1	-1	1.406
16	-1	-1	1	0	1	1	1	0	1	10.1626

17	1	1	1	0	0	1	0	0	0	15.3738
18	0	0	1	1	1	1	1	0	-1	4.3231
19	0	-1	0	0	0	0	0	-1	1	10.0505
20	1	-1	0	0	-1	-1	-1	0	1	6.6228
21	1	0	1	-1	0	-1	0	0	1	9.4316
22	-1	0	-1	0	-1	1	1	1	0	3.8853
23	1	0	-1	-1	1	1	-1	0	0	12.8852
24	-1	0	-1	-1	0	-1	-1	-1	1	2.7331
25	-1	1	1	0	1	-1	-1	0	-1	1.6183
26	-1	0	1	1	0	1	-1	1	1	8.1579
27	-1	1	-1	0	1	1	0	-1	-1	1.0582
28	1	-1	0	-1	1	-1	0	-1	-1	4.3068
29	1	1	-1	1	0	-1	1	-1	-1	2.5104
30	-1	1	0	-1	1	0	1	-1	0	4.0124
31	0	1	1	0	0	0	1	1	-1	3.672
32	1	-1	0	0	0	1	1	1	-1	4.761
33	1	-1	-1	0	0	-1	-1	1	0	7.1593
34	-1	1	1	-1	-1	1	0	-1	0	5.1449
35	0	0	-1	-1	1	0	0	0	-1	2.1749
36	0	-1	-1	0	1	0	1	-1	0	7.1671
37	0	-1	1	1	-1	-1	0	1	-1	3.0084
38	1	1	-1	-1	-1	0	0	1	-1	2.7471
39	-1	-1	1	-1	-1	0	0	1	1	6.6467
40	0	1	0	-1	-1	-1	-1	1	0	4.2251
41	-1	-1	-1	-1	-1	-1	1	0	-1	0.9637
42	-1	1	-1	0	-1	0	-1	0	1	3.8843
43	1	-1	-1	1	0	1	0	0	1	15.8726
44	1	0	1	0	-1	0	-1	-1	-1	4.4323
45	0	-1	0	-1	1	1	-1	1	1	17.7434
46	-1	0	0	-1	0	0	-1	1	-1	1.5073
47	-1	-1	-1	1	1	0	-1	1	-1	1.0998
48	0	-1	-1	1	-1	1	-1	-1	0	6.1027
49	1	0	1	-1	1	-1	1	1	0	12.7236
50	1	0	-1	1	-1	0	1	1	1	8.5339
51	-1	-1	1	1	0	-1	0	-1	0	3.913

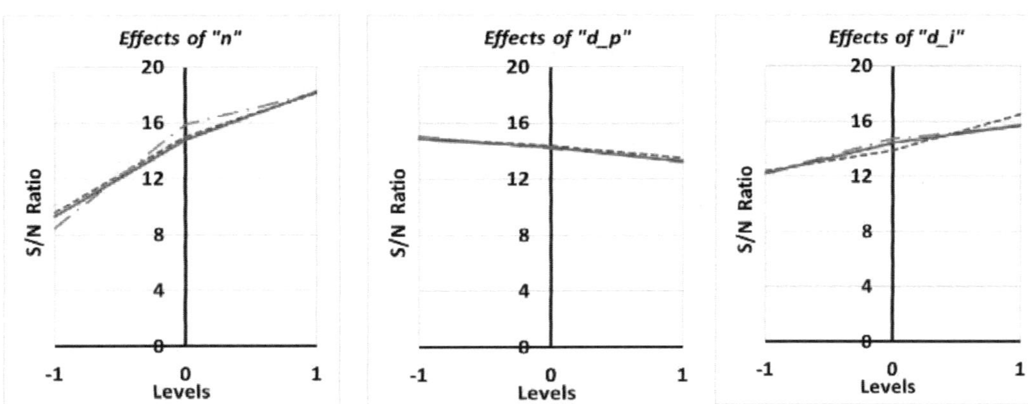

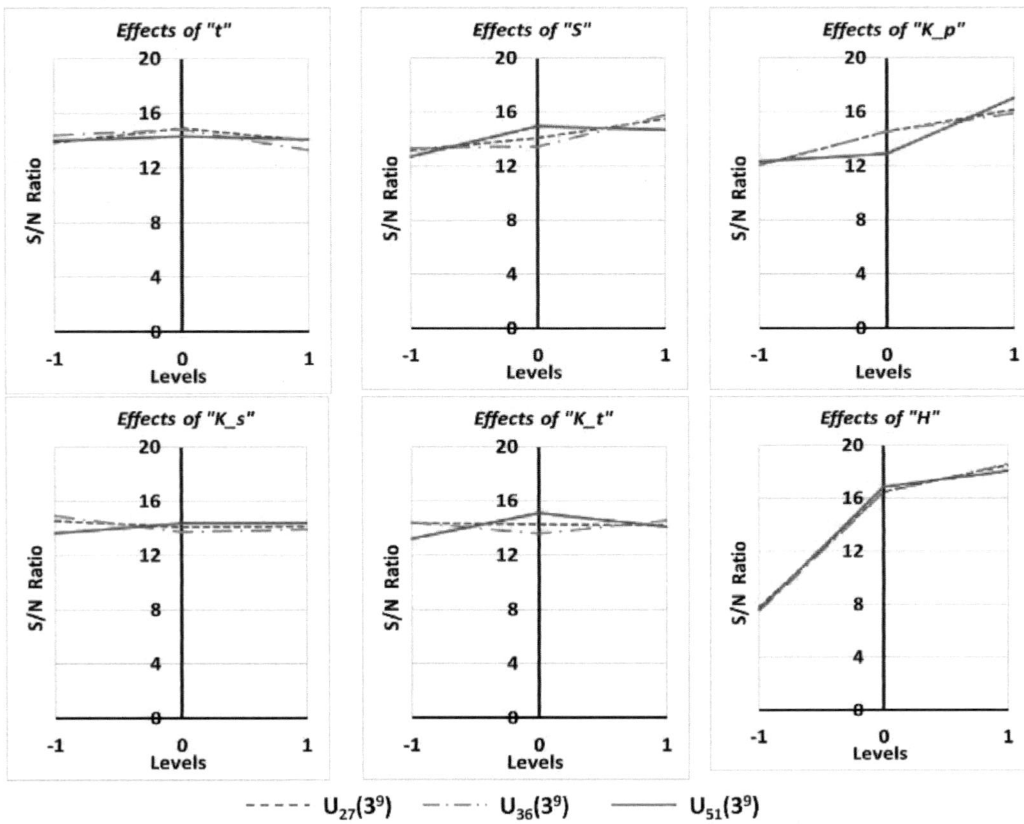

Figure 6. Signal to noise ratios of the energy pile thermal conductance predicted at different uniform designs runs.

4.4. Regression Model Verification

Three cubic regression models have been created from the three designs. To examine these models, the thermal conductance of a base case with all factors are at mid-level is calculated. Then each factor is changed to the extreme levels separately whilst other factors are at the mid-level. The energy pile conductance of these cases are compared to that predicted by the three cubic regression models. As shown in figure 7, the models achieved by designs $U_{36}(3^9)$ and $U_{51}(3^9)$ are in better correlation with the simulation experiment results. However, model achieved by design $U_{27}(3^9)$ is slightly deviated.

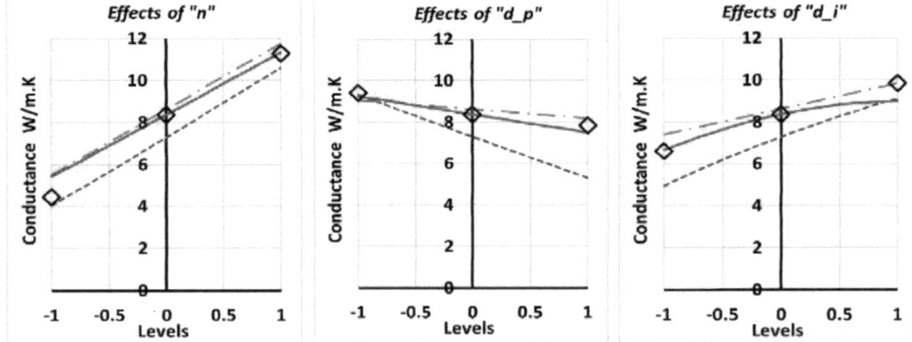

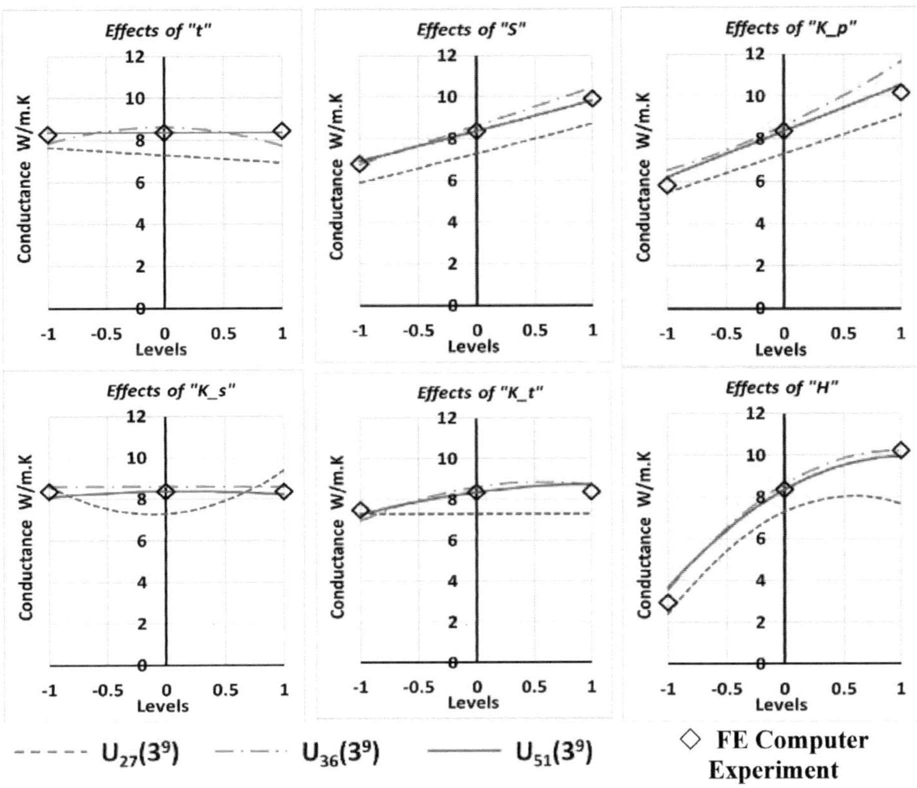

Figure 7. Predicted energy pile conductance using the cubic regression models and FE experiments.

4.5. Optimized Thermal Conductance

Each of the uniform designs have shown different maximum conductance within the investigated combinations. The combinations related to the maximum conductance for each design is shown in table 4. By using the three measures; mean of means, Signal to noise ratio, and cubic regression model, applied to the three designs $U_{27}(3^9)$, $U_{36}(3^9)$, and $U_{51}(3^9)$ we can optimize the energy pile factors for maximum thermal conductance to the values shown in table 4. The energy pile with these configurations has achieved thermal conductance of 34.52 W/m.K. To examine this result, a set of 27 simulation experiments is carried out with the optimum combination achieved as base combination. Then each factor is examined separately whilst keeping the other factors at the optimum value. As shown in table 5, the optimum combinations shows the maximum energy pile conductance in almost all cases. A slight increase (<2%) is observed with choosing thicker tube. This test also shows that using Stainless steel tube (K=16 W/m.K) instead of HDPE tubes (K=0.5 W/m.k) increases the pile conductance (27%). However, using galvanized steel (K=32 W/m.K) increases the pile conductance (1.7%). It is worth noting that these results are limited to the investigated boundaries of each factor.

Table 4. Maximum conductance achieved by each design and the optimum factors for all cases.

	$U_{27}(3^9)$	$U_{36}(3^9)$	$U_{51}(3^9)$	Optimum
Number of Tubes (n)	+1	+1	0	+1
Pile Diameter (d_p)	-1	+1	-1	-1
Tube Inner Diameter (d_i)	+1	0	0	+1
Tube Thickness (t)	-1	-1	-1	0
Distance between Tubes (S)	+1	+1	+1	+1
Pile Thermal Conductivity (K_p)	+1	+1	+1	+1

Ground Thermal Conductivity (K_s)	-1	0	-1	0
Tube thermal Conductivity (K_t)	0	0	+1	+1
Heat Transfer Coefficient (H)	0	+1	+1	+1
Energy Pile Thermal Conductance (C_p)	24.4	21.33	17.75	34.52

Table 5. Maximum thermal conductance verification.

n	d_p	d_i	t	S	K_p	K_s	K_t	H	C_p
-1	-1	1	0	1	1	0	1	1	9.896524
0	-1	1	0	1	1	0	1	1	22.66916
1	-1	1	0	1	1	0	1	1	34.5194
1	-1	1	0	1	1	0	1	1	34.5194
1	0	1	0	1	1	0	1	1	28.37823
1	1	1	0	1	1	0	1	1	25.5727
1	-1	-1	0	1	1	0	1	1	20.5455
1	-1	0	0	1	1	0	1	1	27.63778
1	-1	1	0	1	1	0	1	1	34.5194
1	-1	1	-1	1	1	0	1	1	33.8404
1	-1	1	0	1	1	0	1	1	34.5194
1	-1	1	1	1	1	0	1	1	35.17938
1	-1	1	0	-1	1	0	1	1	15.69314
1	-1	1	0	0	1	0	1	1	23.04178
1	-1	1	0	1	1	0	1	1	34.5194
1	-1	1	0	1	-1	0	1	1	19.84264
1	-1	1	0	1	0	0	1	1	28.39127
1	-1	1	0	1	1	0	1	1	34.5194
1	-1	1	0	1	1	-1	1	1	34.25466
1	-1	1	0	1	1	0	1	1	34.5194
1	-1	1	0	1	1	1	1	1	34.72366
1	-1	1	0	1	1	0	-1	1	26.92667
1	-1	1	0	1	1	0	0	1	34.14625
1	-1	1	0	1	1	0	1	1	34.5194
1	-1	1	0	1	1	0	1	-1	6.734525
1	-1	1	0	1	1	0	1	0	25.07089
1	-1	1	0	1	1	0	1	1	34.5194

5. Conclusions

Energy piles are crucial member of GSHP system to reject or pump heat into ground to reduce the consumption of fossil fuel and CO_2 emission. The efficiency of the energy pile increases with the increase of its steady state thermal conductance. The current work present a statistical approach to define the optimum condition with the least number of experiments using uniform design. Uniform design $U_{36}(3^9)$ has shown acceptable level of error with significantly low number of experiments. The maximum energy pile steady state thermal conductance is achieved with the highest number of tubes, largest tube diameter, largest distance between tubes, highest pile thermal conductivity and highest heat transfer coefficient. Although, smaller pile diameter slightly increase the energy pile thermal conductance, the constructional limitation might stand against reducing these factors with slight insignificant decrease in the energy pile thermal conductance.

Acknowledgement

This publication was made possible by grant No. **NPRP 7-725-2-270** from the Qatar National Research Fund (a member of Qatar Foundation). The statements made herein are solely the responsibility of the authors.

References

[1] Kharseh M, Al-Khawaja M, Suleiman MT. Potential of ground source heat pump systems in cooling-dominated environments: residential buildings. *Geothermics*. 2015 Sep 30; 57: 104-10.
[2] Brandl H. Energy foundations and other thermo-active ground structures. *Geotechnique*. 2006 Mar; 56 (2):81-122.
[3] Fadejev J, Simson R, Kurnitski J, Haghighat F. A review on energy piles design, sizing and modelling. *Energy*. 2017 Jan 24; 122: 390-407.
[4] Cecinato F, Loveridge FA. Influences on the thermal efficiency of energy piles. *Energy*. 2015 Mar 15; 82 :1021-33.
[5] Hu P, Zha J, Lei F, Zhu N, Wu T. A composite cylindrical model and its application in analysis of thermal response and performance for energy pile. *Energy and Buildings*. 2014 Dec 31; 84 :324-32.
[6] He M, Rees SJ, Shao L. Applications of a dynamic three-dimensional numerical model for borehole heat exchangers. *InProceedings of Effstock, the 11th International Conference on Thermal Energy Storage*, Stockholm International Fairs, Stockholm, Sweden 2009.
[7] Lamarche L, Kajl S, Beauchamp B. A review of methods to evaluate borehole thermal resistances in geothermal heat-pump systems. *Geothermics*. 2010 Jun 30; 39 (2):187-200.
[8] Śliwa T, Gołaś A, Wołoszyn J, Gonet A. Numerical Model of Borehole Heat Exchanger in Ansys CFX Software/Numeryczny Model Otworowego Wymiennika Ciepła W Pakiecie Ansys Cfx. *Archives of Mining Sciences*. 2012; 57 (2):375-90.
[9] Wołoszyn J, Gołaś A. Modelling of a borehole heat exchanger using a finite element with multiple degrees of freedom. *Geothermics*. 2013 Jul 31; 47 :13-26.
[10] Nikishkov G. *Finite Element Equations for Heat Transfer. Programming Finite Elements in Java*™. 2010:13-9.
[11] Weissman SA, Anderson NG. Design of experiments (DoE) and process optimization. A review of recent publications. *Organic Process Research & Development*. 2014 Aug 29; 19 (11):1605-33.
[12] Fang KT, Lin DK, Winker P, Zhang Y. Uniform design: theory and application. *Technometrics*. 2000 Aug 1; 42 (3):237-48.
[13] Yang J, Li L, Yang L, Li J. Uniform design for the parameters optimization of pin-fins channel heat sink. *Applied Thermal Engineering*. 2017 Jun 25; 120: 289-97.
[14] Fang KT, Wang Y. *Number-theoretic methods in statistics*. CRC Press; 1993 Dec 1.
[15] Roy RK. *Design of experiments using the Taguchi approach: 16 steps to product and process improvement*. John Wiley & Sons; 2001 Feb 13.

IOP Conf. Series: Materials Science and Engineering **383** (2018) 012035 doi:10.1088/1757-899X/383/1/012035

The study on optimization of characteristic parameters of electromagnetic abrasive jet pressurize

Xie-Huaibei*, Wang-yan, Cui-Song and Li-Changpeng

Anhui University Of Science And Technology, School of Mechanical Engineering, Huainan, Anhui, China

*E-mail: austxhb@163.com

Abstract. In view of the poor continuity of the pre-mixed abrasive jet equipment and the non-uniform mixing of the abrasive particles in the post-mixed abrasive jet system, this thesis is based on electromagnetic abrasive slurry jet technology. The affecting factors of jet compression characteristics of electromagnetic abrasive slurry are studied through the combination of numerical simulation analysis and experimental research. According to previous research results, the influence of two parameters of magnetic flux density and voltage on the outlet velocity and kinetic energy of conductive medium are predicted through numerical simulation. Combined with orthogonal test characteristics, we set up a test platform and carry out some experiments. The experimental results show that the biggest influence on jet supercharging characteristics is magnetic flux density and the next one is the voltage. The results of numerical simulation are verified. When the flux density is 0.7T, the voltage is 72V, the concentration of conductive medium is 18% and the characteristics of jet pressurize are best.

1. Introduction

Through the concerted efforts of several generations, high pressure water jet technology has extended a number of different jet varieties on the basis of pure water jet, about abrasive jet, ultra supercritical carbon dioxide jet, excited jet and so on. Water jet are widely used in mechanics, marine engineering and coal mining etc. It is used for machining, hull cleaning and drilling of coal and rock etc [1]. However, traditional abrasive jet technology still has some technical problems that need to be overcome. The working pressure is very high in post mixing abrasive jet equipment, and the damage degree of the seal and the frequency of replacement are high, which also prone to leakage. The pre-mixed abrasive jet equipment has a high frequency of start and stop and poor work continuity [2-5].

In 1961, James B.Friauf first carried out the study of MHD propulsion unit [6].In 1971,our country developed the MHDT, and carried out experiments in water [7].Yang Yong-yin [8] etc studied the effect of high polymer concentration on jet performance at high pressure. Liu Xiao-jian [9] etc uses sodium based bentonite to improve the suspension properties of the slurry, Liu Ting-cheng [10] etc successfully applied abrasive jet to the field of cleaning. Zhang Dong-su [11-12] etc proposed the technology of electromagnetic abrasive slurry jet and made this technology possible.Hu Ya-lin [13] carried out the research on the performance of conductive medium in the electromagnetic pressurize device and the ratio of the abrasive paste is obtained. In this paper, this ratio is used in the slurry solution. Jiang Hong [14] pointed out that the circular cross section of the flow channel could obtain

Content from this work may be used under the terms of the Creative Commons Attribution 3.0 licence. Any further distribution of this work must maintain attribution to the author(s) and the title of the work, journal citation and DOI.
Published under licence by IOP Publishing Ltd

IOP Conf. Series: Materials Science and Engineering **383** (2018) 012035 doi:10.1088/1757-899X/383/1/012035

better jet pressure characteristics easier and this structure is adopted in this paper.

However, there is no relevant report on the influence of key factors on the supercharging characteristics of electromagnetic abrasive slurry jet. How to determine the main key factors has not been carried out about the effect of electromagnetic field driving conductive medium on the characteristics of jet pressurize. In this paper, the influence of two parameters of magnetic flux density and voltage on the out velocity and kinetic energy of conductive medium is predicted by numerical simulation. It has constructed a test platform of electromagnetic abrasive slurry jet, combined with orthogonal test characteristics. Finally, an experiment of jet pressurize is carried out. Under the current test conditions, the optimum parameters of the key factors are obtained to determine the main influencing factors.

2. Numerical simulation

It is similar to the principle of MHDT, electromagnetic abrasive jet technology utilizes medium conductivity and the conductive medium moves under the action of electromagnetic force of orthogonal electromagnetic field [15].

USSR Academy of Sciences, the institute of high temperature physics, used superconducting spiral MHDT channels as models to carry out experiments of MHD performance analysis and pressure distribution parameters [16].According to the literature [17], the relation among the speed of the conductive medium, the density of the conductive medium, conductivity of conductive medium, voltage of electrode plate, runner length, the distance between the plates and the magnetic flux density of the magnetic source is as shown in equation 1.

$$v = \sqrt{\frac{2\eta\sigma UBL}{\rho D}} \tag{1}$$

In equation (1), η is conversion rate; σ is electrode plate conductivity, $s \cdot m^{-1}$; U is voltage value between electrode plates, V; B is magnetic flux density, T; L is runner length, m; ρ is density of conductive medium, $kg \cdot m^{-3}$; D is runner section diameter, m.

According to the equation (1), the K value is:

$$K = \sqrt{\frac{2\eta\sigma L}{\rho D}} \tag{2}$$

According to the equation of kinetic energy,

$$E = \frac{1}{2}mv^2 \tag{3}$$

In equation (3), m is the quality of conductive medium, kg; v is the speed of conductive medium, $m \cdot s^{-1}$.

According to the equation (3), change the formula (4) to

$$E = \frac{1}{2}mK^2UB \tag{4}$$

According to the existing experimental conditions, parameters are shown in table 1.The theoretical reference values of each influencing factor are listed and K values are obtained.

Table 1. Theoretical reference value of velocity influence factors of conductive medium.

Influence factor	Conversion rate η	Conductivity σ (s/m)	Length L (m)	Density ρ (kg/m^3)	Separation distance D(m)	Concentration V (%)
Numeric al value	40%	8.8	0.6	1.132×10^3	0.025	18

According to the equation (3), within certain range, we obtain the relationship diagram of outlet velocity of conductive medium, dielectric voltage and flux density, Specifics are as shown in figure 1 and figure 2.

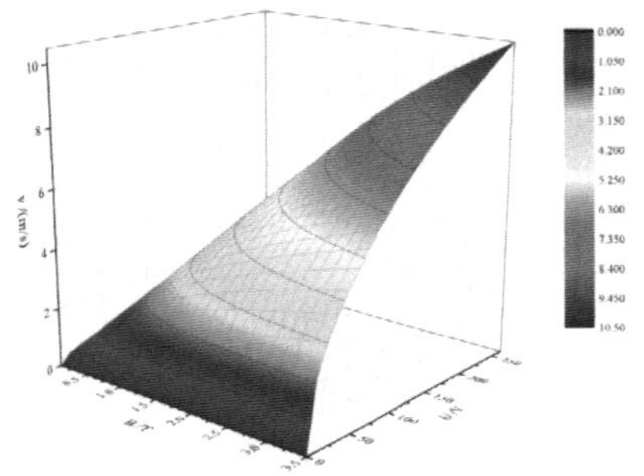

Figure 1. The relationship diagram of outlet velocity of conductive medium and voltage and flux density.

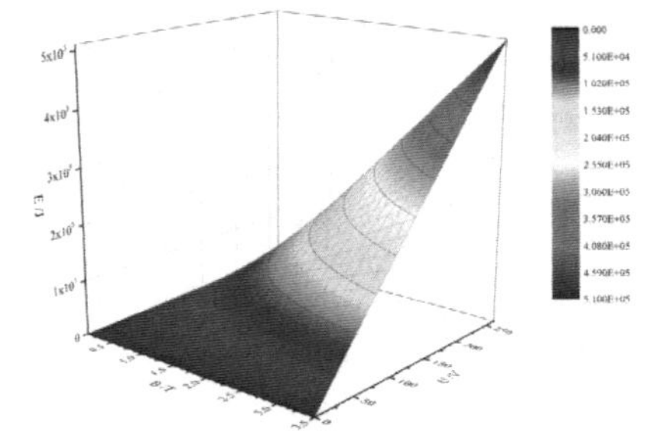

Figure 2. The relationship diagram of outlet kinetic energy of conductive medium, voltage and flux density.

According to figure 1, as for the outlet velocity of the conductive medium, it increases with the increase of voltage and flux density. However, with the increase of voltage value, the outlet velocity speed of the conductive medium gradually decreases.In the late part, the main influence factor of the outlet velocity of the conductive medium is the magnetic flux density. According to figure 2, as for the outlet kinetic energy of the conductive medium, the kinetic energy of the conductive medium greatly increases with the increase of the voltage and flux density. In a range, the effects of voltage and flux density parameters on the kinetic energy of the conductive medium can be predicted.

3. Pressurization characteristic parameter test

The experiment uses three different types of conductive material, namely abrasive slurries mixed solutions, saturated NaCl solutions, and diluted NaCl solutions. The platinum electrode is selected as the electrode plate, three-stage magnetic field strength is used as magnetic field source, and we build an experimental platform of electromagnetic abrasive slurry pressurization characteristics. According to the experiment, the height difference of liquid column is recorded.

In order to explain the experiment better, the concentration, voltage, magnetic flux density and plate spacing are chosen as experimental factors. Each factor takes three levels, the orthogonal test table designed by pseudo level method is shown in table 2. The main index of evaluation is the height of liquid column. 9 sets of experiments are designed according to the orthogonal test table L9 (34). The results are analyzed by range analysis and the results of orthogonal tests are shown in table 3.

Table 2. Factor level diagram of orthogonal test.

Level Factor	Concentration *M*/%	Voltage U / V	Flux density B / T	Separation distance D/*m*
1	10	6	0.1	0.09
2	18	30	0.4	0.05
3	26	72	0.7	0.02

Table 3. Orthogonal test results.

Level Factor	Concentration M/%	Voltage U / V	Flux density B / T	Separation distance D/m	Liquid height H/mm
1	10	6	0.1	0.05	2
2	10	30	0.4	0.025	10
3	10	72	0.7	0.01	26
4	18	6	0.4	0.01	7
5	18	30	0.7	0.05	21
6	18	72	0.1	0.025	12
7	26	6	0.7	0.025	14

8	26	30	0.1	0.01	8
9	26	72	0.4	0.05	16
K	38	33	22	39	$K_1+K_2+K_3=1$
K_2	40	39	33	36	16
K_3	38	54	61	41	
k_1	12.7	11.0	7.3	13.0	$k_1+k_2+k_3=38$
k_2	13.3	13.0	11.0	12.0	.7
k_3	12.7	18.0	20.3	13.7	
Rang	0.6	7	13	1.3	

Notes: The K_1, K_2 and K_3 in the table are respectively the sum of the liquid levels at the level 1, 2 and 3. R is the difference between the maximum and minimum values in K_1, K_2, and K_3.

The experimental results show as follows: The magnetic flux density has the great influence on the supercharging characteristics of MHD jet, followed by voltage ; The effect of concentration and plate spacing is small and the result is the same as numerical simulation. The greater the magnetic flux density is, the greater the supercharging characteristics show. The flux density 0.7T is appropriate. The greater the voltage is, the better the jet supercharging characteristics show. But in the experiment, the increase in voltage will exacerbate the electrolysis, the experimental results increase interference. In order to get better magnetic fluid jet supercharging effect, the voltage value is selected at 72V.When the concentration is increased, the electrolysis phenomenon is intensified, and the interference degree of the experimental effect is enhanced. Therefore, the concentration can not be too high, while too low concentration will lead to poor conductivity of liquids. It reduces the supercharging characteristics of MHD jet, and the concentration of 18% is appropriate. With the increase of the plate spacing, the electric field strength between the electrode plates becomes weaker, the smaller the plate spacing can improve the jet supercharging characteristics. But in experimenting, too small plate distance makes it possible to cause a short circuit between the electrodes. This is not conducive to the experiment, so, the distance between plates is selected at 0.025m.

4. Conclusion

According to the defects of traditional abrasive jet equipment, the research status of electromagnetic abrasive slurry jet technology is analyzed. According to previous research results, The influence of two parameters of magnetic flux density and voltage on the velocity and kinetic energy of the conductive medium is analyzed by numerical simulation.Combined with orthogonal test characteristics, a test platform for electromagnetic abrasive slurry jet has been constructed. The jet pressurization experiment has been carried out. Experimental results show as follows: when flux density is 0.7T, voltage value is 72V, the concentration of conductive medium is 18% and plate distance is 0.025m, the supercharging effect is better. The two key factors of magnetic flux density and voltage that affect the jet pressure characteristics of the electromagnetic abrasive slurry jet have been identified.

References

[1] Xue Shengxiong. High pressure water jet technology project[M]. Hefei University of Technology Publishing House, 2006
[2] Meng Yunqing,Nie Baisheng. Application of abrasive water jet cutting technology in mine safety[J].*China Coal*, 2008(6) pp 45–46
[3] Huang Yifei.Study on surface propulsion mechanism of electromagnetic fluid [D]. Nanjing University of Science and Technology, 2010
[4] Wang Chengyong, Wang Rongjuan,Yang Peixuan, etc. Slurry ratio and drilling performance of

micro abrasive slurry jet[J].*Chinese mechanical engineering*, 2015,16 (11) pp 2191–96
[5] Zeng Wen Liu, R.Y. Liu.Study on Pre-Mixed Micro Abrasive Water Jet Machining System[J].*Applied Mechanics and Materials*, 2014,10 (618) pp 475–79
[6] Friauf J B.Electromagnetic ship propulsion[J].*Journal of the American Society for Naval Engineers*,1961,73 (1) pp 139–142
[7] Zhu Hong, Zhang Feng Qin. MHD design [J]. invigilation system in Shanghai Shipping Transport *Science Research Journal*, 2014, 37 (4) pp 18–22
[8] Yang Yongyin, Wang Rui, and Zhou Weidong. PAM Study on the influence of abrasive fluid jet velocity [J]. *petroleum drilling technology*, 2001, 29 (1) pp 4–6
[9] Liu Xiaojian, Yu Tao. Experimental study on abrasive jet technology and its cutting performance[J]. *Journal of mechanical engineering*, 2006, 42 (4) pp 157–60
[10] Liu Tingcheng, Liu Yan, Fan Xiaohong. Energy storage type pre mixing continuous abrasive water jet system [J]. *cleaning world*, 2009, 25 (10) pp 32–38
[11] Zhang Dongsu, Xie Huaibei, Jiang Hong, etc. Investigation of suspended abrasive fluid jet technology [J]. *fluid machinery*, 2015 (12) pp 47–49
[12] Zhang Dongsu,Miao Sizhong,Xie Huaibei.Research of Abrasive Suspension Magnetic Fluid Jet Technology[C]. *3rd International Conference on Materials Science and Mechanical Engineering (ICMSME)*, Windsor, ENGLAND, DEStech Publicat Inc, 2016 pp 1–4
[13] Hu Yalin. Research on medium of jet driven pneumatic jet device [D]. Anhui University Of Science And Technology, 2016
[14] Jiang Hong. Design and parameter analysis of runner of electromagnetic pressure unit for abrasive slurry [D]. Anhui University Of Science And Technology, 2016
[15] Li Changpeng, Zhang Yangkai. Sizhong seedlings, a feasibility analysis of magnetic fluid abrasive jet [J]. *science technology and engineering*, 2016, 1 (17) pp 215–19
[16] Chen Shuling. Research on integrated optimization design technology of high speed ship intelligent propulsion system [D]. Jiangsu: Jiangsu University of Science and Technology, 2005
[17] Chen Shuling. Research on integrated optimization design technology of high speed ship intelligent propulsion system [D]. Jiangsu: Jiangsu University of Science and Technology, 2005

Influence of electromagnetic compound treatment on microstructure and properties of cemented carbide

WEI Liao[1,2,*] and YANG Yi[1]

[1]Sichuan University, Chengdu, 610065, China
[2]Guangxi Radio and TV University, Nanning 530022, China

*E-mail: weiliao1986@163.com

Abstract. In order to study the effect of external field treatment on mechanical properties, cutting performance, microstructure of cemented carbide, the cemented carbide samples were treated by different magnetic field frequency (f=0.3 Hz, 0.4 Hz, 0.5 Hz) processing methods using self-made electromagnetic compound treatment device. On the basis of the study of this method, the electromagnetic compound treatment method was proposed to couple pulsed magnetic field and pulsed current. The results show that after the treatment of magnetic field frequency, the hardness of the sample slightly higher than that untreated sample, the increase value of percent between 0.5% and 1.3%, the value of hardness was 1594.72 HV, 1587.76 HV, 1582.45 HV, the longer the magnetization time, the higher the hardness; the cutting performance was increased, which means the tool life was increased, the increase value of percent from 4.5% to 7.0%. After electromagnetic compound treatment, the increase value of percent was 1.7% and 2.3%, the value of hardness was 1601.6 HV, 1609.82 HV; the tool life was increased 7.5% and 8.4%. After external field treatment, the angular state of wolfram carbide (WC) particle was transformed into smooth state and was more evenly distributed.

1. Introduction

The cutting tool can withstand higher temperature and mechanical shock during machining, the residual stresses was generated during heat processing and grinding process will accelerate tool wear and influence the surface machining quality of the workpiece [1~3]. The improvement of tool has been an important research topic in the whole machinery manufacturing industry.

In recent years, the research on the improvement of material properties has been carried out by single action of pulsed magnetic field at home and abroad, and the results have been obtained. For example, the practice of American Innovex company shows that pulsed magnetic field treatment can improve the service life of the tool by improving the residual stress distribution in the tool [4].

In recent years, pulsed current treatment as a new technology to improve the microstructure and properties of materials has the advantages of clean, high efficiency and low cost, which has attracted the attention of the material workers [5~7]. The pulsed current treatment has been reported in the field of materials electro plasticity, amorphous, crack healing and fatigue performance improvement [8~12].

In this paper, the effect of electromagnetic compound treatment on the microstrucure and performance of cemented carbide was analyzed and studied, and with the effect of magnetic field treatment alone under the same parameters has carried on the comparative analysis, provided support for further systematic study the change law of cemented carbide microsructure and performance under the external field.

Content from this work may be used under the terms of the Creative Commons Attribution 3.0 licence. Any further distribution of this work must maintain attribution to the author(s) and the title of the work, journal citation and DOI.
Published under licence by IOP Publishing Ltd

IOP Conf. Series: Materials Science and Engineering **383** (2018) 012036 doi:10.1088/1757-899X/383/1/012036

2. Experiment

The experimental objects are cemented carbide cylinder sample and cemented carbide tool, and the workpiece material is 45# quenched and tempered steel.

The sample was put into a self-made electromagnetic compound treatment device. The electromagnetic compound treatment system mainly consists of three parts: pulsed magnetic field system, pulsed current system and control system, the schematic diagram was shown in figure 1 [13].

The pulsed magnetic field system consists of power supply, controller and magnetizing device. In electrical processing system generated electrical pulse by thyristor controlled alternating current transformer. When the electromagnetic compound processing is applied, the excitation power supplies the exciting current to the coil, and the magnetic induction intensity is generated in the closed magnetic circuit formed by the magnetic pole and the sample, while the pulse current is provided by the pulse power supply.

In this experiment, the magnetic field treatment (MFT) has three technics (MFT-A, MFT-B and MFT-C), and parameters were shown in table 1, the electromagnetic compound treatment (ECT) has two technics (ECT-A and ECT-B), parameters were shown in table 2.

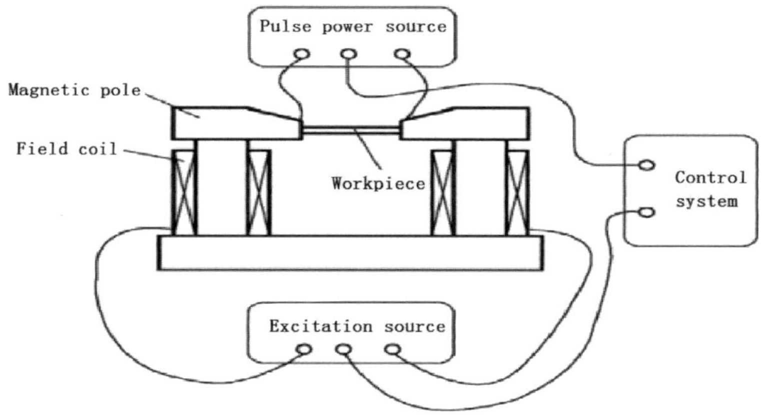

Figure 1. The diagram of electromagnetic compound treatment device.

Table1. Magnetic field treatment parameters.

Technics	Time (S)	Frequency (Hz)
MFT-A	60	0.3
MFT-B	45	0.4
MFT-C	36	0.5

Table 2. Electromagnetic compound treatment parameters.

Technics	Current density (A/mm^2)	Frequency (Hz)	Time (s)
ECT-A	10	0.3	60
ECT-B	20	0.3	60

Then the experiment of turning excircle was done on the NC lathe by the unprocessed tool and the tool processed by magnetic field and electromagnetic compound treatment. During the experiment, the wear extent of flank surface was measured by image measuring instrument and the cutting force was measured by the dynamometer. The microstructure of the sample was observed by scanning electron microscope.

3. Results and discussion

3.1. Effect of external field treatment on microstructure of cemented carbide

Both magnetic field treatment and electromagnetic compound treatment can change the microstructure of the cemented carbide. Figure 2 showed the back scattered electron (BSE) photo of YG8 cemented carbide with untreated, magnetic field treatment and electromagnetic compound treatment.

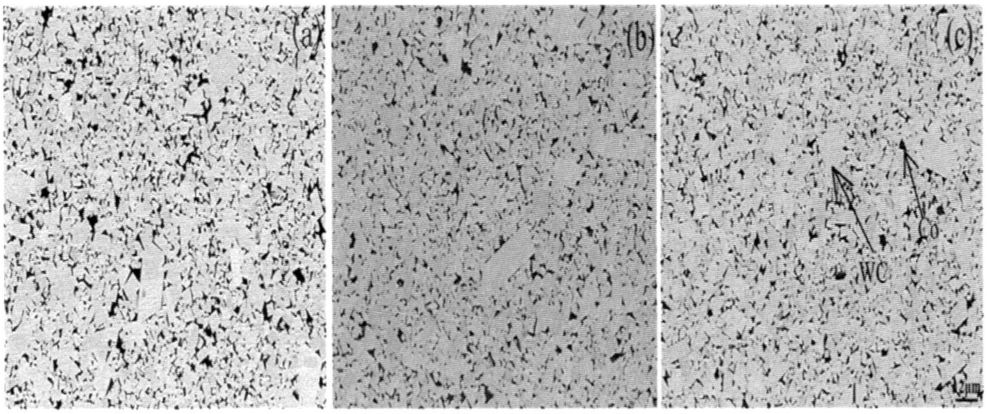

Figure 2. The BSE images of YG8: (a) untreated, (b) magnetic field treatment, (c) electromagnetic compound processing.

It can be seen from figure 2 that: in the untreated, the WC (wolfram carbide) particle of cemented carbide is sharp and angular, as shown in figure 2 (a). The existence of angular result in the stress distribution is extremely uneven, which leads to stress concentration and is unfavorable to toughness of alloy. After magnetic field treatment and electromagnetic compound processing, the angular state of WC particle was transformed into smooth state, which reduces the stress concentration of contact points and improves the toughness of WC [14], as shown in figure 2 (b) and figure 2 (c). At the same time, it can be seen from figure 2 (b) and figure 2 (c) that the WC particle of cemented carbide was more evenly distributed.

3.2. Effect of magnetic field frequency on the mechanical properties of cemented carbide

When the magnetic field strength was fixed, the values of hardness by different magnetic field frequency (f=0.3 Hz, 0.4 Hz, 0.5 Hz) were obtained. The values were shown in figure 3.

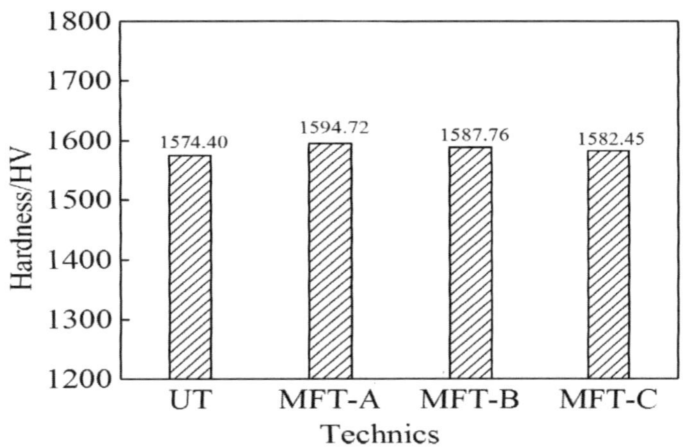

Figure 3. The values of different magnetic field frequency on hardness.

It can be seen from the figure 3 that the hardness of the sample treated with magnetic field frequency slightly higher than that untreated sample. The increase range of percent between 0.5% and 1.3%, the value of hardness was 1594.72 HV, 1587.76 HV, 1582.45 HV, the longer the magnetization time, the higher the hardness. Preliminary analysis may be with the increase of magnetic field frequency, the lower the magnetization time, the lower the dislocation in the sample, the deformation ability of the sample was weakened, and the hardness decreased.

When magnetic field strength was fixed, the tool wear was obtained by changing different magnetic field frequency, the wear curve was done nonlinear fitting by Origin software, the cutting distance of flank surface average wear value of 0.3 mm were obtained, the values were shown in table 3.

Table 3. Tool cutting distance data of different magnetic field frequency.

Technics	Cutting distance (m)
UT	112.76
MFT-A	120.62
MFT-B	119.29
MFT-C	117.78

It can be seen from table 3, the cutting performance of the tool after the magnetic field frequency was increased compared with the untreated tool, which means the tool life was increased from 4.5% to 7.0%. The influence of different magnetic field frequency on the mechanical properties of cemented carbide was analyzed combined with the front, it may be that the hardness of the material was increased after magnetic field treatment .

Figure 4 showed the influence of different magnetic field frequency on the main cutting force and wear extent.

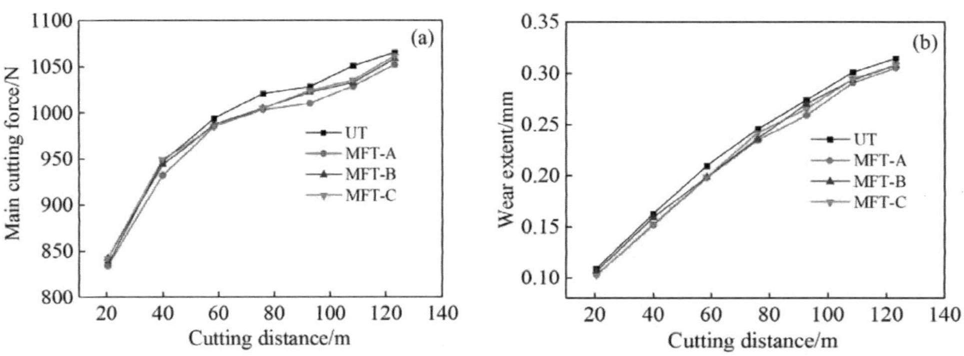

Figure 4. The effect of magnetic field frequency: (a) main cutting force, (b) wear extent.

It can be seen from figure 4 (a) and figure 4 (b), the main cutting force and the wear extent curve are consistent. Therefore, the change of wear extent can be predicted by measuring the change of cutting force so as to study the change of tool cutting performance.

At the same time, after the treatment of magnetic field frequency, the change of treatment effect was not obvious. The preliminary analysis may be due to unsuitable selection of magnetic field treatment parameters, which result in no significant improvement in cutting performance, proper parameters and mechanism analysis were needed further research.

3.3. Effect of electromagnetic compound treatment on the performance of cemented carbide

In the case of not changing the magnetic field parameters, the electromagnetic compound treatment on the sample is carried out by means of coupling pulsed magnetic field and pulsed current, the hardness was obtained. The values were shown in figure 5.

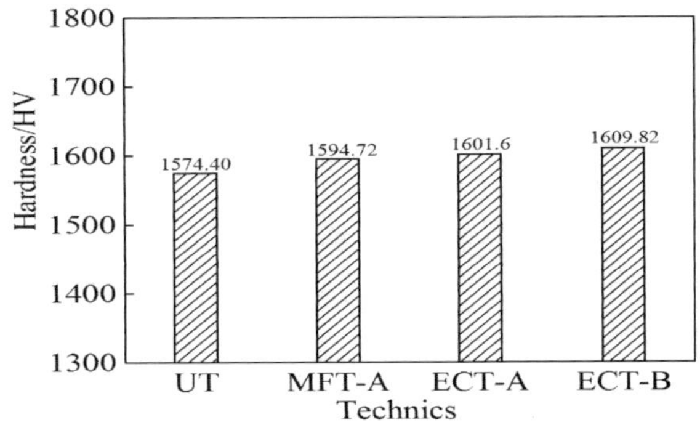

Figure 5. The hardness values of electromagnetic compound treatment.

It can be seen from the figure 5 that the hardness of the sample treated with electromagnetic compound treatment increase than that untreated sample and magnetic field frequency treatment. The increase value of percent was 1.7% and 2.3%. The hardness value was 1601.6 HV, 1609.82 HV. Preliminary analysis because of the effect of pulsed current treatment in the electromagnetic compound treatment.

After the electromagnetic compound treatment, the wear curve was done nonlinear fitting by Origin software, the cutting distance of flank surface average wear value of 0.3mm were obtained, the values were shown in table 4.

Table 4. Tool cutting distance data of electromagnetic compound treatment.

Technics	Cutting distance (m)
UT	112.76
ECT-A	121.16
ECT-B	122.18

It can be seen from table 4, the increase percent of cutting distance for cutting tool after electromagnetic compound treatment was 7.5% and 8.4%. It may be that the hardness of the material was increased after magnetic field treatment.

Figure 6 showed the influence of electromagnetic compound treatment on the main cutting force and wear extent.

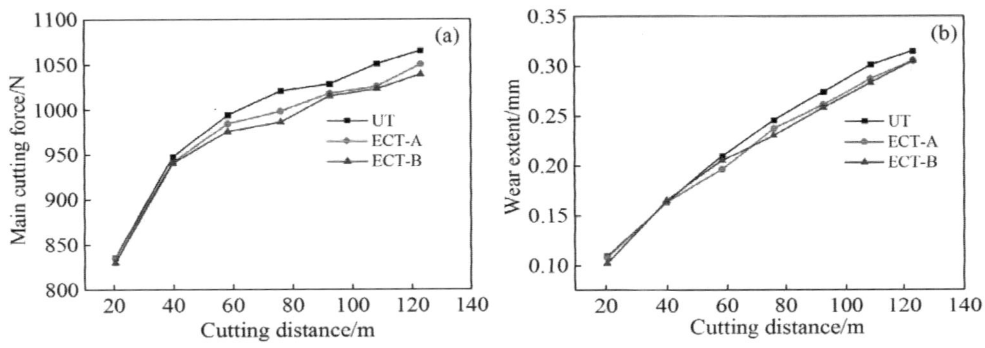

Figure 6. The effect of electromagnetic compound processing: (a) main cutting force, (b) wear extent.

It can be seen from the figure 6 (a) and figure 6 (b), the main cutting force of the electromagnetic compound treatment tool smaller than that of the untreated tool, and the wear extent also less than that of the untreated tool. The main cutting force and the wear curve are consistent.

4. Conclusions

(1) This paper proposed a novel approach to improvement the properties of materials, that was: electromagnetic compound treatment. It was an important research direction and topic in the field of material processing.

(2) After the treatment of magnetic field frequency, the hardness of the sample slightly higher than that untreated sample, the value of hardness was 1594.72 HV, 1587.76 HV, 1582.45 HV, the longer the magnetization time, the higher the hardness; the cutting performance was increased, which means the tool life was increased, the increase value of percent from 4.5% to 7.0%. After electromagnetic compound treatment, the increase value of percent was 1.7% and 2.3%, the value of hardness was 1601.6 HV, 1609.82 HV; the tool life was increased 7.5% and 8.4%.

(3) After external field treatment, the angular state of wolfram carbide (WC) particle was transformed into smooth state and was more evenly distributed.

References

[1] XU Jianming, HU Heng, FU Wei, et al. Behaviors and Mechanisms of Friction and Wear of Special Alloys at High Temperature[J]. *Journal of Steel Research*, 2014, 26 [11] p 60

[2] LI Jian, ZHAO Haiyan, CAI Zhipeng, et al. Reducing residual stresses in engineering structures by pulsed magnetic treatments[J]. *Tsinghua Univ(Sci & Tech)*, 2007, 47 (2) pp 161–64

[3] JIANG Zenghui, WANG Linlin, SHI Li, et al. Study on Tool Wear Mechanism and Characteristics of Carbide Tools in Cutting Ti6Al4V[J]. *Journal of mechanical engineering*, 2014, 50 (1) p 178

[4] MILLER P C.Look at magnetic treatment of tools and surface[J]. *Tooling &Production*, 1990, 55 (12) pp 100–03

[5] LI F, REGEL L L, WILCOX W R. The influence of electric current pulses on the microstructure of the MnBi/Bi eutectic[J]. *Journal of Crystal Growth*, 2001, 223 (1) pp 251–64

[6] BASMAK J P, SPRECHER A F, CONRAD H. Colony(grain)size reduction of eutectic Pb-Sn castings by electropulsing[J]. *Scr Metall Mater*, 1995, 32 (6) pp 879–84

[7] LIAO X, ZHAI Q, LUO J, et al. Refining mechanism of the electric current pulse on the solidification structure of pure aluminum[J]. *Acta Materialia*, 2007, 55 (9) pp 3103–09

[8] SHEN Yifu, GUO Xiaonan, ZHANG Kun, et al. Effect of electric current pulse on metal materials[J]. *Materials Science and Engineering*, 1998, 16 (3) pp 4–7

[9] ZHANG Wei, SUI Manling, ZhOU Yizhou, et al. Electropulsing-induced evolution of Microstructures in metals[J]. *Acta Metallrugica Sinica*, 2003, 39 (10) pp 1009–18

[10] LV Baochen, ZhOU Yizhou, WANG Baoquan, et al. Effect of electropulsing on the fatigued heat-rolled 30CrMnSiA steel[J]. *Chinese Journal of Materials Research*, 2003, 17 (1) pp 16–18

[11] LEVITIN V V, LOSKUTOV S V. The effect of a current pulse on the fatigue of titanium alloy[J]. *Solid State Communications*, 2004, 131 pp 181–83

[12] ZHOU Yizhou, LUO Shen, HE Guanhu, et al. Experimental study on crack healing in steel using electrie current pulse technique[J]. *Chinese Journal of Materials Research*, 2003, 17 (2) pp 169–72

[13] HUANG Xinquan, CAI Zhipeng. Enhancement Effect of Reducing Welding Residual Stress by Composite Electromagnetic Treatment[J]. *JOURNAL OF MECHANICAL ENGINEERING*, 2011, 47 (4) pp 88–92

[14] LI Jian, GUO Xin, YAO Jianjun, et al. Effect of magnetic quenching on performance and structure of cemented carbides[J]. *Journal of Central South University (Science and Technology)*, 2012, 43 (7) pp 2513–18

IOP Conf. Series: Materials Science and Engineering **383** (2018) 012037 doi:10.1088/1757-899X/383/1/012037

Validation of a 5-zone-car-cabin model to predict the energy saving potentials of a battery electric vehicle's HVAC system

Kurt Kruppok[1,*], Felix Claret[1], Peter Neugebauer[1], Reiner Kriesten[1] and Eric Sax[2]

[1]Institute of Energy Efficient Mobility (IEEM), University of Applied Sciences Karlsruhe (HsKA), Moltkestr. 30, 76133 Karlsruhe, Germany
[2]Institute for Information Processing Technologies (ITIV), Karlsruher Institute of Technology (KIT), Engesserstraße 5, 76131 Karlsruhe, Germany

*E-mail: Kurt.Kruppok@hs-karlsruhe.de

Abstract. This paper presents a 5-zone-car-cabin model which is able to simulate the car cabin's thermal condition depending on several influencing parameters. This includes the solar radiation, to which special attention is paid in this paper. In addition, a generic methodology for parameter optimization considering measurements on a reference vehicle is presented. Thus, a very high degree of determination of the model was achieved. This paper is motivated by the impact of auxiliary loads on overall energy consumption in battery electric vehicles. The further use of the model is intended to calculate the energy saving potentials of the heating, ventilation, and air conditioning system by reducing or increasing the target interior temperature. This is necessary for a predictive control of secondary consumers.

1. Introduction

Most of the potential car buyers do not trust in the vehicle range of battery electric vehicles (BEVs), as they are often smaller than the manufacturer indicates. In addition, the so-called range anxiety (The fear that a vehicle cannot reach its destination because of insufficient range.) is still widespread. In order to counteract the range disadvantage, an efficient use of the available energy is necessary. This requires the knowledge of the overall energy requirement to cover the present route. Each route has different energy loads in terms of propulsion and secondary consumers. The heating, ventilation, and air-conditioning (HVAC) system can be the largest auxiliary load. It is shown that the range of electric vehicles significantly reduces when the HVAC system is activated [1], [2]. Depending on the weather conditions, up to 25% of the required energy is used by secondary consumers [3]. Thus, the HVAC system offers a considerable energy saving potential by reducing or increasing the target interior temperature in case of a predictive control. However, possible energy savings quickly lead to the conflict with passenger's comfort. To identify the energy saving potentials of an HVAC system, some preparation steps are necessary, according to figure 1.

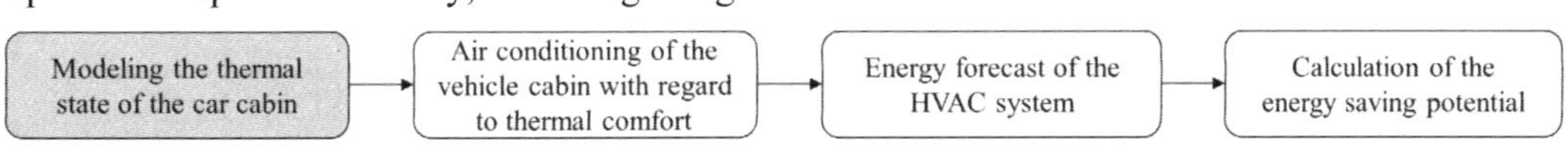

Figure 1. The objective of the research project is shown in partial steps. In this paper, only the first step is discussed.

Content from this work may be used under the terms of the Creative Commons Attribution 3.0 licence. Any further distribution of this work must maintain attribution to the author(s) and the title of the work, journal citation and DOI.
Published under licence by IOP Publishing Ltd

In the first step, detailed knowledge of the condition within the car cabin depending on relevant influencing variables is acquired. Therefore, the thermal behavior is simulated by a 5-zone-car-cabin model considering relevant variables. In the second step, an automatic climate control will be used to estimate how much thermal energy must be supplied to the vehicle interior by the HVAC system. In the next step, the required electrical energy can be calculated using the necessary heat and mass flow. Finally, the energy saving potential of the HVAC system can be calculated for a given route. This paper covers only the first step by validating a 5-zone-car-cabin model.

2. Related work

In previous work the advantages and disadvantages of empirical, single-zone, multi-zone, and CFD (computational fluid dynamics) simulation methods have been discussed [4]. The model presented in this paper is intended to predict the energy demand of an HVAC system, considering the passenger's comfort aspects. This requires a classification of the car cabin into different volumes is necessary since each body part has a different temperature sensation. The temperature distribution and heat flows in the car cabin can be investigated in detail with CFD, but the modelling effort is very high and the simulation speed for fast predictions is too slow [5][6]. As a compromise between computation time and accuracy, a 5-zone model was chosen. In literature, several different models for car cabin thermal calculations have been presented. An overview of the models has been shown in [7] and [8]. Other related works mainly have focused on a single-zone car cabin model [9][10]. When considering solar radiation, a multi-volume model provides advantages compared to a single zone. The partial heating of a zone results in thermodynamic effects that cannot be represented by a single-volume model. Especially in summer scenarios, the resulting predictive energy calculation can achieve a higher accuracy. In other related works only the ambient temperature have been used as input variable [11], [12]. Lajunen et. al has used the simulation software AMESim and considered the flow velocity, but has not taken solar irradiance into account, which results in a less powerful model [13].

Based on all relevant heat transfer mechanisms, a MATLAB/Simulink model was developed, which captures the car cabin as a 5-volume model and calculates the average air volume temperature, see Chapter 3. Further, the model is validated by measurements on a reference vehicle. In addition, a generic methodology for parameter optimization has been applied to increase the accuracy of the model using special selected reference measurements, see Chapter 4. Finally, we summarize our work in Chapter 5, where simulation results of the model are compared with the measurements on the vehicle.

3. Structure of the 5-zone-car-cabin model

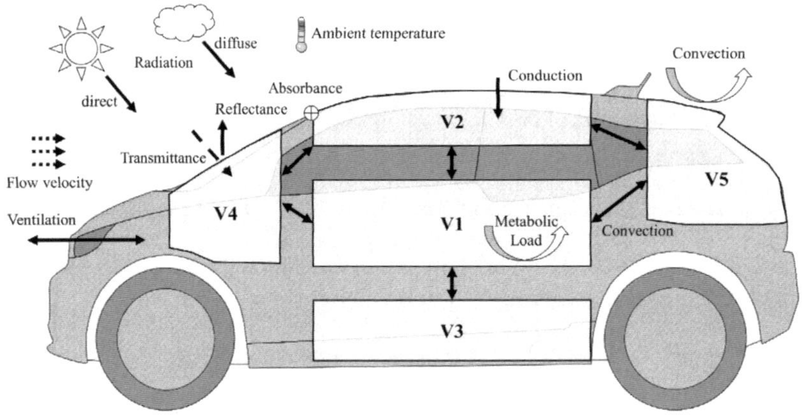

Figure 2. Classification of the vehicle interior into five zones and representation of the influences (ambient temperature, sun, HVAC system, ...) and heat transfer mechanisms which are considered in the model.

3.1. Environmental influences

In the model, ambient temperature, solar radiation, and the exchange of the cabin with the air-conditioning system of the vehicle are considered. This results in heat flows, which influence the condition of the "car cabin, according to figure 2. '*Ventilation*' represents the bidirectional heat flow of the air conditioning system. The general equation describing the heat flow of a body part $\dot{Q}_{body\,part,i}$, is calculated by the sum of heat transfer as well as diffuse and direct radiation:

$$\dot{Q}_{body\,part,i} = S \cdot U \cdot (T_{out} - T_{in}) + \varepsilon \cdot \sigma \cdot S \cdot (T^4 - T^4_{ambient}) + \tau \cdot S \cdot \theta \cdot \dot{I}_{sun} \quad (1)$$

The heat transfer $\dot{Q}_{conduction}$ depends on ambient T_{out} and internal T_{in} temperatures, the surface area S, as well as the overall heat transfer coefficient U which depends on the convection coefficients h_i, h_o, the material thickness λ, and the thermal conductivity k. The solar radiation enters the system through absorption and transmittance. The vehicle's windows hardly absorb radiation energy, which is why thermal energy is transmitted directly in the car cabin.

3.2. Coupling of the 5 zones and chassis parts

The chassis parts roof, underbody, side plates and the windows describe the system boundary. They are considered separately and are connected to the vehicle interior by an associated heat flow. The following assumptions were made for the model: All body surfaces are assumed to be flat surfaces with a simple geometry. The heat input, which enters through the boot and the front wall, is neglected. The air volumes in the interior are connected to each other by convection alone, as are the body panels and thermal masses with the adjacent air volumes. The body elements are considered as layered flat panels. All mentioned components are similarly modeled, except the vehicle windows. These consist only of a single layer. Figure 3 shows the heat transfer mechanisms convection, radiation and transmittance during the passive heating or cooling, but without consideration of the forced convection.

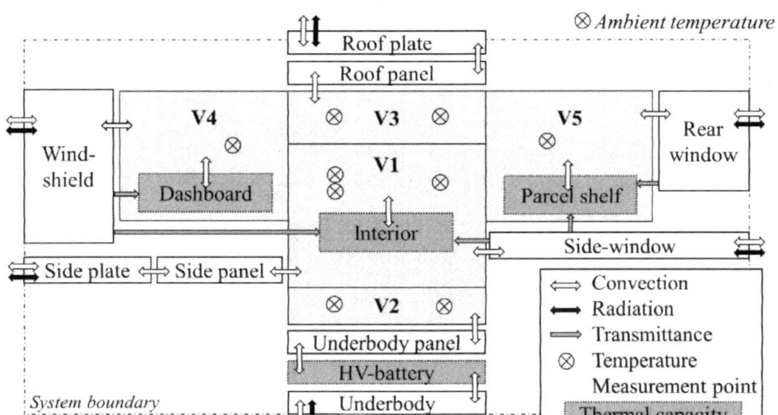

Figure 3. All chassis parts considered in the model and their coupling among each other. Representation of the occurring heat transfer mechanisms with passive heating or cooling. The forced convection is not considered here.

The components with a thermal capacity (also referred to as thermal mass), like the parcel shelf, the dashboard and interior are handled identical except for material parameters and considered ambient influences. Sun radiation acts on the parcel shelf only through the rear window and on the dashboard only through the windshield. The interior, on the other hand, will be irradiated through all windows. Each thermal mass is connected by a heat convection to the volume in which it is located. This is described by the following formula [14]:

$$\dot{Q}_{convection} = \alpha(Nu(Re, Pr)) \cdot S \cdot \Delta T \quad (2)$$

The lithium-ion battery in the reference vehicle's underbody represents an additional thermal mass with 230 kg and affects the thermal condition of the car cabin, in particular by the delayed temperature rise of the footwell during the heating process [15]. In addition, the reference vehicle's body shape (hatchback sedan) influences the heat input into the parcel shelf zone, since heat is also exchanged via side windows. Therefore, heat flow, which enters through the side windows, is divided between the main zone (V1) and the parcel shelf zone (V5). In a conventional sedan, this heat flow completely enters the main zone (V1).

In case of the convective coupling of the volumes six virtual contact surfaces exist, according to figure 2. In the passive state, i.e. heating and cooling of the car cabin only by external conditions, the coupling is exclusively based on natural convection. The heat flow is calculated according to equation 2. In case of an active ventilation of the car cabin by the HVAC system, the coupling of these six contact surfaces by forced air-mass flows is relevant. The resulting heat flow is dependent on the inflow temperature and the interior temperature. The choice of the ventilation mode determines how the mass flows are blown through the car cabin. Since a constant pressure of the car cabin is assumed, the in- and output of air of each volume has to be equal. The amount of the heat flow depends on the temperature difference ΔT of adjacent volumes and can be described by the following formula:

$$\dot{Q}_{ventilation} = \dot{m} \cdot c \cdot \Delta T \qquad (3)$$

Upward flows above the dashboard and the parcel shelf result in circulation flows, which cause an air exchange in the entire car cabin. [4] They arise above the thermal mass because the components heat up faster than the adjacent air volumes due to solar radiation. The airflow is deflected from the windows into the head zone (V3), from there into the main zone (V1) and finally back into the parcel shelf (V5) or dashboard zone (V4). The magnitude of the heat flows depends on the temperature difference between adjacent volumes and on resulting air mass flow above the thermal masses. A further influence caused by the sun is also considered: the solar radiation passes through the windshield and is absorbed by the dashboard. The heat radiation of the heated dashboard, however, does not transmit back through the windshield due to the wavelength of the radiation. This effect is negligible when solar radiation is not considered. All heat flows mentioned are added to a resulting heat flow for each volume $\dot{Q}_{V,i}$, which is generally calculated as follows:

$$\dot{Q}_{V,i} = \dot{Q}_{windows} + \dot{Q}_{body\,part} + \dot{Q}_{thermal\,masses} + \dot{Q}_{convection} + \dot{Q}_{ventilation} \qquad (4)$$

From this, the temperature difference can be calculated in each volume using the general gas equation:

$$\Delta T = \int_0^t \frac{\dot{Q}_{V,i}}{n \cdot c_p} dt = \int_0^t \frac{\dot{Q}_{V,i} \cdot R \cdot T}{p \cdot V \cdot c_p} dt \qquad (5)$$

4. Reference measurements and parameter optimization

Measurements on a reference vehicle are used for the comparison between model and reality. Most of the variables are assumed to be constant (at 20°C and 1 bar air pressure) including environment parameters, material-specific properties, and geometric quantities. The choice of vehicle and interior colors also has an influence on the thermal behavior. The reference vehicle uses carbon, the natural fiber kenaf and eucalyptus, [16]. Most of the vehicle components are composed of several materials, so that lumped material properties are assumed for these components. The parameter optimization applied here is intended to compensate the deviations from the real value. The geometric dimensions of the reference vehicle were measured and transferred to simple geometric shapes in the model. All specified parameters are listed in the measurement record [17].

In addition to the measuring points in the vehicle, a further temperature sensor records the ambient temperature. The mean value of the reference vehicle's left and right solar sensors and the vehicle speed were recorded. The output variables, which were recorded for a target-actual comparison, are the five average volume temperatures.

After recording a reference measurement, it is compared with the results of the model. If measured values and simulation are close enough, the next process step can be continued, which means a further measurement with an additional variable will be compared with the model. However, if the model quality is unacceptable, a parameter optimization is performed. Afterwards measurement and model are compared again. This procedure works iterative until the evaluation criterion is fulfilled, which means the reality is represented sufficiently well, see figure 4.

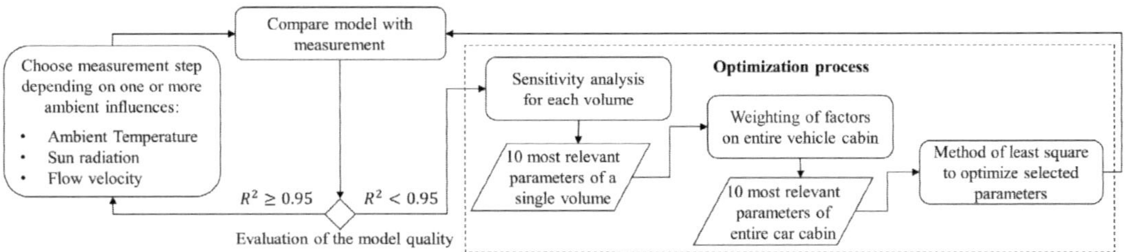

Figure 4. Methodology of optimization process to increase model accuracy. In each step, the measurement is extended by an additional environmental influence and is compared with the model. Finally, the model quality is evaluated.

In order to evaluate the model quality, the coefficient of determination R^2 is used, which indicates the relationship between declared variance and observed variance. R^2 is always between 0[1] (no relationship between the results of the model and the reality) and 1 (the model perfectly reflects reality). For $R^2 \geq 0.9$, there is a strong link between the model and the reality. The objective criterion of optimization was $R^2 \geq 0.95$. The coefficient of determination is calculated as follows:

$$R^2 = 1 - \frac{\sum_{i=1}^{n}(y_i - \hat{y}_i)^2}{\sum_{i=1}^{n}(y_i - \bar{y})^2} = \frac{\sum_{i=1}^{n}(\hat{y}_i - \bar{y})^2}{\sum_{i=1}^{n}(y_i - \bar{y})^2} \quad (6)$$

For the general setting of the reference measurements, the three input variables can be combined with two selected states to a total of eight (2^3) cases, which are shown in table 1. A consideration of both simple cases (Case 1 and 2) makes sense, since different thermodynamic effects come into effect. Therefore, both were considered and summarized. For each iteration, an additional environmental influence (flow velocity or solar radiation) should be added.

Table 1. 8 combinations of the three environmental influences: ambient temperature, flow velocity and solar radiation.

	Flow velocity $[km/h]$	0		available	
	Solar radiation $[W/m^2]$	0	available	0	available
$T_{interior} - T_{ambient}$ $[°C]$	Positive	Case 1	Case 3	Case 5	Case 7
	Negative	Case 2	Case 4	Case 6	Case 8

If the model quality is not sufficient, a sensitivity analysis with all factors used in the model will be performed so that their respective share of the variance of the output variable can be determined. Each parameter is hereby multiplied with a correction factor, which is initialized with the value 1 and receive upper and lower limits (± 10%). Subsequently, for each correction factor datasets are generated from random values in the permitted interval. For each dataset, a simulation is performed and the results are evaluated using the correlation between the correction factor and the curve of the

[1] R^2 is intended to evaluate a linear correlation so that negative values can be calculated for a very large deviation between model and measurement.

volume temperature. The correlation coefficient R between $x(i)$ and $y(j)$, which is in the interval $[-1\ 1]$, is calculated from the covariance matrix $\underline{C}$ [18]:

$$R(i,j) = \frac{\underline{C}(i,j)}{\sqrt{\underline{C}(i,i)\underline{C}(j,j)}} \tag{7}$$

with

$$C = cov(x,y) = E[(x-\mu_x)(y-\mu_y)] \tag{8}$$

where $\mu_x = E[x]$ and $\mu_y = E[y]$.

All factors are ranked according to their influence [19]. A change in the parameters can have the effect that the temperature profile of one volume is influenced positively, while a different volume is negatively influenced. Therefore, the five volumes are weighted according to their importance in terms of thermal comfort: V1 (0.4), V2 + V3 (each 0.2) and V4 + V5 (each 0.1). The ten most important factors of each volume are given the scores 10 to 1, all others are scored with 0. As a result, the volume-specific influences can be transferred on the entire car cabin. From the weighted parameters, the ten most important are selected for optimization. The cost function used within the optimization is the method of least squares, in which the sum of deviations between the model curve $f(x_i)$ and the values of the measurement $\hat{y}_i$ are minimized [20]:

$$min \sum_{i=1}^{n} (f(x_i) - \hat{y}_i)^2 \tag{9}$$

5. Simulation results

Including the general criteria for selecting reference measurements according to table 1, four reference measurements were selected, see table 2, which are discussed here.

Table 2. Setup for the four selected reference measurements.

Reference measurement:	Case 1	Case 2	Case 3	Case 8
Flow velocity $[km/h]$	0	0	0	$v_{veh.} + v_{wind}$
Solar radiation $[W/m^2]$	0	0	400-700	200-800
Initial vehicle temperature $[°C]$	20	4	20	20
Ambient temperature[2] $[°C]$	0 to -8	20	20	20
Measurement duration $[h]$	15	22	3	2
Number of passengers $[-]$	0	0	0	2

5.1. *Results of heating (Case 1) and cooling (Case 2) without sun radiation and flow velocity*

Except for the specific heat capacity of the parcel shelf and the external heat transfer coefficients, the top 10 of the heating process (case 1) and the cooling process (case 2) have the same parameters, so that 12 parameters were relevant for the optimization. A detailed description of all optimization results is available in the measurement record [14].

After about 15 hours of cooling, a constant temperature level in all 5 volumes was achieved, see figure 5. The temperature of the dashboard zone (V4) drops to the lowest level. The other volume temperatures are about 1-1.5°C above this level. During the heating, the highest temperature is achieved in the head zone (V3). The difference in temperature levels is caused by temperature stratification in the car cabin, since warm air has lower density and rises as much as possible. After about 10 hours of heating, constant temperatures in all volumes occur. They are only slightly increased until the end of the measurement. The process of cooling is significantly faster compared to the heating process, since the effect of natural convection in the cabin is fully utilized. The cooling of the air on the component's cold surfaces causes a direct increase in density. As a result, the air drops

[2] The ambient temperature cooled down, because Case 1 was measured overnight.

quickly to the lower parts of the car cabin. This decrease causes a continuous air mass exchange. Whereas, in case of a heating process without sun radiation, there is almost stagnant air.

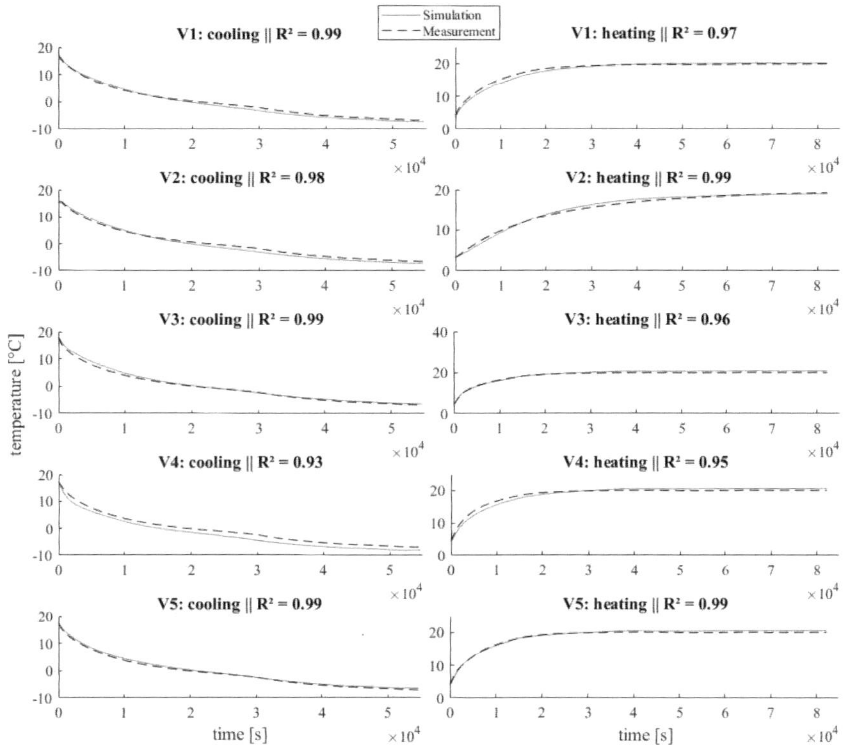

Figure 5. Comparison of simulated and measured temperatures for each volume. The left column shows the cooling process (Case 1) during an ambient temperature of 0°C, which falls to -8°C during the measurement period of 15 hours. The right column shows the heating process (Case 2) during an ambient temperature of 20°C over a period of 22 hours. Both cases are without solar radiation and flow velocity. The accuracy of the model is given by the coefficient of determination R^2.

Through the optimization of parameters for the heating and cooling process without sun and flow influence, all volumes reach the objective criterion with the exception of the dashboard (V4) during the cooling process. However, this can be tolerated, since all other volumes have a very high coefficient of determination and the dashboard zone has only the fourth priority.

5.2. Results involving solar radiation (Case 3 and Case 8)

The second step focused on the factors that specifically reflect the influence of the sun on the car cabin. In this case, a heating process under the influence of solar radiation was considered. The measurement was recorded in Germany in the spring around noon.

The mean value from the two intensity curves was used as input variable for the simulation via zones V1, V4 and V5 by transmittance as well as indirectly through absorption via V3. The target temperatures were almost reached after 1.5 hours with the exception of V5, but heating up was much too slow, according to figure 6. On the one hand, this is due to the division of the heat flows between the chassis parts and the car cabin, and, on the other hand, due to the consideration of the irradiated surfaces. These miscalculations clearly show that the correct representation of the car cabin condition is only possible under consideration of the sun's position. This depends on the driving direction, but also on location, time of day and season. In addition, the used measurement equipment (the reference vehicle's two built-in solar sensors) is not sufficient to validate the effect of the sun on the 5 zones.

The solar radiation angle has a major influence. Therefore, the sun intensity cannot be considered as a lumped value.

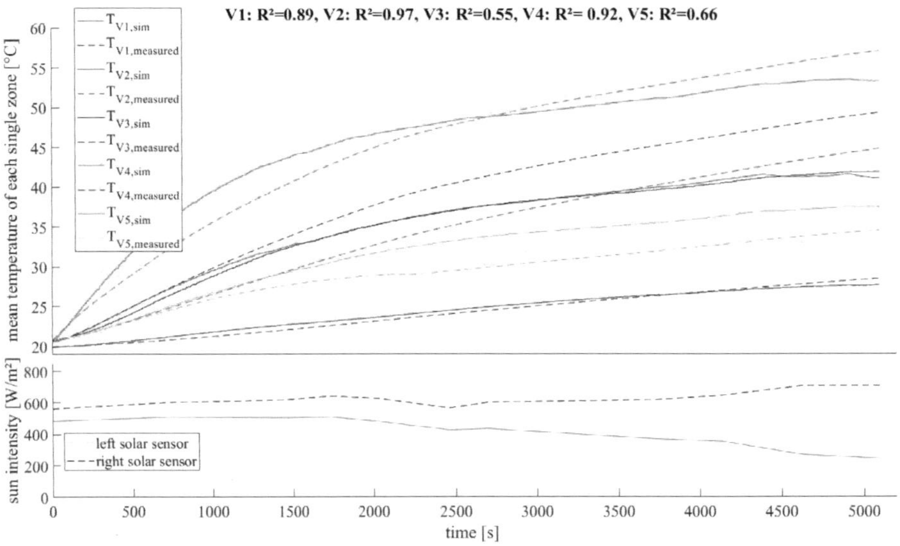

Figure 6. Top: Comparison of the heating curves of simulation and measurement at 20°C ambient temperature and negligible low flow velocity. Bottom: Reference vehicle's recorded solar intensity during the heating process.

This is also confirmed by the observation of the recorded test drive. The simulation results differ from the measured temperatures, as in the first part of the trip the sun was mostly shining in V5 and in the second part in V4 and partly in V1. For this reason, the recorded test drive (case 8) was not displayed. Similarly, the parameter optimizations for case 3 and case 8 were not carried out because it makes no sense to vary the relevant parameters without a correct representation of the sun's influence.

6. Conclusions and future work

The 5-zone interior model presented in this paper can depict the thermal condition of the car cabin depending on the outside temperature, the flow velocity and solar radiation. All heat transfer mechanisms such as heat radiation, heat conduction and heat transitions are taken into account and even natural phenomena caused by the sun's influence are implemented into the model. The achieved model accuracy for the heating and cooling process of the car cabin, which was achieved by parameter optimization, is very satisfactory since they reproduce the reality now very well. Nevertheless, the model is kept so generic, so that an adaptation to a reference vehicle can easily be implemented. The presented methodology allowed to adapt the model step by step to reality by including reference measurements, sensitivity analyzes and subsequent parameter optimization. Due to the realistic representation of the thermal condition in the car cabin, the model will be used as a basis for further investigations, e.g. how much air-conditioning is necessary in order to ensure the passengers' thermal comfort for a given route.

In future work, an automatic climate control for the car cabin, which was already implemented, will be validated. Based on the required heat and mass flow, the electrical energy of a HVAC system is calculated. The simulations results will be compared with the recorded energy consumption of reference vehicle's HVAC system. The 5-zone-car-cabin model can thus be used for further investigations with regard to potential energy saving potential of the HVAC system, which can lead to a direct increase in the range of BEVs. The investigation regarding the sun radiation has shown, that the model and the measurement of sun radiation has some weak points. Therefore, the model will be extended by the calculation of the sun's movement. Then, it will be validated with solar intensity sensors for each volume instead of a left and right sensor, which could not correctly record the radiation direction and intensity of the sun for each volume.

References

[1] K. Umezu und H. Noyama, *Air-conditioning system for electric vehicles (i-MiEV)*, in SAE Automotive Refrigerant & System Efficiency Symposium 2010

[2] S. Gerson, K. v. Rueden und B. Voss, *Optimized energy management and components for electric vehicles under different environmental conditions*, 2010

[3] K. Kruppok, C. Gutenkunst, R. Kriesten und E. Sax, *Prediction of energy consumption for an automatic ancillary unit regulation*, in 17. Internationales Stuttgarter Symposium, Wiesbaden; Springer Fachmedien, 2017, pp. 41–56

[4] J. Hofhaus und S. Wagner, *Idealisierte energetisch-analytische Abbildungsmethode der Temperaturschichtung bei der passiven Aufheizung in der Fahrzeugkabine*, in PKW-Klimatisierung VI, Renningen, Expert Verlag, 2010, pp. 94-110

[5] S. C. Pang, M. A. Kalam, H. H. Masjuki und M. A. Hazrat, *A review on air flow and coolant flow circuit in vehicles' cooling system*, International Journal of Heat and Mass Transfer, Bd. 55, Nr. 23, pp. 6295–306, 2012

[6] T. Ye, *A Multidisciplinary Numerical Modeling Tool Integrating CFD and Thermal System Simulation for Automotive HVAC System Design*, SAE International, 2012

[7] M. A. Fayazbakhsh und M. Bahrami, Comprehensive modeling of vehicle air conditioning loads using heat balance method

[8] H. Lee, Y. Hwang, I. Song und K. Jang, *Transient thermal model of passenger car's cabin and implementation to saturation cycle with alternative working fluids*, Energy, Bd. 90, pp. 1859–68, 2015

[9] R. Valentina, A. Viehl, O. Bringmann und W. Rosenstiel, *HVAC system modeling for range prediction of electric vehicles*, 2014, pp. 1145–50

[10] A. Lajunen, Energy Efficiency and Performance of Cabin Thermal Management in Electric Vehicles

[11] S. Yang, C. Deng, T. Tang und Y. Qian, *Electric vehicle's energy consumption of car-following models*, Nonlinear Dynamics, Bd. 71, Nr. 1-2, pp. 323–29, 2013

[12] J. van Roy, N. Leemput, S. Breucker, F. Geth, P. Tant und J. Driesen, *An availability analysis and energy consumption model for a flemish fleet of electric vehicles*, 2011

[13] Lajunen, T. Hadden, R. Hirmiz, J. Cotton und A. Emadi, *Thermal energy storage for increasing heating performance and efficiency in electric vehicles*, 2017 IEEE Transportation Electrification Conference and Expo (ITEC), 2017

[14] W. M. Rohsenow, J. P. Hartnett und Y. I. Cho, Handbook of heat transfer, Bd. 3, McGraw-Hill New York, 1998

[15] *BMW i3 - Official Specs*, [Online]. Available: http://www.bmwblog.com/2013/07/10/bmw-i3-official-specs/. [Zugriff am 6 9 2017]

[16] M. Usta, *No fumes, kenaf and olive leaves | Magazyn Usta*, [Online]. Available: http://ustamagazyn.pl/en/2016/04/zero-spalin-kenaf-i-liscie-oliwne/. [Zugriff am 6 9 2017].

[17] K. Kruppok, *Measurement Document for Validation of 5-Zone-Car-Cabin-Model*, 20 09 2017. [Online]. Available: http://www.mmt.hs-karlsruhe.de/downloads/IEEM/NEVVE2017_5-Zone-Car-Cabin-Model_Measurement.pdf.

[18] *Analyze Relation Between Parameters and Design Requirements*, MathWorks, [Online]. Available: https://de.mathworks.com/help/sldo/ug/interpreting-analysis-results.html.

[19] K. Siebertz, D. van Bebber und T. Hochkirchen, *Sensitivitätsanalyse*, in Statistische Versuchsplanung, Berlin Heidelberg, Springer-Verlag, 2010, p. 247

[20] Weisstein und E. W, *Least Squares Fitting*, [Online]. Available: http://mathworld.wolfram.com/LeastSquaresFitting.html. [Zugriff am 6 9 2017]

A novel decoupling control method based on neural network for EV's Driving PMSM

Wanbang Zhao[1], Qiang Song[1,*] and Yishan Huang[2]

[1]School of Mechanical Engineering, Beijing Institute of Technology, Beijing, China
[2]Hunan CRRC TIMES ELECTRIC VEHICLE CO., LTD, Zhuzhou, China

*E-mail: songqiang@bit.edu.cn

Abstract. Permanent Magnet Synchronous Motor (PMSM) is often used in Electrified Vehicle (EV), while its I_d (direct-axis current) and I_q (quadrature-axis current) are coupled. Traditional FOC (Field Oriented Control) method can't get an accurate decoupling control of them when the motor speed changed with the vehicle's driving condition, especially at high speed. This coupling issue leads an unstable torque output. And further, it deteriorates the NVH (noise vibration and harshness) performance of EV. This paper focuses on the decoupling solution and puts forward a new control strategy which combines the neural network control idea with FOC method. The novel method gives a neural network controller based on four single neuron PID controllers, which its function is to realize the d-q axis interacted adjustment and decoupling. Four single neurons PID controllers achieve the negative feedback control of I_d to U_d (direct-axis voltage), I_q to U_d, I_q to U_q (quadrature-axis voltage), and I_d to U_q respectively. Creatively, it takes PMSM speed as one of the neuron inputs to adjust the feedback weight of I_d and I_q dynamically. A comparing simulation which is set up in the Simulink platform is given in this paper. Simulation results show that this method gives a good self-adaptiveness and decouples the influence of I_d and I_q, as well as improve the motor control quality at high speed.

1. I_d/I_q coupling in PMSM

Nomenclature

I_d	direct-axis current (A)
U_d	direct-axis voltage (V)
I_q	quadrature-axis current (A)
U_q	quadrature-axis voltage (V)
T_e	electromagnetic torque (Nm)
I_s	the current of stator (A)
R_s	stator resistance (Ω)
L_d	direct-axis equivalent excitation inductance (H)
L_q	quadrature-axis equivalent excitation inductance (H)
ψ_f	the rotor flux (Wb)
β	the angle between rotor flux and the current of stator (rad)
p_0	the number of poles

The torque equation for PMSM is followed.

Content from this work may be used under the terms of the Creative Commons Attribution 3.0 licence. Any further distribution of this work must maintain attribution to the author(s) and the title of the work, journal citation and DOI.
Published under licence by IOP Publishing Ltd

$$T_e = p_0 \psi_f I_s \sin\beta = p_0 \Psi_f \times I_s \tag{1}$$

FOC control method decomposes the stator current into the rotational coordinates which is orthogonal to the rotor permanent magnet flux linkage.

$$\begin{bmatrix} i_d \\ i_q \end{bmatrix} = \sqrt{\frac{2}{3}} \begin{bmatrix} \cos\theta_r & \cos\left(\theta_r - \frac{2\pi}{3}\right) & \cos\left(\theta_r - \frac{4\pi}{3}\right) \\ -\sin\theta_r & -\sin\left(\theta_r - \frac{2\pi}{3}\right) & -\sin\left(\theta_r - \frac{4\pi}{3}\right) \end{bmatrix} \begin{bmatrix} i_A \\ i_B \\ i_C \end{bmatrix} \tag{2}$$

After decoupling, the value of I_d and I_q can be adjusted by changing U_d and U_q respectively. Here I_d is excitation current, while I_q produces electromagnetic torque. To control I_d and I_q will realize the weak magnetic and torque control respectively.

In the FOC control method, the PMSM's voltage and current has the following relationship [1].

$$\begin{aligned} U_d &= L_d \tfrac{dI_d}{dt} + R_s I_d - \omega_r L_q I_q \\ U_q &= L_q \tfrac{dI_q}{dt} + R_s I_q + \omega_r L_d I_d + \omega_r \psi_f \end{aligned} \tag{3}$$

According to formula (3), it can be seen that I_d and I_q are mutual coupled when we set the U_d as a constant. It is the same for the U_q. The coupled items are $\omega_r L_q I_q$ and $\omega_r L_d I_d$, which are related to the speed. When motor at low speed, $\omega_r L_q I_q$ and $\omega_r L_d I_d$ can be ignored. I_d and I_q are mainly decided by U_d and U_q respectively. But while motor at high speed, ω_r increased significantly, the value of $\omega_r L_q I_q$ is much bigger than $R_s I_d$, and the value of $\omega_r L_d I_d$ is much bigger than $R_s I_q$. When the motor rotated at a certain high speed, I_d is directly correlated with the voltage U_q, I_q is mainly influenced by U_d. In the middle speed, the U_d and U_q have a complex relationship with I_d and I_q.

The complicated working condition of EV motor make this problem even worse. Researchers put forward many methods to solve the influence of the cross coupling of I_d and I_q in the torque control. The commonly used techniques include feedforward compensation and diagonal compensation. One method is to compensate the cross coupling by feedforward, which takes a compensation factor which is equal to two currents cross coupling item into the transfer function[2]. Another method is to give a diagonal matrix into the current loop controller to cross compensate the voltage[3]. Other researches also gives another single PID controller to make the PID parameters of U_d and U_q controller self-adaptive[4-5].

2. Novel neural network control method

A single neuron PID controller can be obtained by combining the artificial neuron with the PID controller. PID transfer function can be written as follow.

$$\Delta V(k) = A_1 \Delta e(k) + A_2 e(k) + A_3 (e(k) - 2e(k-1) + e(k-2)) \tag{4}$$

In the formula (4), three inputs error of PID (including $\Delta e(k)$, $e(k)$ and $(e(k) - 2e(k-1) + e(k-2))$) are set to be the inputs.

$$\begin{aligned} &\Delta e(k) = U_1 \\ &e(k) = U_2 \\ &e(k) - 2e(k-1) + e(k-2) = U_3 \end{aligned} \tag{5}$$

The structure single neuron PID controller is shown in figure 1. Here, A_1, A_2, A_3 are the weights of the inputs.

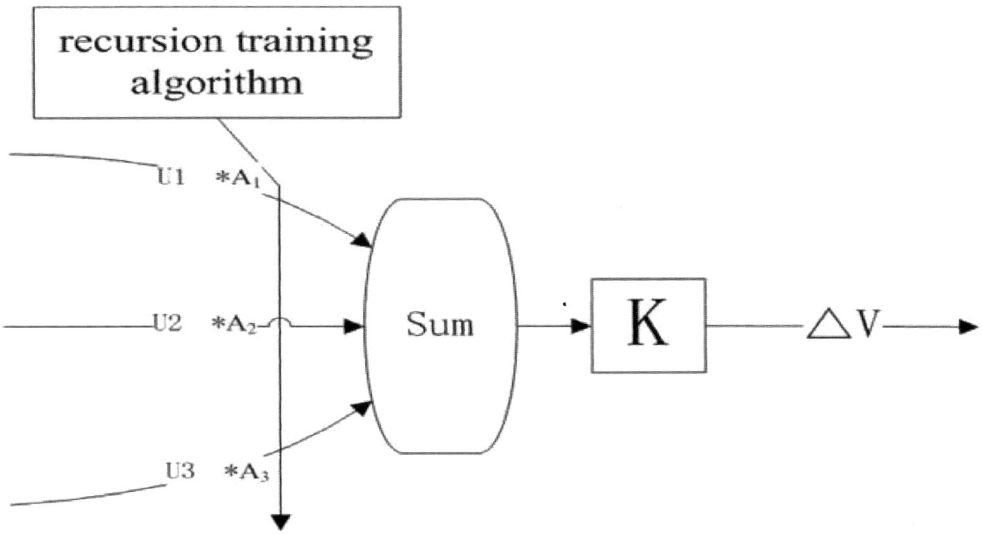

Figure 1. Single Neuron PID controller.

In figure 1, the Single Neuron controller's Recursion Training Algorithm (RTA) is used to adjust the PID parameters dynamically in order to realize a self-adaptive control. The principle of RTA method is showed in formula (6). Here, v (k) is the output. A_1, A_2, A_3 are the weights of inputs. p_1, p_2, p_3 are the learning rates of the weights.

$$\begin{aligned}
&A_1(k+1) = A_1(k)+p_1 e(k)v(k)U_1(k)\\
&A_2(k+1) = A_2(k)+p_2 e(k)v(k)U_2(k)\\
&A_3(k+1) = A_3(k)+p_3 e(k)v(k)U_3(k)\\
&A_i'(k) = A_i(k) / \sum_{i=1}^{3}|A_i(k)|\\
&\Delta v(k)=v(k)-v(k-1)=K\sum_{i=1}^{3}A_i'(k)U_i(k)
\end{aligned} \tag{6}$$

Based on the single neuron PID controller method mentioned above, and also on the existing neural network control methods for PMSM[6-8], this paper puts forward a novel neural network interaction adjustment control strategy (the NNPID method). The control structure diagram is shown in figure 2. The strategy bases on four single neuron controllers, and use the motor rotated speed as one of the inputs to adjust the coupling ratio（W_d and W_q） of each neuron in the output U_d or U_q. The function of this neural network controller is to realize the d-q axis current decoupling. The four neurons achieve the negative feedback control of I_d to U_d (dd neuron), I_q to U_d (qd neuron), I_q to U_q (qq neuron), and I_d to U_q (dq neuron) respectively. Then the PMSM controller's control signal can be obtained according to the neural network's output (U_d and U_q) using SVPWM modulation method. By the proposed method, the desired I_d and I_q of PMSM can be controlled stably.

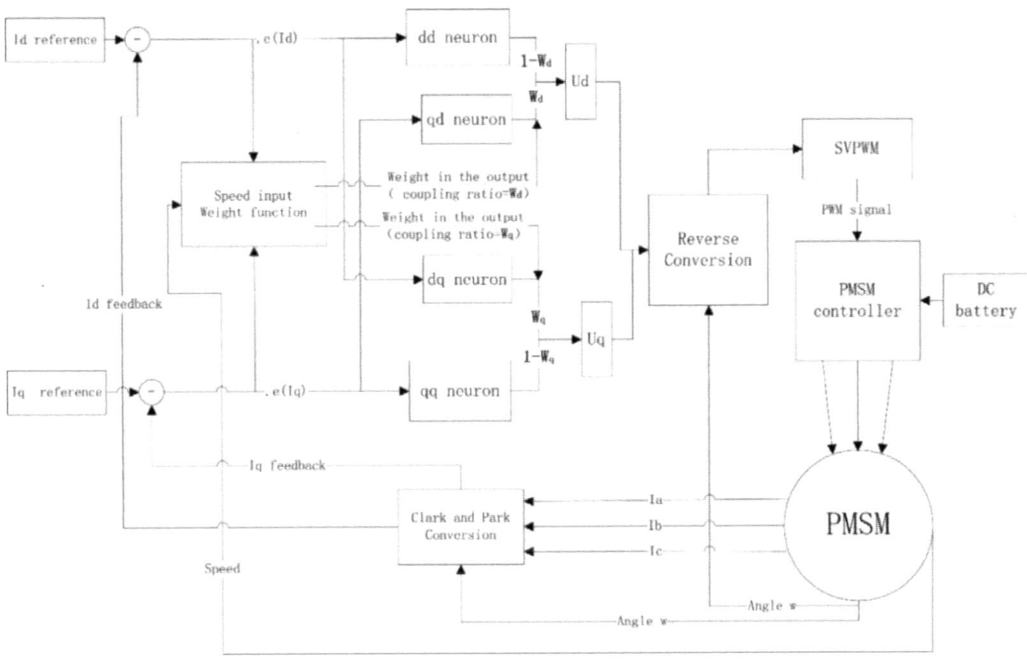

Figure 2. Interactive NNPID method for EV's driving PMSM.

3. Simulation and analysis

The simulation model is created in the Simulink software. The chosen PMSM has the following parameters, R_s is 9.8 mΩ; L_d and l_q are 0.081mH and 0.27mH; ψ_f is 0.0396Wb; 4 pairs of pole. The Simulink sample time of this simulation is set as 1.2×10-6 s. In the simulation, the motor's rotated speed linear increases from 0 to 6000 rpm in 0.2 second. Shaft torque is set as 110Nm from 0rpm to 3000rpm. Then, the torque drops to 50Nm at the time of 0.1s, and keeps at 50Nm from 3000rpm to 6000rpm. According to the torque setting, we get that I_q reference is 580A before 0.1s, and 250A after 0.1s. During the simulation, I_d reference is always set as 5A without loss of generality.

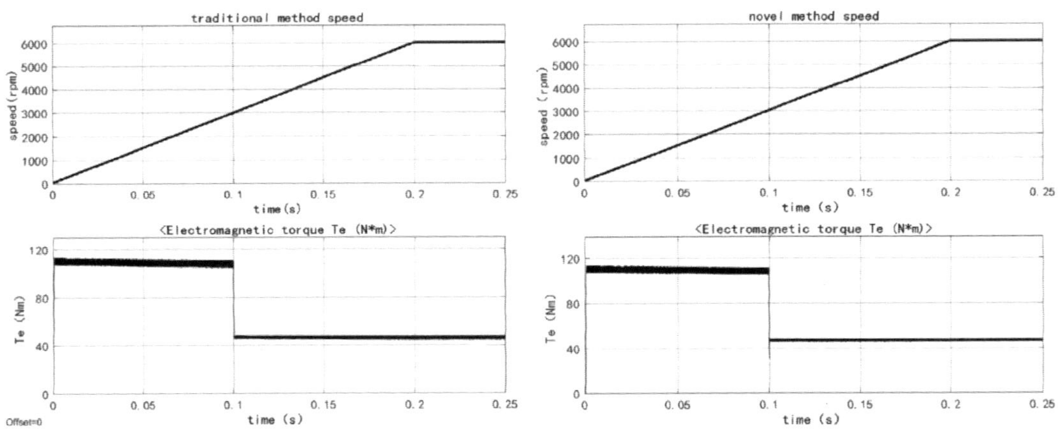

Figure 3. speed/torque for traditional PID method (left); speed/torque the novel NNPID method (right).

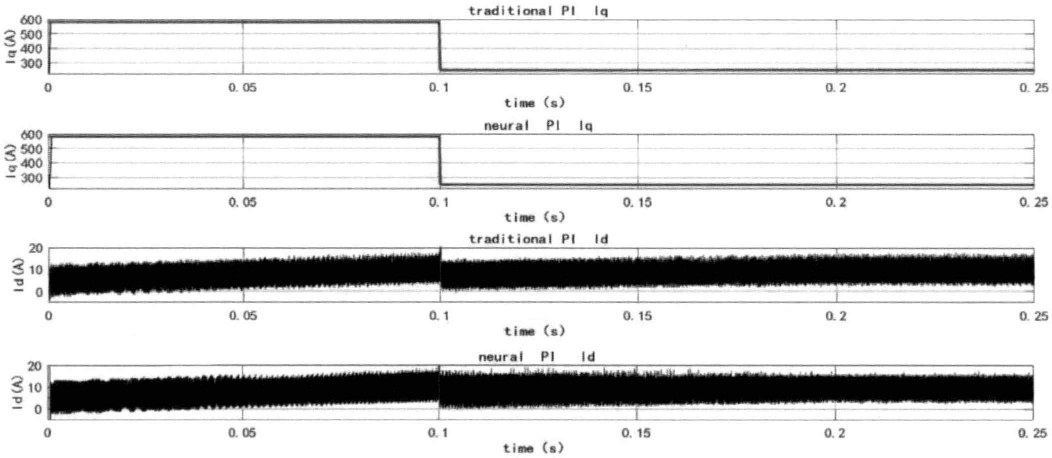

Figure 4. I_d and I_q simulation result (comparison diagram).

Figure 3 gives an illustration on the simulation results of torque output between traditional PID method and proposed NNPID method. The fluctuation of motor shaft output torque is reduced from 105-111 Nm (traditional PID method) to 107-111 Nm under novel NNPID method when motor working in low speed, and from 44.5-48 Nm (traditional PID method) reduced to 46 - 47.5 Nm under novel NNPID method when at high motor speed and low torque reference. The proposed interactive feedback control method in this paper improves the torque control stability about 1%. At the same time, the steady wave error is decreased by 57%. In figure 4, the simulation results show that this proposed control algorithm can track the target currents effectively, and achieve a brilliant current control performance during the whole process, even when the motor speed changes dramatically.

We can draw the following conclusions from the figure 3 and figure 4. At steady state, the NNPID controller's control precision is higher than traditional PID controller, the fluctuation of torque is smaller and the steady state current is more stable. Under the proposed method, the motor's response time of the current and torque are similar to the mature traditional PID control. But, here are big overshoot for torque output during torque drop-off. PID control parameters can be adjusted online, and the coupling ratio of I_q and I_d to U_q can be changed automatically to approach the coupling in the formula (3) (see figure 5).

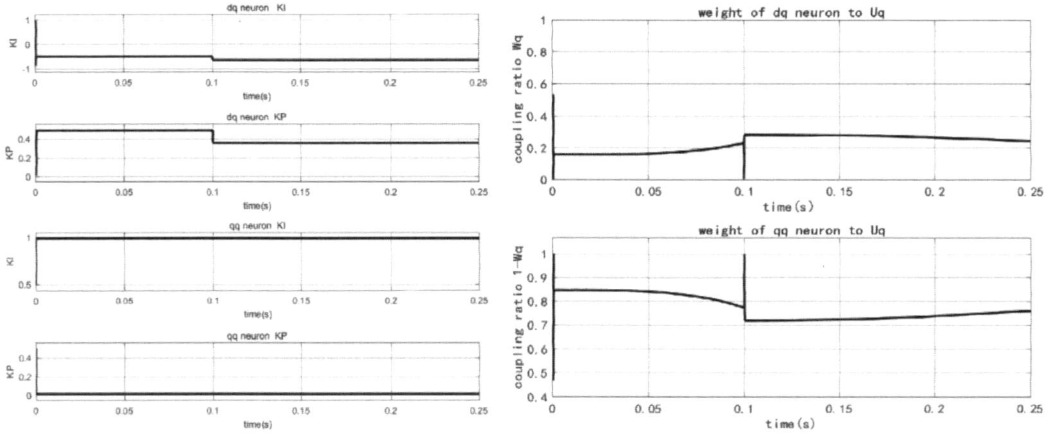

Figure 5. the adjustment of the each one's PI parameters(left) and weight for neuron qq and dq (right) .

IOP Conf. Series: Materials Science and Engineering **383** (2018) 012038 doi:10.1088/1757-899X/383/1/012038

4. Conclusion

This paper puts forward a novel NNPID method to solve the coupling of I_d and I_q in EV's PMSM driving motor. The novel NNPID method and the traditional PID method are compared by simulation. Results show that the novel method can decouple the d-q current, and give a better performance during the speed changed. But the shortage of this NNPID method is that the neuron weigh adjustment algorithm needs more calculation time. The learning rate can only decide by testing, too fast learning rate will lead to non-convergent and out of control.

Acknowledgements

This work was supported by the National Key R&D Program of China (2016YFB0100605).

References

[1] WANG Chengyuan, XIA Jiakuan. Motor Modern Control Technology [M]. Beijing: House of Chinese Machine, 2006

[2] YANG Nan-fang LUO Guang-zhao LIU Wei-guo. (2011) Decoupling control of current loop for permanent magnet synchronous motor drives using error compensation. *Electric Machines and Control* 2011(10) pp 50-54

[3] Jung J, Nam K. (1999) A Dynamic Decoupling Control Scheme for High-speed Operation of Induction Motors Industrial Electronics, IEEE Transactions on, 1999, 46 (1) pp 100-10

[4] LI Xiao-ning ZHAO, Xian-feng HUANG and Da-gui SHAO Wei. (2012) Decoupling control for permanent magnet synchronous motor based on single neuron. *control theory & Applications*, July 2012

[5] Zuo, S., Song, Y., Wang, L., and Zhou, Z. (2014). Neuron-adaptive pid based speed control of scsg wind turbine system. *Abstract and Applied Analysis*, 2014, (2014-5-19), 2014 (5) pp 1-10

[6] Kumar V, Gaur P, Mittal A P. (2014) "ANN based self tuned PID like adaptive controller design for high performance PMSM position control". *Expert Systems with Applications*. 2014, 41 (17) pp 7995-8002

[7] Chen, H. (2012). Research of the electro-hydraulic servo system based on rbf fuzzy neural network controller. *Journal of Software*, 7(9)

[8] Fayez F.M. El-Sousy. (2014). Adaptive hybrid control system using a recurrent RBFN-based self-evolving fuzzy-neural-network for PMSM servo drives. *Applied Soft Computing* 21 (2014) pp 509–32

IOP Conf. Series: Materials Science and Engineering **383** (2018) 012039 doi:10.1088/1757-899X/383/1/012039

Dynamic signal control strategy for intersection near a highway-railway at-grade crossing

Manman Li[1,2,3], Yingshuai Li[1,2,3] and Jian Lu[1,2,3,*]

[1]Jiangsu Key Laboratory of Urban ITS, Southeast University, Nanjing, China
[2]Jiangsu Province Collaborative Innovation Center of Modern Urban Traffic Technologies, Southeast University, Nanjing, China
[3]School of Transportation, Southeast University, Nanjing, China

*Email: lujian_1972@seu.edu.cn

Abstract. To decrease the influence of train passage on the efficiency of the intersection near a highway-railway at-grade crossing, this paper proposes a signal control strategy that dynamically applies two fixed signal timings and a priority phase based on the different scenarios, including two rules for judging different scenarios and two signal timings. The rule for judging whether a train is crossing a highway-railway at-grade crossing or not is established by using detected information. The rule for judging whether the influence of train passage is eliminated is designed according to train passage time and signal cycle length with the influence of train passage. Signal timing parameters for traffic flow, with and without the influence of train passage are respectively optimized by the ARRB method. The simulation results indicated that the performance of the dynamic strategy is superior to the performance of the two fixed signal schemes.

1. Introduction

Highway-railway at-grade crossings exist in several regions, such as Beijing, Shanghai, Texas, Illinois, and California [1-3]. With the increase in urban mobility, it is difficult to manage their traffic [2, 3]. A number of researches have been done into the safety of traffic control devices used at highway-railway at-grade crossings[1-6]. Because the absolute priority of train passage leads to an intensive increase in vehicle delays of the intersections near highway-railway at-grade crossings, it is important to improve its efficiency. To achieve this goal, this paper proposes a dynamic control strategy that dynamically applies two fixed signal schemes and a priority phase based on different scenarios.

The remainder of the paper is organized as follows. Section 2 presents an overview of the dynamic control strategy. Section 3 explains the installed locations of detectors and the rule for judging whether the train is crossing the highway-railway at-grade crossing or not. Section 4 discusses how to optimize signal timing parameters using the ARRB method for the traffic flow with or without the influence of train passage. The rule used to judge whether the influence of train passage is eliminated or not is stated in section 5. Section 6 presents the simulation experiments and results. The conclusion and future research direction are discussed in section 7.

2. Control Logic

Trains passing through a highway-railway at-grade crossing have absolute priority, therefore, any

Content from this work may be used under the terms of the Creative Commons Attribution 3.0 licence. Any further distribution of this work must maintain attribution to the author(s) and the title of the work, journal citation and DOI.
Published under licence by IOP Publishing Ltd

traffic movements at the intersection near the highway-railway at-grade crossing that conflict with passing trains must be prohibited. As the train passage time is relatively long, a large number of vehicles will gather at the conflicting approaches of the intersection. In order to improve the efficiency of the intersection, conflicting traffic movements should be allocated more time to cross through the intersection after the train passed through. A dynamic control strategy for this purpose is proposed here, as shown in figure 1. The dynamic control strategy includes two rules and two signal timings. The two rules are designed respectively to judge whether the train is crossing the crossing or not and whether the influence of train passage is eliminated or not. Signal timings for traffic flow with or without the influence of train passage are optimized by the ARRB method.

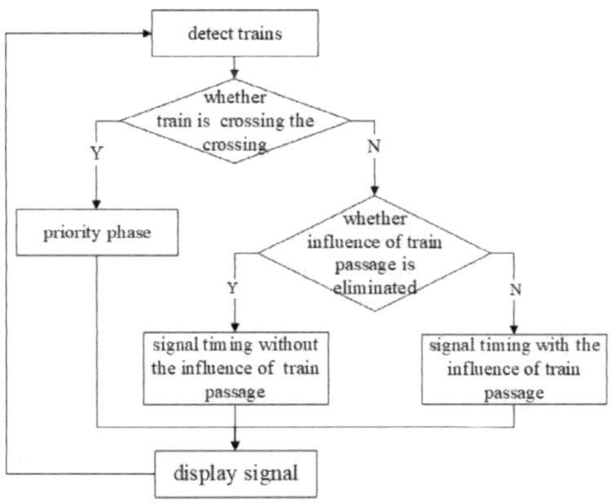

Figure 1. Control Logic.

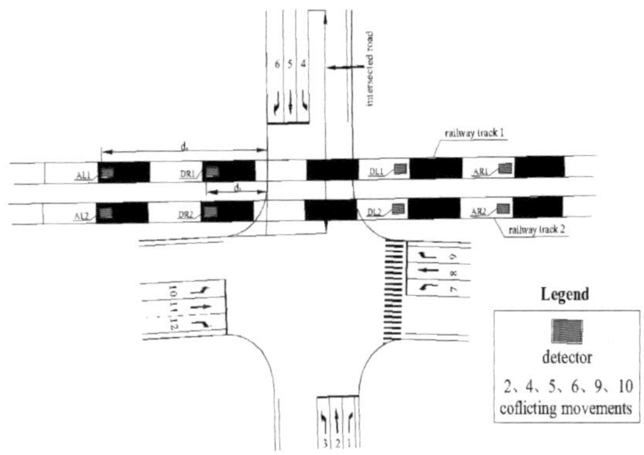

Figure 2. Detector Layout.

The control logic shown in figure 1 is described as follows:

Step 1: Detecting train;

Step 2: Judging whether the train is crossing the crossing or not. If yes, go to step 3; otherwise, go to step 4;

Step 3: Running the priority phase. Then go to step 1;

Step 4: Judging whether the influence of train passage is eliminated. If yes, go to step 5; otherwise, go to step 6;

Step 5: Running the signal timing without the influence of train passage. Then go to step 1;

Step 6: Running the signal timing with the influence of train passage. Then go to step 1.

3. Rule for Judging Whether the Train is Crossing the Crossing

3.1 Detector Layout

In order to judge whether the train is crossing the highway-railway at-grade crossing or not, it is necessary to install four detectors on each railway track, as shown in figure 2.

The four detectors are arranged in two pairs. One of each pair detects the arrival of the train (the arrival detector), such as AR1, while the other is used to detect the departure of the train (the leaving detector), such as DR1. The installation locations of arrival detectors and leaving detectors are calculated as formula 1 and formula 2, respectively.

$$d_s = t_{\max} * V_{\max r} + d_a \tag{1}$$

$$d_l = d_a \tag{2}$$

where d_s is the distance between a train-arrival detector and its nearer edge of the intersected road (m); t_{max} is the longest time a vehicle among conflicting movements takes to cross the crossing from the stop lines (s); V_{maxr} is the maximum speed of the train (m/s); d_a is the minimum gap between the train front (rear) and the vehicle rear (front) (m); and d_l is the distance between a train-leaving detector and its nearer edge of the intersected road (m).

3.2 Rule

When a road intersects with two railway tracks at grade, the rule for judging whether the train is crossing the crossing or not is designed as formula 3. For more or fewer than two railway tracks, the rule only requires to modify the number of variables.

$$\begin{cases} TN(j) \le 0, \ j=1,2,3,4 & \text{the train is crossing} \\ \text{otherwise} & \text{no train is crossing} \end{cases} \tag{3}$$

where $TN(j), j=1,2,3,4$ are variables.

Each variable $TN(j)$ records the information detected by a pair of detectors. The value is updated based on three scenarios:

Scenario 1: Initialization $TN(j)=0, j=1,2,3,4$;

Scenario 2: When the arrival detector of detector group j is activated by the train front, $TN(j)=TN(j)+1, j=1,2,3,4$;

Scenario 3: When the leaving detector of detector group j is activated by the train rear, $TN(j)=TN(j)-1, \mathrm{j}=1,2,3,4.$

Next, the principle of the rule is explained. For a railway track intersecting with a road at grade, if the gaps between two train fronts continuously crossing from one direction is larger than the sum of d_s, d_l, the width of the intersected road and the length of the former train, the two trains will be respectively regarded as one. Otherwise, trains will be regarded as a group, because their passage times partly overlap. Therefore, trains cross the crossing in two forms: a train or a group of trains. For this reason, variation in the recording variables for different forms will respectively be discussed to state the suitability of the rule. Let us suppose that there exists railway track 1. When a train crosses the crossing from right to left, TN(1) will be from 0 to 1 and then from 1 to 0 while TN(2) will be from 0 to -1 and then from -1 to 0. When TN(1) becomes 1 from 0, it indicates that the train is about to cross the crossing. Then, when it becomes 0 from 1, it indicates that the train has just crossed the crossing. At this time, TN(2) is always lower than 0 or equal to 0, although TN(2)=-1 or TN(2)=0 depends on the relation between the train length and d_s-d_l+the width of intersected road. Therefore, if TN(1)$\le$0, j=1,2 no train is crossing the crossing; otherwise, a train is crossing the crossing.

For a group of trains crossing from right to left, TN(1) will firstly be from 0 to 1, then vary above 0 several times, and finally return to 0, such as $0\text{-}1\text{-}2\text{-}1\text{-}2\text{-}3\text{-}\cdots\text{-}1\text{-}0$. Among these, the maximum value of TN(1) depends on d_s, d_l, the width of the intersected road, the length of trains, and the minimum front gap between trains. When it becomes 0 again, it similarly indicates that a group of trains has just crossed the crossing. However, TN(2) will firstly be from 0 to -1, then vary below 0, and finally return to 0, such as $0\text{-}(-1)\text{-}(-2)\text{-}(-3)\text{-}\cdots\text{-}(-1)\text{-}0$, which is always lower than 0 or equal to 0. Therefore, the former rule is also suitable for the form.

In the case of two railway tracks, no train is crossing the crossing only when no train is crossing the crossing on track 1 and simultaneously no train is crossing the crossing on track 2, which means the terms of judging whether no train is crossing the crossing for every railway track must be unified, that is, TN(1) ≤ 0, TN(2) ≤ 0, TN(3) ≤ 0, TN(4) ≤ 0. Therefore, when the value of j is modified as $j \in \{1,2,3,4\}$, the rule for one railway track is also applicable for two railway tracks. Accordingly, the rule will be practical for more than two railway tracks after a similar modification.

4. Signal Timing

Signal timings are optimized by the classic ARRB method that is described in section 3 of Chapter 4 on the literature [7]. For traffic flow without the influence of train passage, when the goal of optimization is set to decrease the average vehicle delay, the signal cycle length, effective green times of all movements, and the green time of phases are calculated by formula 4-6, formula 7-9 and formula 10, respectively [7]:

$$C=\frac{1.4L+6}{1-Y} \tag{4}$$

$$Y=\sum_{i=1}^{n_{cm}} y_i \tag{5}$$

$$y_i=\frac{q_i}{S_i},\ i=1,\cdots,n_{cm} \tag{6}$$

where C represents signal cycle length (s); L represents total loss time (s); Y represents the sum of critical flow ratio of the intersection; q_i, S_i, y_i represent traffic volume (pcu/h), saturation traffic volume (*pcu/h*) and flow ratio of critical movement *i*, respectively; and n_{cm} represents the number of critical movements.

$$t_i=(C-L)*\frac{u_i}{U},\ i=1,0,\cdots,n_{cm} \tag{7}$$

$$u_i=\frac{q_i}{x_i*S_i},\ i=1,\cdots,n_{cm} \tag{8}$$

$$U=\sum_{i=1}^{n_{cm}} u_i \tag{9}$$

where t_i represents the effective green time of critical movement *i* (s); u_i and x_i represent green split and the saturation degree of critical movement *i*, respectively; U represents the sum of green split of the intersection; and other variables are the same as previously defined.

The effective green time of non-critical movements $i\in\{n_{\text{cm}},\cdots,n\}$ can similarly be calculated as critical movements. Finally, the green time of phase *p* can be calculated by formula 9:

$$G_p=(t+l)_p-I_p,p=1,2,\cdots,P \tag{10}$$

where G_p represents the green time of phase *p*; *t+l* represents the sum of the effective green time and loss time of a traffic movement that belongs to phase *p*(s); and I_p represents the interval time of phase *p*.

For traffic flow with the influence of train passage, signal timing parameters are re-optimized on the condition that the traffic volume of conflicting movements is multiplied by a constant *m* that is more than 1. During the period that the influence of train passage is not eliminated (abbreviated influence period), the arrival amount of the conflicting movements that need to be given right of way to cross the intersection equals to the sum of that of the influence period and train passage period. This can be regarded as the increase in the traffic volume of the conflicting movements during the influence period. As the result of the change to the input variable of formula 6 (traffic volume), signal timing parameters should be re-optimized. Considering that the increased percentage of the traffic volume of the conflicting movements during the influence period varies with the length of the train passage period and the influence period, and in order to prevent oversaturation of the intersection because of releasing the arrival amount of the conflicting movements in a signal cycle, it is proposed that a fixed multiple *m* of the average arrival amount of the conflicting movements is given right of way in each signal cycle, but *m* musts make signal cycle length lower than 120s [8].

Moreover, the saturation degree of conflicting movements x_i, *i* belongs to the set n, can be set as a relatively higher value but lower than 0.95, $x_i\leq 0.95$, *i* belongs to the set *n* [7], owing to the reduction of the randomness of conflicting movements during the influence period because of vehicle accumulation during the train passage period.

5. Rule for Judging Whether the Influence of Train Passage is Eliminated

When signal timing for traffic flow with the influence of train passage is re-optimized, the average traffic volume of the conflicting movements is multiplied by a constant *m* that is more than 1. Hence, the number of vehicles of conflicting movements that can be released during a signal cycle with the influence of train passage (abbreviated influence signal cycle) is *m* times the arrival number of those movements. The signal cycles necessary to let all the accumulated vehicles in a train passage period cross is as formula 11:

$$M = \frac{T * q_i}{(m-1) * q_i * C} = \frac{T}{(m-1) * C_{IT}}, i \text{ belongs to the set } n. \quad (11)$$

where M represents the number of signal cycles with the influence of train passage needed; T represents the length of a train passage period(s); and C_{IT} represents the signal cycle length with the influence of train passage(s).

However, because of the randomness of train arrival, the next group of trains or a train is probably about to cross the crossing before all the accumulated vehicles have been given permission. Therefore, formula 11 should be modified to formula 12:

$$M = M + \frac{T}{(m-1) * C_{IT}} \quad (12)$$

Based on the above, the rule for judging whether the influence of train passage is eliminated or not is designed as formula 13:

$$\begin{cases} M = 0 & \text{the influence of train passage is eliminated} \\ M \neq 0 & \text{the influence of train passage is not eliminated} \end{cases} \quad (13)$$

M is updated based on four scenarios in the following:

Scenario 1: Initialization: $M = 0$;

Scenario 2: After a group of trains or a train crosses the crossing, $M = M + \frac{T}{(m-1) * C_{IT}}$;

Scenario 3: After a signal cycle with the influence of train passage is completely displayed, $M = M - 1$;

Scenario 4: If $M<0$, then $M=0$. That is because M of scenario 2 is not always a positive integer.

6. Case Study

The simulation model of the intersection of Jingang Road, Yingang Road, and Chuangye *No.*1 Road in Dalian City, Liaoning Province, China, which is near a highway-railway at-grade crossing and was selected as an experimental intersection, is established by VISSIM 4.3, a professional micro-traffic simulation software that can be used to simulate and evaluate urban traffic[8-10]. The dynamic control strategy proposed is achieved by the VAP[8].

6.1 Test Case

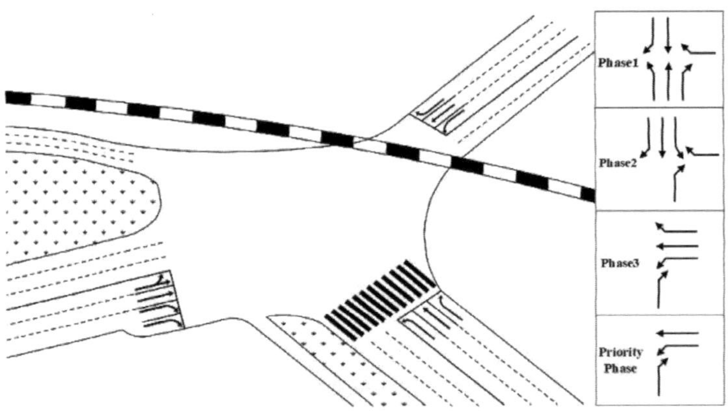

Figure 3. Geometric Characteristics and Signal Phases of the Experiment Intersection.

The geometric characteristics, signal phases, and phase sequence of the experimental intersection are displayed in figure 3.

The saturation traffic volume of all movements is 1468 pcu/h. The start-up lost time and the compensation time after green time are set as 2.4s and 0s, respectively. The conversion coefficients of a small vehicle are set according to the composition of traffic volume: if the proportion of large vehicles is higher than 85%, the conversion coefficient is 1.35; if the proportion is higher than 35% and lower than or equal to 85%, the conversion coefficient is 1.15; otherwise, the conversion coefficient is 1. The conversion coefficient of a container truck is set as 2.34[7].

The priority phase used in the train passage period is shown in figure 3. As a result of simulation experiments only being run for peak hour, the signal timings at peak hour without the influence of train passage and with the influence of train passage are optimized and displayed in table 1. In practical applications, the signal timing at any hour can be designed using the same methodology so that an intersection near a highway-railway at-grade crossing can be efficiently controlled all day.

Table 1. Signal Timing.

without the influence of train passage[a]				with the influence of train passage[b]			
Signal Phase	Green Interval Time (s)	Green Time (s)	Green End	Signal Phase	Green Interval Time (s)	Green Time (s)	Green End
1	5(3+2)	17	22	1	5(3+2)	33	38
2	4(3+1)	10	36	2	4(3+1)	18	60
3	4(3+1)	8	48	3	4(3+1)	8	72

[a]saturation degree of all movements is set as 0.8, x_i=0.8, i=1,2,……,9.
[b]saturation degree of conflicting movements is set as 0.9 and the other is the same as the former, that is, x_i=0.9, i=1,2,3,4,5,6; x_i=0.8, i=7,8,9; traffic volume of conflicting movements is multiplied by 2, m=2.

Two fixed signal timings used in the dynamic control strategy are used as comparative cases, respectively. When the fixed signal schemes are used, a signal controller is added to give priority to the train.

6.2 Analysis of Results

Average vehicle delay times given different train traffic volumes are shown in table 2 and table 3.

Table 2. Average Vehicle Delay (Train Traffic Volume is 6 trains/h).

Random	Average Vehicle Delay Time（s/per vehicle）			Percentage of Difference {②-min(①,③)}/ Min (①,③)
	① fixed signal scheme (m=1)	② dynamic signal strategy (m=2)	③ fixed signal scheme (m=2)	
1	**19.7**	19.1	20.2	-3%
2	**12.5**	9.7	13.4	-22%
3	9.5	7	**8.7**	-20%
4	10.5	8.4	**9.8**	-14%
5	**11**	9.1	11.4	-17%
Average Percentage of Difference				-15%

As shown in table 2, the performance of the dynamic control strategy is superior to the better performance of fixed signal schemes on the condition of different traffic arrival modes. This is because the dynamic control strategy can catch the variability of the influence of the train passage caused by train arrival randomness, train length and the variation of train speed and accordingly adjust the signal timing and the number of signal cycles to let the accumulated vehicles go, whereas both the fixed signal schemes cannot. When the train traffic volume increases, the same results can be found from table 3, which further validates the proposed dynamic control strategy has good performance.

Table 3. Average Vehicle Delay (Train Traffic Volume is 9 trains/h).

Random	Average Vehicle Delay Time（s/per vehicle）			Percentage of Difference
	① fixed signal scheme (m=1)	② dynamic signal strategy (m=2)	③ fixed signal scheme (m=2)	{②-min (①, ③)}/ Min (①, ③)
1	**29**	27.5	33.4	-5%
2	**10.5**	8.7	10.7	-17%
3	17.3	14	**15.6**	-10%
4	17.2	12.8	**15.8**	-19%
5	21.7	19.1	**20.7**	-8%
	Average Percentage of Difference			-12%

7. Conclusion and Discussion

On the condition of non-saturation of the intersections near a highway-railway at-grade crossing, the quick release of the vehicles that accumulate during the train passage period and the reduction of the average vehicle delay time are able to be achieved by increasing the average traffic volume of traffic movements conflicting with the train passage to re-optimize the signal timing and deciding its running times according to the length of the train passage period and the signal cycle length with the influence of train passage.

Future research will focus on designing an adaptive control scheme for the intersection near a highway-railway at-grade crossing to further improve its effectively.

Acknowledgments

This research was funded by the National Natural Science Foundation of China (Grant No: 51478110) and Science and Technology Program of Jiangsu Province (No. BY2016076-05).

References

[1] R. A. Gómez, M. Knodler, D. L. Fisher. An evaluation of traffic control devices and driver distraction on driver behavior at railway-highway grade crossings,*17th International Conference Road Safety On Five Continents*.(Statens väg-och transport for sknings institute, 2016).

[2] H. Z. Liu. Discussion on Design of Highway and Railway Grade Separation. Urban Roads Bridges & Flood Control, 9, 2012: 011.

[3] X Yan, L D Han, S Richards, et al. Train-vehicle crash risk comparison between before and after stop signs installed at highway-rail grade crossings. Traffic injury prevention, 11, 5 (2010):535-42.

[4] S Kadiyala, P Gubbala, S D Schrock. Human Behavior at Railroad Grade Crossings, *2016 Joint Rail Conference*.(American Society of Mechanical Engineers, 2016).

[5] M. G. Lenné, C Rudin-Brown. M., J. Navarro, et al. Driver behaviour at rail level crossings: Responses to flashing lights, traffic signals and stop signs in simulated rural driving. Applied ergonomics, 42,4 (2011):548-54.

[6] W. R. Schulte. Effectiveness of Automatic Warning Devices in Reducing Accidents at Grade Crossings[J]. Transportation Research Record, 611(1976):49-57.

[7] Y. S. Quan. Urban Traffic Control. *(China Education Press,1989).*

[8] Y. M. Bie, D. H. Wang, Q Wei, et al. Conditional active and passive bus signal priority strategies considering saturation degree restriction at an isolated junction. Journal of Jilin University (Engineering and Technology Edition), 41,5 (2011): 1222-27.

[9] M. Fellendorf, P. Vortisch. Microscopic traffic flow simulator VISSIM. *Fundamentals of traffic simulation*. (Springer New York, 2010: 63-93).

[10] J. Stevanovic, A. Stevanovic, P. T. Martin, B. Thomas. Stochastic optimization of traffic control and transit priority settings in VISSIM [J]. Transportation Research Part C: Emerging Technologies, 16, 3 (2008): 332-349.

Optimization of low-floor vehicle dynamic performance based on Kriging model and NSGA-Ⅱ algorithm

YANG Yang, Ding Junjun*, LI Fu and Li Jincheng

School of Mechanical Engineering, Southwest JiaoTong University, China

*E-mail: dumdream0851@163.com

Abstract. Echoing the pressing need to improve optimization efficiency and change the method of vehicle dynamic performance optimization by varying single parameter, a new method based on the Kriging approximate model and NSGA-II genetic algorithm is proposed to obtain more accurate results and shorten the design cycle. Accordingly, taking the suspension parameters as the variables, the dynamic performance indexes as the responses to fit the Kriging approximate model, the dynamic model of the 70% low-floor vehicle has been established. Subsequently, multi-parameter, multi-objective optimization of vehicle dynamic performance is achieved by virtue of the approximate model. The results show that, compared with the original method, all indexes are improved except the vertical comfort and stability. In particular, the optimization efficiency is significantly enhanced, and the effectiveness and feasibility of applying the optimization method to vehicle dynamic performance by virtue of the Kriging model have been proved.

1. Introduction

Increasing transport congestion in urban areas due to excessive motor vehicle traffic can favor the development of passenger rail transport. The corresponding vehicle dynamics has been promoted quickly, and the vehicle-track coupling model has also been widely used.

Due to the complexity of the large wheel rail system, the process of vehicle dynamics analysis is far beyond the scope of theoretical understanding. Therefore, computer simulation should be introduced. The vehicle dynamics model finally boils down to a nonlinear two order differential equation. The railway train is made up of multi-section vehicles, and the degree of freedom is about 500, so the speed of calculation has become most relevant [1]. Due to the advantages of reducing the 'numerical noise', improving the optimization efficiency and shortening the actual solving time by several orders of magnitude among others. Approximate optimization technique has growingly come under the spotlight in recent years. Currently there are some widely-used approximate models, such as the RSM model, RBK model, orthogonal polynomial model, and Kriging model etc.

Kriging model, also known as the space partial interpolation method, is the optimal unbiased estimator for the regional variables in the limited region based on the variation function. It is widely used in optimization design because of its superiority of fitting in high degree of nonlinear and multi-modal issues [2]. Many groups have contributed to the application of Kriging model. For instance, J.Jakumeit [3] applied the Kriging model to the plates punch forming and optimized the pressure-pad-force; ShinkyuJeong [4] used the Kriging model coupled with genetic algorithm to optimize the designs of the two-dimensional and multi-element airfoil flaps, instead of the original complex CFD

Content from this work may be used under the terms of the Creative Commons Attribution 3.0 licence. Any further distribution of this work must maintain attribution to the author(s) and the title of the work, journal citation and DOI.
Published under licence by IOP Publishing Ltd

analysis-solving process; Yue-huaGao [5] introduced the Kriging model into the field of injection molding and carried out the optimization design of injection molding process; and Dawei Yin [6] established the model of low pressure compressor for turbofan engine based on the Kriging model, and drew upon it as an example of the characteristics of the engine.

The dynamic performance optimization of rolling stock is realized by parameter optimization. However, owing to the complexity of the vehicle system, as well as the highly nonlinear wheel rail and the suspension system which give rise to computational complexity, the optimization design is time consuming. Moreover, the occurrence of the optimal failure problem caused by the non-feasible solution of approximate linear programming is frequent. In this paper, the Kriging approximate model is introduced into the optimization process to solve these problems.

2. The Kriging model

The Kriging approximate model was first applied to mining and geographical statistics, and subsequently to aerospace, stamping forming, dynamic analysis and other fields. The mechanism of the Kriging approximate model is to estimate the value of some specific unknown points, by using the weighted average method according to a certain number of sample points, which have been measured in the limited regions of the unknown points. Among them, the purpose of the weights is to minimize variance, which makes the estimation of the unknown points the best linear unbiased estimation [7]. Specific methods are as follows: Let $X^{(i)}=(x_1^{(i)},\dots,x_d^{(i)})(i=1,\dots,n)$ be the measured sample points of size n, d the dimension of the design vector, the corresponding response value is $y(X^{(i)})$, and the estimated value is $f(X)$, which is defined as equation (1), then equation here.

$$f(X)=\sum_{i=1}^{N}\beta_i y(X_i) \tag{1}$$

The relationship between the responses to the function to be fitted $Y(x)$ and the independent variables can be expressed by equation (2), then equation here.

$$y(x)=\sum_{i=1}^{N}\beta_i y(X_i)+z(X)=f(X)+z(X) \tag{2}$$

$Z(x)$ is a random process function indicating the local deviation between the estimated value and the response value and boasting the following statistical characteristics, defined as equation (3), then equation here.

$$\begin{cases} E[z(X)]=0 \\ Var[z(X)]=\sigma^2 \\ Cov[Z(X^{(i)}),Z(X^{(j)})]=\sigma^2 R(X^{(i)},X^{(j)}) \end{cases} \tag{3}$$

where in $R(X^{(i)},X^{(j)})$ is the correlation function to the data points $X^{(i)},X^{(j)}$, and Gaussian correlation function is used in this paper as equation (4), then equation here.

$$R(X^{(i)},X^{(j)})=\exp(-\sum_{k=1}^{d}\theta_k\left|x_k^{(i)}-x_k^{(j)}\right|^2) \tag{4}$$

In the above table θ_k is the unknown correlation parameter employed to fit the model.
Once the correlation function is selected and the best θ_k is estimated, the Kriging model can be used to predict the response $\hat{y}(X)$ as equation (5), then equation here.

$$\hat{y}(X)=\hat{\beta}+r^T(X)R^{-1}(Y-f\hat{\beta}) \tag{5}$$

where $\hat{y}$ is the vector of estimated response values at each sample point, f is the vector with values of the trend function evaluated at each sample point, $\hat{\beta}$ is a constant, and $r^T(X)$ is the vector of correlation values between the untried location x and the sample data points.

The constant $\hat{\beta}$ can be estimated by virtue of the equation (6), then equation here.

$$\hat{\beta} = (f^T R^{-1} f)^{-1} f^T R^{-1} Y \quad (6)$$

The estimate of the variance is equation (7), then equation here.

$$\sigma^2 = \frac{(Y - f\hat{\beta})^T R^{-1}(Y - f\hat{\beta})}{n} \quad (7)$$

The maximum likelihood estimate is obtained by maximizing the θ_k likelihood estimate obtained by equation (8), then equation here.

$$\frac{n\ln(\sigma^2) + \ln(R)}{2} \quad (8)$$

3. Establish the vehicle Kriging approximate model

In this paper, a 70% low-floor tram designed by a company is considered as the object of the research. And according to the multi-body dynamics analysis, the Kriging model is established in accordance with the vehicle indexes, including the primary and secondary stiffness and damping of three direction, and the comfort, stability, wheel rail lateral force, wheelset lateral force, derailment coefficient, and wheel load reduction.

3.1. Vehicle dynamics model

The unit form of the tram is M+T+M, which suggests 'two motors one trail', and sets up the top and bottom hinge joints between the vehicles to form a statically indeterminate structure.The two ends of the tram are supported by the motor bogies, while the intermediate vehicle is supported by the trailer bogie. The motor bogie adopts the structure of the external axle box with the bolster, which can adapt to the large angle between the motor bogie and the vehicle body. In comparison, the trailer bogie uses the structure of the external axle box without bolster. The primary suspension employs conical rubber spring due to its sound performance in noise reduction, while the secondary suspension uses hourglass rubber pad in order to improve the stability of the vehicle. The way of full-suspended hollow axle is applied in the driving device, which can effectively reduce the unsprung mass.

Based on multi-body dynamics theory, the vehicle dynamics model, which couples the lateral-vertical-longitudinal with 69 degrees of freedom, is established in SIMPACK. In the model, the wheel rail contact relationship is considered as nonlinear, which is the same condition as that of the vehicle suspension system. Furthermore, the wheel rail contact force is calculated by the simplified theory FASTSIM, which is proposed by Kalker. To begin with, the curve radius is set as 100m, the length of the transition curve is 20m, the super elevation is 0.075m, and the speed is 25km/h.

To evaluate the curve performance, indexes including wheel rail lateral force, wheelset lateral force, derailment coefficient, and rate of wheel load reduction are taken into account. In the meantime, comfort is evaluated by the spelling index. Stability is analyzed by a RMS-value over 100m track length and 10m step length to the lateral acceleration of the frame. In addition, the peak acceleration of RMS is taken into account in the evaluation, according to UIC518 and EN14363.

3.2. The optimal Latin hypercube sampling

A certain number of sample points are needed to fit the response function when the approximate model is established. The optimal Latin hypercube sampling is used in this paper, which is an improved form of random Latin hypercube[8]. Its mechanism is that in a n-dimensional space, the coordinate interval of each dimension $[x_k^{min}, x_k^{max}], k\in[1,n]$ is evenly divided into m intervals, and each subinterval is referred to as $[x_k^{i-1}, x_k^{i}]$, $i\in[1,m]$. m points are randomly selected, each horizontal plane of each factor has been ensured to be considered only for once, which is a Latin hypercube sampling with m samples in a n-dimensional space. Optimal Latin hypercube sampling method improves the uniformity of random Latin hypercube and makes the fitting of the factor and the response more accurate, and can be robust to the changes of the model.

From past design experience, the value of the primary and secondary suspension stiffness, a total of six directions, as well as the lateral damping are determined. In order to obtain reasonable optimal results, a wider range of sampling listed in table 1 is finally used.

Table 1. Sample range.

	Primary suspension, x&y (kN/m)	Primary suspension, z (kN/m)	Secondary suspension, x&y (kN/m)	Secondary suspension, z (kN/m)	Damper (kNs/m)
Sample upper limit	4500	800	100	1000	10
Sample lower limit	1500	300	40	400	40

3.3. Fitting Kriging approximate model

References In accordance with established ideas and methods of the Kriging model, firstly, a certain number of sample points that meet the requirements by the optimal Latin hypercube sampling should be established, and then the corresponding sample points should be adopted in the vehicle dynamics model and be solved to obtain the corresponding response of the curve performance. Subsequently, the Kriging approximate model of sample points should be established according to the response values, and an error comparison should be drawn between the results of the approximate model and the dynamic model. If the error does not meet the requirements, the number of sample points has to be raised to improve the accuracy of the approximate model.

3.4. Error analysis

The established Kriging model serves to verify the accuracy by taking other sample points in order to ensure the effectiveness of the model. And the Kriging model can be used in the approximate analysis after confirmation. When it comes to the verification of the accuracy, the optimal Latin hypercube sampling is used to take 20 sample points within the sampling range and calculate the response of those 20 sample points respectively in SIMPACK and by the approximate model. In this paper, the average relative error as equation (9) and (10) is used to analyze the accuracy of the approximate model.

$$err = \frac{1}{m}\sum_{1}^{m}\frac{|y(x)-\hat{y}(x)|}{|y(x)+\varepsilon|} \tag{9}$$

$$\varepsilon = \begin{cases} 0 & |y(x) \neq 0| \\ 0.01 & |y(x) = 0| \end{cases} \tag{10}$$

The results are shown in table 2. The data suggest that the approximate model is highly accurate, and the maximum error is 1.30%, which falls into the acceptable range and indicates that the model meets the requirements.

Table 2. Evaluation index error analysis.

Index	Stability	Lateral comfort	Vertical comfort	Rate of wheel load reduction	Wheel rail lateral force	Wheel axle lateral force	Derailment coefficient
err	1.30%	0.55%	0.33%	1.30%	0.2 6%	0.46%	0.25%

4. Dynamic performance optimization

The trademark characteristic of urban rail transit is full of small radius curve. For railway vehicles, sound curve performance can decrease the force between wheel and rail so as to reduce wear, and achieve the purposes of energy-saving, noise reduction and operating cost cutting. On the contrary, poor curve performance, lack of guidance, will produce great force between wheel and rail giving rise to more wear, and even derailment or overturning or other accidents. Hence, good dynamic performance is the common goal pursued in vehicle production and operation department. For the tram used in this paper, the primary and secondary suspension parameters were optimized by the traditional method in the design process. Now, based on the Kriging approximate model, a more reasonable parameter matching relationship is obtained by genetic algorithm.

4.1. Optimization method

The NSGA- II , which is non-dominated sorting genetic algorithm with elite strategy and also the improved version of NSGA published in 1994, is introduced to optimize the curve passing performance. It draws upon a fast non-dominated sorting algorithm to reduce complexity of the calculation. Furthermore, it employs the elite strategy, expands the sampling space, and compares the arithmetic operators between the degrees of congestion. As a result, it overcomes the defects of manually specifying the shared parameters in the NSGA.

In the non-dominated sorting genetic algorithm with elite strategy, the degree of congestion calculation is significant in ensuring the diversity of the population. Its calculation steps are as follows:

- Each degree of congestion i_d is set as 0;
- For each objective, non-dominated sorting is conducted on the population, the degree of congestion of the two individuals on the boundary is infinite, which is $o_d=I_d=\infty$.
- Calculate the degree of congestion to other individuals as equation (11) , then equation here.

$$i_d = \sum_{j=1}^{m} (|f_j^{i+1} - f_j^{i-1}|) \tag{11}$$

where in, i_d represents the degree of congestion of the i point, f_j^{i+1} represents the j-th objective function value of the i+1 point, and f_j^{i-1} represents the j-th objective function value of the i-1 point.

When it comes to multi-objective optimization, due to the contradiction between the objective functions, it is difficult to find the solution to make each objective function reach their optimal value. At present, the common solution would be the Pareto solution, which indicates the value of any objective function cannot be further improved, otherwise, it would worsen the value of other objective function. The Pareto optimal solutions are continuous and the number of the solutions is usually infinite, which constitutes the Pareto front. The final solution of multi-objective optimization is to choose the best compromised solution from all the Pareto optimal solutions.

4.2. Optimization results

The vehicle dynamic performance is directly affected by the matching relationship of the stiffness of the vehicle suspension system. In this paper, the Kriging approximate model is employed to optimize the vehicle curve performance based on the NSGA- II algorithm, which considers the stiffness of primary and secondary with six directions in total and the damping of shock absorber as the design variables. Moreover, it takes indexes including the wheel rail lateral force, wheelset lateral force, derailment coefficient, wheel load reduction, stability index, lateral stability, and vertical stability as the response values. Table 3 shows the optimized suspension parameters, and table 4 shows the comparison between the calculated results and the original parameters. In the comparison of the results before and after optimization, it shows that the wheel rail lateral force is reduced by 767.4N, the wheelset lateral force is reduced by 581.5N, and all the performance indexes are improved except the vertical comfort and stability. Compared with the traditional optimization method of changing single variable, genetic algorithm has greatly increased the optimization speed For instance, in this case using a genetic algorithm takes only 18 hours instead of approximately 200 hours by optimizing in the original design process, which serves to significantly improve efficiency and shorten the design cycle. Thus, it indicates smaller human consumption, and the workload is greatly reduced.

Table 3. Optimization results.

	Stability	Lateral comfort	Vertical comfort	Wheel rail lateral force (N)	Wheel axle lateral force (N)	Derailment coefficient	Rate of wheel load reduction
Optimized	0.84	1.82	2.23	35214	17854	0.616	0.246
Original	0.78	2.25	2.22	35981.4	18435.5	0.624	0.257
Difference	-0.06	0.43	-0.01	767.4	581.5	0.008	0.011

Table 4. Optimized suspension parameters.

		x stiffness (kN/m)	y stiffness (kN/m)	z stiffness (kN/m)	Damper (kNs/m)
Optimized	Primary	3395.587	2376.82	457.50	13.11
	Secondary	72.28	47.77	613.55	
Original	Primary	3000	3000	500	20
	Secondary	120	120	600	

5. Conclusion

In this paper, a new 70% low-floor tram is selected as the study object. According to its structure, a multi-body dynamics model with 69 degrees of freedom has been established, and the Kriging approximate model is applied to optimize the vehicle dynamic performance based on the NSGA- II algorithm. It takes the primary and secondary suspension parameters with seven in total as the design variables, and takes the vehicle dynamic performance indexes as the response values.

The results show that, NSGA- II optimization algorithm serves to deliver good results, moreover, compared with the original design parameters, all the performance indicators are improved except the vertical comfort and the stability, so the optimization method can be used directly in the design of the vehicle suspension parameters. The method based on the Kriging model can increase the optimization efficiency without significant sacrifice of accuracy. Meanwhile, the optimization target has been remarkably improved, an indicator of its effectiveness and feasibility applied to the optimization of rolling stock dynamic performance.

6. Reference

[1] Jianhua H, Jing Z, Pingbo W 2005 Vertical vibration isolation and suspension parameter optimization of railway vehicle *J. Journal of Traffic and Transportation Engineering* 5 10-4

[2] Xu R, Song W 2010 Application of Improved Kriging-Model-Based Optimization Method in Airfoil Aerodynamic Design *J. Journal of Northwestern Polytechnical University* 28 503-10

[3] Jakumeit J, Herdy M, Nitsche M 2005 Parameter optimization of the sheet metal forming process using an iterative parallel Kriging algorithm *J. Structural & Multidisciplinary Optimization* 29 498-507

[4] Jeong S, Murayama M, Yamamoto K 2015 Efficient Optimization Design Method Using Kriging Model *J. Journal of Aircraft* 42 413-20

[5] Gao Y, Wang X 2008 An effective warpage optimization method in injection molding based on the Kriging model *J. The International Journal of Advanced Manufacturing Technology* 37 953-60

[6] Yin D, Li B, Wang Y, et al 2011 Aeroengine compressor characteristics metamodeling using Kriging method *J. Acta Aeronautica Et Astronautica Sinica* 32 99-106

[7] Sakata S, Ashida F, Zako M 2003 Eigenfrequency optimization of stiffened plate using Krigingestimation *J. Computational Mechanics* 31 409-18

[8] Zhao M, Cui W 2007 Application of the optimal Latin hypercube design and radial basis function network to collaborative optimization *J. Journal of Marine Science and Application,* 6 24-32

IOP Conf. Series: Materials Science and Engineering **383** (2018) 012041 doi:10.1088/1757-899X/383/1/012041

Matching performance of CHN60N rail with various wheel profiles for Chinese high-speed railway

Wang Jian, Wang Ping and Ma Dao-lin*

MOE Key Laboratory of High-Speed Railway Engineering, Southwest Jiaotong University, China

*Email: madaolin@swjtu.edu.cn

Abstract. The matching of wheel profile with rail profile is the key issue in design of high-speed railway, it directly affects the vehicle dynamics performance during travelling and the wear behavior of wheel/rail. The matching performance of wheel profiles LMA, S1002CN and XP55 widely used in Chinese high-speed railway with CHN60N rail was analyzed in the study in terms of wheel/rail contact geometry, wheel/rail rolling contact and vehicle-track coupling dynamics. The results indicate that: By comparing the matching performance of the three wheel profiles with CHN60 rail, S1002CN profile matches best with the CHN60N rail. XP55 profile is not applicable to CHN60N rail.

1. Introduction

Correct wheel-rail profile matching can improve the dynamic performance, riding stability and curving performance of traveling vehicle, reduce wheel/rail wear, prolong fatigue life of wheel and rail, and ensure riding safety[1-8]. There are mainly three types of electric multiple unit (EMU) available in Chinese high-speed railway, which may differ in wheel tread (as show in figure 1). For instance, CRH1 and CRH2 EMU adopts LMA tread, CRH3 EMU adopts S1002CN tread, while CRH5 EMU prefers XP55 tread.

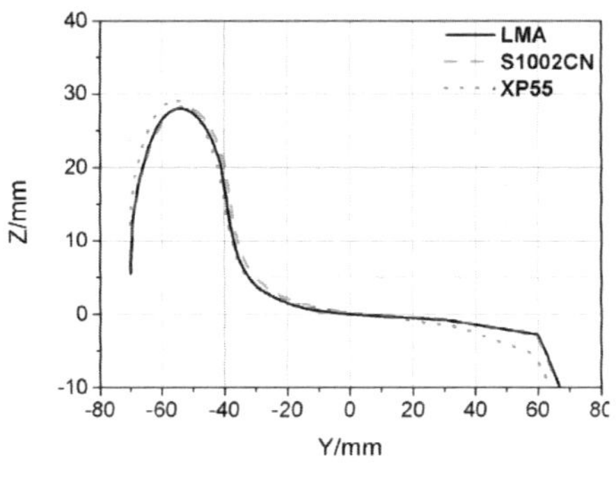

Figure 1. Typical wheel treads for Chinese high-speed railway.

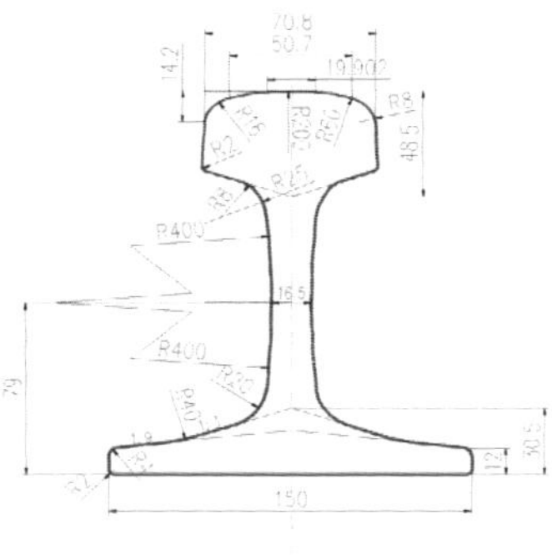

Figure 2. CHN60N rail used in Chinese high-speed railway.

In this study, the matching relationship of CHN60N rail (as show in figure 2) with the various wheel treads available in Chinese high-speed railway was analyzed comprehensively in terms of

Content from this work may be used under the terms of the Creative Commons Attribution 3.0 licence. Any further distribution of this work must maintain attribution to the author(s) and the title of the work, journal citation and DOI.
Published under licence by IOP Publishing Ltd

wheel/rail contact geometry, wheel/rail rolling contact state and vehicle-track coupling dynamic performance.

2. Wheel/Rail Contact Geometry

The distribution of wheel/rail contact points on rail top surface at different lateral displacements for three wheel profiles is plotted in figure 3. The figure can reflect wheel/rail wear trend to some extent. As can be seen from figure 3 that, LMA/CHN60N contact points are densely distributed and concentrate within about 0.4mm from rail centerline; at lateral displacement of -9~1.8mm, LMA/CHN60N contact points move to the interior of the rail with distribution width of 6.08mm. XP55 is a conical profile (profile conicity=5.5%), so XP55/CHN60N contact points are denser than LMA/CHN60N contact points, concentrating within 6~7mm from rail centerline; this may produce certain unbalanced loading and lead to uneven wear of rail. S1002CN/CHN60N contact points are distributed much even than the above two types and have the largest distribution width, which is contributive to the uniformity of wear on rail top surface.

Equivalent conicity reflects the running smoothness in straight line and the curving performance of wheel-set. Figure 4 presents the variation of equivalent conicity with wheel-set lateral displacement when CHN60 rail matches with different wheel profiles. For LMA profile, at lateral displacement of below 2mm, the equivalent conicity remains the same, and this profile produces good lateral stability and low wheel/rail dynamic effect; while at lateral displacement of above 2mm, the equivalent conicity increases steadily. For S1002CN profile, at lateral displacement of below 2.2mm, the equivalent conicity fluctuates significantly but increases on the whole; at lateral displacement of 2.2mm~5.4mm, the equivalent conicity drops remarkably from 0.14 to 0.10; and at lateral displacement of greater than 5.4mm, the equivalent conicity becomes stable at about 0.10. For XP55 profile, the equivalent conicity has small fluctuation at lateral displacement of below 1.2mm, but changes greatly at lateral displacement of 4.6mm and 5.8mm, from 0.045 to 0.067 and from 0.061 to 0.071 respectively, which may cause abrupt change in lateral force of wheel-set, affecting the lateral stability of vehicle.

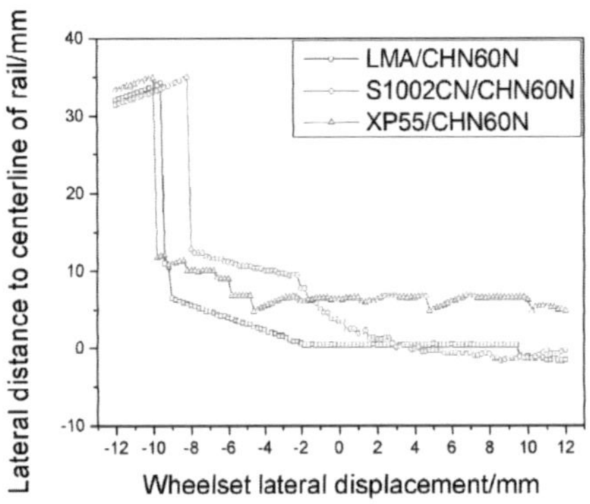

Figure 3. Distribution characteristics of wheel/rail contact points on rail top surface for three profiles.

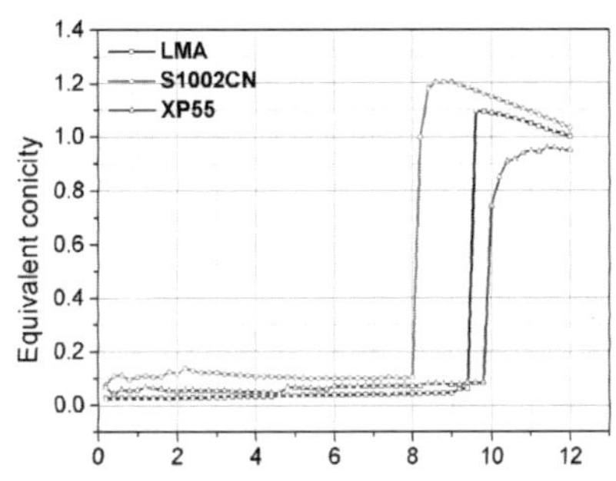

Figure 4. Equivalent conicity for the matching of three profiles with CHN60N rail.

3. Wheel/Rail Rolling Contact

The size of contact area directly influences the distribution of contact stress on wheel/rail contact area. figure 5 provide the size of contact area when different wheel profiles are in contact with CHN60N rail at different lateral displacements. As can be seen from figure 5 that, when the wheel-set is aligned with the track centerline, the contact area is 86.72mm2 for LMA profile, 113.6mm2 for S1002CN profile and 80.56mm2 for XP55 profile, respectively. When the wheel-set moves slightly to the left (within 2.5mm), S1002CN profile has much greater contact area than the other two types. As the wheel-set moves to the left further till the left flange clings to the rail, the left wheel/rail contact area increases gradually, but the right wheel/rail contact area remains basically the same for LMA profile; the left and right wheel/rail contact area decreases gradually for S1002CN profile, with the minimum contact area of 64.8mm2 for left wheel and rail and 84.32 mm2 for right wheel and rail, respectively; the left wheel/rail contact area increases first and then decreases, while the right wheel/rail contact area remain unchanged for XP55 profile.

If the wheel/rail normal contact force remains the same, the normal contact stress of wheel and rail depends mainly on the characteristics of wheel/rail profiles. The distribution of the maximum normal stress in the contact patch between three wheel profile types and CHN60N rail at different lateral displacement is plotted in figure 6. The distribution of wheel/rail contact area is contrary to the distribution of the maximum normal contact stress of wheel and rail.

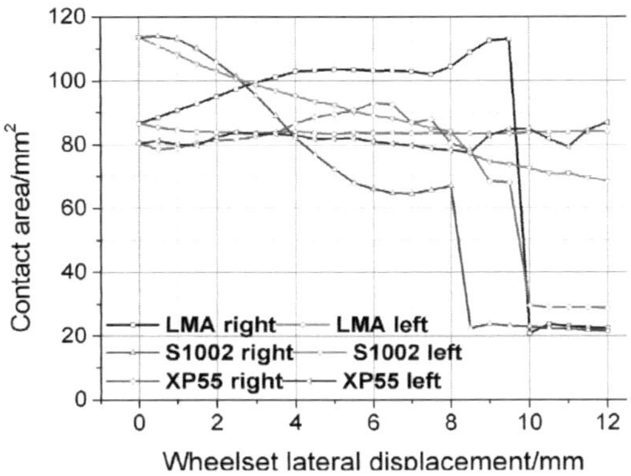

Figure 5. Area of wheel/rail rolling contact.

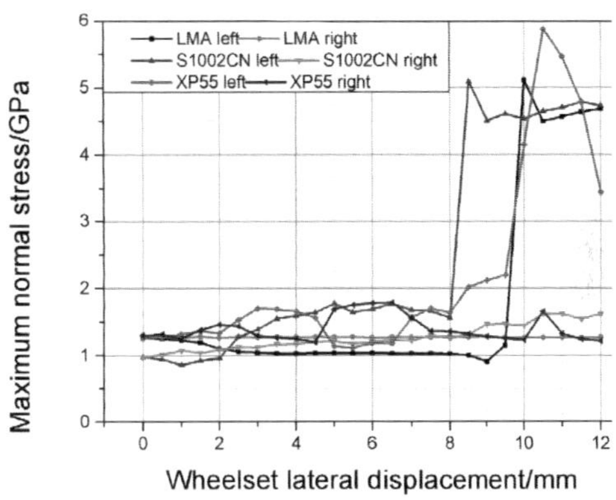

Figure 6. Maximum normal stress of wheel/rail rolling contact.

4. Assessment of Dynamic Performance

4.1. Curving Performance

Figure 7 demonstrates the variation of lateral displacement of front guide wheel-set (the first wheel-set) with the traveling distance during curving movement. As can be seen in the figure that, S1002CN profile has the smallest lateral displacement of guide wheel-set and the best curve negotiation ability; LMA profile and XP55 profile produce much greater lateral displacement of guide wheel-set than S1002CN profile, and have poor curve negotiation ability. This lies in that LMA profile and XP55 profile have smaller equivalent conicity, and the wheel-set lateral displacement is practically irreversible.

Figure 8 shows the variation of wear index of outer wheel of guide wheel-set with the traveling distance during curving motion. When the wheel-set approaches the transition curve, the maximum wear index of outer wheel of guide wheel-set is 7.42N·m for S1002CN profile, 13.66N·m for XP55 profile, as well as 21.05N·m for LMA profile since the flange closely clings to the rail. Therefore, the wheel/rail wear will be worsened when the wheel of LMA profile travels in the transition curve. When the wheel-set rolls on to the circular curve, the wear index of outer wheel of guide wheel-set for S1002CN profile and XP55 profile is almost the same, about 2.50 N.m, while the wear index of LMA profile may have a peak of 7.50 N.m in the circular curve.

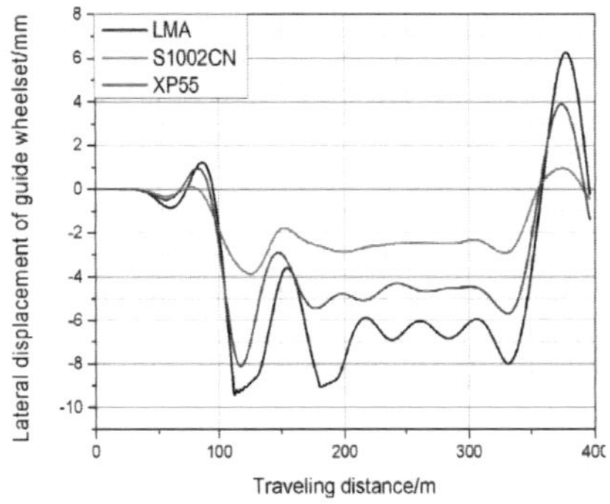

Figure 7. Lateral displacement of guide wheel-set during curving.

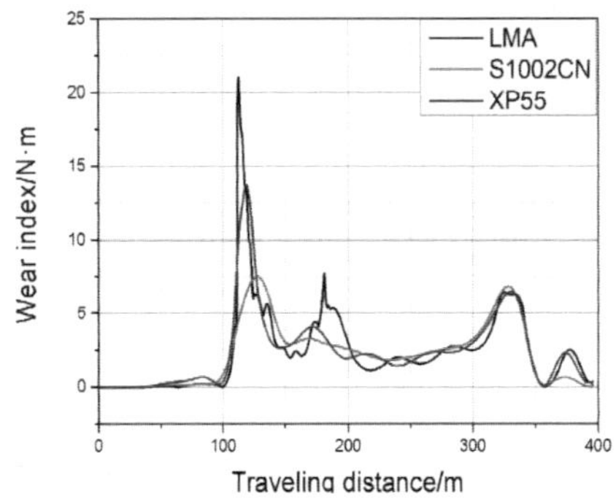

Figure 8. Wear index of outer wheel of guide wheel-set during curving.

4.2. Lateral Riding Stability

A vehicle-track coupling dynamic model was used in the study. Applying the fourth, fifth and sixth grade track irregularity PSD of US railways to the straight track, and letting the vehicle travel at 250km/h produces the lateral and vertical vibration acceleration of carbody. The vibration response of vehicle of different wheel profiles was evaluated with Sperling index. The evaluation results are as shown in table 1.

Tablel. Sperling Index for Carbody Vibration Response Under the Excitation of Irregularities.

Profile Type	Grade 6 PSD of US Railway		Grade 5 PSD of US Railway		Grade 4 PSD of US Railway	
	Lateral	Vertical	Lateral	Vertical	Lateral	Vertical
LMA	1.65	0.92	2.01	1.15	2.20	1.38
S1002CN	1.68	0.92	2.07	1.15	2.28	1.38
XP55	1.82	0.92	2.21	1.15	2.47	1.38

Table 1 indicates that, wheel profile type mainly affects the lateral running smoothness of carbody, not the vertical running smoothness of carbody. Under the effect of irregularities, the vehicle of LMA profile has the best lateral running smoothness of carbody, followed by S1002CN profile that is slightly inferior to LMA profile, and the XP55 profile which has the worst lateral running smoothness of carbody.

5. Conclusions

(1) For the matching of LMA profile with CHN 60N rail: The wheel/rail contact area is large and uniform, the maximum normal contact stress is small, however, the excessive distribution of wheel/rail contact points on rail top surface may cause uneven wear. This matching case can provide the best running smoothness of carbody under random excitation, but poor curve negotiation ability due to smaller equivalent conicity.

(2) For the matching of S1002CN profile with CHN 60N rail: At small lateral displacement of wheel-set, the wheel/rail contact area is large; the maximum normal contact stress is the smallest; the wheel/rail contact points are distributed uniformly on rail top surface. This matching case can provide good running smoothness of carbody under random excitation and best curve negotiation ability.

(3) For the matching of XP55 profile with CHN 60N rail: The wheel/rail contact area is small; the maximum normal contact stress is the greatest; the wheel/rail contact points are concentrated within 6~7mm from the center of rail top surface, which may lead to severe uneven wear of rail. This matching case can provide the least running smoothness of carbody under random excitation and poor curve negotiation ability.

To sum up, the matching case of S1002CN profile with CHN60N rail is the best. XP55 profile is not applicable to CHN60N rail.

References

[1] Leary J F, Handal S N, Rajkumar B. Development of freight car wheel profiles - a case study. *Wear*, Vol.**144**, 1991, pp. 353-62.

[2] Wu H M. *Investigations of Wheel/rail Interaction on Wheel Flange Climb Derailment and Wheel/rail profile Compatibility*. Illinois Institute of Technology Press, Chicago, 2000.

[3] Shevtsov I Y,Markine V L, Esveld C. Optimal design of wheel profile for railway vehicles. *Wear*, Vol. **258**, 2005, pp. 1002-30.

[4] CUI Da-bin, LI Li, JIN Xue-song, XIAO Guang-wen. Numerical Optimization Technique for Wheel Profile Considering the Normal Gap of the Wheel and Rail. *Journal of Mechanical Engineering*, Vol. **45**, 2009, pp. 205-11.

[5] SUN Shan-chao, WANG Cheng-guo, LI Hai-tao. Analysis of wheel/rail contact geometric parameters' effect on the dynamic behavior of high-speed passenger car. *China railway Science*, Vol. **27**, 2006, pp. 93-98. (in Chinese)

[6] JIN Xue-song, WEN Ze-feng, ZHANG Wei hua. Analysis of contact stresses of wheel and rail with two types of profiles. *Chinese Journal of Mechanical Engineering*, Vol. **40**, 2004, pp. 5-11. (in Chinese)

[7] HELLER R, LAW E H. Optimizing the wheel profile to improve rail vehicle dynamic performance. *Vehicle system dynamics*, Vol. **15**, 1979, pp. 179-97.

[8] JIN Xue-song. *Creep theory and experimental investigation of wheel and rail system.* Southwest Jiaotong University Press, Chengdu, 1999.

Intelligent classification model for railway signal equipment fault based on SMOTE and ensemble learning

Lianbao YANG*, Ping LI, Rui XUE, Xiaoning Ma, Xinqin LI and Zhe WANG

China Academy of Railway Sciences, Beijing, 100081, China

*E-mail: yanglianbao_121@163.com

Abstract. In this paper, we propose a novel intelligent classification model to classify the railway signal equipment fault based on SMOTE and ensemble learning. To tackle the imbalanced fault text data, the model uses SMOTE algorithm to generate the minority railway signal equipment fault class data randomly, making the data balanced. Then the model adopts the base classifier, such as Logistic Regression, Multinomial Naive Bayes, SVM and the ensemble classifier, such as GBDT, Random Forests to classify the data processed by SMOTE. To combine the advantages of various classifiers, the model integrates multiple classifiers by way of voting. Based on the experiment analysis of railway signal equipment fault text data from 2012 to 2016, the result shows that the model has a significant improvement in fault classification accuracy, recall rate and f-score.

1. Introduction

For the various mechanism of different kinds of railway signal equipment, the frequency of fault occurrence is not the same, which causes the problem of imbalanced fault classification. It means that only a small amount of data belong to a kind of railway signal equipment fault, and the vast majority of the data belong another kind of railway signal equipment fault. To avoid the imbalanced fault classification problem, this paper proposes a novel railway signal equipment fault intelligent classification model, which mainly includes two steps. Firstly, the data balance, which means representing the unstructured text data in vector and balance the transformed vector data. Secondly, the EL (Ensemble Learning) [1], which means training and integrating classifiers by the processed data to accomplish intelligent classification of railway signal equipment fault.

The data balance is mainly to balance the data by changing the sample distribution of the data set, which is mainly divided into over-sampling and under-sampling [2]. Over-sampling is to generate small class data automatically, and under-sampling is to remove data from the large category. In 2002, Chawla proposed the SMOTE (Synthetic Minority Oversampling Technique) [3] algorithm which is commonly used in the over-sampling methods. The SMOTE algorithm includes several improved versions, such as the Borderline-SMOTE [4], SVM-SMOTE. The basic idea of SMOTE is to synthesize some new data in minority classes to achieve the balance. The ensemble learning is mainly to make full use of the difference of classifiers through training multiple classifiers and implement the integration of different classifiers through Voting. Traditional classification is primarily based on a single classifier model, such as LR (Logistic Regression), DT (Decision Tree), SVM (Support Vector Machine) and Multinomial NB (Multinomial Naive Bayesian) [5]. However, these classifier models are mainly suitable for training on balanced data. Ensemble classifiers mainly consist of bagging and

Content from this work may be used under the terms of the Creative Commons Attribution 3.0 licence. Any further distribution of this work must maintain attribution to the author(s) and the title of the work, journal citation and DOI.
Published under licence by IOP Publishing Ltd

boosting methods. RF (Random Forest) [6] is the main representative algorithm of bagging method and GBDT (Gradient Boost Decision Tree) [7] is the representative algorithm of the boosting method.

In view of experts and scholars research of imbalanced data classification, and combined with the characteristics of railway signal equipment fault data, this paper puts forward a novel multiple classifiers ensemble learning railway signal equipment fault intelligent classification model based on the SMOTE and Voting. The model utilizes SMOTE algorithm to implement small categories of railway signal equipment fault data automatically, and integrates base classifiers such as LR, Multinomial NB, SVM and ensemble classifiers such as RF, GBDT by means of Voting to achieve railway signal equipment intelligent classification. In order to validate the correctness and effectiveness of the model, a total of 10 classed 641 records for the railway signal equipment fault in the railway station from 2014 to 2016 are analyzed.

2. Imbalanced fault text data processing based on SMOTE

SMOTE is a kind of over-sampling technique which is used to synthesize the minority class data to meet the balance of the data set, which can enhance the effect of classification. Its basic principle is: by selecting a minority sample x_i and k nearby similar samples, select x_j from nearby randomly of its class, and by random linear interpolation, constructs the new minority sample x_{new} :

$$x_{new} = x_i + u\left(x_i - x_j\right), 0 \le u \le 1 \tag{1}$$

Since traditional SMOTE does not consider the distribution characteristics of its adjacent samples, it may cause that repetition occurs between classes. In recent years, some improved algorithms based on SMOTE have been proposed, and the representative algorithms include the Borderline-SMOTE algorithm and SVM-SMOTE algorithm, etc. Borderline-SMOTE only makes linear interpolation on a few samples of the boundary, thus enhancing the impact of boundary samples. SVM - SMOTE is based on the proximity ratio of different types of samples, and the classification boundary can be constructed through SVM, which can be interpolated according to the actual sample data distribution, making the distinction between categories more obvious. Therefore, this paper selects SVM - SMOTE algorithm to generate data in minority classes. The effect of different SMOTE algorithms on the generation of samples in minority is shown in figure 1. The blue dots in figure 1 represent the minority samples, and the red dots represent the majority samples.

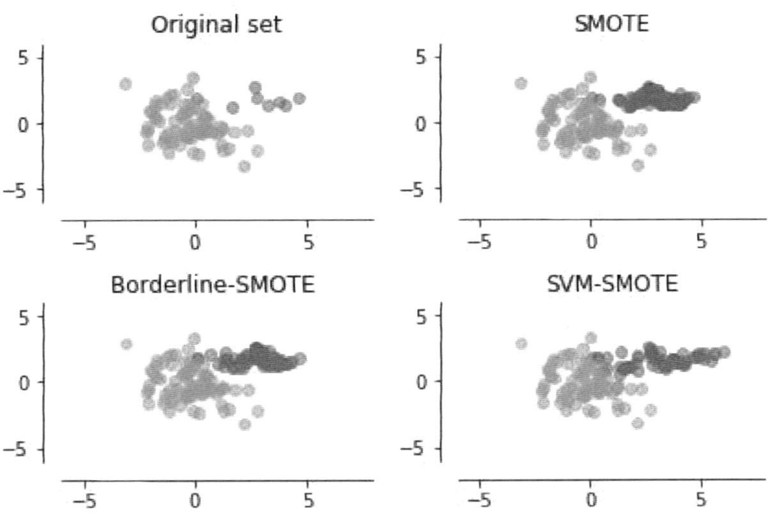

Figure 1. Different SMOTE algorithms used to synthesize the minority samples.

3. Voting based intelligent classification model of multi-classifier ensemble learning

3.1. Base classifier

The commonly used text classifiers are LR, DT, SVM and Multinomial NB, etc. Suppose n vector data set with a m dimension feature, which has c classification, can be expressed as: $D=\{x_i,y_i\}, i=1,2,...,n, x_i=(x_i^1,x_i^2,...,x_i^m), y_i\in\{0,1,2,..c-1\}$

3.1.1. LR.

LR is a kind of classification method based on statistical analysis.

$$p(y=k\mid x)=\frac{e^{-g_k(x)}}{1+\sum_{j=0}^{c-1}e^{-g_j(x)}}, k=0,1,2,\cdots,c-1 \tag{2}$$

Therefore, the corresponding Logistic regression model can be obtained:

$$g_k(x)=\beta_{k0}+\beta_{k1}x^1+\beta_{k2}x^2+\cdots+\beta_{km}x^m \tag{3}$$

The calculation of parameters β is usually estimated by the maximum likelihood method

3.1.2. DT.

DT is a special tree structure, mainly used for classification and decision making. A decision tree contains three types of nodes: decision nodes, usually represented by rectangular boxes; opportunity nodes, usually represented by a circle; endpoints, usually represented by triangles. The commonly used decision tree generation algorithms are ID3, C4.5 and C5.0.

3.1.3. SVM.

SVM is constructed by constructing a hyperplane $f(x)$ that allows the function to represent the relationship between the class y and the sample x vector. Definition of linear x insensitive loss function is:

$$|y-f(x)|_\varepsilon=\begin{cases}0, & |y-f(x)|\le\varepsilon \\ |y-f(x)|-\varepsilon, & |y-f(x)|>\varepsilon\end{cases} \tag{4}$$

If there is a hyperplane:

$$f(x)=\omega^Tx+b=0, \omega\in R^n, b\in R \tag{5}$$

Let:

$$|y-f(x)|\le\varepsilon \tag{6}$$

The sample set D is called a ε -linear, $f(x)$ is regression estimation function. The distance from the sample point $\{x_i,y_i\}$ to the hyperplane is:

$$d_i=\frac{|\omega^Tx_i+b-y_i|}{\sqrt{1+\|\omega\|^2}}\le\frac{\varepsilon}{\sqrt{1+\|\omega\|^2}} \tag{7}$$

To get the optimal hyperplane classification, it can transform to an optimization problem, as let the $\|\omega\|^2$ to be minimum. But for nonlinear problems, SVM maps samples to a high-dimensional feature space through nonlinear mapping function $\varphi(x)$, and computes inner product through kernel function. The optimization problem can be expressed as:

$$\min(\frac{1}{2}\|\omega\|^2+\frac{C}{n}\sum_{i=1}^{n}(\xi_i+\xi_i^*))$$
$$s.t\begin{cases}\omega^T\varphi(x_i)+b-y_i\le\varepsilon+\xi_i\\ y_i-\omega^T\varphi(x_i)-b\le\varepsilon+\xi_i^*\\ \xi_i,\xi_i^*\ge 0\end{cases} \tag{8}$$

In (8), ξ_i, ξ_i^* are the slack variable; C is the penalty factor. The larger penalty factor is, the larger the penalty for sample classification errors, and adjusting C can improve SVM's generalization ability.

3.1.4. Multinomial NB.

Multinomial NB is naive bayes model which is adapted to discrete characteristics. It takes documents as a collection of frequency words, and when calculating the prior probability and conditional probability, smooth will be processed so as to solve posteriori probability is 0 if a characteristic didn't appear in the training sample.

The calculation of prior probabilities $p(y=k)$ is as follows:

$$p(y=k)=\frac{N_{y=k}+\alpha}{N+c\alpha} \tag{9}$$

The condition probabilities $p(x_i|y=k)$ is as follows:

$$p(x_i|y=k)=\frac{N_{y=k},x_i+\alpha}{N_{y=k}+m\alpha} \tag{10}$$

In (9), (10), $N_{y=k}$ is the number of samples of the k category, $N_{y=k},x_i$ is the number of samples in category k with the eigenvector is x_i . α is the smoothness value, and when $\alpha=1$ it's called the Laplace smooth; when $0<\alpha<1$, it's called Lidstone smooth, when $\alpha=0$, it's means don't do smooth.

3.2. Ensemble classifier

An ensemble classifier is a classifier that makes common decisions by combining multiple base classifiers with a certain strategy. It mainly includes the boosting algorithm which supposes the interdependence of base classifiers, and the bagging algorithm which supposes the independence of each base classifier. Boosting algorithm is to operate the sample set to get the sample subset, then take advantage of sample subset to train base classifiers, and finally get the ensemble classifier through the weighted fusion of trained base classifiers. Bagging algorithm begins with randomly selecting training data to train base classifier, and then combines the different trained base classifiers to get the ensemble classifier. This article selects a parallel ensemble classifier RF, which is based on bagging and a serial ensemble classifier GBDT, which is based on Boosting for text classification.

The RF uses the CART decision tree as the base classifier, and improves the establishment process of decision tree. Decision tree selects an optimal feature make the left and right subtrees partition on all nodes in a sample of n features, but RF randomly picks n_{sub} ($n_{sub}\le n$) features of a sample characteristics of nodes, then chooses an optimal feature to make the left and right subtrees partition, further enhancing the generalization ability of the model and avoiding over-fitting phenomenon. The main tuning parameters of RF are divided into bagging frame parameters and CART decision tree parameters. The bagging frame parameters include the maximum number of iterations, etc. The CART decision tree parameters include the maximum depth of the tree.

GBDT, also called MART (Multiple Additive Regression Tree), is a kind of iteration of the decision tree algorithm. The algorithm is composed of many CART Regression decision trees, through the gradient promotion algorithm optimizing the loss function, and finally gets the optimal regression tree. GBDT tuning parameters are much more, and can mainly divided into boosting framework and CART tree parameters. And the boosting framework parameters include the maximum number of iterations, learning step length, etc. The CART decision tree parameters include the depth of the tree.

3.3. Voting based multi-classifier integration learning

The basic idea of ensemble learning is to build one strong classifier based on some weak classifiers through different strategies, solving classification problem. The advantage of various classifiers ensemble learning is that it solved some statistical problems in reality when using weak classifiers. Through combining results of weak classifiers, ensemble learning can reduce location risk in classification. Therefore, ensemble learning always has a better performance in generalization.

Supposing error rates of classifiers are independent, as the number of base classifiers increases, error rate of ensemble learning decreases exponentially, and finally trends to zero. However, base classifiers are always somehow dependent with each other. To choose relatively accurate and diverse classifiers for ensemble learning, this paper picks classifiers based on their classification performance in sample datasets. And then through a voting system of the best classifier combined with other classifiers, this paper chooses the combination with the best classification performance as a final ensemble classifier. The workflow of this method is shown in figure 2.

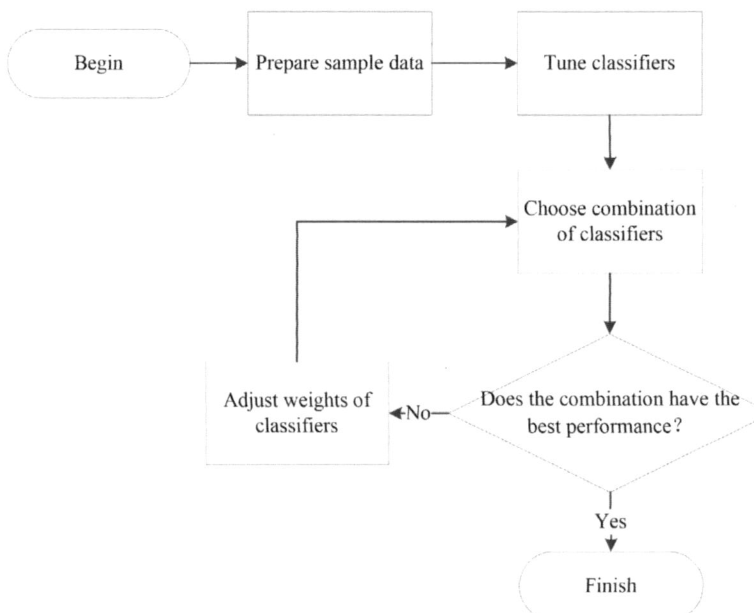

Figure 2. Workflow of Various Classifiers Ensemble Learning Based on Voting.

4. Experiment analysis

This experiment uses imbalanced fault data of signal equipment from one railway bureau signal division, to validate the model proposed above. Experiment data include 643 items covering 10 types of faults. Precision, recall and F-score are used as evaluation index of the model.

Precision is defined as:

$$Precision = \frac{1}{|C|}\sum_{i \in C}\frac{(TP_i + TN_i)\times TP_i}{TP_i + FP_i} \quad (11)$$

Recall is defined as:

$$Recall = \frac{1}{|C|}\sum_{i \in C}\frac{(TP_i + TN_i)\times TP_i}{TP_i + FN_i} \quad (12)$$

F-score is formulated as:

$$F_1 = \frac{2\times Precision \times Recall}{Precision + Recall} \quad (13)$$

TP_i (true positive) means the number of items distributed to this class correctly, TN_i (true negative) means the number of items distributed not to this class correctly, FP_i (false positive) means the number of items wrongly distributed to this class, FP_i (false negative) means the number of items belonged to this class while distributed to other classes, and C represents the number of classes.

The experiment includes two parts. The first part is imbalanced fault text data experiment analysis, and the second part is fault intelligent classification experiment analysis.

4.1. Imbalanced fault text data experiment analysis

This experiment uses Jieba to accomplish word segmentation of signal equipment fault text, and calculates weights and normalizes through TF-IDF, and finally represents fault text as vectors. To validate effects of SMOTE when dealing with imbalanced fault text data, this experiment generates some data in minority through SVM-SMOTE. TDCS equipment fault data are generated from origin 6 items to 172 items, blocking equipment fault data are from 6 items to 84 items, and microcomputer interlocking fault data are from 6 items to 133 items. And the overall data become 1014 items, as shown in table 1.

Table 1. Comparison between raw data and data after SMOTE.

Category	Raw Data	Data after SMOTE
CTC equipment fault	21	21
LKJ equipment fault	15	15
TDCS equipment fault	6	172
Blocking equipment fault	6	84
On-board equipment fault	23	23
Turnover fault	175	175
Power supply panel fault	53	53
Track circuit fault	242	242
Computer interlocking fault	6	133
Signal fault	96	96
Total	643	1014

4.2. Fault intelligent classification experiment analysis

This experiment uses two different datasets (raw data and data after SMOTE) to train and test widely used traditional classifiers (LR, Multinomial NB and SVM), and ensemble classifiers (RF and GBDT). And then compares their performance on precision, recall and F-score, to validate SMOTE's effect of imbalanced fault text data classification. From both two datasets randomly take 80% as training data, and 20% as testing data.

4.2.1. Raw text data classification experiment.

Ensemble classifiers need tuning based on training data to achieve a better performance. This experiment tunes through GridSearchCV. After tuning, the estimator of RF is 180, the estimator of GBDT is 180 as well, learning rate is 0.3, and subsample is 0.8. The performance of tuned ensemble classifiers and base classifiers shows in table 2.

Table 2. Raw data classification performance.

	Algorithm	Precision	Recall	F1
Base Classifiers	LR	0.8001	0.7752	0.7357
	Multinomial NB	0.9136	0.8527	0.8745
	SVM	0.8997	0.8527	0.8489
Ensemble Classifiers	RF	0.9078	0.8837	0.8778
	GBDT	0.9036	0.8760	0.8671

Table 3 shows that for imbalanced data, ensemble classifiers are slightly better than base classifiers. And among base classifiers, logic regression has the worst performance in this experiment, while among ensemble classifiers, RF shows the best performance in classification.

4.2.2. Text data classification experiment after SMOTE.

After using SMOTE to balance data, this experiment retrains and retests classifiers. And then the tuned parameters are the following: RF estimator is equal to 160, while GBDT estimator is 180, learning rate is 0.3, and subsample is 0.6. The classification performance after SMOTE is shown in table 3.

Table 3. Data after SMOTE classification performance.

	Algorithm	Precision	Recall	F1
Base Classifiers	SMOTE + LR	0.8963	0.9015	0.8768
	SMOTE + Multinomial NB	0.9312	0.9261	0.9267
	SMOTE + SVM	0.9538	0.9458	0.9416
Ensemble Classifiers	SMOTE + RF	0.9441	0.936	0.932
	SMOTE + GBDT	0.9253	0.931	0.9294

Table 4 describes that after SMOTE, the performance for both base classifiers and ensemble classifiers is significantly improved, especially SVM.

4.2.3. SMOTE + Voting ensemble classification experiment.

Based on above experiments, choose some classifiers to ensemble based on voting. Among those combinations, ensemble learning based on SVM with other four classifiers voting stands out. And the performance shows in table 4.

Table 4. SMOTE + Voting ensemble classification performance.

	Algorithm	Precision	Recall	F1
SMOTE + Voting Ensemble Learning	SVM + LR + RF	0.9538	0.9458	0.9416
	SVM + LR + GBDT	0.9538	0.9458	0.9416
	SVM + LR + Multinomial NB	0.9538	0.9458	0.9416
	SVM + Multinomial NB+ RF	0.9607	0.9557	0.9528
	SVM + Multinomial NB + GBDT	0.9572	0.9507	0.9474
	SVM + GBDT + RF	0.9517	0.9458	0.9425

Table 4 demonstrates that combinations such as SVM + LR + RF, SVM + LR + GBDT, SVM + LR +Multinomial NB do not improve the final performance, which is just like SVM alone. While ensemble learning with various classifiers such as SVM + Multinomial NB + RF, SVM + Multinomial NB + GBDT, SVM + GBDT + RF has a better performance compared to using SVM alone. And SVM + Multinomial NB + RF performs the best, which improves 1% on precision, recall and F-score.

4.3. Experiment conclusion
Based on above experiment analysis, take SVM classifier as an example, and compare performance of using raw data, data after SMOTE and data after SMOTE + Voting, shown in figure 3.

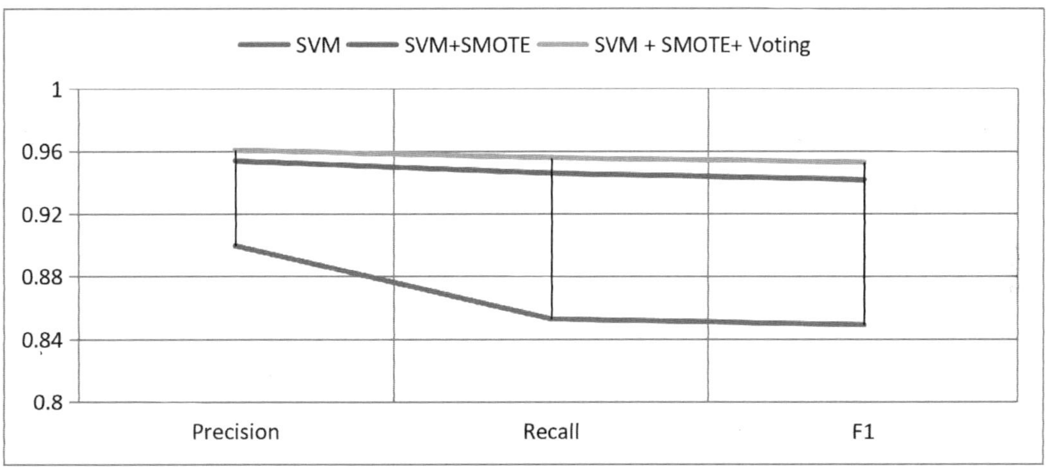

Figure 3. SVM performance comparison of different methods.

From figure 3 it shows that for imbalanced data, SVM+ SMOTE+ Voting method shows the best performance in precision, recall and F1. Therefore, intelligent classification model for railway signal equipment fault text data, which is firstly proposed in this article, has great advantages and can be applied in railway signal equipment.

5. Conclusion
This article uses SVM-SMOTE to automatically generate data in minority at first, balancing fault text data and improving classification performance. And then it applies these relatively balanced new text data to train and test models, and chooses models with competitive performance. Models used as classifiers include base ones such as LR, Multinomial NB, SVM, and ensemble ones such as RF and GBDT. Finally based on those models, this paper proposes a new ensemble learning classification model based on voting method. This article also includes an experiment dealing with imbalanced fault

text data from railway signal equipment of one railway bureau. The experiment validates the model mentioned above through precision, recall and F-score, offering new solution to intelligent classification of railway signal equipment faults.

Acknowledgement
The authors would like to thank the reviewer's great patient to revise the paper and the support of the China Academy of Railway Sciences within the major issues of the fund (no. 2017YJ005 and no.2017YJ02).

References

[1] Dietterich T G 2000 *Mutliple classifier systems*, pp:1-15
[2] He H, Bai Y, Garcia E A, Li S 2008 *IEEE International Joint Conference on Neural Networks*. pp:1322-28.
[3] Chawla N V, Bowyer K W, Hall L O, Kegelmeyer W P 2002 *Journal of Artificial Intelligence Research*, 16: 321-57.
[4] Han H, Wang W Y, Mao B H 2005 *Lecture Notes in Computer Science* 3644:878-887.
[5] Aggarwal C C, Zhai C X 2012 *Mining Text Data*.pp:163-222.
[6] Breiman L 2001 *Machine Learning* 45 (1): 5-32.
[7] Friedman J H 2001 *Annals of Statistics* 29 (5):1189-232.

A new method for short-term traffic congestion forecasting based on LSTM

Ying Zhong[1], Xin Xie[2,*], Jingjing Guo[3], Qing Wang[4] and Songlin Ge[2]

[1]College of Information Science and Engineering, Hunan University, China.
[2]School of Information Engineering, East China Jiaotong University, China.
[3]College of Electronic Information, Wuhan University, China.
[4]Central South University Railway Campus, Central South University, China.

*Email: xiexin@ecjtu.edu.cn

Abstract. Traffic congestion in metropolitan areas such as shenzhen, has become more and more serious. Over the past decades, many academic and industrial efforts have been made to alleviate this issue. In this paper, we propose a novel approach to predicting short-term traffic congestion. At first, we pre-process the data to get the speed, traffic, lane number of these parameters. Second, we carry out statistical data and create training samples. Third, We establish a hybrid neural network prediction model based on LSTM and substitute the generated samples into training. Finally, we use the model to predict the future congestion situation. The experimental results show that our model achieves good predictive results.

1. Introduction

With the steadily growth number of vehicles, road traffic congestion has become an increasingly important problem [1]. In the USA, for example, there were 835 vehicles per 1000 persons in 2015. There are many problems related with this increase, such as noise, pollution, and traffic jams, which lead to time being wasted in travel and productivity loss. Therefore, traffic forecasting, especially 5 short-term traffic forecasting has been paid more and more attention, and has become a key research topic in traffic engineering field.

The rest of the paper is organized as follows. Section 2 reviews the background and related work. Section 3 proposes a new Short-Term Traffic Congestion Forecasting Model. Results and Evaluation is described in Section 4. Finally, Section 5 presents the conclusion.

2. Background and Related Work

As we all know, we can build various models to predict traffic congestion. And we introduce the development of short-term traffic flow forecasting from neural networks and non-neural networks. For non-neural networks, we introduce several representative articles. At first, Lee et al. [2] attempt to find new knowledge between traffic congestion and weather by using big data processing technology. Second, Surya et al. [3] proposed the design of a system that utilizes and efficiently manages traffic light controllers according to the congestion predicted at each line using real time data. At last, Kong et al. [4] proposed a novel approach to estimate and predict the urban traffic congestion using floating car trajectory data efficiently. In this method, floating cars are regarded as mobile sensors, which can probe a large scale of urban traffic flows in real time. As to neural networks, Rohan et al. [5] proposed

Content from this work may be used under the terms of the Creative Commons Attribution 3.0 licence. Any further distribution of this work must maintain attribution to the author(s) and the title of the work, journal citation and DOI.
Published under licence by IOP Publishing Ltd

the significance of Jordan sequential network for prediction of future values, depending upon the current value and aggregate past values. In addition, Fouladgar et al. [6] proposed a decentralized deep learning-based method where each node accurately predicts its own congestion state in realtime based on the congestion state of the neighboring stations. While the related works presented above focus on neural networks and traditional congestion prediction methods, we are the first to combine the advantages of LSTM in timing prediction and the advantages of general neural networks, and using it for congestion prediction.

3. A new Short-Term Traffic Congestion Forecasting Model

3.1. LSTM
In this section, the LSTM model [7], as a particular type of Recurrent Neural Network (RNN), is introduced. LSTM is a special type of RNN that can learn long-term dependency information. It has achieved considerable success and has been widely used. Another variation of the RNN is the Gated Recurrent Unit (GRU), which is proposed by Cho et al. [8].

3.2. Short-Term Traffic Congestion Forecasting Model
We make congestion predictions for each crossroads. The prediction model is shown in figure 1.

Step 1: Clean the original traffic data, and then read the traffic flow P, speed V, lane number data N, and use these data to produce the sample set.

Step 2: we define congestion coefficient $C = k * P / (V * N)$, and k is the constant. According to the congestion coefficient C, the traffic situation could be divided into A: very congested B: congestion C: general congestion D: slightly congested E: not congested.

Step 3: For the designated crossroads, lane number data N is constant. So we model the speed and traffic flow in time and space. For example: we can get the p_{t+1} according to the in front of several moments p_t ,p_{t-1} ,p_{t-2} ,p_{t-3} ,p_{t-4} and the p_{s+1} according to the near the crossroads p_s ,p_{s-1} ,p_{s-2} ,p_{s-3} ,p_{s-4} . In the same way we can get v_{t+1} and v_{s+1} .

Step 4: Get the input layer $x = [p_{s+1} , p_{t+1} , v_{s+1} , v_{t+1}]$, and the five-dimensional vector p_{s+1} , p_{t+1} , v_{s+1} , v_{t+1} could be spliced into a 20-dimensional vector, as the neural network input.

Step 5: The model has two layers of hidden layers which use the sigmoid activation function, with 64 and 128 neurons, respectively.

Step 6: The output layer uses the softmax activation function, and the loss function uses the cross entropy function.

Step 7: Use the batch gradient drop method to training.

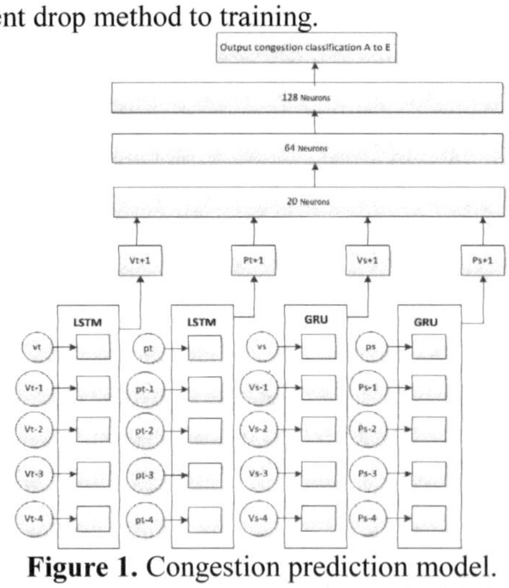

Figure 1. Congestion prediction model.

4. Results and Evaluation

4.1. Experimental Setting

In this paper, we use HUAWEI Official Data set to verify the validity of our proposed model. The data set is collected by the camera at all crossroads in Shenzhen, and consists of 36 fields. Among them, the field includes the license plate number, speed, car color, lane number and other information, which originates from 2017 Huawei Developer Contest [9]. We have uploaded all the experimental code to this site [10] . We use Tensorflow and python to realize our model.

4.2. Experimental Results and Evaluation

The data in the data set is provided with data from all the intersections of Shenzhen May 11 to May 20, 2017. We have to predict the traffic jam at the crossroads numbered 10300202. So, we only need to deal with the designated intersection and its adjacent five junctions of information. We divided the data for 10 days into the first eight days and the next two days. The first eight days of data we used to make samples, and then used to train the model. Two days after the data used to calculate the accuracy rate and other indicators, and to verify our model.

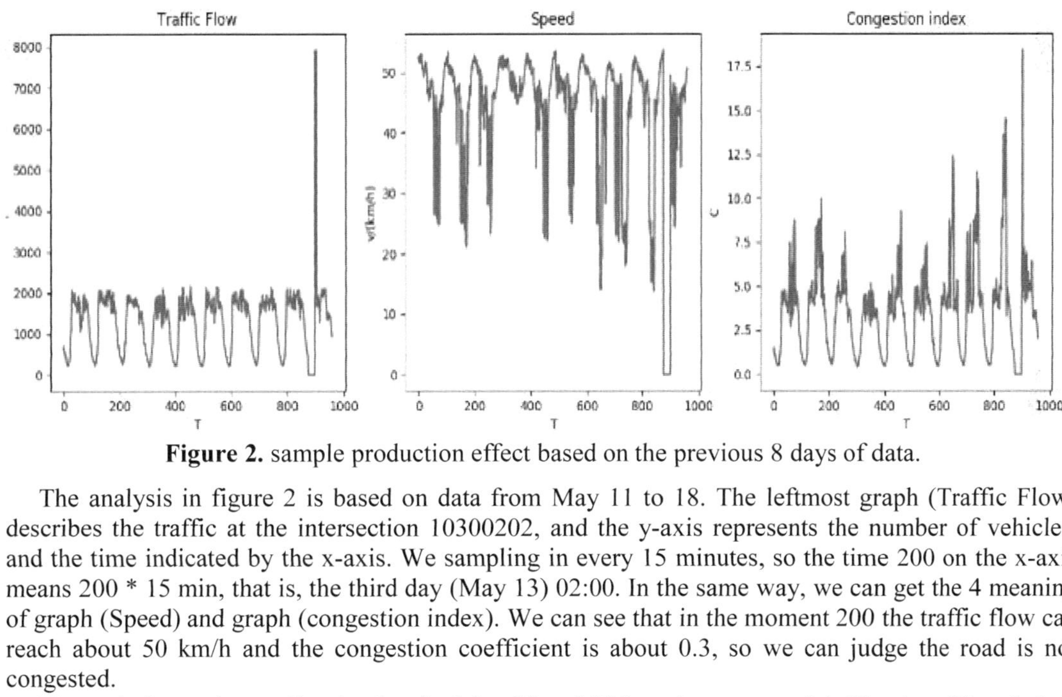

Figure 2. sample production effect based on the previous 8 days of data.

The analysis in figure 2 is based on data from May 11 to 18. The leftmost graph (Traffic Flow) describes the traffic at the intersection 10300202, and the y-axis represents the number of vehicles, and the time indicated by the x-axis. We sampling in every 15 minutes, so the time 200 on the x-axis means 200 * 15 min, that is, the third day (May 13) 02:00. In the same way, we can get the 4 meaning of graph (Speed) and graph (congestion index). We can see that in the moment 200 the traffic flow can reach about 50 km/h and the congestion coefficient is about 0.3, so we can judge the road is not congested.

Figure 3 shows the predicted value for May 19 and 20 based on our model. The time 80 on behalf of 80 * 15 min, that is, 20:00 May 19. In this time, the speed of is 20km/h below, and congestion coefficient of 11 or so. Thus, the road is very congested, since it is the get off work late peak. According to the predicted value and the original value, we can calculate the values of Precision, Recall and F1-Measure are 0.984, 0.975, and 0.979 respectively. The experimental results show that our model achieves good predictive results.

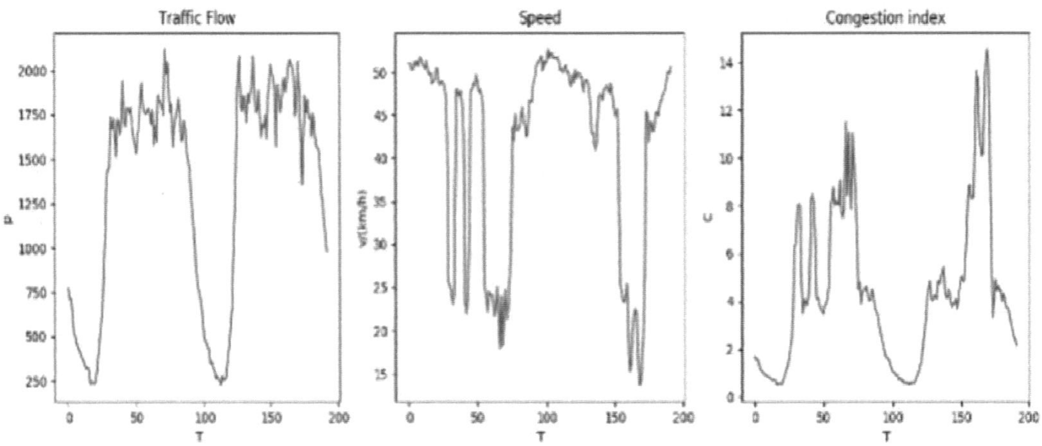

Figure 3. The effect of congestion prediction based on data for the next 2 days.

5. Conclusion

In this paper, we propose a novel approach to predicting short-term traffic congestion. At first, we pre-process the data to get the speed, traffic, lane number of these parameters. Second, we carry out statistical data and create training samples. Third, We establish a hybrid neural network prediction model based on LSTM and substitute the generated samples into training. Finally, we use the model to predict the future congestion situation. The experimental results show that our model achieves good predictive results.

Acknowledgments

The work is supported by the National Natural Science Foundation of China(61762037), Science and Technology Support Program of Jiangxi Province(20151BBE50055), and Science and Technol-ogy Project supported by education department of Jiangxi Province(GJJ150541), and Nanchang City Knowledge Innovation Team(2016T75).

References

[1] Wang F Y. Parallel Control and Management for Intelligent Transportation Systems: Concepts, Architectures, and Applications[J]. IEEE Transactions on Intelligent Transportation Systems, 2010, 11(3):630-638.

[2] Lee J, Hong B, Lee K, et al. A Prediction Model of Traffic Congestion Using Weather Data[C]. IEEE International Conference on Data Science and Data Intensive Systems. IEEE, 2016:81-88.

[3] Surya S, Rakesh N. Flow based traffic congestion prediction and intelligent signalling using Markov decision process[C]. International Conference on Inventive Computation Technologies. IEEE, 2017:1-6.

[4] Kong X, Xu Z, Shen G, et al. Urban traffic congestion estimation and prediction based on floating car trajectory data[J]. Future Generation Computer Systems, 2016, 61(C):97-107.

[5] More R, Mugal A, Rajgure S, et al. Road traffic prediction and congestion control using Artificial Neural Networks[C]. International Conference on Computing, Analytics and Security Trends. IEEE, 2017:52-57.

[6] Fouladgar M, Parchami M, Elmasri R, et al. Scalable Deep Traffic Flow Neural Networks for Urban Traffic Congestion Prediction[C]. International Joint Conference on Neural Networks. 2017:2251-2258.

[7] Gers F A, Schmidhuber J, Cummins F. Learning to forget: continual prediction with LSTM.[C]. Artificial Neural Networks, 1999. ICANN 99. Ninth International Conference on. IET,

2002:2451.
[8] Cho K, Merrienboer B V, Gulcehre C, et al. Learning Phrase Representations using RNN Encoder-Decoder for Statistical Machine Translation[J]. Computer Science, 2014.
[9] http://developer.huawei.com/ict/forum/thread-21745.html
[10] https://github.com/0zhongying0/ICSTTE-paper-zhongying00

Research on airport trailer emergency scheduling model based on genetic simulation annealing algorithm

Zhiqiang Zhou, Shi Liu * and Kunyuan Huang

College of Computer,Inner Mongolia University, Hohhot, China

*Email: liushi@imu.edu.cn

Abstract. With the increase of air transport business year by year, the airport comprehensive service capabilities are increasingly facing severe challenges. As a key part of airport service, ground scheduling is crucial to the departure of flights on time. by analyzing the traditional rules of airport trailer and combining with the service processes and the characteristics of trailers, the paper develops a multi-target trailer emergency scheduling model with rolling windows. When the model is solved, the simulation annealing algorithm is introduced into the genetic algorithm to prevent the optimal result falling into the local optimal solution. The simulation results show that the model can more effectively solves the problem of airport trailer emergency scheduling in the case of flight delays than the traditional scheme,and to relieve the pressure of flight delay for busy airports.

1. Introduction

With the increase of air transport business year by year, the airport comprehensive service capabilities are increasingly facing severe challenges. As a key part of airport service, ground scheduling is crucial to the departure of flights on time.

At present, most of the airport flight scheduling service is still in the manual scheduling phase[1], adopting the First in first out scheduling strategy.Although the operation is simple,efficiency is low. It is obvious no value for solving the flight delay when the weather suddenly changes,flight delays, Special vehicle breakdown and other emergencies,Especially.the ground scheduling of trailers during the flight access to station operation is studied by Jiayan Du [2-3],developing a VRPTW model according to the operation flow of the towing vehicle and Minimizing the total scheduling time cost of the entire process .simultaneously,Xia Feng[4]has established the coordinated scheduling model of refuel tanker and shuttle bus with taking the refueling service and pick-up service time of flights as the constraint,and aimed at the objective of minimizing the number of safeguards and making the total starting time of service earliest, and given model solution based on multi-objective genetic algorithm.

The paper develops a trailer emergency scheduling model, by analyzing the traditional rules of the trailer scheduling based on the existent researches and related papers,and combining with the service flow and time characteristics of the trailer,then gives the optimized scheduling scheme.

2. Traditional trailer scheduling strategy

Now, most airports adopt the First in first out strategy in the airport trailer scheduling [5].The strategy,which is to select a grouping scheme from the scheduling system,and the grouping which can be completed in the shortest time t_i in the GPS system corresponding to the scheduling system is selected for transmission. Among all the grouping schemes waiting to be scheduled (see figure 1).It is

Content from this work may be used under the terms of the Creative Commons Attribution 3.0 licence. Any further distribution of this work must maintain attribution to the author(s) and the title of the work, journal citation and DOI.
Published under licence by IOP Publishing Ltd

confirmed that the time which takes for the packet completion service lags behind the time it takes to transmit the longest packet on the process ,comparing with the corresponding GPS system [6].

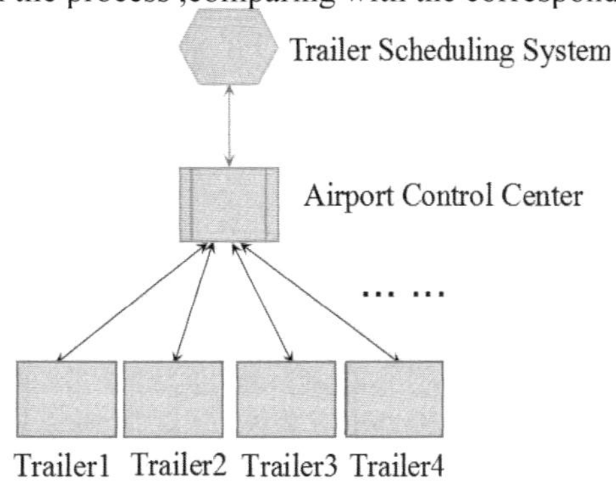

Figure 1. The structure of traditional trailer scheduling.

When a trailer group arrives,it will be sorted into the corresponding queue by classification.the scheduling system will calculate the completion time of the process,at the same time [7].Then select the packet with the shortest completion time among the packets as the next packet transmitted from the output port, and the packet period is T.

Among them, the scheduling time is calculated as follows: $v(0) = 0$, and

$$v(t_{j-1}+T) = v(t_{j-1}) = T / \sum_{i \in B} \phi_i, T \le t_j - t_{j-1}, j = 2,3,\cdots \tag{1}$$

In the above equation(1),the scheduling of the trailer as one event, and t_j represents the time of the j th event finished .

The complexity of the scheduling system is expressed as $O(M)$, and M is the number of connected trailers.In the scheduling process, the (σ_i, r_i) is used as the constraint of scheduling algorithm, and the node trailer scheduling delay as follows equation(2) [8]:

$$f(\sigma_i, r_i) = \frac{\sigma_i}{r_i} + \frac{L_{max}}{C} \tag{2}$$

Among them, C is the service rate of the server,and L_{max} is the maximum of the packet length of vehicles.

3. Trailer emergency scheduling model

The paper develops a trailer emergency scheduling model,and gives the optimized scheduling scheme.

3.1. putting forward of the problem

According to the conventional ground specific vehicle scheduling strategy, the nearest trailer is assigned at the current time, but the actual distance between the current failed trailer and the next trailer in the group is ignored. And the trailer's work flow is chained with strict scheduling rules and time constraints. A change of a link or a group of trailers may cause subsequent trailers to fail to schedule normally and affect the normal launch of subsequent flights. In order to efficiently schedule flights under the condition, it is necessary to optimize the scheduling strategy and improve the level of ground scheduling and security.

So, the paper develops a multi-target trailer emergency scheduling model with rolling windows.and gives a certain number of delayed flights and serviceable trailers, we construct a coordinated emergency scheduling model for all kinds of trailers with flight delays and assign them to service needed flights according to the traction of trailers, and gives the start time of services in the process of scheduling.

3.2. Model building

3.2.1. Model description and symbolic description

In this paper, the rolling window strategy is used to divide the flight delay time into multiple equal time periods. After that, the flights in each time period are optimized according to their planned take time-off time and delay time.In this way, not only can the scheduling plans of delayed flights be developed quickly, but also the real-time processing of flight information can be ensured.

In the actual operation of the airport scheduling, the trailers are divided into three categories called small, medium and large.The specific information as the table1 show.

Table 1. the correspondence between flight models and the type of trailers.

Aircraft type	Representative	Carrying capacity	trailer type
Large airliner	B747/A340/A380	Over 200 people	Large
Medium airliner	B737/C919/A320	Between 100and200	Large/Medium
Small airliner	ARJ21/MA60	less than 100 people	All

Suppose the step size of the rolling window is L, the starting time value is Q, the rolling window is RW_t at time t is taken as[Q+(t-1)*L,Q+t*L].According to graph theory, the multi-target trailer emergency scheduling model can be abstractly represented as a graph G=(FG,EG,SE,NG),when the flight is delayed.Where FG represents the number of delayed flights N in the rolling window RW_t; EG represents the path from the trailer to the delayed flight position in the rolling window RW_t (N scheduling paths are generated for N delayed flights); SE represents the window on which the towing flight is trailed in the rolling window RW_t; NG indicates the number of the trailer in the rolling window RW_t(the total numbers of the trailers is p).For each late flight $i \in$ FG, there is a latest departure time O_i (the flight will be excluded in the current window RW_t beyond which the latest departure time) and the time window $SE_i = [s_i, e_i]$ $(e_i < O_i)$ [8].

For each scheduling path k, s_{kj} represents the time when the trailer j starts to schedule a delayed flight in the flight schedule path k, that is the trailer j arrives at the scheduled downtime. If there is a trailer j in the scheduling path k, then $x_{kj} = 1$, otherwise $x_{kj} = 0$.

3.2.2. Model assumptions

In the following study in this paper, for the delay of departure flights, the establishment is based on the following assumptions:

- Each flight happens to be towed once.
- The flight has been completed refueling, boarding and other work, the trailer towed to the taxiway can be released takeoff.
- During the process of receiving the towing service, the towing vehicle will not malfunction and other events that affect scheduling will occur.
- Once the trailer has started its work, it will not be interrupted until the end of the service.

3.2.3. trailer emergency scheduling model constraints

Trailer emergency scheduling model constraints is extremely limited and in order to operate safely, all kinds of special vehicles must travel at a fixed path and speed.in the flight scheduling process, the use of the trailer must be in accordance with the "flight-trailer" corresponding mode to draw the flight.

Second, the number of trailers, the departure time of the flight, the constraints in the rolling window [Q + (t-1) * L, Q + t * L] are now defined as follows.

Due to delayed flight and scheduling path one by one, so there $x_{kj} = x_{ij}$. For subsequent convenience and verification, the following x_{kj} is indicated by x_{ij} . The number of trailers and delayed flights as follows equations(3)-(4) :

$$\sum_{a=1}^{3} p_a = P \ , \sum_{b=1}^{3} n_b = N \ , i = 1,2,3 \cdots N , j=1,2,3 \cdots P . \tag{3}$$

$$\sum_{i=1}^{N} \sum_{j=1}^{P} x_{ij} \le P \tag{4}$$

Time constraints in the scheduling process are as follows equations (5)-(7):

$$\sum_{i=1}^{N} x_{ij} \le O_i, \ \forall i \in FG \tag{5}$$

$$e_{i-1} + T_{i+1} < s_i < e_i \tag{6}$$

$$Q + (t-1) * L < s_i < e_i < Q + t * L, i \in EG \tag{7}$$

3.2.4. Trailer emergency scheduling optimization goals

According to the actual operation of the airport and the emergency scheduling model of the above trailer, the various types of trailers under the condition of flight delay are optimized to achieve the goal of minimizing the total service time of all delayed flights,as follows equation (8).

$$T_{\min} = \sum_{i=1}^{N} \sum_{j=1}^{P} x_{ij} \cdot T_{ij} \tag{8}$$

3.3. Model solution

According to the above model, the key to solve the delay flight scheduling path problem is to reasonably determine the relationship between the flight and each trailer, to minimize the total delay time when the corresponding types of flights and trailers and the departure time of flight are satisfied, and to avoid genetic algorithm (GA) is easy to fall into the condition of local optimization and slow convergence of simulation annealing algorithm (SAA), and a model based on simulation annealing genetic algorithm to solve the emergency scheduling of trailers is constructed [9-11].

3.3.1. Algorithm steps

The specific idea of simulation annealing genetic algorithm is as follows:

Step 1: Initializing the control parameters. The population size is N , the mutation probability is P_m , the initial annealing temperature is T_0 , the temperature cooling parameter is α .

Step 2: Construct chromosomes $C_h(h = 1,2,\cdots N)$ to generate initial population.For example, chromosomes {6, 2, 5, 4, 1, 3} indicate that flight number 1 is the 6th towing service, flight number 2 is the second towing service, and so on.

Step 3: Calculate fitness.For each chromosome C_h generated in the population, to determine whether the corresponding feasible solution after decoding.If the corresponding solution is feasible, the corresponding objective function value Z_h is obtained.If the chromosome C_h corresponds to a non-feasible solution, a large integer M is assigned.The fitness function $f_h = 1/Z_h$, f_h is the greater the value, indicating that the algorithm is close to the optimal solution.

Step 4: Chromosome cross.Select two individuals C_i and C_j randomly and perform partial matching crossover(PMX)to generate two new individuals C_i' and C_j' , and then calculate their fitness function values f_1' and f_2'. If $\min\{1, \exp(-(f_h - f_h')/T_k)\} > b$, b is a random number between 0 and 1, then the current individual is received;

Step 5: The cross after the above-mentioned individual mutation operation, combined with the fourth step of the method to determine whether to receive the variant solution.

Step 6: to determine whether to meet the convergence conditions. Otherwise, $T_{k+1} = \alpha * T_k$,turn to the step 3,and continue to find the optimal solution.

4. Application examples

Based on the above trailer emergency scheduling model, this section will use the delayed flight data as an example that to verify the reliability of the scheming model,and comparing with the conventional trailer scheduling method. The scheduling time for each flight is assumed to be 10 minutes.Specific flight information as follows table 2.

Table 2. Delay flight information at 19:20 on day.

Flight number	Flight type	Original schedule time	Current Flight schedule	Current Trailer state	Flight number	Flight type	Original schedule	Current Flight schedule	Current Trailer state
CA1302	large	17:30	20:30	large1	CZ2203	large	19:50	20:40	
CZ2201	Medium	17:40	19:25	Medium3	CA1322	Small	20:05	20:45	
CF3105	Medium	17:55	19:30	Medium4	MF1268	Medium	20:20	20:50	
HU6681	Small	18:10	19:40	Small6	CF3109	Medium	20:25	21:00	
CA1308	Small	18:25	19:45	Small7	CA1338	Medium	20:32	21:10	
MF1241	Small	18:40	19:50	Small8	CZ2204	Small	20:45	21:15	
CZ2202	Small	18:55	19:55	Medium5	CA2356	Small	21:10	21:20	
CA1307	Medium	19:15	20:10	large2	CF3111	large	21:30	21:30	
CF3106	large	19:20	20:15		HU6688	Medium	21:35	21:35	
HU6682	Small	19:30	20:35		CZ2205	large	21:40	21:40	

According to the traditional "First in first out" strategy, the trailer number 1-8 is a scheduling group.The actual scheduling time as show in the table 3.

Table 3. The traditional scheduling strategy actual scheduling situation.

Flight number	Schedule trailer	The actual scheduling time			Flight number	Schedule trailer	The actual scheduling time		
		First group	Second Group	Third group			First group	Second Group	Third group
CA1302	large1	20:30			CZ2203	large1		20:50	
CZ2201	Medium3	19:25			CA1322	Small7		20:55	
CF3105	Medium4	19:30			MF1268	Medium3		21:00	
HU6681	Small6	19:40			CF3109	Medium4		21:05	
CA1308	Small7	19:45			CA1338	Medium5		21:10	
MF1241	Small8	19:50			CZ2204	Small8		21:15	
CZ2202	Medium5	19:55			CA2356	Small6			21:25
CA1307	large2	20:10			CF3111	large2			21:30
CF3106	large2		20:40		HU6688	Medium3			21:35
HU6682	Small6		20:45		CZ2205	large1			21:40

According to the table 2 and table 3,we can know that total actual delay time of the flight :75min.

However,according to the proposed cooperative scheduling of various types of trailers based on genetic simulated annealing algorithm for scheduling flights, set L = 30min, N = 20, the mutation probability $P_m = 0.005$, initial temperature $T_0 = 100^\circ C$, the actual scheduling situation is as follows figure 2 and table 4.

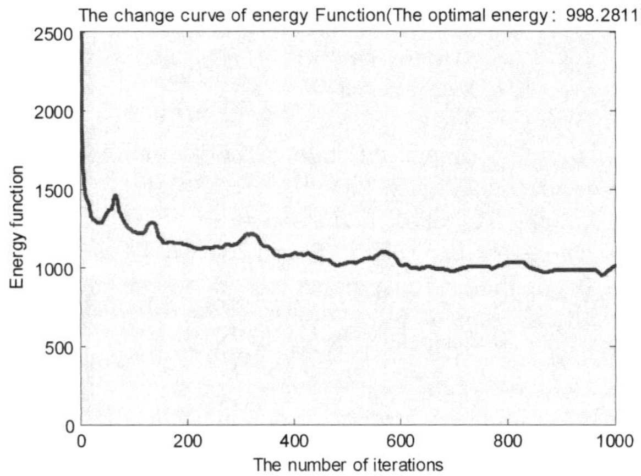

Figure 2. The change curve of energy Function.

Table 4. The trailer coordinated emergency scheduling plan actual scheduling situation.

Flight number	Schedule trailer	Scheduling time Window				Flight number	Schedule trailer	Scheduling time Window			
		1	2	3	4&5			1&2	3	4	5
CA1302	large2			20:30		CZ2203	large1		20:40		
CZ2201	Medium3	19:25				CA1322	Small8		20:45		
CF3105	Medium4	19:30				MF1268	Medium4		20:50		
HU6681	Small6	19:40				CF3109	Medium5			21:00	
CA1308	Small7	19:45				CA1338	Medium3			21:10	
MF1241	Small8	19:50	20:00			CZ2204	Small6			21:15	
CZ2202	Small6		20:05			CA2356	Small7			21:20	
CA1307	Medium3		20:10			CF3111	large1				21:30
CF3106	large1					HU6688	Medium4				21:35
HU6682	Small7			20:35		CZ2205	large2				21:40

According to the table 2 and table 4,can see that the above trailer emergency scheduling method which is only a total delay of 15 minutes on the basis of the current flight schedule.

5. summary

By analyzing the traditional rules of airport trailer and combining with the service processes and the characteristics of trailers,the paper develops a multi-target trailer emergency scheduling model with rolling windows.And the simulated annealing algorithm is introduced into the genetic algorithm to prevent the optimal result falling into the local optimal solution when the model is solved. The simulation results show that the model can more effectively solve the problem of airport trailer emergency scheduling in the case of flight delays.In the future research work,we can refine the relevant factors and go further quantitative analysis to improve the efficiency of airport scheduling.

References

[1] Makhloof M A,Waheed M E and Badawi U A E R 2014 Real-time aircraft turnaround

operations manager *Production Planning&Control* 25 (1) pp 2-25
[2] Du J Y,Jens B 2014 Planning towing processes at airports more efficiently Transportation Research *Logistics and Transportation Review* 70 (1) pp 293-304
[3] Angus C; Ip W H,Lu D 2005 An aircraft service scheduling model using genetic algorithms *Journal of Manufacturing Technology Managemen*t 16 (1) pp 109–19
[4] Feng X, Ren Z Y 2016 Research on collaborative scheduling of refueling trucks and ferry pushes based on genetic algorithm *Journal of Transportation Systems Engineering and Information Technology* 16 (2) pp 155-63
[5] Sun D W, Chang GR; Wang C,Xiong Y,Wang XW 2010 Efficient Nash equilibrium based cloud resource allocation by using a continuous double auction *International Conference on Computer Design and Applications* pp 194-99
[6] Jing H M ,Zhang L J 2007 Modeling and simulation of vehicle scheduling problem with multiple vehicles *Computer simulation* 17(8) pp 261-64
[7] Lu J W 2010 Preferred method of test nodes based on genetic particle swarm compound arithmetic *Computer Measurement & Control* 54 (2) pp 1036-38
[8] Wang J H,Shi M Y,Wang T T 2012 Coverage optimization of mobile nodes based on QPSO *Microelectronics & Computer* 33 (2) pp 96-99
[9] Zhang Y X,Qi Y X 2017 Improved genetic simulated annealing algorithm for solving TSP *Intelligent Computer and Application* 7(3) pp 52-54
[10] He Q,Wu Y L 2017Application of Improved Genetic Simulated Annealing Algorithm in TSP Optimization *Control and decision making* 13 (10) pp 1-7
[11] Lin Z L 2009 Coverage optimization strategy of wireless sensor networks based on particle swarm optimization *Computer simulation* 26 (4) pp 190-93

IOP Conf. Series: Materials Science and Engineering **383** (2018) 012045 doi:10.1088/1757-899X/383/1/012045

Express/local train plan optimization for urban rail transit in condition of full-length and short-turn modes

LUO Qin[1,2], HOU Yufei[2], LI Wei[2,3,*] and ZHANG Xiongfei[1,2]

[1]College of Urban Traffic and Logistics, Shenzhen Technology University, Lantian Road 3002, Shenzhen, P.R.China
[2]Shenzhen Key Laboratory of Urban Rail Transit, Shenzhen University, Nanshan Ave 3688, Shenzhen, P.R.China
[3]Key Laboratory of Optoelectronic Devices and Systems of Ministry of Education and Guangdong Province, College of Optoelectronic Engineering, Shenzhen University, Nanshan Ave 3688, Shenzhen, P.R.China

*E-mail: aliweib1@126.com

Abstract. Operating the express/local train plan can meet the characteristics of imbalance of passenger volume temporal and spatial distribution, and satisfy the demands of long-distance commuters who want to travel fast. Although the program can improve the organizational operating efficiency and passenger service level, it will reduce the carrying capacity of system to some extent. Above all, based on analysing the factors of the express/local train plan and the composition of passenger travel time in detail, the 0-1 integer programming model of stop-schedule plan problem was established by minimizing the passenger travel time. And combined with a simulation case, the immune genetic algorithm was designed to study the effect of line capacity.

1. Introduction

With the continuous expansion of urban scale and the development of urban rail transit network, the round-trip passenger flow between the city centre and suburban grows gradually. As the central city has been saturated basically, more and more companies and residential buildings tend to transfer to the suburbs, which further promotes the swelling of the traffic flow. Therefore, cities need to use the express and local train operating mode to meet the demands for the commuting passenger. In this mode, how to ensure the service level of rail transit, and provide the foundation and safeguard for rail transport operating enterprises working better, is a significant project.

Following are some representative studies about the express/local train plan. Wang L et al.[1] studied the operational efficiency of the rail transit system, and proposed the optimal goal of the express/local train plan on the relationship between the stop scheme and the operational efficiency. Altazin E et al. [2] presented that stops on train lines can be skipped to minimize both the recovery time and the waiting time of passengers. Cheng J [3] considered both the total travel time of the passengers and operating costs to build up a multi-objective optimization model, and designed an algorithm to solve the problem. Hassannayebi E et al. [4] studied the optimal time table of express/local train and built up a model in order to reduce the passenger waiting time by using variance and penalty function Pan H C [5] focused on the analysis of the influence of the carrying capacity of express/slow train multi-operation, then discussed the relations between the proportion of

Content from this work may be used under the terms of the Creative Commons Attribution 3.0 licence. Any further distribution of this work must maintain attribution to the author(s) and the title of the work, journal citation and DOI.
Published under licence by IOP Publishing Ltd

express/slow train and the maximal capacity. Han Y et al. [6] established to determine the reasonable opening train number between two major cities, train passenger capacity, the train stations along the number of stops and parking opportunities mathematical model to make the total time delay to minimum and railway enterprises benefit the most. Liao J Q [7] respectively devised the operation scheme in condition of the single route and the full-length & short-run route aiming at the maximum travel saving time of all passengers. Gao Y et al. [8] aimed to find the relationship among energy consumption, travel time and timetables, and propose a bi-objective programming model to better understand the relationship between lowering energy consumption and reducing travel time.

In this paper, by analysing the different passenger trip time because of choosing different types of trains, we propose to reduce total travel time as the target, and accordingly, make a reasonable stop plan of the express train.

2. Analysis of Passenger Flow Characteristics and Effects

2.1. Passenger Flow Characteristics

The long-distance regional rail transit line is connected in series with urban areas, suburbs and satellite towns, and the intercity passenger flow is not only large, but also imbalanced in time direction. During the morning rush, the traffic flow moves from satellite towns or suburbs to the downtown, and at the evening peak, the situation is opposite. Meanwhile, with different levels of development along the rail transit, the boarding of stations is also inconsistent, and overall, it shows the disproportion of the section flow. Consequently, two types of the fluctuating passenger flow have demands for regional express rails. On the one hand, Commuter flow prefers to take the through train from downtown area to the suburban. On the other hand, short-distance passengers want to achieve that any two stations can be reached directly.

2.2. Travel Time Cost

Although operating express/local train makes the total trip time of passengers who take local train to be prolonged because of overtaking, it saves the time of those who take express train.

2.3. Carrying Capacity

Express and local train scheme will reduce the capacity of lines. In the traditional pattern, the line capacity is only related to the minimum headway I, and its calculating method is

$$n_r = 3600/I \tag{1}$$

However, the carrying capacity of express/local train take the deduction coefficient method for calculation and analysis. The deduction coefficient of the express train refers to the number of the local cars deducted from the parallel operation chart because of drawing an express train, and the formula is

$$\varepsilon_e = 1 + \left(t_d/I\right) - k \tag{2}$$

where, t_d is the difference of travel time of express/local train, and k means the times which are crossed of slow trains [9].

3. Optimization Model and Algorithm

3.1. Problem Description

Operating express/local train combination plan is mainly for long-distance passengers, it can not only shorten the travel time of those who take the express train, but also increase their time benefit. At the same time, this plan also needs to consider the problem that crossed by the express train will increase the waiting time and stop yielding time of people taking local trains at the stations. Based on the above analysis, this paper studies the express train stop scheme of express/local train combination plan in condition of the full-length & short-run route. As well, we make use of the time benefit of passengers and traffic capacity as the key index to evaluate the superiority of the plan.

3.2. Model Assumption

The model selects the morning peak as the optimization period to study. It uses the mixed-routing mode, and the long train routing operates both of local and express trains, while the short train routing only operates local trains. Two kinds of trains own the same technical parameters. Regardless of the circulation of rolling stock, the model only discusses the one-way situation. Every station can support trains to turn back. Passengers arriving can all be considered to obey normal distribution, because people are generally reluctant to transfer other trains, the model define when the passenger's travel starting point and the final station are slow stations, the passengers only wait for the slow, when the passenger's starting point and the end of the station is only a slow station, the passengers only wait for the slow, besides, when the passenger's starting point and the end of the station are fast stations, the passengers can either choose express or slow trains. The last assumption is that every station can support overtaking.

3.3. Modeling

Passenger travel time t includes waiting time t_w and running time t_r. In which, because of arriving time of passengers obey normal distribution, their average waiting time is half of the interval.

$$t = t_w + t_r \tag{3}$$

According to the different nature of OD, the total passenger travel time is divided into four categories, and model parameters are listed in table1.

Table 1. Model Parameters.

Parameters (unit)	Explanation
f_1	The operating frequency of the slow train of the long routing
f_2	The operating frequency of the express train of the long routing
f_3	The operating frequency of the slow train of the short routing
q_{ij-m}^{F-l} (person)	The number of passengers waiting to the mth train which is the slow train of the long routing from the ith station to the jth station
q_{ij-m}^{F-e} (person)	The number of passengers waiting to the mth train which is the express train of the long routing from the ith station to the jth station
q_{ij-m}^{S-l} (person)	The number of passengers waiting to the mth train which is the slow train of the short routing from the ith station to the jth station
T (min)	The length of the optimized period
h_k	The increased delay time because the slow train is crossed by the express train at preceding stations
x_k	Boolean variable. When the slow train is crossed by the express train at the kth station, the value is one, otherwise, the value is zero.
t_{ij} (min)	The interval running time of trains
t_s (min)	The train stopping time
h_w	The increased stopping time because the slow train is crossed by the express train at this station
x_w	Boolean variable. When the slow train is crossed by the express train at the wth station, the value is one, otherwise, the value is zero.
$c_1 / c_2 / c_3$	Boolean variable. When passengers take the slow train of the long routing, the value of c_1 is one, others is zero, when passengers take the express train of the long routing, the value of c_2 is one, others is zero, and when passengers take the slow train of the short routing, the value of c_3 is one, others is zero.
$t_3^{(l)}$ (min)	The travel time of passengers who choose to take slow trains

$t_3^{(e)}$ (min)	The travel time of passengers who choose to take express trains
q_{ij-m} (person)	The number of passengers waiting to the mth train
$\eta_{\max}$	Maximum section load factor
a	Passenger capacity of the train

1) When the passenger travel OD belongs to the following four cases, the passenger can only choose the slow trains of the long route, and the time is as formula (4):

- The starting point i (i=1, 2, 3, …n) is a slow station of the long route, the final station j is any type of station.
- The starting point i is an express station of the long route, the final station j is a slow station of long and short route.
- The starting point i is a slow station of the short route, the final station j is a slow or express station of long route.
- The starting point i is an express station of the short route, the final station j is a slow station of long route.

$$t_1 = \sum_{m=1}^{f_1} q_{ij-m}^{F-l} \cdot \left(\frac{\mathrm{T}}{2f_1} + \sum_{k=1}^{i-1} h_k x_k + + t_{ij} + \sum_{w=i+1}^{j-1} \left(t_s + x_w h_w \right) \right) \quad (4)$$

2) When the passenger travel OD belongs to the following two cases, the passenger can choose the slow or express trains of the long route, and the time is as formula (5).

- The starting point i is an express station of the long route, the final station j is an express station of long and short route.
- The starting point i is an express station of the short route, the final station j is an express station of the long route.

3) When the starting point i and the final station j are both the express station of the short route, the passenger can choose the slow or express trains of any type of station, and the time is as formula (6).

$$t_2 = \left(\sum_{m=1}^{f_1} q_{ij-m}^{F-l} + \sum_{m=1}^{f_2} q_{ij-m}^{F-e} \right) \cdot \left(\frac{\mathrm{T}}{2\left(c_1 f_1 + c_2 f_2\right)} + \sum_{k=1}^{i-1} c_1 h_k x_k + t_{ij} + \sum_{w=i+1}^{j-1} \left(t_s + c_1 x_w h_w \right) \right) \quad (5)$$

$$t_3 = t_3^{(l)} + t_3^{(e)} \quad (6)$$

$$\begin{cases} t_3^{(l)} = \left(\sum_{m=1}^{f_1} q_{ij-m}^{F-l} + \sum_{m=1}^{f_3} q_{ij-m}^{S-l} \right) \cdot \left(\frac{\mathrm{T}}{2\left(c_1 f_1 + c_3 f_3\right)} + \sum_{k=1}^{i-1} h_k x_k + t_{ij} + \sum_{w=i+1}^{j-1} \left(t_s + x_w h_w \right) \right) \\ t_3^{(e)} = \left(\sum_{m=1}^{f_2} q_{ij-m}^{F-e} \right) \cdot \left(\frac{\mathrm{T}}{2f_2} + t_{ij} + \sum_{w=i+1}^{j-1} t_s \right) \end{cases}$$

$$t_4 = \left(\sum_{m=1}^{f_1} q_{ij-m}^{F-l} + \sum_{m=1}^{f_3} q_{ij-m}^{S-l} \right) \cdot \left(\frac{\mathrm{T}}{2\left(c_1 f_1 + c_3 f_3\right)} + \sum_{k=1}^{i-1} h_k x_k + t_{ij} + \sum_{w=i+1}^{j-1} \left(t_s + x_w h_w \right) \right) \quad (7)$$

4) When the passenger travel OD belongs to the following two cases, the passenger can only choose the slow trains of the long or short route, and the time is as formula (7):

- The starting point i is a slow station of the short route, the final station j is the express or local station of short route.
- The starting point i is an express station of the short route, the final station j is the local station of short route.

In summary, we establish the following objective function to achieve the minimization of the total passenger travel time.

$$\min Z = \min \left(\sum_{i=1}^{n-1} \sum_{j=i+1}^{n} t_1 + t_2 + t_3 + t_4 \right) \quad (8)$$

$$s.t. \quad q_{ij-m} \le \eta_{\max} \cdot a$$

IOP Conf. Series: Materials Science and Engineering **383** (2018) 012045 doi:10.1088/1757-899X/383/1/012045

$$f_d \geq 1 \quad (d=1, 2, 3)$$

3.4. Solution Algorithm

The evolutionary algorithm, which is represented by genetic algorithm, is an iterative search algorithm which imitates the biological evolution mechanism. It uses the crossover and mutation operator to realize the information interaction and local search between individuals in the group, and provides optimization opportunities for each individual. By survival of the fittest competition selection mechanism, genetic algorithm guides the population to a better direction of evolution. In this paper, we do the objective function is to solve the minimization problem. Thus, we structure a fitness function as formula (9), and the greater the fitness value, the better the individual.

$$F(x) = m - Z(x) \tag{9}$$

where, m is the total travel time of passengers in the traditional pattern.

Based on the general genetic algorithm, immune genetic algorithm is added the inoculation operator and makes the antibody mutate by utilizing the population information[10]. In addition, it is added the detection operator in order to lead a better evolutionary direction and improve the efficiency of algorithm. And it is also added balance operator to coordinate the concentration between the better individuals and the worse ones. The solution flow chart is shown in figure 1.

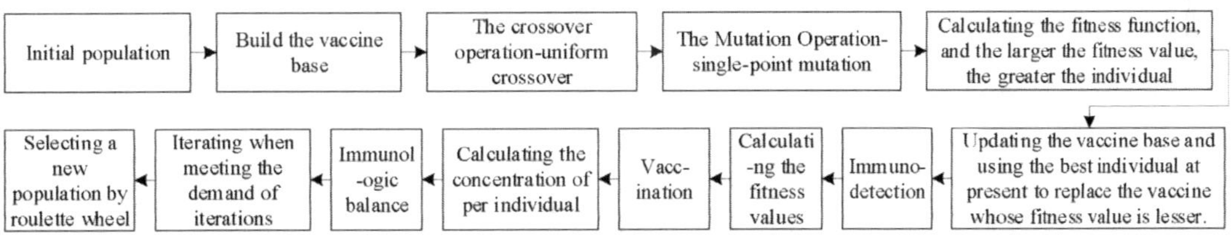

Figure 1. The flow of solving algorithm.

4. Case Analysis

4.1 . Line Profile

Taking a regional express rail as an example, this paper analyses and examines the optimization problem on express/local train stopping schedule under the circumstance of the full-length & short-run route. There are 26 stations and the 7th to 20th stations belong to the short routing section. Based on the flow data of this case, when the short route is from the 7th to the 20th stations and the proportion of the long and short route is 1:1, the demand of passenger flow can be satisfied furthest.

4.2. Solution

We design the immune genetic algorithm to solve model. The algorithm parameter values are shown in table 2. From the case, the departure times during optimized period are 20 pair per hour, and in this situation, we calculate the travel time by changing the proportion of the express/local train which changes from 19:1 to 1:19, and when the proportion is 5:15, in which there are 5 express trains, the objective function value takes the minimum, that is, the passenger travel time is the shortest. At this point, the best express train stop program is shown in table 3, in which the number 1 represents that express trains stop at the station, and the number 0 means that express trains do no stop at the station. And by computation, in this plan, the express train overtake at the 7th and the 19th station.

Table 2. Values of Algorithm Parameters.

Parameters	Value	Parameters	Value	Parameters	Value
Population	50	Crossover probability	0.9	Inoculation probability	0.8
Iterations	100	Mutant probability	0.05	The number of vaccine in vaccine base	8

Table 3. The Best Express Train Stop Program.

1	2	3	4	5	6	7	8	9	10	11	12	13	14	15	16	17	18	19	20	21	22	23	24	25	26
1	0	1	0	1	1	0	0	1	1	1	1	1	1	1	1	1	0	0	1	0	1	0	1	1

4.3. Capacity Analysis
It has an impact on carrying capacity to operate the express/local train. According to the foregoing case, when the minimum headway is 120 s and the ratio of express and local train is 1:3, the capacity depresses from 30 to 25 pairs/h.

4.4. Time Benefit Analysis
Different from the general trains, passengers who take the direct trains save the travel time, while the time of some passengers travel by slow trains is increased. In this case, the passenger travel time of general trains is 38 688 413 min, while in combination mode, the time saves 545 985 min. Comparing with the total travel time before and after the plan, we can find that in the specific passenger flow conditions, although the traffic capacity is reduced, the time benefit is still obvious.

5. Conclusion
This paper sets up a model, based on express and local train operating plan, aiming for minimize travel time of passengers, and by analysing different time composition of the full-length and short-run route. Case analysis results confirm the validity of this model. However, the combined operating mode will make the carrying capacity reduce, in the actual operation; it can obtain the relatively large capacity to change the proportion of express and local train.

Acknowledgements
This paper is supported by the Research Projects of the Social Science and Humanity on Young Fund of the Ministry of Education under grant No. 15YJCZH108, the Research Projects of Natural Science Foundation of Guangdong Province under grant No. 2015A030310341, and the Research Project of Shenzhen Technology University.

References
[1] Wang L and Wei G J 2015 *Stopping Schedules of the Express /Local Train for Urban rail Transit Based on GASA Algorithm* (Shanghai: Urban Mass Transit) pp 96-100
[2] Altazin E, Dauzère-Pérès S and Ramond F 2017 *Rescheduling through Stop-Skipping in Dense Railway Systems* vol 12 (Amsterdam: Elsevier) pp 73-84
[3] Cheng J 2016 *Study on Stop Schedule Plan Optimization of Express/Local Train for Regional Rail Line* (Shanghai: Shanghai University of Engineering Science)
[4] Hassannayebi E, Zegordi S H and Amin-Naseri M R 2016 *Train Timetabling at Rapid Rail Transit Lines: A Robust Multi-Objective Stochastic Programming Approach* vol 17 (Operational Research) pp 1-43
[5] Pan H C. and Yang T 2009 *Express/Slow Train on Urban Rail Transit Line Based on Carrying Capacity* (Shanghai: Urban Mass Transit) pp 48-51
[6] Han Y and Fu Y P 2012 *Optimization of Intercity Rail Train of Mass Transit Type Plan* (Third International Symposium on Innovation & Sustainability of Modern Railway) pp 685-689
[7] Liao J Q 2016 *Research of Regional Rail Transit Express Train Plan* (Beijing: Beijing Jiaotong University)
[8] Gao Y, Yang L X and Gao Z Y 2016 Energy consumption and travel time analysis for metro lines with express/local mode *Transportation Research Part D Transport & Environment.* TRPDT&E10(2016)009
[9] Qu M Y and Huang S M 2012 Study on express/local train plan of urban rail transit vol 34 (Beijing: Railway Transport and Economy) pp 77-82
[10] Zhou X Z 2014 *Application Research of Immune Genetic Algorithm for High-speed Railway Train Operation Adjustment* (Beijing: Beijing Jiaotong University)

IOP Conf. Series: Materials Science and Engineering **383** (2018) 012046 doi:10.1088/1757-899X/383/1/012046

Emergency disposal of urban rail transit based on distributed workflow

LUO Qin[1, 2], LYU Yitong[2], MO Yihong[3], and LI Wei[2, 3,*] and ZHANG Xiongfei[1,2]

[1] College of Urban Traffic and Logistics, Shenzhen Technology University, Lantian Road 3002, Shenzhen, P.R. China
[2] Shenzhen Key Laboratory of Urban Rail Transit, Shenzhen University, Nanshan Ave 3688, Shenzhen, P.R. China
[3] Key Laboratory of Optoelectronic Devices and Systems of Ministry of Education and Guangdong Province, College of Optoelectronic Engineering, Shenzhen University, Nanshan Ave 3688, Shenzhen, P.R. China

*E-mail: aliweib1@126.com

Abstract. Under the background of the network operation of urban rail transit, the rapid expansion of rail network structure in various cities, on the one hand, has promoted the rapid development of rail transit industry. On the other hand, it put tremendous pressure on the daily operation and management of rail transit operators. The traditional centralized disposal mode is not in a position to meet the increasingly complex network and passenger demand any more. In this paper, based on the distributed workflow theory, a distributed multi-aircraft collaborative system framework is established for emergency management of urban rail transit and an emergency response model, using Oracle database management system as the main technical to realize the digital management of contingency plans, the theory is finally applied in practice, and the design of the urban rail transit emergency response system for OCC layer is realized.

1. Introduction

With the development of the society, urban traffic is under tremendous pressure. Under such circumstances, urban rail transit has become the direction and the first choice for the future development of the transportation with its advantages of high safety, high traffic volume and high speed, not occupying the ground space and riding comfortably.

In the case of rapid development of rail transit, the number of rail transit operations is continuously increasing. China's urban rail transit has entered a networked operation era. Compared with single-line operation, routes of networked operation are more complicated and operation equipment is more diversified, which make transport more efficient. But it also makes the capacity of rail transit network traffic volume more prominent contradictions, and the potential safety problems of the facilities are outstanding. Under such circumstances, it is particularly important to enhance the resilience of urban rail transit to emergencies. This requires that the urban rail transit emergency response mechanism must be able to carry out timely and efficiently in advance, and also during and after the incident: the prevention phase of the plan should be perfect and well documented; the disposal phase should achieve timely detection and rapid, correct disposal.

Content from this work may be used under the terms of the Creative Commons Attribution 3.0 licence. Any further distribution of this work must maintain attribution to the author(s) and the title of the work, journal citation and DOI.
Published under licence by IOP Publishing Ltd

The existing mode of emergency treatment of rail transit is centralized disposal mode, with unified command and step-by-step responsibility[1]. In the disposal, the OCC director dispatching, will give instructions in a unified manner and then delegate the command to other departments because this method is not a command at the scene, only through the feedback of two command to understand the scene, leading to difficult to grasp the best handling of emergencies. With the increasingly complicated network structure, the drawbacks of centralized disposal have become increasingly obvious.

Based on the above situation, this paper proposes an urban rail transit emergency disposal system based on distributed workflow, with a view to improving the existing defects of centralized disposal. Based on the standardization and digital management of emergency response plan of rail transit, based on workflow theory, urban rail transit emergency disposal mechanism, organizational structure, disposal workflow and so on to model and redesign, to build distributed collaborative work system, the use of distributed collaborative system to achieve multi-position information, multi-point linkage, co-processing and improve the efficiency and reliability of the entire emergency disposal operations [2], in order to solve the future network operation under the conditions of the safety of urban rail transit proposed a new solution.

Xu G [3] carried out simulation model of passenger flow evacuation at metro stations and established a model using cellular automata technique. The results can be applied to station layout evaluation and optimization, passenger flow evacuation scheme design and optimization. Xu R H, et al [4] at present, the emergency management plan of urban rail transit is deficient, and the emergency plan computer integrated and multi-media management system is designed. This research can solve the problem of decentralized management, such as the system structure, the working mode and the work flow of the system, including the comprehensive inquiry, plan management, linkage and command Event evaluation, decision support and emergency communications. Hoogendoom M [5] demonstrated the method of using TTL language (Temporal Trace Language) to establish contingency plan model. Gadomski A M [6] found other use of case-based reasoning methods and decision-making theory for disposal programs generated based on the research results, Liu S G [7] compared the commonly used document processing method models in the six studies to solve the shortcomings of the existing case processing systems. Based on the improved petri net distribution processing framework and workflow model.

2. Emergency disposal characteristics of rail transit based on distributed workflow

The concept of distributed workflow is relative to earlier centralized workflow management systems. To date, many companies and institutions both at home and abroad have conducted fruitful research and put forward many models and assumptions. The distributed workflow system uses a specific protocol to synchronize the execution of each workflow machine. When choosing different workflow products, each workflow machine needs a common standard to generate process instances, and can transfer processes and activities among heterogeneous workflow machines to support common management functions.

When applying distributed workflow to emergency response process, all participants in each process of emergency response can operate according to the disposal process. When an emergency occurs, all relevant departments and personnel receive the operation instruction of emergency information at the same time, as well as the system to automatically match the solution and disposal programs to ensure that all relevant departments to receive information at the first time and conduct targeted operations, improved rail transit between the various departments and staff information exchange methods have improved the efficiency of collaborative work among members. In this work environment, the members of the group conduct activities such as mutual communication and coordination around the solution of the same emergency. The business process shows geographical distribution, asynchronous synergy, synchronous synergy and process rationality Sexual characteristics, which respectively reflect the geographical space, time space and behaviour space characteristics of the emergency disposal process based on distributed workflow.

2.1. Regional cooperation

Regional synergy reflects the spatial distribution of emergency response process. Including the distribution of the execution location during the disposal and the distribution of the location of the participants. In particular, the distribution of disposal execution positions reflects the characteristics of the distribution of the execution at different locations in multiple disposal processes in an urgent situation. In the process of emergency response of rail transit, when an emergency occurs, different personnel in different geographical locations such as dispatching centre OCC and station need to exchange data, information and knowledge by using network communication technology in different spatial locations for the same disposal event collaborative disposal.

2.2. Asynchronous synergies

Asynchronous synergy reflects the synergy of collaboration and the asynchronous nature of collaboration time. Each participant performs the work process together, but each participant is an independent individual, and each participant has its own goal. For the time characteristics of the collaborative process, the asynchronous nature of the collaborative time reflects the time difference in the collaborative process. The participants in the business process collaborate on the same task at different times. Through the design of disposal methods and disposal process, different staff in different departments of rail transit such as dispatching centre, communication and signal centre and station control desk make different processes and procedures of different time and place at the same time according to the decision instruction given by the different aspects of asynchronous operation in time.

2.3. Synchronization synergies

Synchronization Synergy reflects the synergy of collaboration and collaboration features of the synchronization time. Synergy is described in Section 2.2. Specifically, in synchronous collaboration, multiple participants can work on the same data model in real time, and the data model is consistent, allowing each participant to communicate with each other in real time during the operation. For the emergency treatment of rail transit, the advantages are spatially dispersed participants of each emergency response can communicate face-to-face in a virtual workspace which can reduce the traditional transmission of information such like by telephone, and eliminating the time and Spatial barriers between participants which contribute to the efficiency of emergency response work.

2.4. Process rationality

Process rationality reflects the orderliness of the process operation. Each participant in the disposition process has a clear responsibility to accomplish the identified tasks and there is a definite implementation dependency between the participants. At present, with the rise of the Internet and the development of distributed computing technology, the characteristics of the distribution, heterogeneity and autonomy of modern enterprise information systems have become increasingly obvious. Such an information system environment is referred to as the HAD environment [8]. The centralized workflow model can no longer meet the needs of the ever-increasing number of resources in the enterprise and the urgent need for collaboration among enterprises.

3. Modelling of emergency response system for rail transit based on distributed workflow

3.1. Distributed multi-machine collaborative architecture

Based on the current development of urban rail transit, this paper proposes to apply the concept of "distributed coordination" to the system. Computer-based platform designed and implemented collaborative work system through the computer to achieve the exchange of information to ensure that all small groups within the working group, timely and accurate transmission and sharing to create an efficient and virtual work space, improve the accuracy of information transmission and enhance the coordination of emergency response work, so as to ensure that the urban rail transit operation and management department can meet the requirements of handling "safe" and "efficient" emergency operations [9].

For the four-layer and eight-level distributed multi-machine collaborative architecture established in this paper, controlled-level government, enterprises and the network of three control levels through the cloud access to data, command-level lines and on-site command level through the cloud access to the accident and the disposal of instructions issued to the cloud, after the implementation of the rescue level downloaded emergency treatment, as show in figure 1.

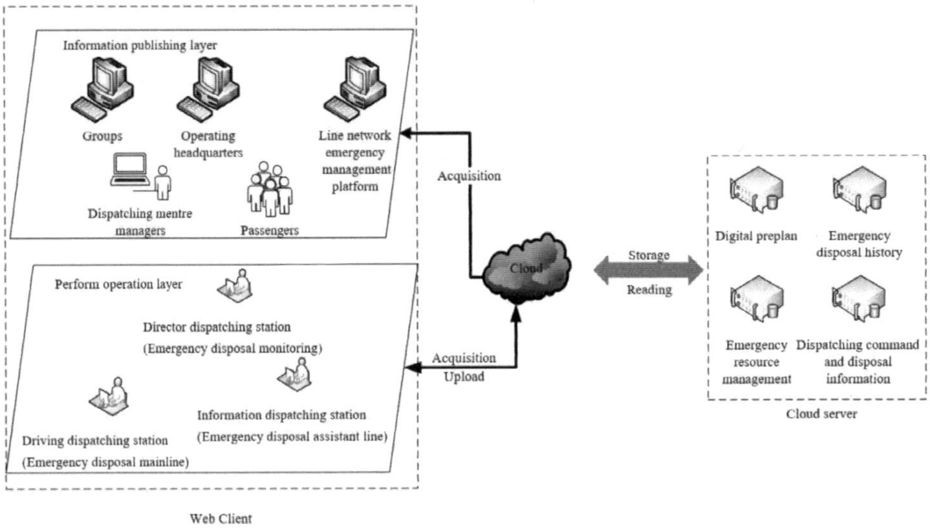

Figure 1. Data transfer between web client and cloud server.

3.2. Rail transit emergency system organizational structure

In this paper, the UML language is used to model the organizational structure of a rail transit emergency response system based on distributed workflow. As shown in figure 2, the organizational structure of the emergency system to rail transit is divided into the control command module and the emergency execution module.

In the control and command module, the network operation control center is responsible for overseeing the operation status of the entire rail transit network, supervising the emergency disposal status and reporting the emergency disposal status. When an emergency occurs, the emergency operation leading group is mainly responsible for arranging emergency response work for emergencies and coordinating emergency response work among various departments. In order to improve the ability to cope with emergencies and unexpected incidents, the leading group organizes emergency drills and publicity and education work. Line network operation control center corresponds to a number of affiliated operations control center, compared to the network operation control center, its monitoring range is small, usually a line or a regional operating status.

The emergency management module is connected with the control command module through the operation control center. The on-site command department is in charge of directing the emergency work at the scene. The comprehensive coordination group is responsible for coordinating the on-site emergency treatment.

3.3. Rail transit emergency system disposal process

At the time of the accident, the station pre-treated the accident and at the same time, the witness alerted the OCC through the emergency disposal system. After dispatching the alarm, the characteristic information of the accident was sorted out and the situation of the accident was monitored. If the accident did not lead to traffic delays, start the relevant emergency plan and accept the scheduling order until the disposal of the end. If the driving delay has been caused, the emergency plan will continue to be activated within three minutes and the site will be treated for disposal. If the disposal has not been completed within three minutes, the traffic dispatch will start to be adjusted and the station will continue to dispose of the accident along with the traffic adjustment. So far the emergency treatment of rail transit basically ended.

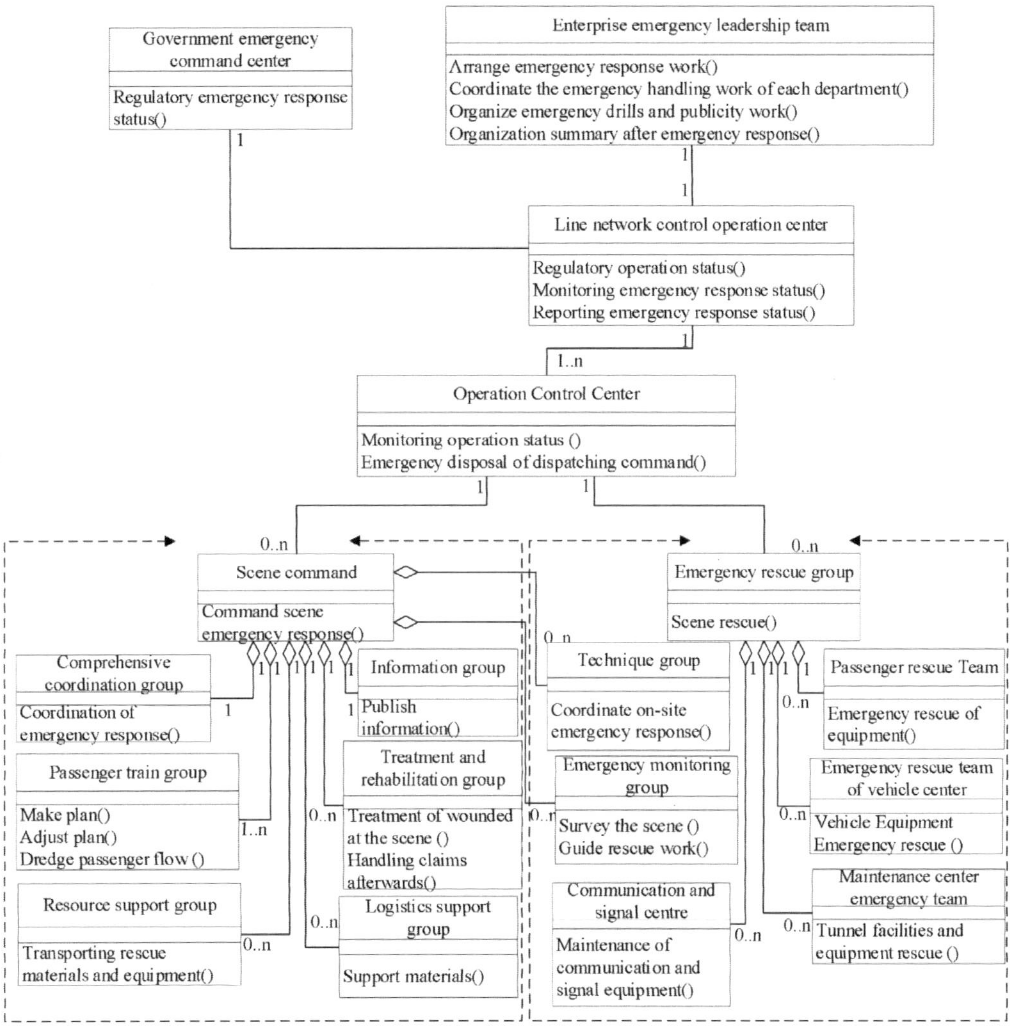

Figure 2. Urban rail transit emergency response organizational structure.

4. Design and implementation of emergency response system

The system mainly faces the OCC scheduling level to realize the digitalization, standardization and standardized integration and management of the existing contingency plans. The carding and clarification of the emergency disposal process, the intelligent quick connection alarm, the accurate judgment of the event grade, the disposal plan and the automatic scheduling command Generate and issue, the process of recording and post-summary evaluation and other key functions.[10]

At the time of the accident, the witness first carried out the alarm through the mobile phone directly, and then through the alarm window of the emergency disposal system, selected the line where the current accident occurred, the direction of the up and down direction and the specific accident contents and released the information and displayed it in the system Interface. Station control room and dispatch command center receive the accident information through the emergency system interface, the dispatcher command station control room to deal with the accident.

Based on the system, any participant can log in to the system to know the specific content and location of the emergency in real time. When the accident occurred, the expert system established based on the historical data command the dispatcher and the station control room. Disposal orders, the completion of the processing time and other information will be displayed in the module below in the system. The system defines the disposition time for each incident as 180 seconds, when the time is over, all departments and personnel stop the operation immediately and start the next step to rescue.

5. Conclusion

In this paper, we analyze the working principle and working characteristics of distributed workflow and establish a distributed multi-machine collaborative architecture framework with four levels and eight levels. We apply the idea of distributed workflow and established a model of the emergency disposal process of rail transit. Finally, it is applied to the practical application, combined with the expert database, developed an emergency response decision support system of urban rail transit, the traditional emergency measures have been improved and perfected.

Acknowledge

This paper is supported by the Research Projects of the Social Science and Humanity on Young Fund of the Ministry of Education under grant No. 15YJCZH108, the Research Projects of Natural Science Foundation of Guangdong Province under grant No. 2015A030310341, and the Research Project of Shenzhen Technology University.

References

[1] Yu H F, Qin Y, et al 2013 *Traffic Information and Safety. Study on Auxiliary Decision System of* Urban *Rail Transit Emergency Disposal* [J]. p 163-68.

[2] Wang Z Q, Zhang M, et al 2007 School of Transportation Engineering, Tong Ji University, Shanghai, China. Emergency Treatment Assistant Decision System of Urban Rail Transit [A]

[3] Xu G 2003 *Modern tunnel technology. Computer Simulation of Evacuation of Subway Station* [J]. (02) p 32-34.

[4] Gu J H, Xu R H, et al 2006 Urban Rail Transit Research, Design of Emergency Management Information System for Urban Rail Transit [J]. 9 (2): p 57-59

[5] Hoogendoorn M, Jonker C M, et al. 2007 Computational and Mathematical Organization Theory, Modeling centralized organization of organizational change[J]. 13(2): p 147-84.

[6] Gadomski A M, Bologna S, et al. 2001 International Journal of Risk Assessment and Management, Towards intelligent decision support systems for emergency managers: the IDA approach[J]. 2(3): p 224-42.

[7] Liu S Y and Liu Z G 2015 *Application Research of Computers, A distributed processing system model based on improved Petri net for workflow case* [J], (02): p 475-78.

[8] Du S Z. 2004 Zhejiang University, Research on Time Workflow Modeling in Distributed Collaborative Work Environment [D].

[9] Zhang Y, Shi M L, et al 2003 *Computer Research and Development, HCM: A collaborative work description model for virtual enterprise*[J] 40(5): p 752-56.

[10] Wang W J and Meng F K, et al 2006 Computer Engineering, Research on Emergency Plan Based on Ontology [J]. 32(19): p 170-72

IOP Conf. Series: Materials Science and Engineering **383** (2018) 012047 doi:10.1088/1757-899X/383/1/012047

The comparison on resistance performance and running attitude of asymmetric catamaran changing shape of tunnel stern exit region

Sang-Won Kim*, Gyeong-Woo Lee and Kwang-Cheol Seo

Dept. of Naval Architecture and Ocean Engineering, Mokpo National Maritime University, Mokpo 58628 Korea

*E-mail: swkim5834@naver.com

Abstract. Catamaran vessels have been used for various purposes such as leisure craft and research vessel. Especially, asymmetric catamaran has a half of the symmetrical hull and flat side of facing inward demi-hull, and has feature of planing hulls such as hard chine, spray rail. Therefore, these vessels can travel at high speed with better stability and resistance performance. However, these features cause asymmetric catamaran to have large trim angle at high-speed condition, which cause problem on vessel's operation performances such as visibility and dynamic instability motion. To overcome these negative effects, tunnel shape in stern has been introduced. By applying this shape, trim angles are also suppressed because large dynamic lift developed on the exit region of tunnel stern. In this research, tunnel stern was applied on the asymmetric high-speed catamaran to evaluate vessel's hydrodynamic performance by numerical method, and the tunnel stern types are distinguished by slope of tunnel exit region: flat or narrow types. Consequently, it is confirmed that the total resistance of tunnel types is lower about 1.6~5.0% than the bare hull in the wide speed range while showing almost same phase with the bare hull after $FnV=2.3$. Also, narrow type improves resistance performance further than flat type. On the other hand, trim angles show different trend with each tunnel type. In case of narrow type, the trim angle is stabilized after $FnV=1.8$ while that of flat type show an increasing trend as ship speed increase. These indicated that the narrow type has better resistance and operation performance than others.

1. Introduction

Recently, there has been a significant increase on the use of catamaran vessels for various purpose such as sail yacht, ferries, leisure craft and research vessel. The advantages of this vessel compared to mono-hull are that they have better transverse stability, more attractive layout accommodation [1]. Furthermore, catamarans often have a shallower draft and better resistance performance than comparably-class mono-hulls. There are two types of catamaran hull, such as Symmetric and Asymmetric Hull. Symmetric hulls have symmetric demi-hull and usually are used sail yacht and leisure boat which need to be comfortable. On the other hand, asymmetric hulls have a half of the symmetrical hull and flat side of facing inward demi-hull with hard chine, spray rail like planing hulls. Therefore, these vessels can travel at high speed with better stability and have been widely used for fast vessels. However, these vessels usually build with some degrees of shaft angle due to the engine and gearbox inside the hull. This has known to cause loss of propeller. Moreover, the large trim angle at high-speed condition also cause problem on vessel's operation performances such as visibility and

Content from this work may be used under the terms of the Creative Commons Attribution 3.0 licence. Any further distribution of this work must maintain attribution to the author(s) and the title of the work, journal citation and DOI.
Published under licence by IOP Publishing Ltd

IOP Conf. Series: Materials Science and Engineering **383** (2018) 012047 doi:10.1088/1757-899X/383/1/012047

dynamic instability motion. To overcome these negative effects on operation performance of vessel, tunnel stern which is round shape in stern has been introduced. By applying this shape in stern, shaft angle can be reduced, and trim angles are suppressed because large dynamic lift developed on the stern.

In this research, we applied tunnel stern on the asymmetric high-speed catamaran and evaluated resistance performance and running attitude. Also, we monitored flow characteristic according to changes exit region of tunnel sterns. The tunnel stern types are distinguished by slope of tunnel exit region: flat or narrow type. The performance of catamaran was estimated by using commercial CFD software, STAR-CCM+. Reynolds averaged Navier-stokes equations with an SST turbulence model was used along with the volume of fluid method to describe the two-phase flow of water and air around the hull. Furthermore, running altitude of the planing hull was used via an overset method. Validation of this simulation was performed with Warped hull.

2. Numerical simulation

The governing equations of fluid were discretized by Finite Volumes Methods[2][3]. The discrete form is shown in equation 1, and continuous integral form of the RANS equation is represented in equation 2.

$$\frac{d}{dt}(\rho x\phi V)_0 + \sum_f [\rho\phi(\upsilon \cdot a - G)]_f = \sum_f (\Gamma\nabla\phi \cdot a)_f + (S_\phi V)_0 \tag{1}$$

$$\frac{d}{dt}\int_V \rho x\phi dV + \int_A \rho\phi(\upsilon - \upsilon_g)da = \int_A \Gamma\nabla\phi da + \int_V S_\phi dV \tag{2}$$

where ρ and ϕ are a density and scalar quantity in continuum, respectively. υ and υ_g are velocity and grid velocity, respectively. Γ and S are the face diffusivity and source coefficient, respectively. This equation is composed of unsteady term, convection flux, diffusion flux and the volumetric source term, in sequence.

The unsteady term represents growth rate of variation ϕ over time on control volume and is only included in transient calculation. There are two temporal discretization: First-order and second-order temporal scheme. Generally, second-order scheme with relatively coarse grid ensures same accuracy to first-order scheme with relatively fine grid, and second-order scheme is used in this research.

2.1. Computational domain

The mesh was divided into one stationary background region and one moving overset region close to the hull. Both parts were meshed with trimmed, predominantly hexahedral cells with local refinements at the free surface and the wake. In the overset region, the mesh was same size with background region where the hull is located, and prism layer mesh constructed around the hull. To resolve the free surface accurately, a high mesh density was used in the region around the free surface in vertical direction where induced wakes by hull were expected to be present. The whole mesh consists of a total of about 2.16 million of cells, where the inner region has about 898,000 cells and the outer region has about 1,268,000. The height of the first cell layer around the hull was set to $Y^+ \approx 75$. The computational domain and boundary conditions are illustrated in figure 1.

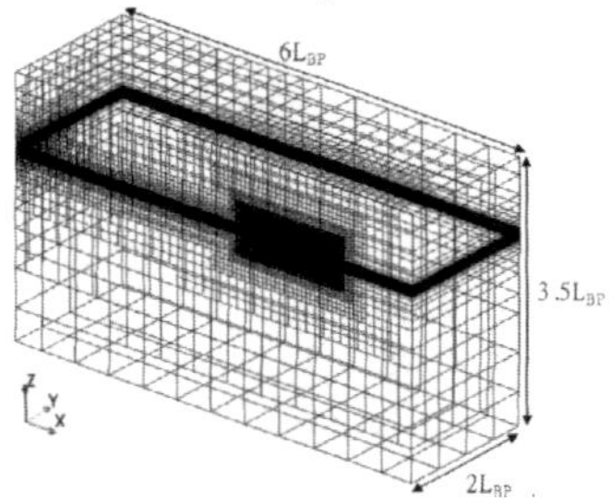

Figure 1. Computational domain.

2.2. Numerical method

We applied established coordinate system and numerical methods[4] for incompressible and viscosity of fluid to calculate performance of planing hull, and table 1 listed key features of numerical models. The commercial CFD code used in this study was STAR-CCM+.

Table 1. Key features of numerical models.

	Model
Governing Equation	Reynolds-Averaged Navier-stokes
Time	Implicit Unsteady
Turbulent Model	SST (Menter) *k-w*
Wall Treatment	All Y^+ Treatment
Spatial Discretization	Cell Centered FVM
Velocity/Pressure Coupling	SIMPLE Algorithm
Free Surface Model	Volume of Fluid method High-Resolution Interface Capturing
Body Motion	Overset Grid Method

3. Principle dimension of asymmetric catamaran and tunnel stern

Table 2 lists the principle dimension of asymmetric catamaran hull applied in this study while figure 2 illustrates the model profiles. The model has a hard chine and flat side on inward of demi-hull.

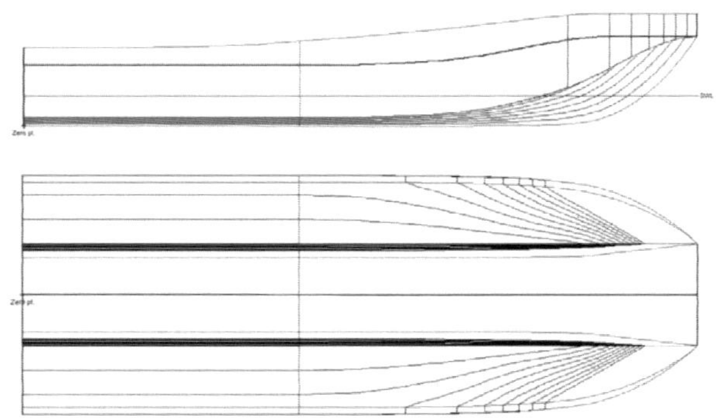

Figure 2. Model profiles of Asymmetric High-Speed Catamaran.

Table 2. Principal dimensions of Asymmetric Catamaran.

Dimension	Value
$Length_{BP}$ (m)	1.803
Breath (m)	0.69
Draft (m)	0.0905
Displacement (kg)	48.7
Speed(Fn_V)	0.85~2.57

The rear and longitudinal views of tunnel stern are illustrated in figure 3. Generally, tunnel stern is composed of three regions; entrance, propeller and exit region[5]. Among these areas, the exit region starts from edge of propeller region to transom and has an important impact on the trim angle of the vessel. The dynamic lift developed on the exit region can result in the bow-down trim[6][7]. The principle of tunnel stern in the present study is shown in table 3. The bare hull represents base hull without tunnel stern while same shape and height is applied on the exit and propeller region in flat type, the inclination angle was set so that tunnel become narrows throughout exit region.

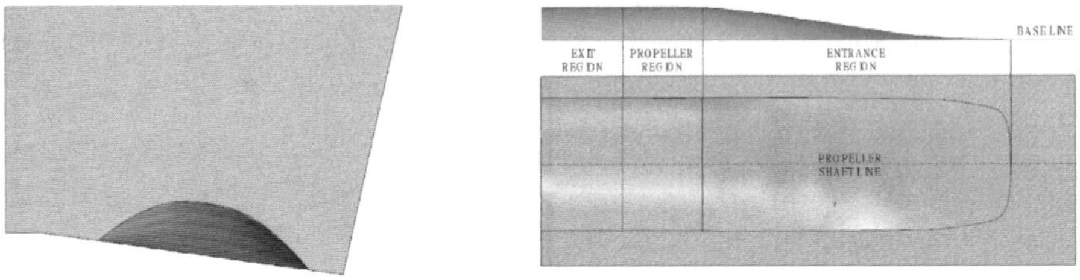

Figure 3. Conceptual design of tunnel stern (Left: Rear view, Light: Bottom and side view).

Table 3. Principle of the tunnel stern models.

Shape		Dimensions	
		Flat type	Narrow type
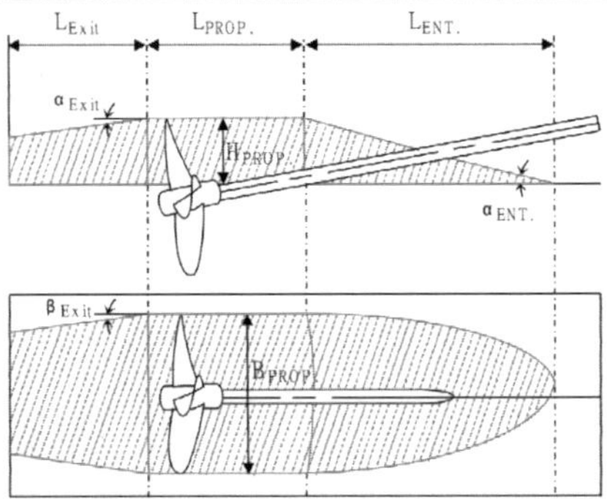	$L_{ENT.}$	12.2%L_{BP}	12.2%L_{BP}
	$L_{PROP.}$	5.0%L_{BP}	5.0%L_{BP}
	L_{Exit}	8.9%L_{BP}	8.9%L_{BP}
	$H_{PROP.}$	33.6%T	33.6%T
	$B_{PROP.}$	18%B	18%B
	α_{ENT}	6°	6°
	α_{Exit}	0°	3°
	β_{Exit}	0°	3°

4. Result and discussion

4.1. Validation of the numerical method

In this paper, the experiment and numerical results from the Warped hull 2, were used to validate a CFD model[8]. Table 4 lists the principle dimensions of the Warped hull 2 while figure 4 illustrates the concept and geometry of the model.

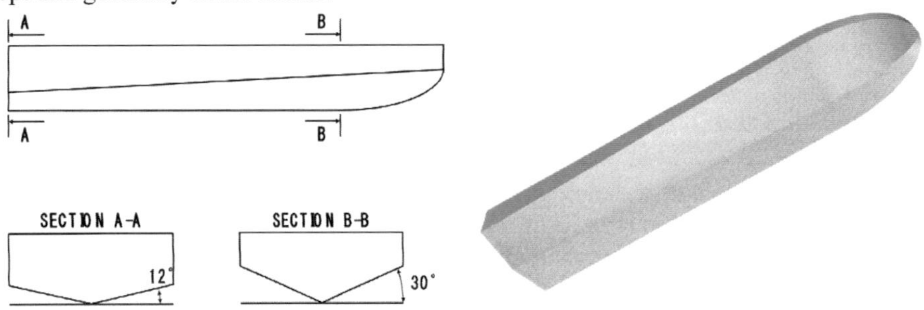

Figure 4. Geometric view of the Warped Hull 2 [9].

Table 4. Principal dimensions of Warped Hull 2.

Dimensions	Values
$Length_{OA}$ (m)	1.9
$Length_{A\text{-}B}$ (m)	1.5
Breath (m)	0.424
$Draft_{AP}$ (m)	0.11
Displacement (kg)	32.73
Speed(Fn_V)	1.92~3.57

The results were compared with experimental and CFD data for the total resistance, trim and wetted surface area as given in figures 5~7. Qualitatively Overall numerical predictions were reliable with experimental values but there were slightly underestimations about 2~8% of the total resistance, 3.5~4% of the trim angle and 5~7.2% of the wetted surface for all ranges of *Fn$_V$*, quantitatively.

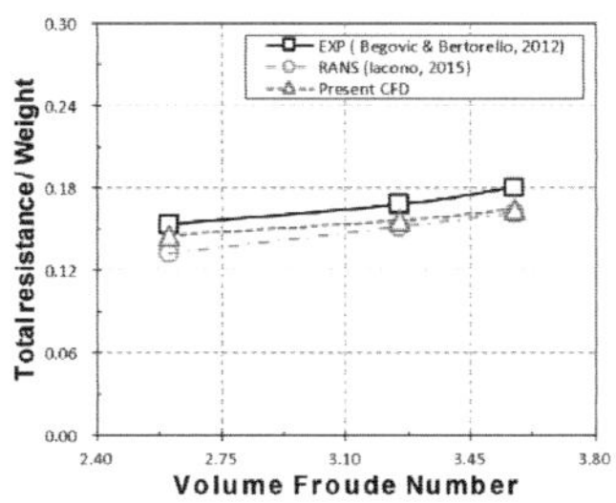

Figure 5. Comparison of total resistance.

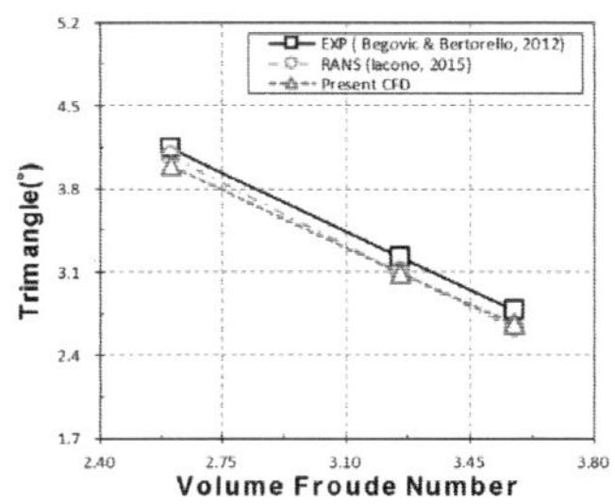

Figure 6. Comparison of trim angle.

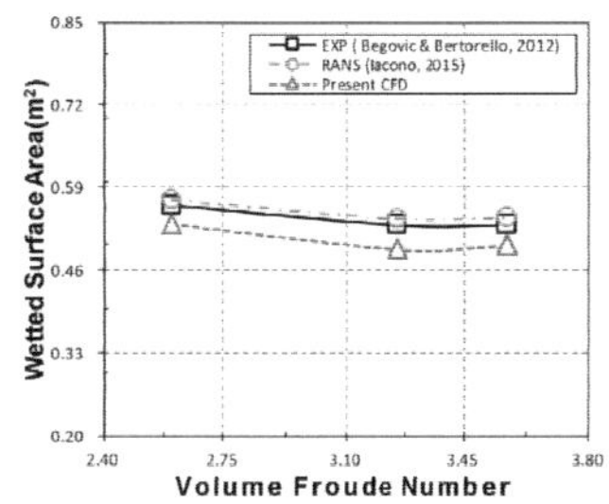

Figure 7. Comparison of Wetted surface area.

4.2. Resistance performance

Numerical analysis was performed with bare hull and two types of tunnel stern. The frictional, residuary and total resistances were shown as dimensionless form by weight and compared in figures 8~10. In total resistance performance, the tunnel types demonstrate good performance about 1.6~5.0% than bare hull over wide-range of *FnV* while showing same performance after *FnV=2.3*. For comparing details of the resistance performance, the total resistance was divided into the frictional and residual resistance. The residual resistance was improved when applied the tunnel types, and it was improved by 5.2~7.3% for the narrow type and 3.1~4.6% for the flat type. The frictional resistances show equivalent performances except that the narrow type shows poor performance after *Fn$_V$=2.0*.

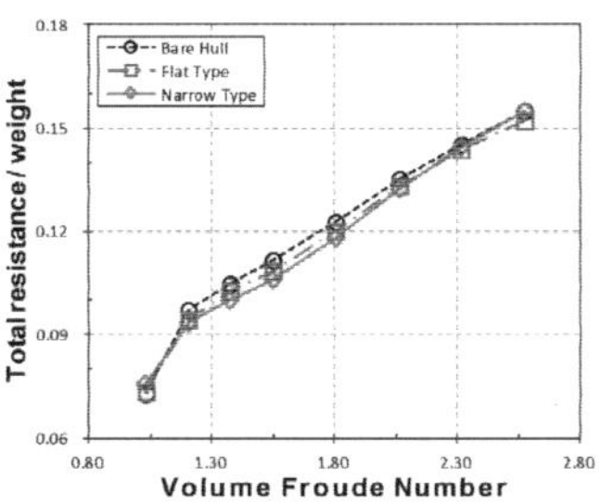

Figure 8. Total resistance with tunnel types.

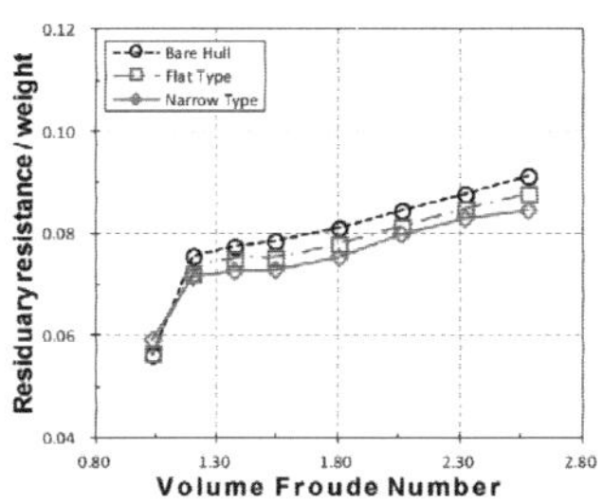

Figure 9. Residuary resistance with tunnel types.

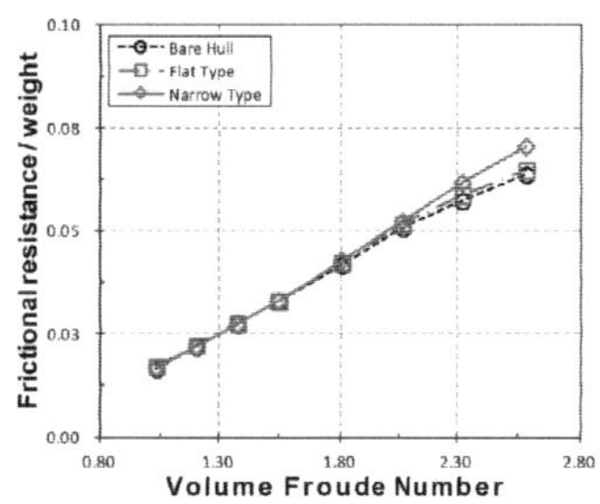

Figure 10. Frictional resistance with tunnel types.

4.3. Running attitude and wetted surface area

The wetted surface, Sinkage and trim angle were shown in figures 11~13, and sinkage was represented as dimensionless form by initial draft. figure 12 shows that tunnel types have similar draft and more submerge than bare hull over all *Fn$_V$* range. However, Trim angle of tunnel types are different tendency with each other after *Fn$_V$=1.37*. The flat type shows similar trend with bare hull but trim angles increase from 4.9% to 25.8% in general. However, the trim angles of narrow type peaks at *Fn$_V$=1.37* and has stabilized trend after *Fn$_V$=1.54*. This indicates that changing tunnel exit region cause trim angle to suppress largely. Also, because of Sinkage and trim angle trend, the wetted surface area of tunnel stern increase from 0.97% to 2.89% than bare hull while flat type comes close to bare hull after *Fn$_V$=1.8*.

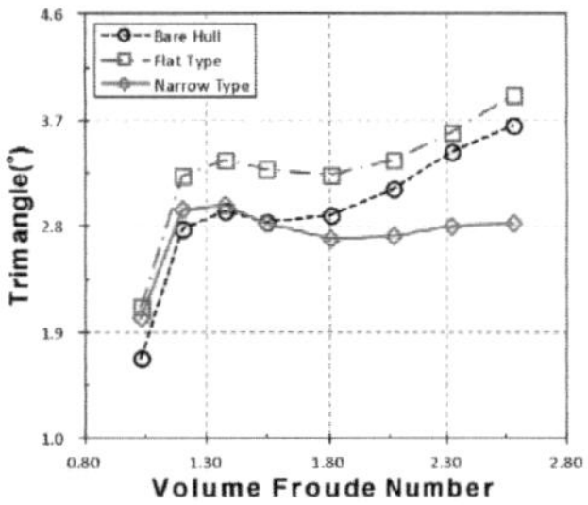

Figure 11. Trim angle with tunnel types.

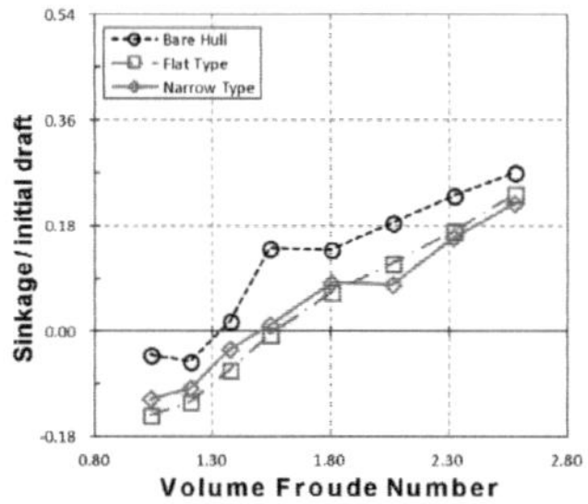

Figure 12. Sinkage with tunnel types.

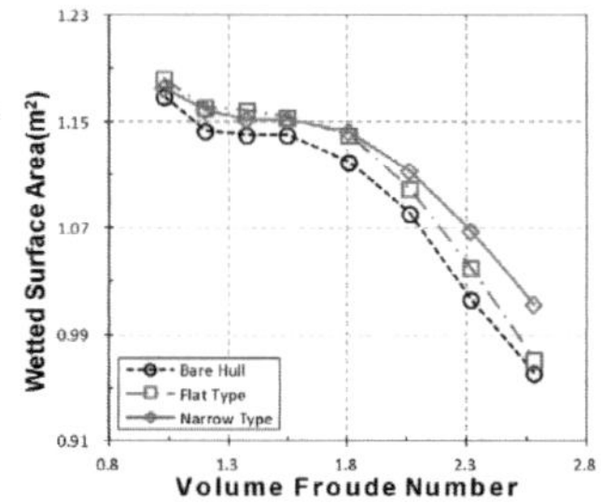

Figure 13. Wetted surface area with tunnel types.

4.4. Longitudinal and transverse pressure distribution

The longitudinal and transverse pressure distributions were evaluated for verifying the characteristic of tunnel stern, and its values were measured at 0.23m in transverse direction from centreline and 0.09~0.30m longitudinal direction from transom, shown as figure 14.

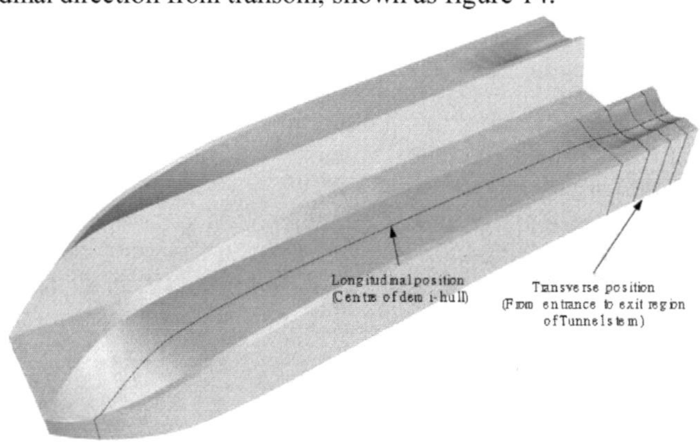

Figure 14. Measured positions of longitudinal and transverse pressure distribution.

Figures 15~18 show longitudinal pressures of tunnel types around *Fn*$_V$=*1.20~2.58*. The bare hull and tunnel types has peak pressure point at bow. The areas where occurred maximum pressure distribution on the bow are from free surface to spray area. However, there are relatively low pressure distribution from tunnel entrance at the stern of tunnel type. These phenomena become more pronounced as the *Fn*$_V$ increases. Also, the pressures of tunnel types increase significantly after certain position of tunnel exit region and propeller and peak almost same with that of the bow.

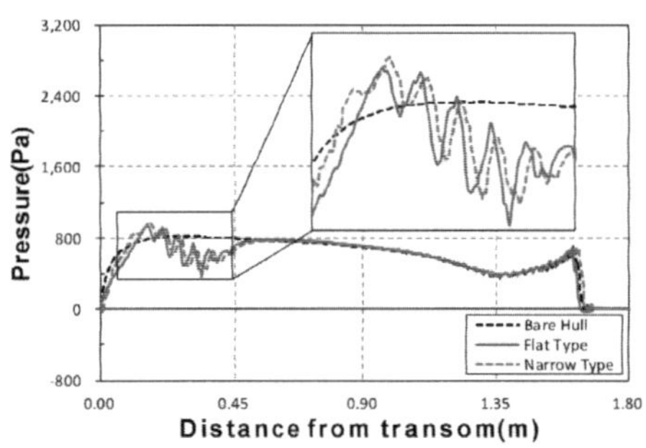

Figure 15. Comparison of longitudinal pressure distributions at Fn$_V$=1.20.

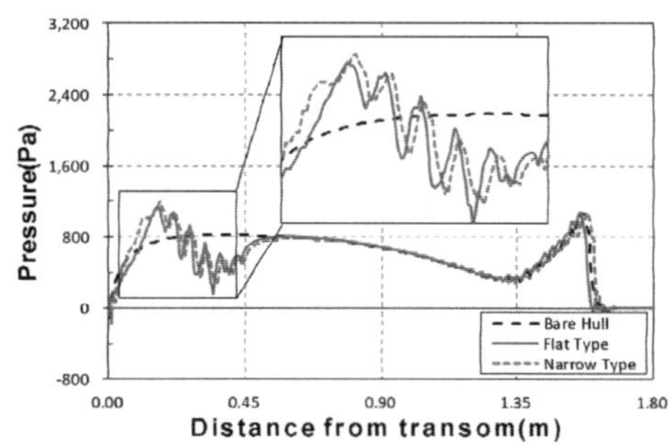

Figure 16. Comparison of longitudinal pressure distributions at Fn$_V$=1.54.

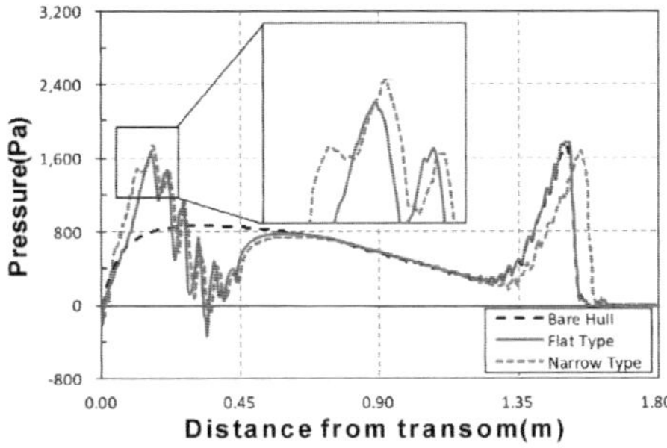

Figure 17. Comparison of longitudinal pressure distributions at Fn_V=2.06.

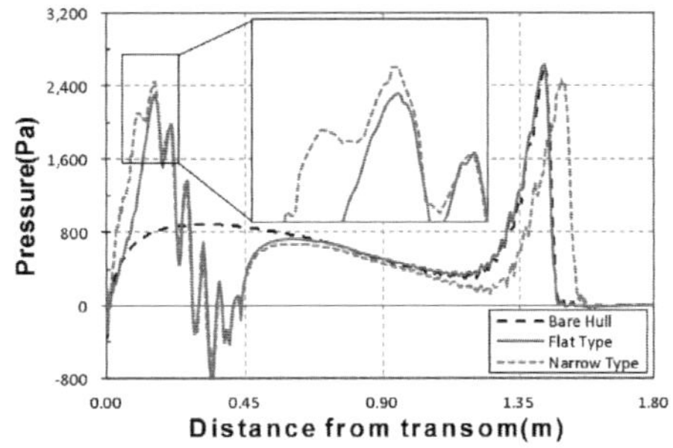

Figure 18. Comparison of longitudinal pressure distributions at Fn_V=2.58.

These differences of pressure distribution can be confirmed clearly from transverse direction around the tunnel stern and shown as figure 19. The transverse pressure distributions were measured at four point from transom where located at 0.09, 0.19m in tunnel exit region, and 0.23, 0.3m where located in propeller region and tunnel entrance respectively.

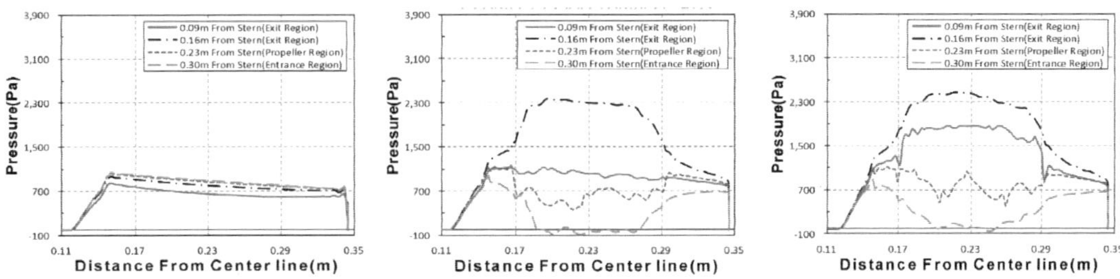

(a) (b) (c)

Figure 19. Transverse pressure distribution at Fn_V=2.57 / (a) Bare hull, (b) Flat type (c) Narrow type.

While the bare hull has uniform pressure distribution, it can be seen clearly that the entrance and propeller region have low pressure sections, and tunnel exit region have high pressure sections. In figure 19(b), exit region of flat type has high pressure section at 0.16m from stern, but the pressure distribution at 0.09m from stern is similar with that of bare hull. However, in cases of the narrow type, the pressure distributions at 0.09m from stern sharply increase than that of bare hull and flat type. This indicates that, increasing the slope of tunnel exit region, the overall pressure distributions of exit region also increased, and it suppress trim angle largely.

Figure 20 shows longitudinal wave profiles at stern. The amplitudes of first wave period of tunnel types are slightly lower than that of bare hull because the separation points of tunnel stern are located upper that of bare hull. figure 21~22 show Transverse wave profiles of 0.85 and 1.25m from the tunnel stern where are in first wave period of stern, this indicates that wave-making resistance of tunnel types is lower than that of bare hull.

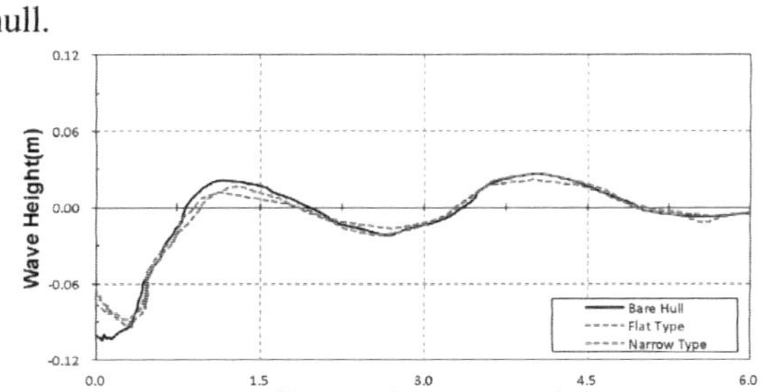

Figure 20. Longitudinal wave profile of position 0.23m from the centreline at Fn_V=2.58.

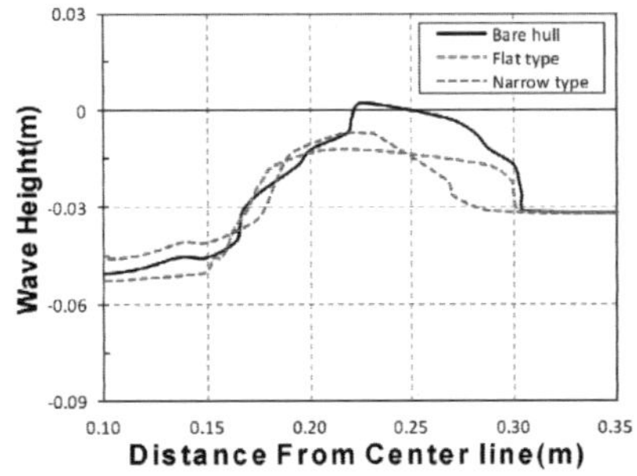

Figure 21. Transverse wave profile of 0.85m from the transom at Fnv=2.58.

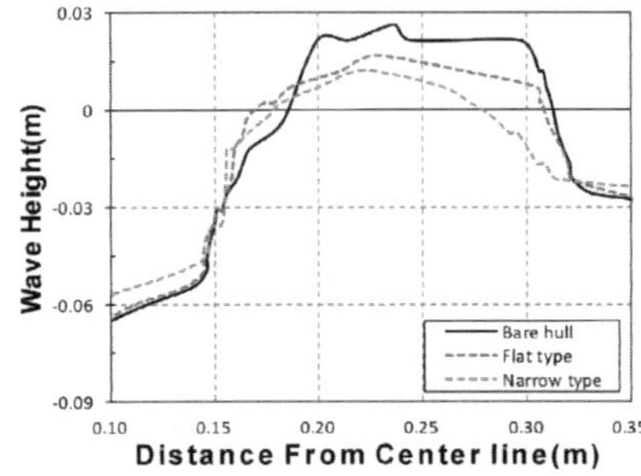

Figure 22. Transverse wave profile of 1.25m from the transom at Fnv=2.58.

5. Conclusion and Future Work

In this study, a detailed analysis of the flow characteristics on stern tunnel type and the performances of asymmetric catamaran have been carried out via commercial software, STAR-CCM+. The validation was performed, and results were adequate for the prediction of performances in calm water. The following conclusions were drawn.

1) The numerical investigation indicated that there was a close dependency between resistance performance or running attitude and shape of tunnel stern. This implied that vessel performance could be adjusted by shape of tunnel stern.

2) Tunnel stern used in the study showed good performance of wave-making resistance reduction, the vessel applied tunnel stern are submerged slightly than bare hull because there is some loss of volume at stern.

3) The investigation showed that tunnel types produced better resistance performance than bare hull. In case of the running attitude, trim angle of each tunnel type shows the different trend. These trends could be confirmed by analysing the pressure distributions at the tunnel stern. It was showed that the pressure increased greatly in exit and propeller region with increasing inclination angle of tunnel exit region.

Future studies should further research the flow characteristics and performance of the stern tunnel with propeller. Furthermore, the parametric study on slope and length of tunnel exit region and needs for optimization on resistance performance and running attitude. To achieve this, it will be necessary to research the propeller of small crafts.

Acknowledgments

This research was performed with the support of the Ministry of Education and the National Research Foundation of Korea's Element design technology development of Marine Tourism Control Patrol Type High-speed government vessel for modernization Promotion (2015H1C1A1035813).

References

[1] Setyawan D, Utama I K. A. P., Murdijanto M, Sugiarso A and Jamaluddin A 2013 Development of Catamaran Fishing Vessel *J. Technology and Science* Vol **21** No 4 pp 167-73

[2] Rosenfeld M and Kwak D 1991 Time-dependent solutions of viscous incompressible flows in moving coordinates *Int. J. Numerical Methods in Fluids* Vol **13** No 10 pp 1311-28

[3] Orihara H and Miyata H 2003 Evaluation of added resistance in regular incident waves by computational fluid dynamics motion simulation using an overlapping grid system *J. Marine Science and Technology* Vol **8** No 2 pp 47-60

[4] Kim S.W, Seo K.C, Lee D.K and Lee G.W 2017 A Numerical Study on Dynamic Instability Motion Control of Wave-Piercing High-Speed Planing Craft in Calm Water using Side Appendages *J. Korean Society of Marine Environment & Safety* Vol **23** No 3 pp 320-29

[5] Blount D.L 1997 Design of Propeller Tunnels for High-Speed Craft *Proc. FAST Conf.* (Sydney:

Australia) p 151-56
[6] Hankley D.W 1980 Developmental Tests of a Pre-Prototype Landing Craft Personnel – Large LCP (L) MK 12 Combatant Craft Engineering Department Report 6660-50
[7] Hankley D.W 1986 U.S. Patent Number 4,622,016 assigned to United States of America Secretary of Navy (Propeller Tunnels)
[8] Iacono M 2015 Hydrodynamics of Planing Hull by CFD Thesis for the Degree of Master of Science Univ. Naples Federico II pp 1-202
[9] Begovic E and Bertorello C 2012 Resistance assessment of warped hull form *J. Ocean Engineering* **56** pp 28-42

IOP Conf. Series: Materials Science and Engineering **383** (2018) 012048 doi:10.1088/1757-899X/383/1/012048

Deduction and prevention of container overboard accident during loading and unloading

Zhibiao Chen[1,] *, Xianling Zhu[2] and Jingzhou Zhang[1]

[1] CCS Wuhan Rules & Research Institute, Wuhan China, 430022
[2] Wuhan Institute of Shipbuilding Technology, Wuhan China, 430050

*E-mail: zbchen@ccs.org.cn

Abstract. To prevent the recurrence of containers overboard accident during loading and unloading, taking the accident at Yueyang Chenglingji Teminal as the study object, the mechanical deduction is carried out. A kind of method to determine whether a container will overturn is put forward, including the criterion and the quantitative algorithm of the corresponding indicators, i.e. the allowable heel angles. By verifying, the estimated results according to the algorithm described in this article, are basically consistent with the Yueyang accident. Thereout a set of measures to prevent such accidents are provided, including adding allowable heel angles during loading and unloading in the loading manual, carrying out stowage plan and evaluating and controlling the ship floating state before operating, monitoring and adjusting of the ship floating state during operating.

1. Introduction

Container waterway transportation has the characteristics of large transport capacity, small energy consumption, low cost, less land occupation and less pollution, so developing vigorously over the years. At the same time, accidents of containers overturning and falling into water also happen frequently, not only in transit, but also in loading and unloading occasionally. "7.28" containers overboard accident in 2011 of Yueyang Chenglingji Songyanghu International Container Terminal, resulted in 26 FEU scattered in the river, and the captain of the ship on the container missed in water at the same time[1], caused great loss and far-reaching influence.

Through qualitative analysis in macroscopic view, the cause of the Yueyang accident was the uneven weight distribution of the goods carried in the containers, and the stowage calculation wasn't carried out, and the random loading and uneven stowage, all of these led to the container falling into water. However, the author believes that it is necessary to carry out a mechanical quantitative analysis for the accident, to obtain the key quantitative algorithms and preventive measures, so as to prevent the recurrence of such accidents.

2. Accident process

By known information, the container ship was loading at the terminal before the accident. The ship had heeled by the starboard side before the last container lower down, and this container's slot was located on starboard side, see figure 1(1). After lower down, the heel by starboard was aggravated, see figure 1(2). And then the container columns overturn to the starboard side, see figure 1(3). At last, Domino effect appeared, and the containers overturned and fell into the water, see figure 1(4).

Content from this work may be used under the terms of the Creative Commons Attribution 3.0 licence. Any further distribution of this work must maintain attribution to the author(s) and the title of the work, journal citation and DOI.
Published under licence by IOP Publishing Ltd

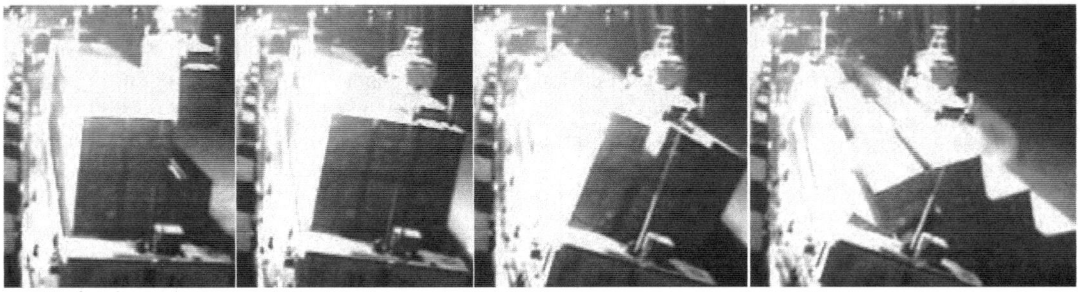

Figure 1. Process of overboard accident.

3. Accident deduction

3.1. Deduction condition

The ship is a typical inland waterway container ship. The securing scheme adopts the splayed lashing to the whole container stack [2] [3]. The container was being loaded at the terminal during the accident. Therefore, the accident scene can be simplified as follows:

- During loading, bridge locks with transverse connectors had not yet been installed on the top of the containers, and neither the lashing wire nor lashing band for container stacking had been installed. Each container column was independent, there was no interaction force between columns. Therefore, the single column of the container stack can be taken as the study object.
- Stack cones were placed between adjacent tiers in the same container column (when loading the twist lock was opened, it was also considered as a stack cone). Therefore, the transverse constraint existed between the container tiers, but no vertical constraint.
- The ship berthed at the terminal. There was no the rolling, pitching and heaving motions caused by waves. Therefore, the analysis can be carried out according to the static situation, considering the container gravity, and centre of gravity located in container centre.
- When the accident occurs, the ship was being in the maximum heeling. The heel angle should be the initial heel angle before the last container lower down, plus the increment caused by the last container lower down.

3.2. Force analysis

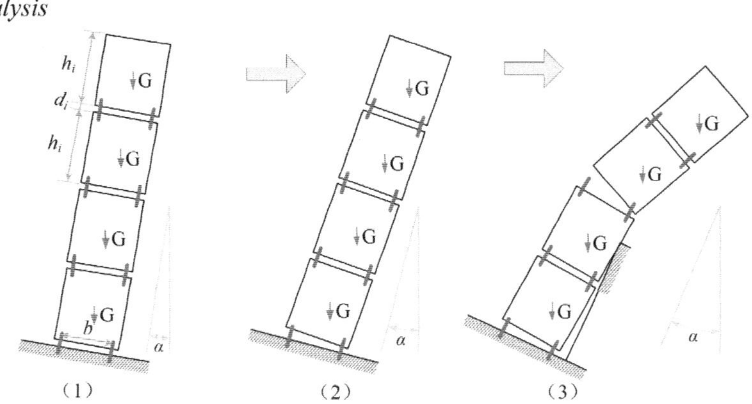

Figure 2. Overturning mechanical model.

By further analysis of container overboard details, the author found that the top three containers were overturned at first, see figure 1(3), but then only the top two containers overturned together because obstructed by the hatch or the container of the adjacent column, see figure 1(4), but in either case, they all overturned around the stack cone of the container corner. In combination with the above 3.1, considering a container column as an object, the overturning mechanics model as shown in figure 2 is established.

Overturning moment of the container column around the stack cone, *M*, can be calculated as follows:

$$M = nG\cdot\left[\frac{n(h_i+d_i)}{2}\sin\alpha-\frac{b}{2}\cos\alpha\right]=\frac{1}{2}nG\cdot\left[n(h_i+d_i)\sin\alpha-b\cos\alpha\right] \quad (1)$$

where α——the ship heel angle;

b——the space between the container left stack cone and the right, generally taken as 2.259m;

h_i——the container height, generally taken as 2.591m[4];

d_i——the space between the container tiers, generally taken as 0.028m;

n——the total number of the container tiers;

G——the average weight of the containers, taken as no more than the rated weight.

When the overturning moment M>0, n tiers from top down will overturn.

Let M=0, the formula of critical heel angle after lower down is obtained as follows:

$$\tan\alpha=\frac{b}{n(h_i+d_i)} \quad (2)$$

If n taken respectively as 1, 2, 3 and 4, the corresponding critical heel angles are shown in table 1. From the table, it can be concluded that the more stacking tiers are, the more prone to overturn.

Table 1. Critical overturning heel angle α

n	**tan*α***	***α* (deg)**
1	0.863	40.779
2	0.405	22.028
3	0.271	15.163
4	0.205	11.578

The heel angle α, consists of the initial heel angle before lower down, α_0, and the incremental angle after lower down, θ_h, that is:

$$\alpha=\alpha_0+\theta_h \quad (3)$$

where θ_h can be taken as the incremental angle on the stability arm curve[5] corresponding to the heeling arm l_h caused by lower down, shown as figure 3.

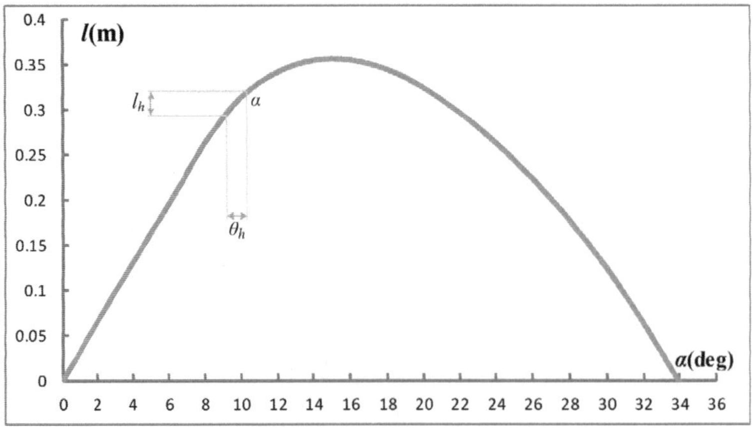

Figure 3. Stability arm curve and heel angle.

The heeling arm, l_h, can be calculated as follows:

$$l_h = \frac{cG}{\Delta}\left[Y\cos\alpha + (Z - \frac{d}{2})\sin\alpha \right] \tag{4}$$

where c——the dynamic coefficients while lower down, generally taken as 1.3;
G——the weight of the operating container;
Δ——the ship displacement;
Y,Z——the respective distance from the ship center line plane and base plane to the center of the operating container after lower down;
d——the ship draft;

At last, the heel angle before lower down, α_0, can be obtained as:

$$\alpha_0 = \alpha - \theta_h \tag{5}$$

It is worth pointing out that, the above analysis also applies to the situation of lifting up a container from the ship. For lifting up, the heeling arm l_h is caused by lifted up, at this time the container weight is taken as negative value; the heel angle α_0 is the heel angle after lifted up; the heel angle α is the heel angle before lifted up.

3.3. Overturning criterion

Some simplified assumptions are adopted in the above analysis process:

- Each container weight in the same column is even;
- No wind load when loading and unloading is considered;
- No ship swaying when loading and unloading is considered.

The partial safety factors are taken into account for the above uncertainties, the critical heel angle in 2.2 of this article can be converted to the allowable heel angle as follows:

$$[\alpha] = \alpha / \gamma_\alpha \tag{6}$$

$$[\alpha_0] = [\alpha] - \gamma_\theta \theta_h \tag{7}$$

where $[\alpha]$——the allowable heel angle after lower down or before lifted up;
$[\alpha_0]$——the allowable heel angle before lower down or after lifted up;
γ_α——the partial safety factor of the uneven weight and the wind load etc., generally not less than 1.3;
γ_θ——the partial safety factor of ship swaying, generally not less than 1.1;
θ_h——the incremental angle caused by the heel moment l_h, determined according to the method of figure 3, α in the figure corresponding to $[\alpha]$ here;

Now, it is possible to determine whether the container will overturn by these two indicators, the allowable angle $[\alpha]$ and $[\alpha_0]$, as follows:

- The allowable heel angle before lifted up or after lower down $[\alpha]$: if the ship heel angle exceeds this indicator before lifted up or after lower down, the overturning may appear; to be determined by the formula (2) (6) of this article.
- The allowable heel angle before lower down or after lifted up $[\alpha_0]$: if the ship heel angle exceeds this indicator before lower down or after lifted up, the overturning may appear.; to be determined by the formula (4) and (7) of this article, taking the operating container as the variable.

4. Verification of criterion

According to the collected information of the accident ship, at the accident time, the maximum tiers number of container stacks was n=4,the columns number is 3; the ship moulded depth is 3.3m; the design draft is 2.5m; the ship displacement is 1600t; the stability arm in the design full loading is shown in table 2.

Table 2. Stability arm l in full loading.

θ(deg)	**0**	**2**	**4**	**6**	**8**	**10**	**12**	**14**
l(m)	0	0.065	0.130	0.196	0.263	0.313	0.341	0.354
θ(deg)	**16**	**18**	**20**	**22**	**24**	**26**	**28**	**30**
l(m)	0.355	0.344	0.324	0.297	0.262	0.222	0.177	0.123

The allowable heel angle after lower down $[\alpha]$ is obtained according to the formula (2) and (6); the heel arm after lower down l_h is obtained according to the formula (4); the incremental angle θ_h is obtained according to the method of figure 3 and table 2; the allowable heel angle before lower down $[\alpha_0]$ is obtained according to the formula (7). The results are shown in table 3. From the table, the larger container weight corresponds the smaller allowable heel angle.

Table 3. Allowable heel angle $[\alpha]$ and $[\alpha_0]$ during loading and unloading.

n	G (t)	$[\alpha]$ (deg)	l_h (m)	θ_h (deg)	$[\alpha_0]$ (deg)
4	30	8.906	0.125	4.036	4.466
4	25	8.906	0.105	3.456	5.104
4	20	8.906	0.084	2.846	5.775
4	15	8.906	0.063	2.216	6.468
4	10	8.906	0.042	1.546	7.205

From the figure 1(2) and 1(3), the heel angle before lower down is about 5~6 degrees at the accident time, and the heel angle after lower down is about 8~10 degrees.

Where the operating container weight is about 18~30t, the estimated results according to the algorithm described in this article, are basically consistent with the ship states at the accident time.

5. Preventive measures

Based on the above analysis, it is necessary to take the following precautionary measures to prevent the recurrence of this kind of accident of containers overturning at terminal:

- In the loading manual the allowable heel angle $[\alpha]$ and $[\alpha_0]$ should be added, shown as Table 2. A series value of the maximum tiers n and the container weight G in the table can be present according to the ship situation.
- Before loading and unloading, the stowage plan and the floating state simulation should be carried out, to evaluate the ship heel angle, and to control the allowable indicators $[\alpha]$ and $[\alpha_0]$ not to be exceeded, respectively corresponding to the heel angle after lower down and the heel angle after lifted up.
- During loading and unloading, the ship heel angle should be monitored, and the heel angle before lower down should be controlled not greater than the allowable heel angle $[\alpha_0]$, or the heel angle before lifted should be controlled not greater than the allowable heel angle $[\alpha]$. If necessary, adjust the floating state and then continue the operation. Where, the ship heel angle can be seen in wheelhouse, and the current allowable heel angle can be interpolated in Table 3, according to the current operating container weight which can be can be obtained from the terminal or the consignor.

General principles are: during loading, if ship has heel angle, the next container should not be lower down as far as possible on the heel side; during unloading, if ship has heel angle, the next container should not be lifted up as far as possible from the opposite side of the heeling; for example, if the ship heel by starboard before lower down, the container cannot be located on the starboard side after lower down.

6. Conclusion

In this article, through the analysis and deduction of the container overboard accident in loading, a criterion and the corresponding quantitative calculation algorithm about the indicators, is concluded to determine whether the container is possible to overturn through the ship heel angle during loading and unloading. The feasibility of this method is preliminarily verified by the study accident. Measures to prevent such accidents are put forward, including adding allowable heel angles during loading and unloading in the loading manual, carrying out stowage plan, and evaluating and controlling the ship heel angle before operating, monitoring and adjustment of the ship heel angle during operating.

Acknowledgments
Foremost, I would like to show my deepest gratitude to my wife, Mrs. Zhu, who is also my partner, has provided me with great encouragement and support in the studying and writing of this article.

References

[1] Lu Jiqing and Qu Shengguo 2012 On loading and securing of inland waterway container ships Inland River Container Transport Forum (Qingdao: China Institute of Navigation)
[2] Chen Zhibiao 2013 Status and improvement of container securing design in inland waterway shipping J. Marine Tech. 1 29-32
[3] China Classification Society 2016 Rules for the construction of inland waterways steel ships vol 1, China Communications Press Co.,Ltd. (Beijing) p 1-104
[4] ISO 1496-1:2013 Series 1 freight containers
[5] Sheng Zhenbang 2015 Ship Theory, SHANGHAI Jiao Tong University Press (Shanghai) p 95-98

IOP Conf. Series: Materials Science and Engineering **383** (2018) 012049 doi:10.1088/1757-899X/383/1/012049

Method on continuous ice-breaking capability assessment of icebreaker

Wang Chuan[1,*], LI Hong-wei[1], LIU Zhi-hui[1] and Yue Qi[1]

[1] School of Petroleum Engineering, China University of Petroleum, Qingdao, China

*Email: wangchuan@upc.edu.cn

Abstract. The opening of the Ice Silk Road is a great significant for the realization of the close connection with the European, and it is a promotion for China's economic development. The icebreaker will play an important role in the Arctic strategy. More accurate ice load calculation and prediction of ice breaking capacity methods will provide important technical support for the design and construction of icebreaker. This paper aims to solve the above problems and carry out the following research work. The ice load in continuous icebreaking type is calculated in 3D numerical method. Then, the motion differential equation in time history is established to calculate the motion response. In addition, the structure strength under ice breaking condition is evaluated based on FEM. At last, the evaluation method of continuous ice breaking capacity is proposed in this paper. Compared with the results of literature, the accuracy of the ice load calculated by this method meets the requirements and could further guide the design and development of ice breaking vessels.

1. Introduction

Global warming has become the common consensus, with the Arctic ice melting, to build "Ice Silk Road" is a new point to promote the Eurasian Economic Union docking based on The Belt and Road Initiative. The icebreaker operate in high latitude and severe cold regions, and ice load becomes the key factor to restrict ship structure safety and ice breaking capacity.

Continuous ice breaking means that icebreaker moves forward with a steady speed, which is the main operation mode. The ice load and the structure strength are important concerns in this condition. The research in this field is relatively limited, many problems need to be solved. Considering the situation, the purpose of this paper is to establish a theoretical prediction method for ice load and ice breaking capability, and make a solid theoretical foundation for model tests in the future. The research results also provide technical support for independent research and development of icebreaker design in China.

2. Rules for icebreaking capacity

This section will focus on the ice breaking capacity evaluation method. In order to calculate the capacity, the following factors should be taken into account[1]:

a. The thrust of the propulsion system is sufficient to meet the ice breaking demand.

b. The structural safety of icebreaker is sufficient to satisfy the ice breaking condition.

The continuous ice breaking capacity is depend on the more accurate environment load calculation. Considering the mass distribution and thrust force, it is the key to solve icebreaker motion response which could show the curve between ice thickness and breaking velocity in maximum thrust force, and

Content from this work may be used under the terms of the Creative Commons Attribution 3.0 licence. Any further distribution of this work must maintain attribution to the author(s) and the title of the work, journal citation and DOI.
Published under licence by IOP Publishing Ltd

then, it is necessary to check the safety of the ship structure in each design ice conditions. Through the above steps, it is obtained that the envelope of the icebreaker continuous icebreaking capability curve. The figure.1 shows the flow-process diagram of the calculation.

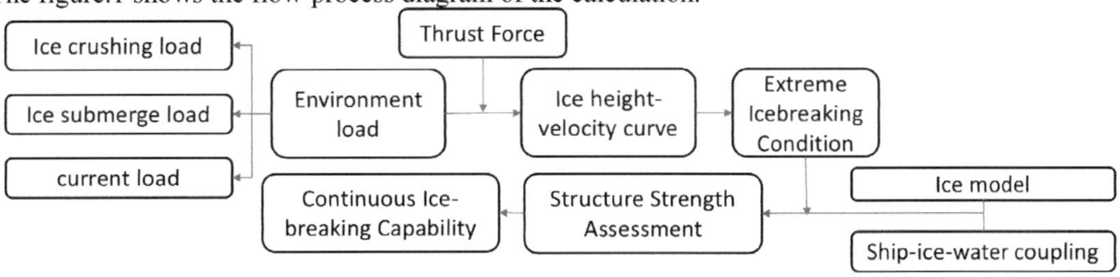

Figure 1. Flow-process diagram of the calculation.

3. Environment Load Calculation Method

The ice crushing load, ice submergence and current are the main loads of icebreaker during operation. Among these components, the ice crushing load is the control load of icebreaker in the process. The calculate method would be proposed in this part.

3.1. *Ice crushing load*

The action of ship ice is the process of contact-extrusion-bending failure[2]. Therefore, the ice breaking load is the integral of the action load of the sea ice which is squeezed by the contact with the hull.

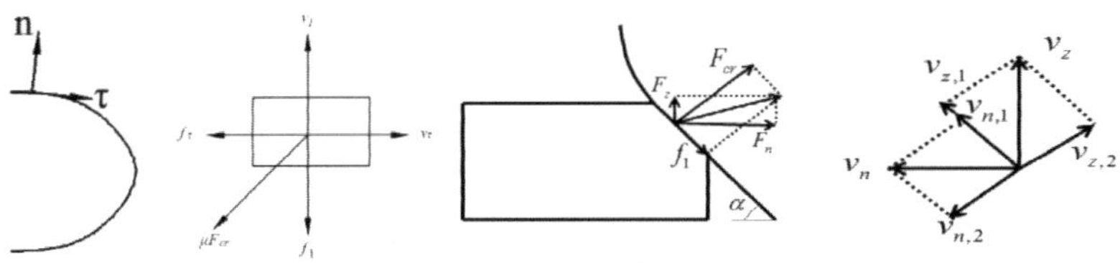

Figure 2. Schematic of the whole contact force system.

The coordinate system and section form of ship ice contact are shown in figure 2. The ice breaking load mainly contains two kinds of calculation components:

a. Extrusion caused by contact between sea ice and hull,

b. Friction caused by the relative motion between the hull and sea ice;

The crushing force F_{cr} is normal to the contact surface. And determined by the ice crushing strength σ and contact area A.

$$F_{cr} = \sigma A \tag{1}$$

On the contact surface, the relative velocity between ship and sea ice could be decomposed into the horizontal relative velocity component v_τ and v_1, and the normal velocity component v_n. f_1 and f_τ are the frictional force in opposition to motion. So the force component could be expressed as:

$$f_1 = \mu F_{cr} \frac{v_1}{\sqrt{v_1^2 + v_\tau^2}} \tag{2}$$

$$f_\tau = \mu F_{cr} \frac{v_\tau}{\sqrt{v_1^2 + v_\tau^2}} \tag{3}$$

$$F_n = F_{cr} \sin\alpha + f_1 \cos\alpha \tag{4}$$

where μ is the frictional coefficient between ice and hull, α is the normal angle between ship and ice surface.

The ice crushing force of six degrees in the global coordinate system is as follows:

$$F_x = -F_n \cos\beta - f_\tau \sin\beta \tag{5}$$

$$F_y = F_n \sin\beta - f_\tau \cos\beta \tag{6}$$

$$F_z = F_{cr} \cos\alpha - f_1 \sin\alpha \tag{7}$$

The expression of ice breaking load in ship coordinate system is obtained by further deformation:

$$F_1^{brk} = F_x \cos\psi \cos\theta + F_y \sin\psi \cos\theta - F_z \sin\theta \tag{8}$$

$$F_2^{brk} = F_x\left(-\sin\psi\cos\varphi + \cos\psi\sin\theta\sin\varphi\right) + F_y\left(\cos\psi\cos\varphi + \sin\varphi\sin\theta\sin\psi\right) + F_z\cos\theta\sin\varphi \tag{9}$$

$$F_3^{brk} = F_x\left(\sin\psi\sin\varphi + \cos\psi\sin\theta\cos\varphi\right) + F_y\left(-\cos\psi\sin\varphi + \cos\varphi\sin\theta\sin\psi\right) + F_z\cos\theta\cos\varphi \tag{10}$$

$$F_4^{brk} = y_b F_3^{brk} - z_b F_2^{brk} \tag{11}$$

$$F_5^{brk} = y_b F_1^{brk} - x_b F_3^{brk} \tag{12}$$

$$F_6^{brk} = x_b F_2^{brk} - y_b F_1^{brk} \tag{13}$$

where (x_b, y_b, z_b) is the coordinate of ship ice contact point in the fixed ship coordinate system, and the above is the expressions of ice resistance caused by extrusion during ice breaking process in six degrees.

3.2. *Ice submerge load*

The resistance caused by ice floe clearing and submergence could be considered according to the formula proposed based on Lindqvist[3].

$$\begin{aligned} R_s = (\rho_w - \rho_{ice})\cdot g\cdot h_i\cdot B\big(T(T+B)/(B+2T)+ \\ \mu_i\left(0.7L - T/\tan\phi - B/4\tan\beta + T\cos\phi\cos\gamma\sqrt{1/\sin\phi^2 + 1/\tan\beta^2}\right)\big) \end{aligned} \tag{14}$$

where ρ_{ice} is the sea ice density, L is the ship length, B is the beam, T is the draft. ϕ is the slop angle of the ship hull, β is the waterline angle.

$$\gamma = \arctan\left(\frac{\tan\phi}{\sin\beta}\right) \tag{15}$$

3.3. *Current load*

The resistance of the current is also one of the load components concerned. As this component is not the focused force, this paper uses the calculation method in the literature [4] to calculate.

$$F_1^c(t) = \frac{0.075}{(\log_{10} R_n - 2)^2}\cdot\frac{1}{2}\rho_w S u|u| \tag{16}$$

$$F_2^c(t) = \frac{1}{2}\rho_w \int_L C_D(x) D(x) v(x)|v(x)| dx \tag{17}$$

$$F_6^c(t) = \frac{1}{2}\rho_w \int_L C_D(x) D(x) v(x)|v(x)| x dx \tag{18}$$

where S is the wet surface area, $C_D(x)$ is the drag coefficient and $D(x)$ is section draft.

4. Structure Assessment Method

The dynamic analysis method is used to evaluate the structural strength of icebreaker based on the finite element software. This part focuses on the coupled dynamic analysis method of ship-ice-water.

4.1. *Ice Model*

Sea ice is a nonlinear and extremely strong material [5]. In order to match the ice layer properties, the sea ice modeling criteria are summarized as follows by comparing the calculation results of various working conditions with the relevant technical literature:

a. Six body units is chosen to simulate the ice unit.

b. The size of sea ice unit could meet the requirements of engineering and precision.
c. Sea ice radiation region needs to meet the non-reflection condition.
d. Ice boundary conditions set for rigid fixation for large enough ice sheet.

Figure 3. Nonlinear characteristics of sea ice.

Based on the previous research results of the authors, multiple failure criteria could be used to simulate the failure modes of sea ice. The damage pattern is shown in the figure.3.

4.2. *Second development of dynamic analysis*

The second development of dynamic analysis is mainly embodied in the following two aspects, and the program is compiled in the figure.4.
a. The new sea ice materials and failure criteria.
b. The coupling analysis model.

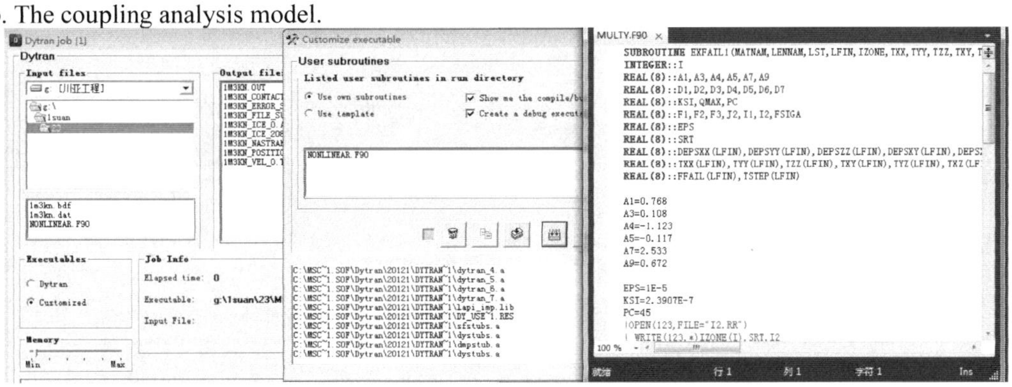

Figure 4. Second development of dynamic analysis.

5. Results and Discussion

According to the evaluation method of continuous ice breaking capacity proposed in this paper, a complete calculation of the ice breaking capacity of an icebreaker is carried out in this section, and the results are discussed in further. First, ice load is calculated in some ice thickness. And the motion response is solved through the motion differential equation. The preliminary h-v curve is contained according to the calculation results which meet the thrust requirements. Then the strength of hull structure under ice breaking condition is checked. Thus the ice breaking capacity of icebreaker is determined

5.1. *The parameter of icebreaker and ice*

The main parameter of the icebreaker and ice are shown in table1 and table 2.

Table 1. Main particulars of the icebreaker.

Length (m)	Beam (m)	Draft (m)	Displacement (t)	Mass inertia ($t \cdot m^4$)		
				I_{xx}	I_{yy}	I_{zz}
96.0	18.4	5.2	4890	1.39e8	2.44e9	2.57e9

Table 2. Main particulars of ice.

Density (kg/m^3)	Young Modulus (MPa)	Poisson ratio	Bending Strength (MPa)	Friction Coefficient
915	5400	0.33	0.75	0.15

5.2. Ice load calculation

In this part, the time domain ice load calculation program is used as a tool to predict the stable ice breaking speed under a fixed thrust force. When the speed is stable, it indicates that the ship could continue to break ice under this ice condition.

Figure.5 is the stable ice breaking speed and ice breaking load under the thickness of the 0.4m. The results shows that the mean ice force is about 2.15×10^5N, and the ice breaking velocity is 11kn. Compared with the Lindqvist [3] result, the accurate of this paper is satisfied. The ice load curve shown in figure.6 reflects its characters. When the ice contact occurs, the ice force increases gradually, and the ice force drops suddenly when the ice layer is destroyed. The fluctuating ice force at the bottom reflects the effect of ice floe on the structure. The variation of ice load presents periodicity. The thickness of ice layer under different ice thickness can be calculated in this method.

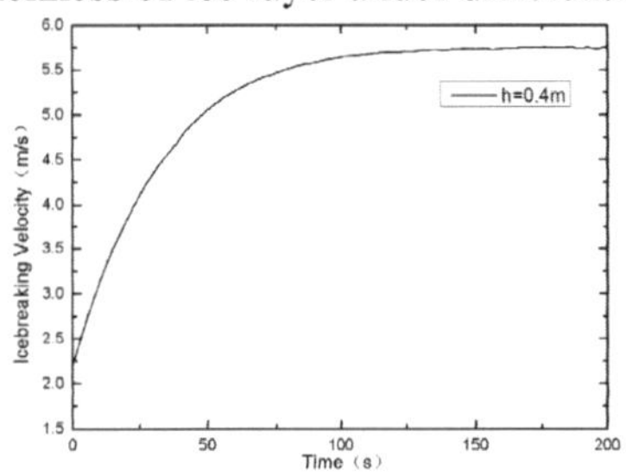

Figure 5. Velocity of icebreaking.

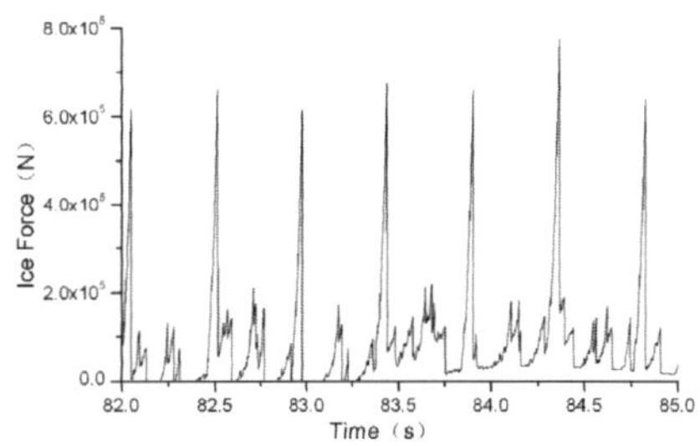

Figure 6. Ice Loads in numerical Method.

Table 3 gives the comparison of the ice load calculation results of the two ice thicknesses between this paper and the literature, and the accuracy is satisfactory.

Table 3. Results of breaking force in straight way.

Ice Thickness	0.4m	0.6m
Steady Velocity	11 kn	9 kn
Results in this paper	2.15×10^5N	4.02×10^5N
Results in Lindqvist[3]	2.53×10^5N	3.93×10^5N

5.3. Assessment of icebreaker structure

According to the calculation results of the previous section [6], the following cases are selected to check the structure strength. The load cases are listed in table 4.

Table 4. Load case of bow part in continuous-breaking way.

Case ID	Icebreaking Velocity	Ice Thickness
1	3.5kn	1.0m
2	8.9kn	0.6m
3	12.2kn	0.3m

The model is established according to the finite element model criterion. Coupling ship, ice and water, the dynamic analysis of the structure is carried out which is shown in figure.7. The analysis results are shown in the figure.8 and figure.9.

Under the ice breaking condition, the stress concentration area of the bow is concentrated in the transverse frame, the side plate and the transverse bulkhead area. As the ship ice breaking process needs not only to further resist the longitudinal ice breaking load, but also to withstand the side extrusion from the sea ice to the ship body mechanism.

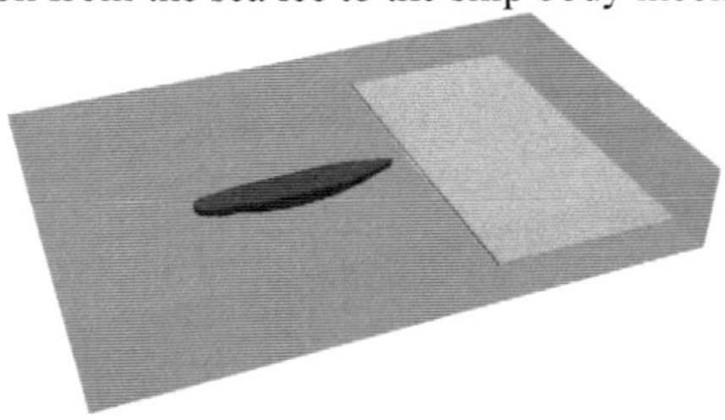

Figure 7. The couple model.

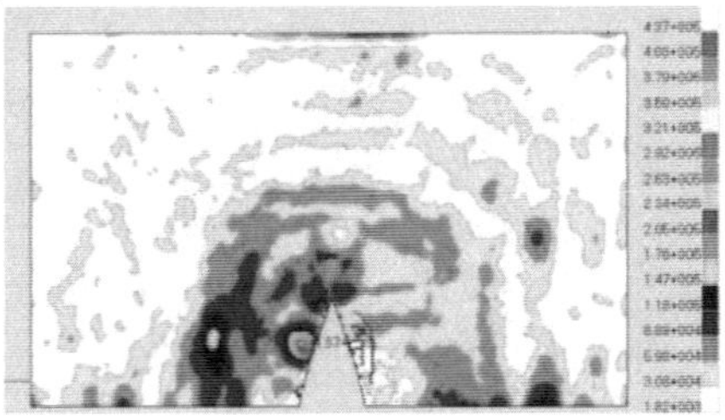

Figure 8. The stress distribution of ice.

The largest stress area appears in the transverse frame. For different ice thickness conditions and different ice breaking speed, it is found that the stress of the ship structure does not only increase with the ice thickness, but has a certain relationship with the ice breaking speed. The safety stress level is 330MP. So the ice breaking capacity curve is shown in figure.10.

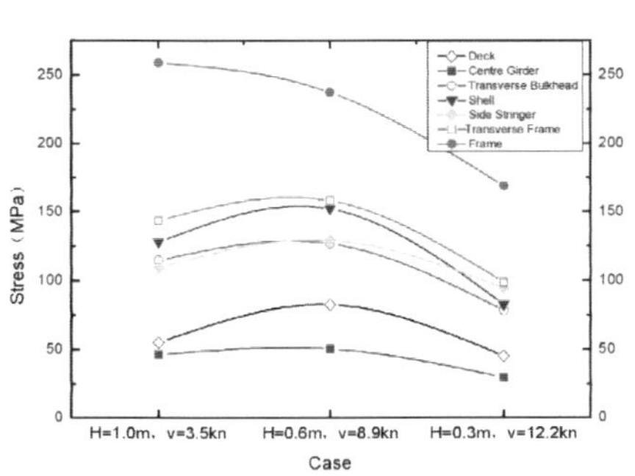

Figure 9. The structure stress summary.

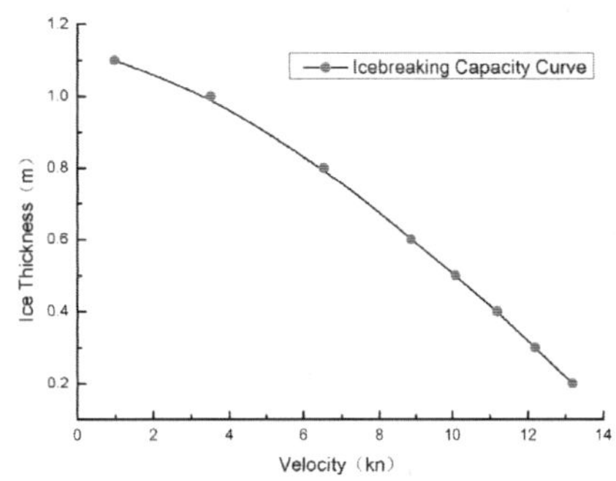

Figure10. Ice breaking capacity curve.

6. Conclusions

The calculation and analysis of ice load and the evaluation method of ice breaking capacity are the theoretical basis and core research contents. Through the study of this paper, the following conclusions are obtained:

The time domain calculation method of continuous ice breaking load and ship motion is developed. And the calculation accuracy is satisfied.

Put forward a set of evaluation method of ice ship icebreaking capability. And structure and load are considered.

Acknowledgments

Supported by National Key R&D Program of China (2016YFC0303800);

The Fundamental Research Funds for the Central Universities(15CX02122A);

Shandong Provincial Natural Science Foundation, China (ZR2016EFP08)

References

[1] Chuan Wang and Huilong Ren 2015 *Journal of Coastal Research* **73** 548
[2] Biao S.,Riska K. and Moan T. 2011 *Cold Regions Science and Technology* **60** 177
[3] Lindqvist G.1989. *Proceedings of POAC*,1989 722
[4] Xiang T. 2013 *Cold Regions Science and Technology* **8** 1
[5] Timco G.W.and Weeks W.F. 2010 *Cold Regions Science and Technology* **60** 107
[6] Ren Huilong and Wang Chuan 2014 *Journal of Ship Mechanics* **18** 1109

IOP Conf. Series: Materials Science and Engineering **383** (2018) 012050 doi:10.1088/1757-899X/383/1/012050

Large eddy simulation of flow past a 3D cylinder at Re=3900

Han Liu*
University of Michigan, Ann Arbor, USA

*E-mail: hanliu@umich.edu

Abstract. Unsteady flow past a cylinder (Re=3900) was computed using large eddy simulation (LES) by OpenFOAM. The study includes the prediction of drag and lift coefficient, and pressure distribution around the cylinder. During the simulation, the one-equation sub-grid scale (SGS) model was applied. LES numerical results were compared with published RANS k- ε model, k- ω model and experimental data. Results indicate that LES one-equation SGS model provides better agreement in force coefficients and pressure distribution than above mentioned RANS turbulence models. Besides, effect of spanwise discretization to numerical results is also investigated, showing that medium gird is fine enough to yield accurate results compared to the fine grid, which is very computationally expensive.

1. Introduction

The flow past circular cylinders has been extensively studied due to its importance in many practical applications, such as hydrodynamic loads on marine piles, offshore platform risers and support legs. In fact, the flow around circular cylinders includes a variety of fluid dynamics phenomena, like separation, vortex shedding and the transition to turbulence. The mechanisms of vortex shedding and its suppression have significant effects on the various fluid-mechanical properties of practical interest: flow-induced forces such as drag and lift forces and pressure coefficient [1].

Currently the numerical simulation methods to predict turbulent flows are limited. This is mainly due to their three-dimensional, unsteady and irregular characteristics. The most accurate method of solving turbulence flow is Direct Numerical Simulation (DNS), in which Navier-Stokes equations are solved on a fine grid using a small time-step. All eddy sizes (including the smallest turbulence scales) will be captured. DNS performs well for simple flows, however, it becomes computationally expensive when complicated flow is needed. The overall cost, including time step, of the computational effort is proportional to Re, which is not suitable to industrial applications with CPU resources available today.

Another numerical simulation method for turbulent flows is Reynolds-Averaged Navier-Stokes (RANS) models. RANS focuses on the mean flow and the effects of turbulence on mean flow properties. Prior to the application of numerical methods the Navier–Stokes equations are time averaged and the turbulence equations are closed by using viscosity terms. Common models of RANS including k- ε method and k- ω method. Due to its modest computational requirement, this approach has been applied successfully in the industrial for the last few decades. However, it tends to fail in predicting the transient behavior of the flow and for flows involving large unsteady vortical structures. [2]

An alternative approach is called large-eddy simulation (LES) which was proposed in 1963 by Joseph Smagorinsky. In LES approach, large eddies are retained, and 3D time-dependent

Content from this work may be used under the terms of the Creative Commons Attribution 3.0 licence. Any further distribution of this work must maintain attribution to the author(s) and the title of the work, journal citation and DOI.
Published under licence by IOP Publishing Ltd

Navier-Stokes equations are solved directly. Small eddies are removed and modeled using a sub-grid scale model (SGS), which results in a significant reduction in computational cost compared to DNS. LES provides higher accuracy degree than the RANS approach since the large eddies contain most of the turbulent energy and are responsible for most of the momentum transfer and turbulent mixing [3]. No model is required for the large flow structures that depend on the individual geometry of the problem. Only the fine–scale turbulence, which has more universal character, needs to be modelled. Therefore, LES method is found to be a more promising and viable tool for simulating turbulent flow and chosen for this study [2].

2. Governing equation

As mentioned above, in LES, only the influence of the small eddies has to be modeled by sub-grid scale model, whereas the large energy-carrying eddies are computed directly. Small eddies are more universal, random, homogeneous and isotropic, which simplifies the development of appropriate models.

In order to separate the large and small scale motions, the three-dimensional, time dependent Navier–Stokes equations are filtered. In the present study a box filter is applied as a filter kernel and an incompressible fluid is assumed. The governing equations are given by equation (1)-(3):

$$\frac{\partial \overline{u_i}}{\partial x_i} = 0 \tag{1}$$

$$\frac{\partial \overline{u_i}}{\partial t} + \frac{\partial}{\partial x_j}(\overline{u_i u_j}) = -\frac{\partial \overline{p}}{\partial x_i} + \frac{\partial}{\partial x_i}[\frac{1}{\text{Re}}(\frac{\partial \overline{u_i}}{\partial x_j} + \frac{\partial \overline{u_j}}{\partial x_i})] - \frac{\partial \tau_{ij}}{\partial x_j} \tag{2}$$

$$\tau_{ij} = \overline{u_i u_j} - \overline{u_i u_j} \tag{3}$$

where $\overline{u_i}$ is the velocity component of the resolved scales, $\overline{p}$ is the corresponding pressure. The filtering procedure provides the governing equations for the resolvable scales of the flow field. Although the continuity equation (1) of the resolved quantities is equal to the original unfiltered one, the filtered momentum equation (2) includes an additional term for the non-resolvable sub-grid scale stresses τ_{ij}, which results from filtering the non-linear convective fluxes. τ_{ij} describes the influence of the small-scale structures on the larger eddies. Only this effect has to be modeled by a sub-grid scale model [4].

For one-equation SGS Model, the eddy viscosity is calculated from equation (4):

$$\mu_{T,a} = \rho_a C_k \Delta k_{sgs}{}^{1/2} \tag{4}$$

where k_{sgs} represents the SGS kinetic energy, as show in equation (5), which is obtained by solving for transport equation for k_{sgs},

$$\frac{\partial}{\partial t}(\rho_r \alpha_r k_{sgs}) + \nabla(\rho_r \alpha_r u_r k_{sgs}) = -\nabla(\alpha_r \frac{\mu_{eff,r}}{\sigma_k} \nabla k_{sgs}) + \alpha_r (G_r - C_\varepsilon \frac{k_{sgs}^{3/2}}{\Delta}) \tag{5}$$

where, G is the production term, defined as equation (6):

$$G_r = \mu_{T,a} \left| \overline{S_{ij}} \right| \tag{6}$$

The value of model constants (C_ε =1.05 and C_k =0.07) in equation are considered on the basis of recommendation by Davidson [5].

3. Numerical implementation

3.1. Computational domain

The computational domain and grid are shown in figure 1 and figure 2. The diameter of the cylinder is D. For selection of spanwise length, according to Ref.[6], when the spanwise length of the cylinder $Z \geq \pi D$, the cylinder's flow will present obvious 3D effect. Three spanwise discretizations are chosen in order to investigate the influence of Δz to numerical results. Table 1 shows the total number of grid sizes:

Table 1. Grid Size.

Grid Type	Nx	Ny	Nz	Total
Coarse	175	114	12	239,400
Medium	175	114	18	359,100
Fine	175	114	24	478,800

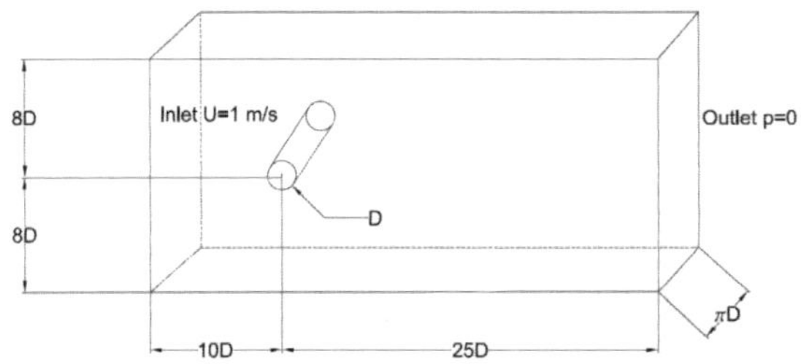

Figure 1. Computational Geometry.

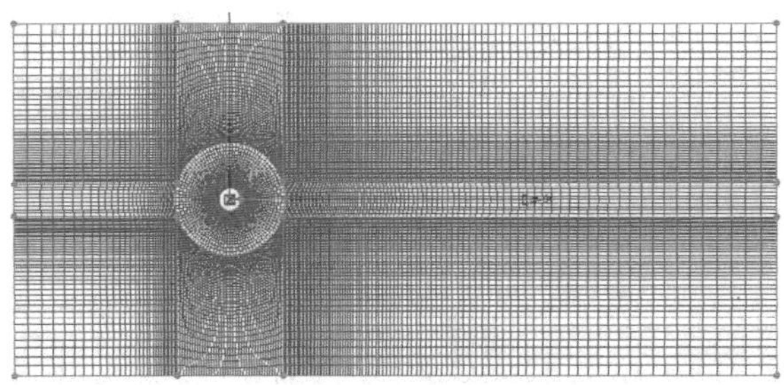

Figure 2. Computational Domain.

3.2. Boundary condition

The boundary condition for inlet is free-stream velocity U=1 m/s with fluctuation scale (0.02, 0.01, 0.01) and zero gradient for pressure. The pressure for outlet is specified as 0. For top, bottom and cylinder, no slip wall is applied. Symmetry Plane is set for the front and back of the domain.

3.3. Wall treatment

When LES is used to simulate the flow at high Reynolds number, a special near-wall treatment (wall-function) has to be introduced with the shortcoming that the near-wall regions cannot be properly resolved. For this project, as a relatively low Reynolds number (3900), the wall-function can be avoided by implementing the near-wall model. Specially, in order to fully resolve the viscous sub-layer, it is necessary to have y+<1 for the first cell adjacent to the cylinder.

3.4. Computational details

Table 2 below lists the computational details for this study.

IOP Conf. Series: Materials Science and Engineering **383** (2018) 012050 doi:10.1088/1757-899X/383/1/012050

Table 2. Computational Details.

Solver	pisoFoam
Time discretization scheme	Euler backward
Δt	0.005s
Time length	30s
Pressure-velocity coupling	PISO

4. Results and discussion

4.1. Drag coefficients

The comparison of drag coefficient between different methods and experimental data is listed in table 3:

Table 3. Comparison of Drag Coefficient.

	Cd		
Exp		0.98 ± 0.05	
LES	**Coarse**	**Medium**	**Fine**
	1.1075	1.0524	1.0447
RANS($k-\varepsilon$)		0.7446	
RANS($k-\omega$)		0.6208	

From figure 3 and 4, it is observed that the predicted drag and lift coefficients using LES agree well with the experimental data. While for the RANS, both $k-\varepsilon$ and $k-\omega$ underestimate the Cd value. Besides, the grid refinement from coarse to medium improves the result significantly, whereas for medium to fine, the improvement becomes less obvious.

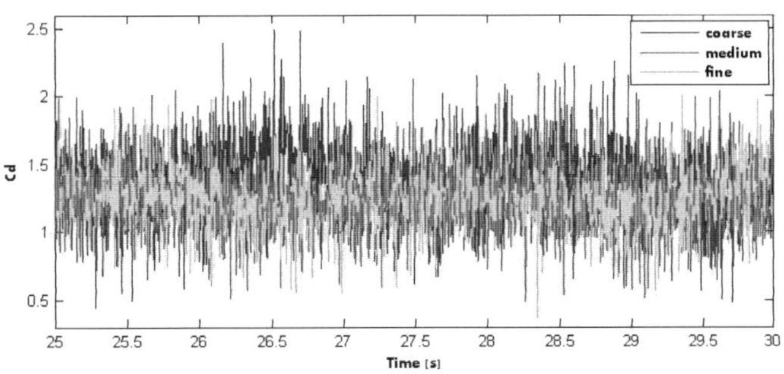

Figure 3. Time history of drag coefficient for different grid when Re=3900.

4.2. Lift coefficient

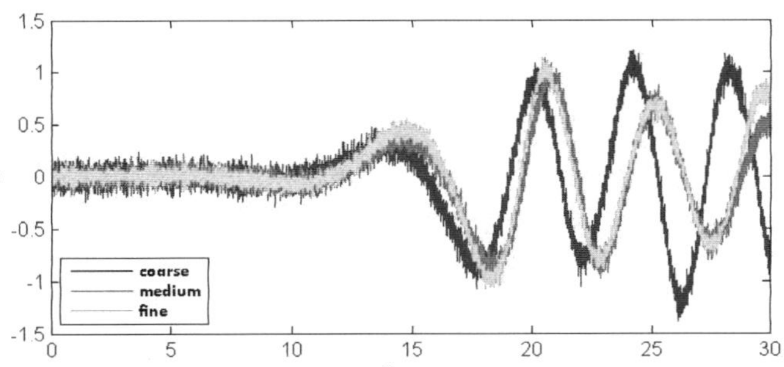

Figure 4. Time history of lift coefficient for different grid when Re=3900.

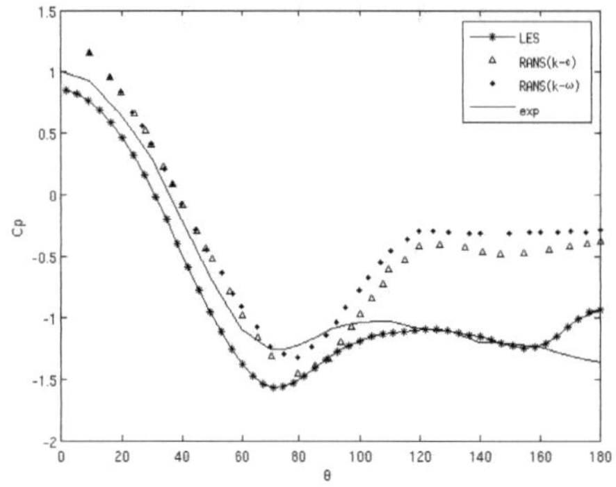

Figure 5. Comparison between different turbulence models and experimental data for pressure distribution around cylinder.

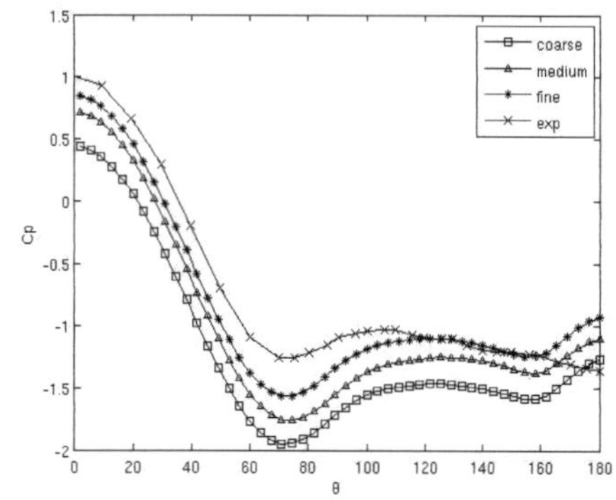

Figure 6. Comparison between different grids for pressure distribution around cylinder.

4.3. Pressure distribution

Figure 5 presents the distribution of the pressure coefficient around the cylinder at Re =3900 predicted by LES, RANS ($k-\varepsilon$) and RANS ($k-\omega$) models, which are compared to the experimental data. It can be observed that all three turbulence models predict the pressure distribution well from stagnation point to 60 degree. However, RANS models failed to accurately predict the pressure distribution after separation point (about 80 degree) of the flow occurred with increasing pressure coefficients. For LES, the pressure still distributes almost the same trend as experimental data, which shows that LES yields a better result than other two RANS models.

As for figure 6, it is obvious that fine grid give better prediction than medium and coarse grid. But the improvement from medium to fine is not that significant.

4.4. Mean stream-wise velocity

For mean stream-wise velocity in figure 7, it can be seen that the overall trend predicted by LES agrees the experimental data well except for the value at about x/D=1, which numerical results overestimate the velocity.

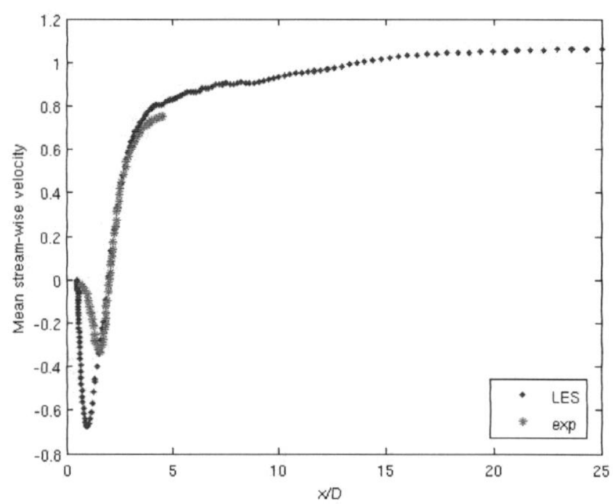

Figure 7. Mean stream-wise velocity in the wake centerline for flow over a cylinder.

5. Conclusion

Unsteady flow past a cylinder (Re=3900) was computed by large eddy simulation (LES). One-equation SGS model was applied, whose results validate against experimental data. Results show that the one-equation SGS LES model predicts the force coefficient, pressure distribution and mean stream-wise velocity better than the RANS ($k-\varepsilon$ and $k-\omega$) models. For investigation of spanwise discretization, it can be seen that if restricted by time and machine condition, medium grid is accurate enough to yield good results.

References

[1] Patel Y., *Numerical Investigation of Flow Past a Circular Cylinder and in a Staggered Tube Bundle Using Various Turbulence Models*, Master's thesis, Department of Mathematics and Physics, Lappeenranta University of Technology, 2010

[2] Jisheng Zhang & Yakun Guo, *Large Eddy Simulation of Flow around Circular Cylinders on Structured and Unstructured Grids*, European Conference on Computational Fluid Dynamics, 2006

[3] Zhiyin Yang, *Large-eddy simulation: Past, present and the future*, Chinese Journal of Aeronautics, (2015), 28 (1): 11–24

[4] Breuer M., *Large Eddy Simulation of the Sub-critical Flow Past a Circular Cylinder: Numerical and Modeling Aspects*, International Journal for Numerical Methods in Fluids, Vol. **28**, pp. 1281-1302, 1998

[5] Tabib M., & Schwarz P., *One-Equation SGS Modeling for Eulerian Large Eddy Simulation of Two Phase Solvent Extraction Pump Mixer Unit*, 17th Australasian Fluid Mechanics Conference, 2010

[6] Chen H. L., *Three-dimensional Numerical Simulation of the Flow Past a Circular Cylinder based on LES Method*, Journal of Marine Science and Application, Vol. **8**, pp.110-16, 2009

IOP Conf. Series: Materials Science and Engineering **383** (2018) 012051 doi:10.1088/1757-899X/383/1/012051

Ultimate strength of typical stiffened panels in container ships with comparison of hull girder bending analysis

Jinju Cui[1,2] and Deyu Wang[1,2,*]

[1] State Key Laboratory of Ocean Engineering, Shanghai Jiao Tong University, Shanghai, China
[2] Collaborative Innovation Center for Advanced Ship and Deep-Sea Exploration, Shanghai Jiao Tong University, Shanghai, China

*E-mail: dywang@sjtu.edu.cn

Abstract. Large container ship's structural strengths are the major concerns of owners and designers, and local stiffened panels within the mid-ship scope are similar due to the typical arrangements adopted in these ships. In this paper, investigations on longitudinal ultimate strength are made for these typical stiffened panels where considerations of modelling extent, boundary conditions and initial deflections are included. Additionally, the resulting force-load curves and collapse modes are compared with hull girder bending analysis (both hogging and sagging), thus validation are made in the more practical background. The study in this paper can be of reference values to similar container ships.

1. Introduction

Container ships are getting larger and are used extensively due to their high transportation efficiency and large capacity, and their structural strengths are the major concerns of owners and designers, especially after the "MOL Comfort" 8100TEU container ship crash accident in 2013 which broke into two and finally sank in deep sea. As long as hull girder ultimate strengths are concerned, the ultimate strengths of local structures and failure modes also need further investigations to help designers better understand the structural behaviours in specific situations.

The ultimate strength for local stiffened panels is a critical problem in practical structural designs, and many researchers have devoted to these studies, e.g. Fujikubo and Yao [1] studied the elastic buckling strength by introducing torsional rigidity to simulate the interaction between stiffener and plate, and the results have been referenced by classification rules; The ultimate strengths of continuous plates and continuous stiffened panels under combined transverse thrust and lateral pressure are also extensively studied [2, 3]; Tanaka, Yanagihara, Yasuoka, Harada, Okazawa, Fujikubo and Yao [4] studied 720 cases with different numbers, types and sizes of stiffeners using Nonlinear Finite Element Analysis (NFEA), and compared the results with predictions by several existing methods, such as CSR, PULS and FYH; Tekgoz, Garbatov and Guedes Soares [5] analysed the effect of residual stress on the ultimate strength of a thin rectangular stiffened and single plate, and studied the sensitive factor as well as best welding sequence etc.; Cui, Wang and Ma [6] studied ultimate strength of typical stiffened panels in container ships under localized pitting corrosion, and investigated the probabilistic characteristics of ultimate strengths for corroded structures; Kim, Lim, Kim, Hwang and Park [7] proposed an empirical formulation for predicting the ultimate strength of stiffened panels subjected to longitudinal compression, which employs a function of plate slenderness ratio and column slenderness

Content from this work may be used under the terms of the Creative Commons Attribution 3.0 licence. Any further distribution of this work must maintain attribution to the author(s) and the title of the work, journal citation and DOI.
Published under licence by IOP Publishing Ltd

IOP Conf. Series: Materials Science and Engineering **383** (2018) 012051 doi:10.1088/1757-899X/383/1/012051

ratio with two correction coefficients; Lindemann and Kaeding [8] applied the idealized structural unit method for ultimate strength analyses of stiffened plate structures, and compared the results with Finite Element Methods; Xu, Song, Pan and Soares [9] studied the ultimate strength of continuous stiffened panels under combined longitudinal compressive load and lateral pressure; Zhang [10] made a review and study on ultimate strength analysis methods for steel plates and stiffened panels in axial compression.

In this paper, we studied the ultimate strength of typical stiffened panels for a typical 10000TEU container ship by Nonlinear Finite Element Analysis, and compared with hull girder ultimate strength results in our previous research [11], so the validation can be made for boundary conditions and modelling extent etc. Additionally, the effects of initial deflections are studied. The results are of reference values for similar container ships.

2. Fundamental methods and model description

2.1. Ultimate strength by Nonlinear Finite Element Analysis

Nonlinear Finite Element Analysis (NFEA) is a unified method and can be applied to a wide range of engineering problems especially for complex structures like ships, aero planes and etc. Generally, for ship structures, the ultimate strength problems can be classified as hull girder bending and torsional ultimate strengths, and the local stiffened panels' (e.g. bottom panels, upper deck panels and etc.) ultimate strengths under uniaxial or bi-axial compression. For container ships, the most critical ultimate strength for hull girder is the hogging case because the loads of static water moment, wave induced moment and cargo loads are combined as the largest according to practical loading manuals, so the corresponding most dangerous local structures are the ones below neutral axis of mid-ship cross section, especially the bottom and inner bottom panels. In this paper, however, we studied both the hogging and sagging cases, and the ultimate strengths of stiffened panels under compression in either case are studied.

The hull girder ultimate strength results for both hogging and sagging results can be found in our previous research [11], and this is used as reference.

To study the ultimate strengths of local stiffened panels, the whole cross section of container ship is divided into several stiffened panels according to typical structural arrangements. Generally, the bottom panel, inner bottom panel, longitudinal girder panel, bilge panel, platform panel, deck panel, longitudinal bulkhead panel and outer shell panel etc. are selected and modelled individually with specific modelling extent and boundary conditions.

2.1.1. Modelling extent and boundary conditions of stiffened panels

The selection of modelling extent of local stiffened panels is according to structural arrangements and mechanics simplifications between connected structures as shown in figure 1. The longitudinal extent is determined by the "1/2+1+1/2 bay" model with periodic boundary conditions at both longitudinal ends, i.e. the displacements except the longitudinal translational ones are the same for paired nodes at the aft and fore ends, thus both symmetric and anti-symmetric longitudinal modes can be captured in the analysis. On the other hand, the transverse or vertical extent of the stiffened panels are determined by their connections with primary support members (PSMs) such as platforms or longitudinal bulkheads etc. Take the bottom panel as an example, the periodic boundary conditions are assigned between edges '*AI*' and '*BJ*' while simply supported boundary conditions are assigned at longitudinal girders and transverse frames locations, i.e. '*KL*', '*MN*', '*AB*','*CD*', '*EF*', '*GH*' and '*IJ*' while the edges are free to be pulled in to release the normal displacements within the '*xz*' plane.

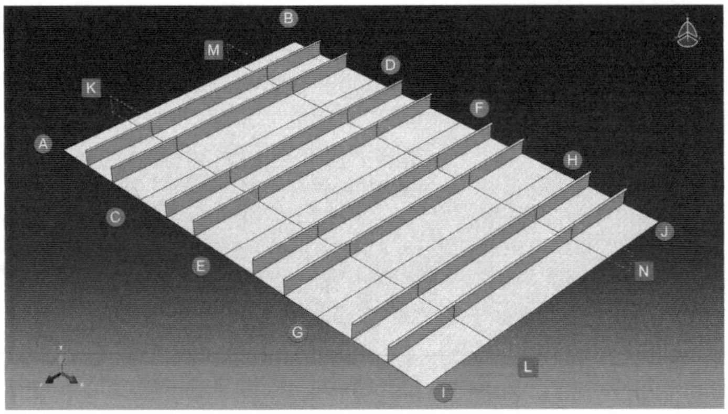

Figure 1. Illustration of modelling extent.

2.1.2. Initial deflections

As discussed by Ueda and Yao [12], both initial deflections and residual stresses exist in ship structures; they also investigated the modes and magnitudes of initial deflections according to measured results and actual practice. Thus, initial deflection is a practical consideration associated with manufacturing, and it is also included in the IACS Unified Requirements S11A; however, due to the lack of relevant information in practical cases, we adopted the initial deflections recommended by ISSC [13], and 3 kinds of initial deflections are proposed for structural analysis with Abaqus:

(a) Buckling mode initial deflection of plating:

$$w_{opl} = A_0 \sin\left(\frac{m\pi x}{a}\right)\sin\left(\frac{\pi y}{b}\right) \tag{1}$$

where x, y and z denote the longitudinal, transverse and vertical directions respectively; a denotes the span between adjacent transverse frames and b denotes the spacing between adjacent longitudinal stiffeners; m denotes the buckling mode of plating defined as the minimum integer satisfying $a/b \le \left[m(m+1)\right]^{1/2}$; A_0 denotes the magnitude of the above initial deflection, and is recommended as $0.1\beta^2 t_p$ where $\beta = b/t_p \bullet (\sigma_Y / E)^{1/2}$ is the slenderness ratio (t_p - plate thickness; σ_Y - material yield stress; E - Young's modulus).

(b) Column type deformation of stiffened panel:

$$w_{oc} = B_0 \sin\left(\frac{\pi x}{a}\right)\sin\left(\frac{\pi y}{B}\right) \tag{2}$$

where B denotes the spacing between longitudinal girders and B_0 denotes the magnitude of the column type mode, and is recommended as $0.0015a$.

(c) Initial tripping distortion of stiffener:

$$w_{os} = C_0 \frac{z}{h_w} \sin\left(\frac{\pi x}{a}\right) \tag{3}$$

where h_w denotes the height of the stiffener and C_0 denotes the magnitude of the stiffener tripping mode, and is recommended as $0.0015a$.

The above initial deflections are adopted in present study while residual stresses are not considered.

3. Model description

A typical 10000TEU container ship is chosen in this paper, the principle particulars are shown as table 1. The thickness distributions at mid-ship section are shown in figure 2 (left), and we can see that the thicknesses of longitudinal structures range from 9 mm to 70 mm, while high strength steels with different yield strengths (AH32, AH36, EH40, etc.) are adopted.

Table 1. Principle particulars for the 10000TEU container ship.

Principle particulars	Values
Overall length: L_{OA}	337.0 m
Length between perpendiculars: L_{PP}	320.0 m
Moulded breadth: B	48.2 m
Moulded depth: D	27.2 m
Scantling draft: T_{SC}	15.5 m
Block coefficient: C_b	0.642

4. Results and discussion

4.1. Results of hull girder ultimate strength

Firstly, the results of hull girder ultimate strength under both hogging and sagging bending moments are introduced as shown in figure 2 (right) according to the authors' previous research [11]. We can conclude that results of hogging and sagging from both Smith's method and NFEA are very close with each other, and the results from NFEA are selected for comparison in the following discussions.

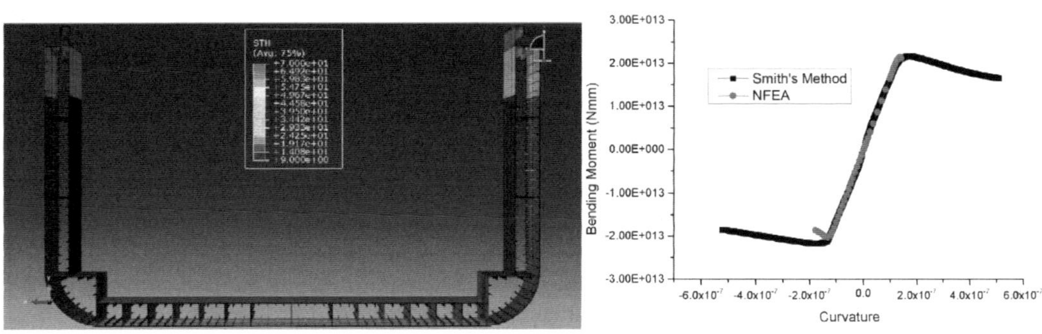

Figure 2. Thickness distributions at mid-ship section (left); Bending moment vs. curvature for Smith's method and Nonlinear Finite Element Analysis method (right).

4.2. Results of local stiffened panels' ultimate strengths

Based on model setups described before, ultimate strengths of local stiffened panels can be calculated by NFEAs in Abaqus, and the results for individual stiffened panels are shown as table 2.

Table 2. Ultimate strength results of stiffened panels for the 10000TEU container ship.

Stiffened panel	Stress distribution at ultimate strength	Force-displacement curves
Bottom		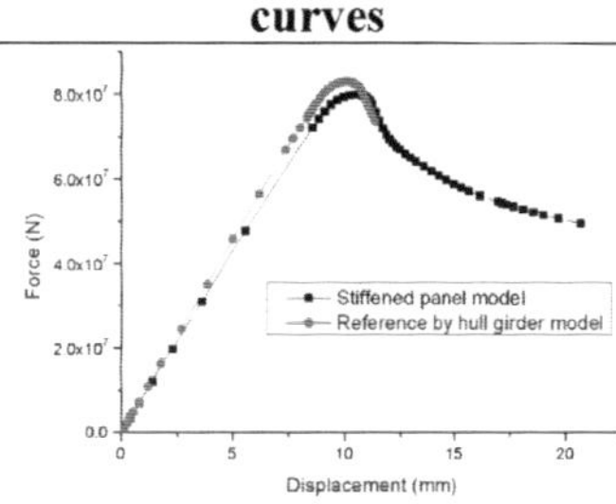

Inner bottom

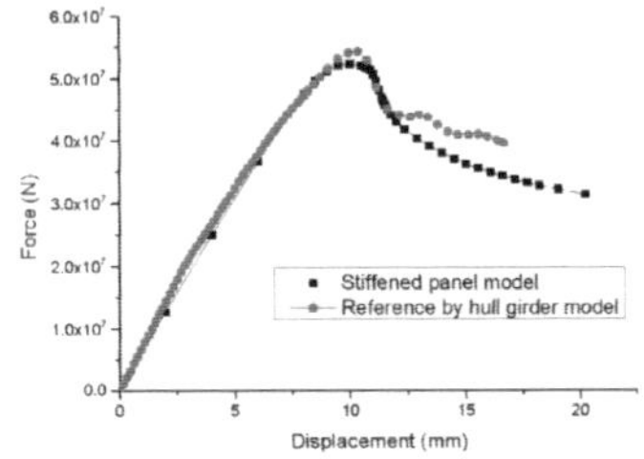

Bilge

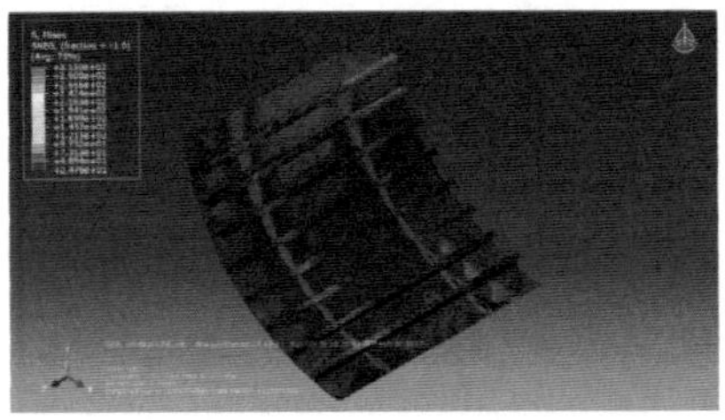

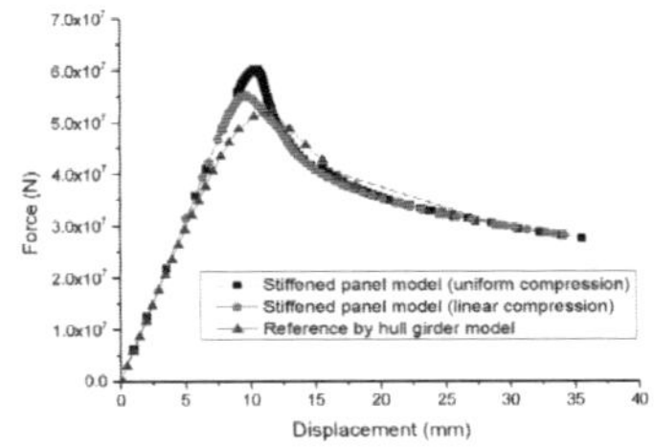

Longitudinal girder

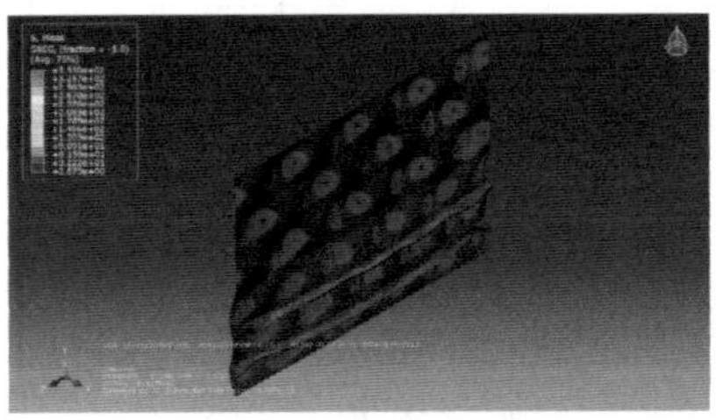

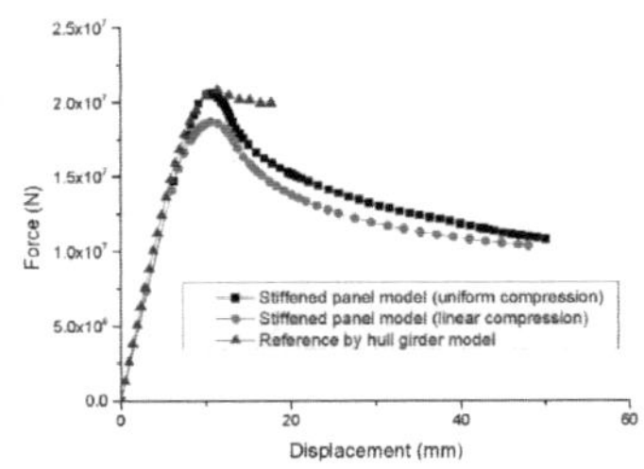

Step platform

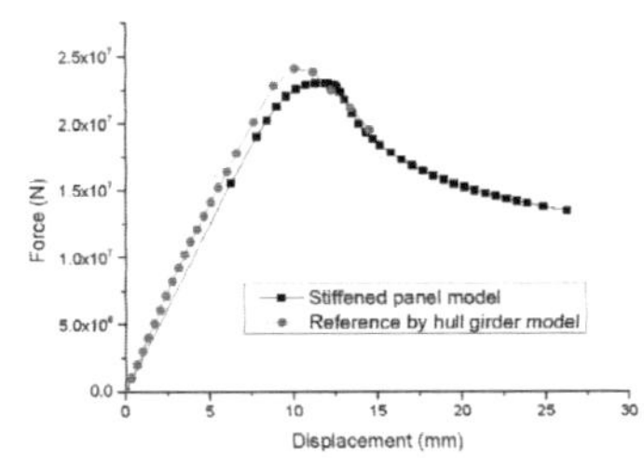

Second deck

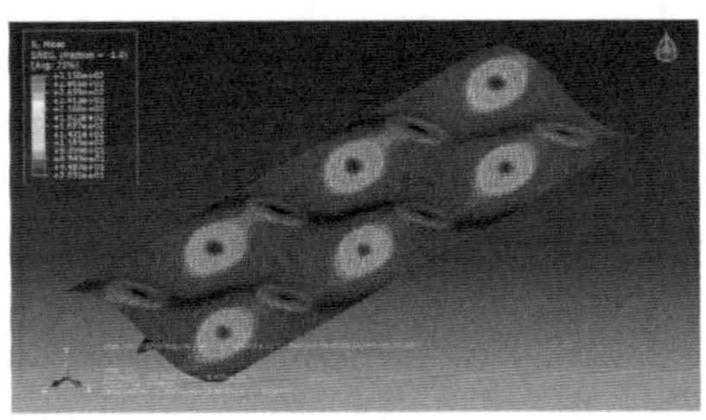

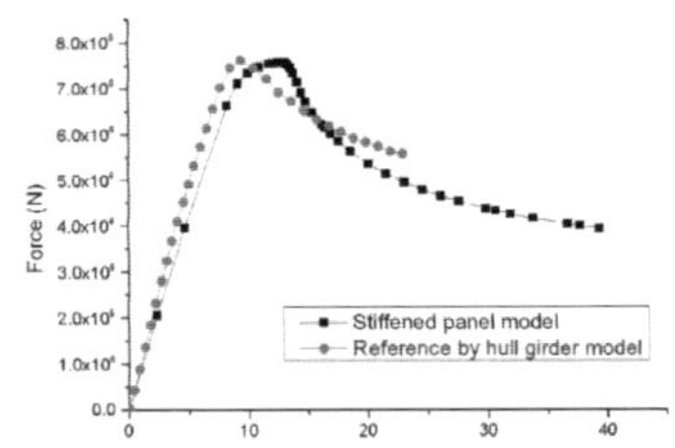

IOP Conf. Series: Materials Science and Engineering **383** (2018) 012051 doi:10.1088/1757-899X/383/1/012051

Longitudinal bulkhead (up)

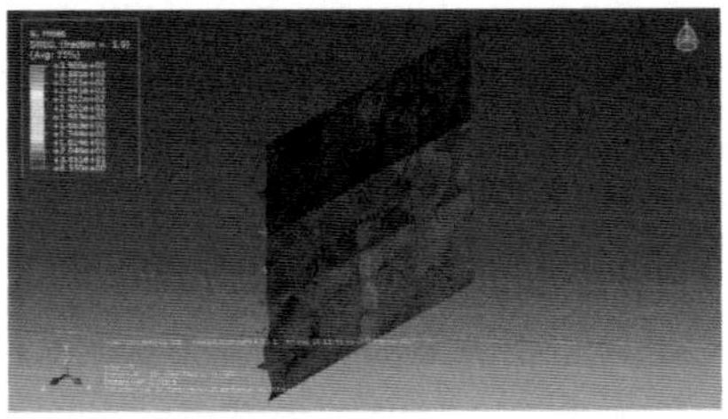
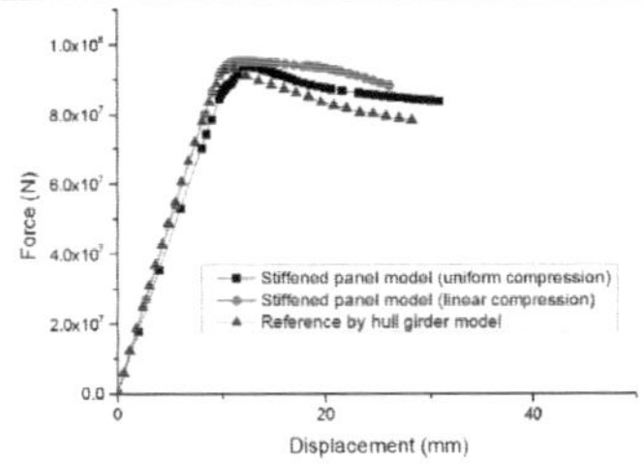

Side shell (up)

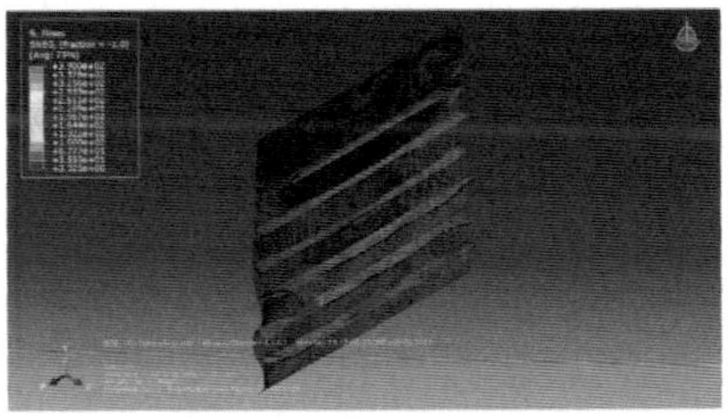
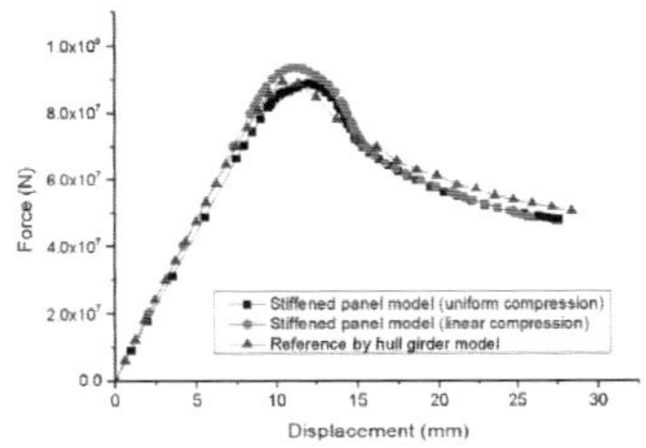

Upper deck

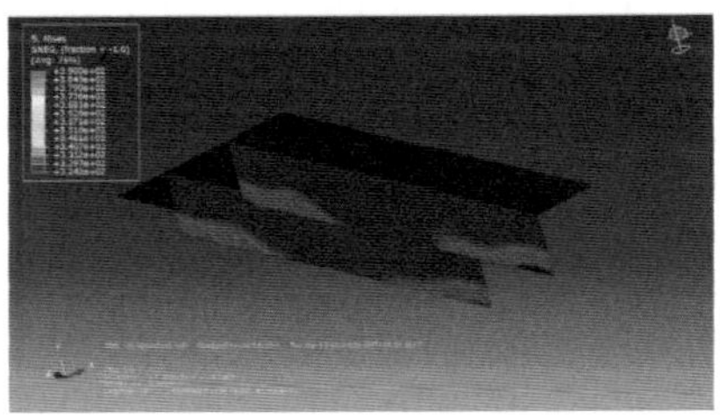
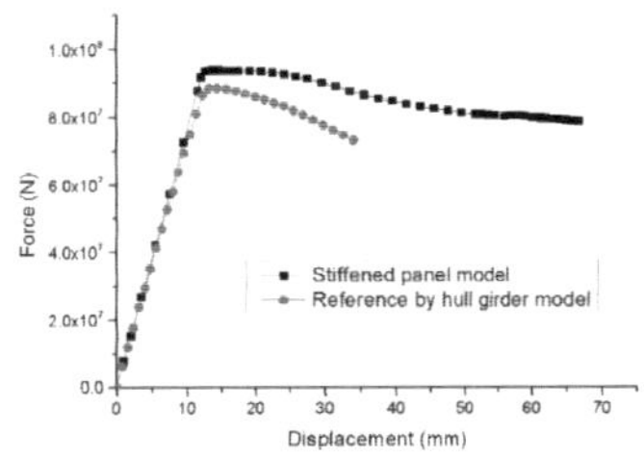

Hatch coaming

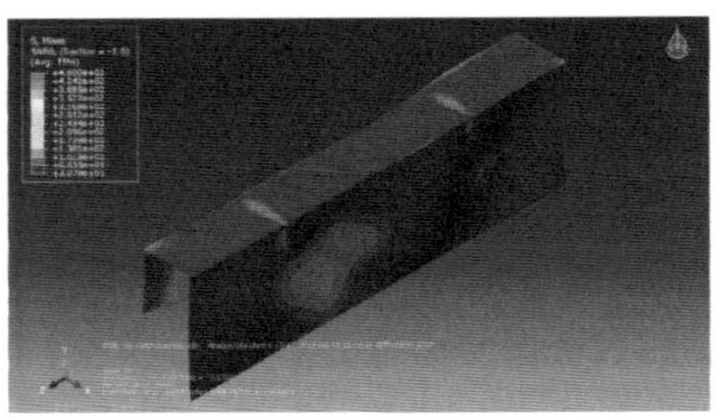
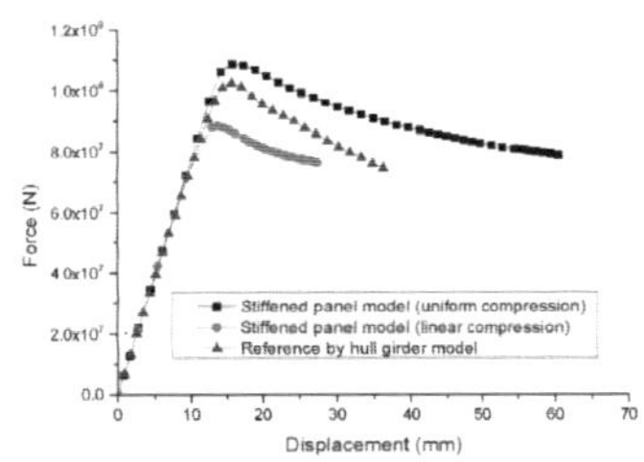

4.2.1. Bottom

Ultimate strength of bottom panel under longitudinal compression is shown in table 2. Results from adopted stiffened panel are very close with that from hull girder analysis, and collapse modes belong to local collapse between stiffeners. The difference may lie in mesh size effect (different mesh sizes are adopted between stiffened panel and hull girder models, and former mesh is much finer than latter) as described in our previous research [14], boundary condition effect and etc.

4.2.2. Inner bottom

Similarly, the results for inner bottom can be found in table 2 which shows local collapse modes and good agreement between stiffened panel and hull girder results.

4.2.3. Bilge

Initial deflections are not considered in bilge panel due to the complicated curved surface whose initial deflections are different from the plane panel and have been rarely reported so far. Additionally, the bilge panel covers a certain vertical extent which means the loading are different between top and bottom in hull girder bending, so we applied both uniform and non-uniform (linear in depth) compression. Apparently, both models overestimate the ultimate strengths, and linear compression is more accurate (5.8% overestimation); linear compression yields smaller ultimate strengths because local collapse happens, develops and expands due to non-uniform loading, i.e. earlier collapse happens rather than sudden overall collapse in uniform compression, and this conclusion can also be applied to other stiffened panels (unless with different thickness distributions).

4.2.4. Longitudinal girder
The #23 longitudinal girder extending from bottom to step platform is chosen, and results are shown in table 2, and we can see that uniform compression yields results of higher agreement with hull girder model, and linear compression yields lower ultimate strengths as analysed before. Additionally, the neutral axis changes rapidly especially after hull girder ultimate bending moments are reached, coinciding with the collapse of #23 girder. However, the linear compression with fixed neutral axis cannot reflect this change, which may cause worse results even than uniform compression.

4.2.5. Step platform
The step platform connecting previous longitudinal girder and side shell is chosen in this study, and the results show local collapse modes and good agreements between two models, and the stiffened panel model slightly underestimates the ultimate strength by hull girder model.

4.2.6. Second deck
The second deck is under compression in sagging bending, and its connections with double shells are not allowed to be pulled in due to double shell's strong rigidity. The results can be found in table 2, and it's evident that the stiffened panel model slightly underestimates the rigidity and ultimate strength. Additionally, the collapse mode belongs to local collapse rather than panel collapse.

4.2.7. Longitudinal bulkhead (up)
The results for longitudinal bulkhead locating beneath upper deck are shown in table 2, which shows good agreement between the stiffened panels and hull girder model. Apparently, collapse happens first in thinner plates, and then expands upwards.

4.2.8. Side shell (up)
The results are very similar with previous longitudinal bulkhead, and uniform compression yields results closer to hull girder model results.

4.2.9. Upper deck
The collapse mode immediately at ultimate strength is shown in table 2. Additionally, the stiffened panel model overestimates the ultimate strength by 5.9% than hull girder model.

4.2.10. Hatch coaming
Hatch coaming structures are also studied as shown in table 2. We can see that the collapse happens in the local plate collapse mode, while the ultimate strength by hull girder model lies between the uniform compression and linear compression models, and uniform compression is more accurate which overestimates the ultimate strength by 6.0% than hull girder model.

4.3. Influence of initial deflection on stiffened panel's ultimate strength
We also considered the effects of initial deflection magnitude on the stiffened panels' ultimate strengths. A series of NFEAs are calculated by varying the magnitudes and direction of initial deflection, and the results are normalized to unity by the ultimate strength values without initial deflection for more clear comparison between stiffened panels. Also note that bifurcation exists in the

intact structures without initial deflections, so buckling and post-buckling phenomenon cannot be captured in some cases, thus small initial deflections are adopted instead in these cases.

To demonstrate how initial deflection affects ultimate strength, we can take the bottom panel as an example, whose stress distributions immediately at ultimate strength are shown in figure 3 for the cases of zero and 1.5 times magnitude of initial deflection, respectively. Apparently, out-of-plane deformations develops much faster with larger initial deflection, thus the ultimate strength is degraded.

The results for different stiffened panels across the mid-ship cross section are shown in figure 4. Apparently, the ultimate strength decreases almost linearly with increased magnitude of initial deflection while sign or direction is not a sensitive factor for a given type of initial deflection combination, resulting in almost symmetric 'umbrella shaped' curves. Additionally, weaker or thinner stiffened panels are more likely influenced by the magnitude of initial deflections, and vice versa.

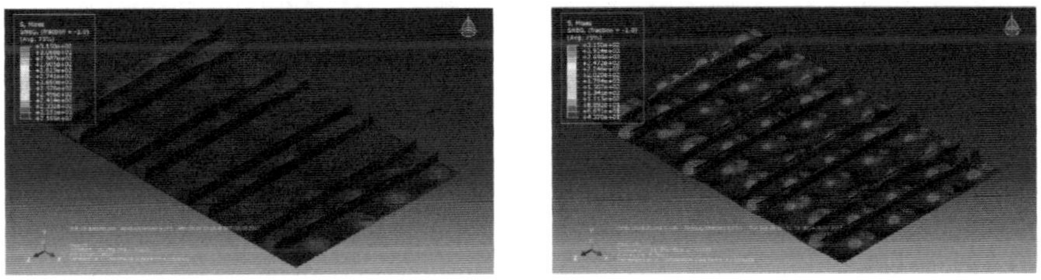

Figure 3. Bottom: stress distribution and deformation shape immediately at ultimate strength (0 x initial deflection, left; 1.5 x initial deflection, right).

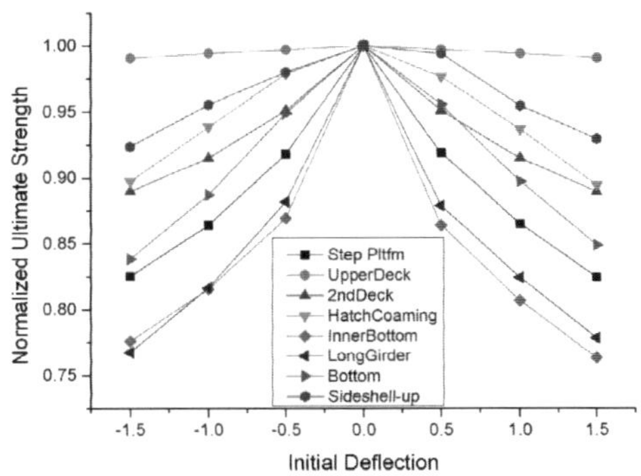

Figure 4. Influence of initial deflection magnitude on typical stiffened panel's ultimate strength (normalized).

5. Conclusions

In this paper, we investigated the ultimate strengths of typical stiffened panels in container ships, and compared the results with hull girder analyses. According to the modelling extent with boundary conditions in this paper, the ultimate strength results of typical panels under longitudinal compression matches well with hull girder model (difference within 6.0%), and the collapse modes belong to local collapse between stiffeners, which fulfils design principle that stiffeners should not fail before plates do. Thus the proposed methods are validated, and are supposed to find applications in other container ships or similar double-shelled ships. Additionally, the following conclusions can also be drawn:

(1) Generally, for vertically-spanned stiffened panels with equal thickness distributions, ultimate strengths of linear compression are smaller than uniform compression; however, due to the change of neutral axis in hull girder bending especially after ultimate strengths, the uniform may even yield more reasonable results than the fixed linear compression in specific cases.

(2) In the analysis of weak stiffened panels (e.g. 2nd Deck) between double shells, considering the rigidity of shell structures, it is more reasonable that their connections cannot be pulled in freely.

(3) The ultimate strength decreases almost linearly with increased magnitude of initial deflection while sign or direction of initial deflection is not a sensitive factor for a given type of initial deflection combination. Additionally, weaker or thinner stiffened panels are more likely influenced by the magnitude of initial deflections, and vice versa.

Acknowledgements
This paper is funded by both the Chinese Government Key Research Project KSHIP-II Project (Knowledge-based Ship Design Hyper-Integrated Platform, No.201335) and High-tech Ship Research Projects Sponsored by MIIT(2016-25).

References
[1] M. Fujikubo, and T. Yao, "Elastic local buckling strength of stiffened plate considering plate/stiffener interaction and welding residual stress," Mar. Struct., **12(9)**, pp.543-64 (1999)
[2] M. Fujikubo, T. Yao, M. R. Khedmati, M. Harada, and D. Yanagihara, "Estimation of ultimate strength of continuous stiffened panel under combined transverse thrust and lateral pressure Part 1: Continuous plate," Mar. Struct., **18(5-6)**, pp. 383-410 (2005)
[3] M. Fujikubo, M. Harada, T. Yao, M. Reza Khedmati, and D. Yanagihara, "Estimation of ultimate strength of continuous stiffened panel under combined transverse thrust and lateral pressure Part 2: Continuous stiffened panel," Mar. Struct., **18(5-6)**, pp. 411-27 (2005)
[4] S. Tanaka, D. Yanagihara, A. Yasuoka, M. Harada, S. Okazawa, M. Fujikubo, and T. Yao, "Evaluation of ultimate strength of stiffened panels under longitudinal thrust," Mar. Struct., **36**, pp. 21-50 (2014)
[5] M. Tekgoz, Y. Garbatov, and C. Guedes Soares, "Ultimate strength assessment of welded stiffened plates," Eng. Struct., **84**, pp. 325-39 (2015)
[6] J. Cui, D. Wang, and N. Ma, "Ultimate Strength of Typical Stiffened Panels in Container Ships Under Localized Pitting Corrosion," Proc. ASME 2016 35th International Conference on Ocean, Offshore and Arctic Engineering, American Society of Mechanical Engineers, V003T002A092-V003T002A092 (2016)
[7] D. K. Kim, H. L. Lim, M. S. Kim, O. J. Hwang, and K. S. Park, "An empirical formulation for predicting the ultimate strength of stiffened panels subjected to longitudinal compression," Ocean Eng., **140**, pp. 270-80 (2017)
[8] T. Lindemann, and P. Kaeding, "Application of the idealized structural unit method for ultimate strength analyses of stiffened plate structures," Ship Technol Res, **64(1)**, 15-29 (2017)
[9] M. C. Xu, Z. J. Song, J. Pan, and C. G. Soares, "Ultimate strength assessment of continuous stiffened panels under combined longitudinal compressive load and lateral pressure," Ocean Eng., **139**, pp. 39-53 (2017)
[10] S. M. Zhang, "A review and study on ultimate strength of steel plates and stiffened panels in axial compression," Ships Offshore Struc, **11(1)**, pp. 81-91 (2016)
[11] J. Cui, D. Wang, and N. Ma, "A Study of Key Influencing Factors in Ultimate Strength Analysis for Container Ships," The 27th International Ocean and Polar Engineering Conference, International Society of Offshore and Polar Engineers, San Francisco, California, USA(2017)
[12] Y. Ueda, and T. Yao, "The Influence of Complex Initial Deflection Modes on the Behaviour and Ultimate Strength of Rectangular Plates in Compression," J. Constr. Steel Res., **5(4)**, 265-302 (1985)
[13] J. Paik, H. Amlashi, B. Boon, K. Branner, P. Caridis, P. Das, M. Fujikubo, C. Huang, L. Josefson, and P. Kaeding, "Ultimate strength-final report of ISSC III. 1," Proc. Germany: International Ship and Offshore Structures Congress (ISSC 2012)
[14] J. Cui, D. Wang, and N. Ma, "A study of container ship structures' ultimate strength under corrosion effects," Ocean Eng., **130**, pp. 454-70 (2017)

IOP Conf. Series: Materials Science and Engineering **383** (2018) 012052 doi:10.1088/1757-899X/383/1/012052

Comparative analysis on the methods of preliminarily estimating resistance of yachts

Wu Pan, Chen Chaohe* and Zheng Xiao

College of Civil and Transportation Engineering, South China University of Technology, Guangzhou 510641, China

*E-mail: chenchaohe@scut.edu.cn

Abstract. The analysis of yacht resistance performance using empirical methods has become an accepted approach over the last decade. Which are based on existing statistical material of yachts to obtain the empirical equations or map methods. There are not many simple methods to obtain the resistance of yachts, and the accuracy of the results is not always uncertain for different types of yachts. The study is focused on determining the appropriate empirical equation to estimate resistance for different kinds of yachts. And based on computational fluid dynamics (CFD), XFlow software to calculate resistance of yachts under turbulent state is selected. This paper provides a review of these methods and two real yachts are chose as examples according to the classification of yachts .It discusses the resultant implications for practical applications. Simulations on an identical yacht using empirical methods and softwares are compared. Good agreements are achieved for both Daisumi Mihiro's empirical methods and numerical simulation methods, which shows that Daisumi Mihiro's empirical equation to estimate resistance of the planning yacht is relatively feasible and Daisumi Mihiro's map method is reliable for the transition and drainage boats.

1. Introduction

Most studies of performance of a ship are based primarily on the calm water resistance of the ship hull, and the same is true of yachts.The current methods for estimating resistance of yachts, using empirical equation estimation methods, or using Maxsurf software with empirical equation as calculation theory, is more efficient than predicting resistance of yacht by ship model test. But the drawback is that the results of different empirical equations have some differences, and the accuracy is uncertain. Looking at the literature on resistance calculation, most scholars focus on optimizing the CFD method of calculating resistance [1-3], but the quantitative analysis on the accuracy of the empirical equations is not sufficient. The existing estimation methods for resistance of yachts are finished to calculate the real yachts. The calculation results are compared with the numerical results to analyze the accuracy of these empirical equations.

The sailing state of yacht has certain influence on its performance, especially on the resistance performance. The applicability of the empirical equation is closely related to the navigation state, and the complex flow pattern often leads to the decrease of the accuracy of the empirical equation. The study focuses on the comparative analysis of the fast methods to estimate resistance of yachts.

2. The classification of yachts

Content from this work may be used under the terms of the Creative Commons Attribution 3.0 licence. Any further distribution of this work must maintain attribution to the author(s) and the title of the work, journal citation and DOI.
Published under licence by IOP Publishing Ltd

According to the volume Froude number Fr_∇, the yachts are divided into three categories, as show in equation (1).

$$Fr_\nabla = \frac{V_S}{\sqrt{g\nabla^{1/3}}} \tag{1}$$

where V_S is designed speed, knot; ∇ is displacement volume of yacht, m^3.

When Fr_∇ is less than1.0, the speed is slow. The hull floating in the water is basically relying on the hydrostatic buoyancy. If the volume Froude number of yacht is in the range, it is called a drainage yacht.

When Fr_∇ is more than 1.0 but less than 3.0, the head of the yacht will rise more obviously as the speed increases. However, at first there is a slight decline occurring in the tail and then the hull gradually tilts. When the yacht is hull-borne, the power of fluid increases but the drainage volume is decreasing. When the volume Froude number of yacht is in the range, it is called the transitional yacht.

When Fr_∇ is more than 3.0, the yacht is at a high speed. At that time, its head and tail will show great changes in the draft, eventually the whole hull is above surface of water. In the case, it can be called a planing boat. Yachts are basically supported by fluid power due to the small static buoyancy.

According to the three categories of yachts, the appropriate empirical equations are selected to estimate resistance of yachts.

3. The empirical equation method to estimate resistance of the planing boat

Among methods to estimate the resistance of planning crafts, the empirical equation put forward by Daisumi Mihiro is a very typical method. Therefore, that is used to estimate the resistance of planing boat. Daisumi have collected the data about brake horsepower from low power to high power ship and effective horsepower values of bare boats. There are more than100 ships. Finally, the brake horsepower and the effective horsepower spectrum are built on the basis of different ship parameters. At the same time, according to the data, Daisumi Mihiro put forward his own unique semi-empirical equation. The equation is mainly used to calculate brake horsepower required by the main engine BHP:

$$BHP = 1.45 * \beta * \Delta^{1.14} * V_S^{1.611} / L_{WL}^{0.806} \tag{2}$$

In the equation,L_{wl} L_{WL} is the length of waterline when the yacht is static, m; Δ is Displacement, t; V_S is designed speed, knot;*BHP* is brake horsepower, ps; β is The correction factor of ship length, $\beta = 1.12 - 0.0048 * L_{WL} \leq 1$, if β is greater than one, β is one.

After the *BHP* obtained by the equation (2), according to the relationship between horsepower and power, the propulsive power *P* is determined. Then the resistance *F* can be determined by the equation *P*=*F***V* between power and resistance.

The 26 'yacht is chose as an example for the planning craft, which is a small power yacht.

The principal dimension parameters are as following: Δ=2.561 t; B_M =2.20m; B_T =1.96m.

The model of the planing boat was completed by Rhinoceros software. As shown in figure 1, the model surface is smooth, which can reduce the error of the simulation result.

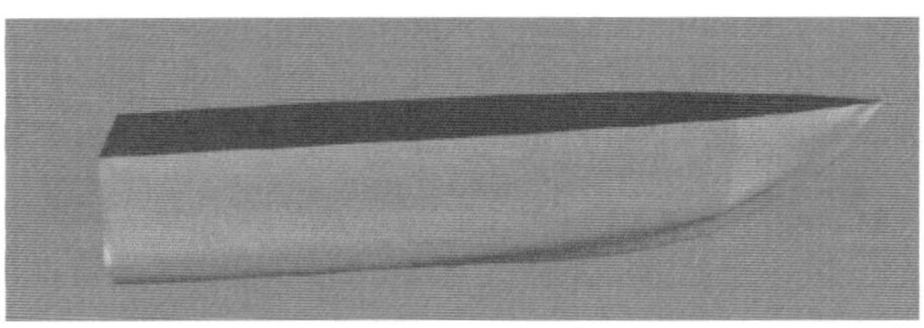

Figure 1. The 26-foot planing boat model.

Using the above method, resistance of the yacht under different volume Froude numbers is calculated .The results are shown in table 1:

Table 1. The result calculated by Daisumi Mihiro's empirical equation.

Fr_∇	3.1	3.2	3.3	3.4	3.5	3.6	3.7	3.8	3.9	4.0
R(kN)	9.00	9.18	9.35	9.52	9.69	9.86	10.03	10.19	10.36	10.52

4. The ways to estimate resistance of the transitional and displacement crafts

In the process of finishing the methods on calculating resistance of yachts, it is found that in most of ship model tests and theoretical methods analysis, the transition yachts and the drainage yachts are put into the same set of experiments, and the volume Froude number of the mother ship is mostly between 0.6~2.8. Therefore, in the classification of resistance calculation methods for yachts, it is envisaged that the resistance estimation methods of the transition and displacement yachts can be put into a group to analyze.

By reading and studying the relative references, it is found that Daisumi Mihiro's estimation method about brake horsepower at slow speed is used more frequently in estimating resistance of the transition and displacement yachts. As shown in figure 2, this is the brake horsepower estimation chart, the abscissa represents the ratio of the speed to the square root of the design waterline, that is V_S/L_{WL}; the ordinate is $BHP/\beta * \Delta^{1.14}$; the curve in the figure represents $L_{WL}/\Delta^{1/3}$. Based on the above information, the resistance of transition and drainage yachts can be calculated by the medium-low-velocity brake horsepower chart.

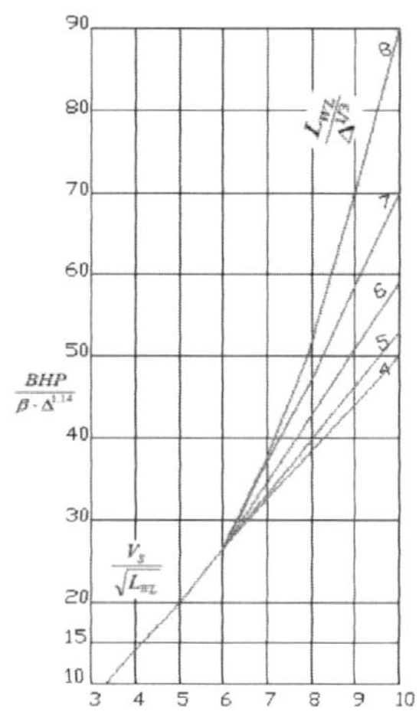

Figure 2. The estimation chart of Daisumi Mihiro's Brake horsepower.

Transitional and drainage yachts are usually medium-low speed yachts,so the 60'yacht is chose as an example. The principal dimension parameters are as follow:

L_{WL} =19.45m, B=5.76m, D=2.56m, d=0.962m, $\Delta = 39.96$t, LCB=1.244m

This is a 60'yacht. The model built by Rhinoceros is shown in figure 3.

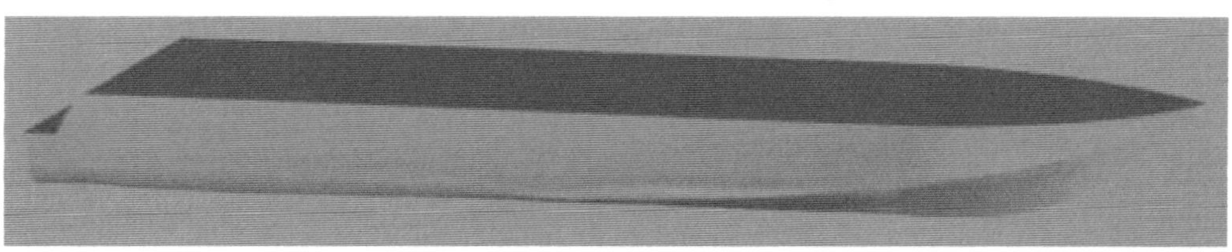

Figure 3. The 60' yacht model.

For Froude numbers Fr_∇ =0.8, 1.3, 1.8, 2.3, 2.8, the resistance of yacht is calculated. The results are shown in the table 2:

Table 2. Results calculated of the Daisumi Mihiro's Brake Horseforce Chart.

Fr_∇	0.8	1.3	1.8	2.3	2.8
V_s (knot)	11.59	18.83	26.07	33.31	40.56
$L/\nabla^{1/3}$	2.63	4.27	5.91	7.55	9.20
$BHP/\Delta^{1.14}$	15.10	18.30	26.20	35.10	51.20
R(kN)	124.79	93.06	96.23	100.89	120.89

5. Numerical simulation based on XFlow software

5.1. Computational theory

XFlow is the simulation software which applied in the new generation of computational fluid dynamics (CFD). It is based on the particle and the whole Lagrangian function, which can simply deal with the traditional complex computational fluid dynamics (CFD) problems. XFlow software has adopted Lattice Boltzmann method and large eddy simulation (LBM-LES), without dividing the mesh, which is a numerical method based on mesoscopic frame work. The method uses the distribution function to count the real variables, and that also make sure the conservation of the mass, momentum and energy in the process of fluid calculation.

5.1.1. Lattice Boltzmann method (LBM).

Boltzmann transportation equation [4] is as equation (3):

$$f_i(r+c_i\Delta t,t+\Delta t)=f_i(r,t)+\Omega_i^B(f_1,\dots f_b) \tag{3}$$

where f_i is the distribution function in the direction i, Ω_i^B is the collision operator, t is the discrete time, r is a position in the lattice and c_i is a velocity.

In the Boltzmann calculation method, the collision operator is simplified by BGK[4], which can solve the problem of low Mach number flow in fluid dynamics.

The operator is defined as equation (4):

$$\Omega_i^{BGK}=\frac{1}{\tau}(f_i^{eq}-f_i) \tag{4}$$

where f_i^{eq} represents the local equilibrium function, and τ is the relaxation characteristic time (which is related to the macroscopic viscosity).

Generally, the expression of equilibrium distribution function [5-6] is as equation (5):

$$f_i^{eq}(r,t)=t_i\rho\left[1+\frac{c_{i\alpha}v_\alpha}{c_s^2}+\frac{v_\alpha v_\beta}{2c_s^2}\left(\frac{c_{i\alpha}c_{i\beta}}{c_s^2}-\delta_{\alpha\beta}\right)\right] \tag{5}$$

In the above equation, c_s is the sound velocity; v_α and v_β are macro viscosities; $\delta_{\alpha\beta}$ is the Kronecker function, t_i is the parameter to ensure the spatial isotropy and ρ is the macroscopic density.

5.1.2. Turbulence model

In XFlow software, the turbulence model is simulated by the large eddy simulation (LES)[7-9]. LES solves the problem that turbulence sizes are larger than or less than given filters .LES is based on local numerical methods to simulate microscopic scales. Its analysis focuses on small scale turbulence simulation, which is close to the real physical model, and that does not require any subjective parameters to describe turbulence phenomena.

That is the approach employed in XFlow. In predicting resistance of yachts, the Wall-Adapting Local Eddy-viscosity (WALE) model is selected to describe the minimum scale of turbulence. WALE model has good performance. For laminar and turbulent flow, it is close to and away from the wall. The model recovers the asymptotic behavior of the turbulnt boundary layer when the layer can be directly solved and it does not add artificial turbulent viscosity in the shear regions out of the wake. The model directly simulates the boundary layer, which can reflect the gradual change of the turbulent boundary layer and also do not need to add artificial turbulent viscosity at the shear zone outside the tail vortex.

5.1.3. Surface info

After completing the resistance prediction of yachts, the surface information of the geometric models can be obtained by post-processing. The correlation coefficients are defined as follows:

c_p : Pressure coefficient, defined as equation (6):

$$c_p = \frac{2P_{static}}{\rho V_{ref}^2} \tag{6}$$

where P_{static} is local static pressure, ρ is the reference density and V_{ref} is the specified reference velocity.

c_f :Skin friction coefficient, defined as equation (7):

$$c_f = \frac{2T_w}{\rho V_{ref}^2}, T_w = \mu\left(\frac{\partial_v}{\partial_y}\right)_{y=0} \tag{7}$$

where T_w is the wall shear stress.

5.2. Numerical calculation of yachts

XFlow software is used to calculate the resistance of yachts.

5.2.1. Numerical calculation of the planing yacht

In the case of different Froude number, the corresponding resistance is shown in the table 3:

Table 3. The results for the planning yacht are calculated by XFlow.

Fr_V	3.1	3.2	3.3	3.4	3.5	3.6	3.7	3.8	3.9	4.0
R(kN)	9.31	9.55	9.82	10.11	10.57	11.09	12.15	12.57	12.81	12.91

When the Froude number is 3.1, the free liquid surface of yacht changes as shown below: Changes in velocity field are shown in figure 4 and figure 5.

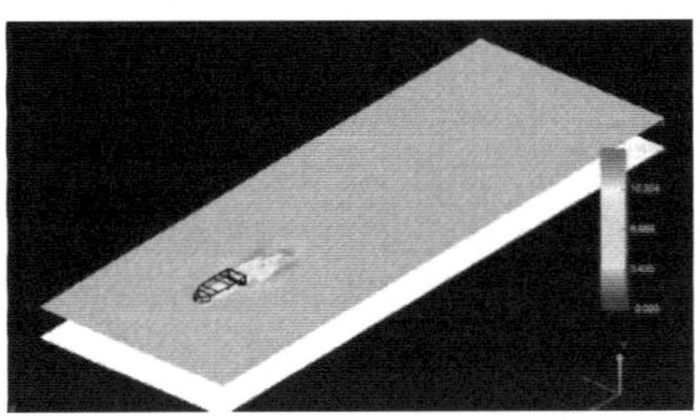

Figure 4. When *t*=1s, the free surface velocity field.

Figure 5. When *t*=30s, the free surface velocity field.

It can be seen that the velocity of the free liquid surface is small when t =1s, and the change of the free liquid velocity distribution becomes more and more obviously over time. When t =30s, the yacht tail produces a large number of velocity field, free liquid surface velocity distribution gradually spreads. After 30 seconds, the liquid level changes little and tends to be steady. The velocity distribution at each time is substantially symmetrical. The variation of the velocity field over time is close to the distribution of the flow field in the actual working condition.

Changes in free surface swirl current field are shown in figure 6 and figure 7.

Figure 6. The free surface eddy current field at t=1s.

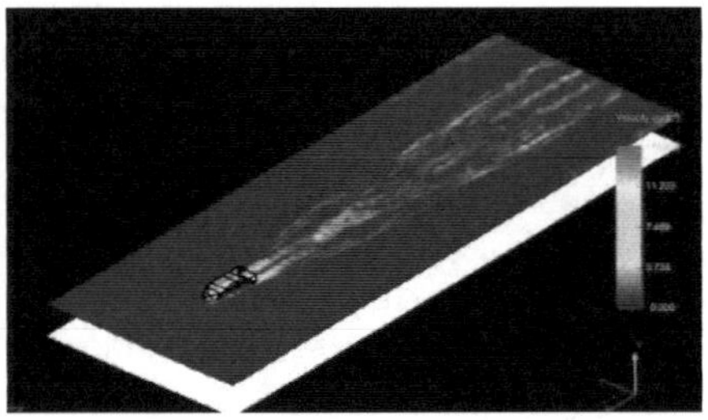

Figure 7. The free surface eddy current field at t=30s.

From the above two free surface vorticity graphs, it can be seen that when t =1s, smaller vortices are generated at both sides of the yacht and at the head, the vorticity is larger at the end of the yacht. When t=30s, the vortices at the rear of the yacht are noticeably large and the vorticity generating area becomes larger. As time increases, the vorticity distribution in the computational domain becomes larger and larger until the entire region is covered, and the vorticity distribution tends to be in a relatively stable state.

5.2.2. Numerical calculation of the transitional and displacement yacht

The resistance of yacht is calculated under different Froude numbers. The calculation data is shown in the table 4:

Table 4. Resistance of the 60' yacht is calculated by XFlow software.

Fr_{∇}	0.8	1.3	1.8	2.3	2.8
R(kN)	109.86	97.65	112.41	135.31	156.32

6. Prediction on the resistance of yachts based on Maxsurf

6.1. The theoretical method for calculating resistance by Maxsurf

In sailing, the resistance of a ship consists mainly of the following three parts: frictional resistance R_f, viscous pressure resistance R_{pv} and wave resistance R_W .Hughes held that anything related to viscidity including viscous resistance and frictional resistance should be combined together [10]. The total resistance of ship above static water can be integrated into the viscous resistance R_V which related to Reynolds number, and the wave resistance R_W related to Froude number as equation (8):

$$R_t = R_{pv} + R_f + R_W = R_V + R_W \tag{8}$$

The ratio of viscous pressure resistance coefficient C_{pv} and friction resistance coefficient C_f is the constant k. The total resistance can be expressed as equation (9):

$$R_t = (1+\kappa)R_V + R_W \tag{9}$$

Viscous resistance includes frictional resistance and viscous pressure resistance. The Hull speed module of Maxsurf uses the ITTC-57 friction resistance equation recommended by ITTC to calculate the viscous resistance R_V .The viscous resistance equation can be seen as equation (10):

$$R_V = \frac{1}{2} c_f (1+\kappa) \rho V^2 S \quad (10)$$

where V is the speed, kn; S is wet area, m^2.

6.2. The results about resistance of the planing craft

The example resistance of the planing craft under different volume Froude number has been calculated. The results are shown in the table 5:

Table 5. The results of the planing craft are calculated by Maxsurf.

Fr_∇	3.1	3.2	3.3	3.4	3.5	3.6	3.7	3.8	3.9	4.0
R(kN)	10.0	10.3	10.5	10.8	11.0	11.4	11.7	12.1	12.4	12.7

When the volume Froude number is 3.1, the waveform of the free surface is shown in figure 8 below:

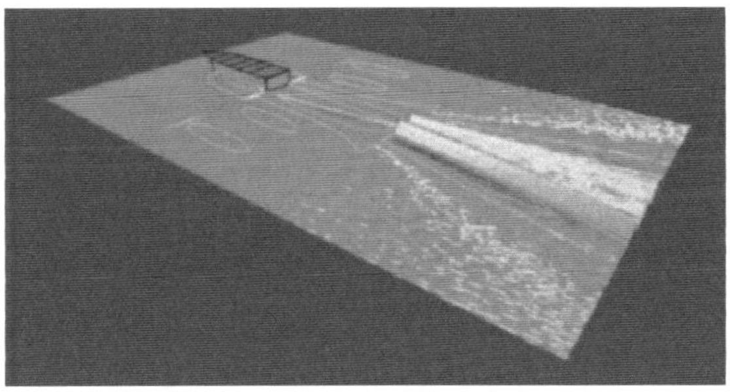

Figure 8. The waveform of the planing craft in free surface.

As can be seen from the above figure, when Fr_∇ =3.1, the waveform of free liquid is symmetrical on the side of the yacht and is concentrated at the end of yacht. The peaks and valleys have changed dramatically, that concentrated in the rear of the yacht. The waveform of the yacht in the free liquid surface also corresponds to the variation of the waveform under the actual conditions.

6.3. The calculation results about resistance of the transition and drainage yacht

Resistance of the transition and drainage yacht under different volume Froude number s has been calculated by Maxsurf.

The results are shown in the table 6:

Table 6. Resistance results of the transition and drainage yacht calculated by Maxsurf.

Fr_∇	0.8	1.3	1.8	2.3	2.8
R(kN)	33	57	107	137	180

7. The Comparative analysis on the resistance results of yachts calculated by different methods

The results of various methods for calculating the resistance of yachts are summarized, and the estimation methods for resistance of planning boats and drainage yachts are compared and analyzed respectively.

7.1. Comparative analysis on resistance of the planing boat by different methods

Based on numerical simulation, the Maxsurf software and the empirical equation, the resistance versus the volume Froude number is investi-gated by using these methods to calculate the resistance of planning boat. As shown in the following figure 9:

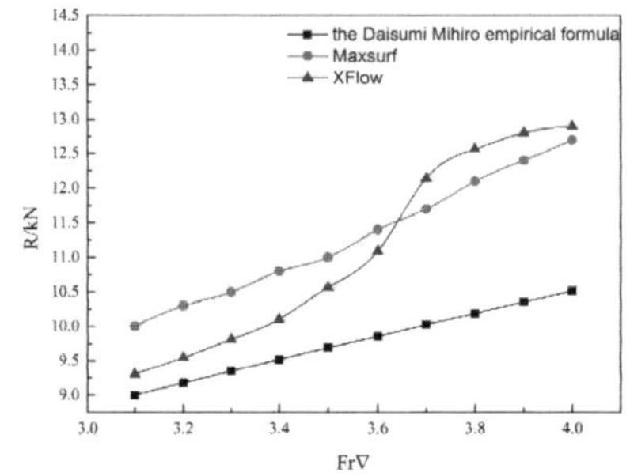

Figure 9. Resistance comparison analysis chart for the planing boat.

As can be seen from the diagram, the results of the three methods to calculate resistance of planing boat are different. The results of XFlow software and Maxsurf software are relatively close, and that are larger than the resistance obtained by the Daisumi Mihiro empirical equation. The tendency of the resistance curve increases with the increase of the volume Froude number, which agrees with the change tendency in actual conditions.

7.2. Comparative analysis on resistance of the transition and drainage yacht by different methods

By comparing the results of XFlow with the resistance of Maxsurf and empirical equation, the analysis curve can be drawn, as shown in figure10.

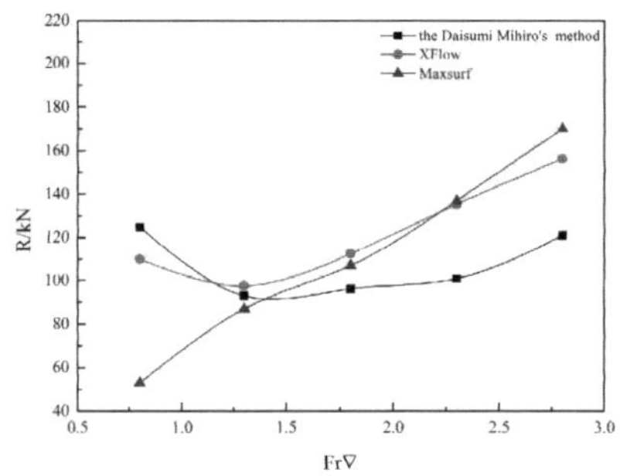

Figure 10. The comparative analysis chart of resistance for the transition and drainage boat.

From the diagram above, the results of the Daisumi Mihiro's map method and XFlow software are relatively close under the same condition. The trends of their resistance curves are the same. Which go down first and then rise. The change is consistent with the trend of resistance for the transition and drainage yacht under actual operating conditions. The results of Maxsurf are quite different from the XFlow when the volume Froude number is small, but with the increase of volume Froude number, the resistance calculated by the two methods is closer.

8. Conclusions

For single yachts, starting from resistance forecast, it has analyzed the predicting methods of the resistance for displacement yachts, transitional yachts and planning boats. The results are compared and analyzed. The following conclusions are drawn:

8.1. The comparison and analysis on resistance estimation methods for the planning boat

The results show that resistance curves obtained by the three methods have the same trend and are consistent with the actual situation. Quantitative analysis shows that the results calculated by Maxsurf and XFlow are relatively close and are larger than the Daisumi Mihiro empirical equation. With the increase of the volume Froude number, the difference between the resistance results obtained from the software and the empirical equation is greater.

Therefore, in the preliminary analysis of resistance, Daisumi Mihiro empirical equation to estimate resistance of the planning craft is relatively credible. These computational experiences provide an effective reference for resistance prediction of planning boats in the future.

8.2. The comparison and analysis on the ways to estimate resistance for the transition and drainage boats

The results of three methods for calculating resistance of the transition and drainage boats are presented. From qualitative analysis, the resistance of Maxsurf software varies with volume Froude number is not in conformity with the actual yacht. But in quantitative analysis, when the volume Froude number is greater than1.3, the calculation result is closer to the result of XFlow software and the Daisumi Mihiro's map method. Because XFlow is more comprehensive in theory compared with Daisumi Mihiro's map, so the final results in the later high speed part is are higher than the results of Daisumi Mihiro's map method. But the overall trend is consistent with the actual resistance trend of the transition and displacement type yacht with the volume Froude number. To sum up, for estimating resistance of the displacement and transitional crafts, Daisumi Mihiro's map method is more helpful and credible. When the volume Froude number is greater than 1.3, it is feasible to estimate the resistance of transition and drainage yacht by Maxsurf. That has some guidance and reference for the resistance prediction of transitional and drainage yachts in the future.

Above all, the XFlow software efficiently calculates the resistance of yachts. Which is easy to handle free surface, but the stability of free surface needs to be improved. In terms of time and cost, there is no model test to determine the resistance of yachts. Therefore, further research is needed on the forecast for the resistance of yachts.

References

[1] Yoo J and Kim HT 2006 Computational and experimental study on performance of sail of a sailing yacht *Ocean Eng.* 33 1322–42
[2] Venkata Subbaiah B, Santosh G Thampi and Mustafa V 2015 Modelling and CFD analysis of traditional snake boats of Kerala *Aquatic Procedia* 4 481–91
[3] Salina Aktar, Goutam Kumar Saha and Md Abdul Alim 2010 Drag analysis of different ship models using computational fluid dynamics tools *Proc.of Int. Conf. on Marine Technology(MARTEC)*Dec-2010 pp.123-29
[4] McNamara G and Zanetti G 1998 Use of the boltzmann equation to simulate lattice-gas automata *Physical Review Letters* 61 p. 2332
[5] Chen S and Doolen G 1998 *Lattice Bolzmann method for fluid flows* (Annu. Rev.Fluid Mech. vol 30) pp. 329-64
[6] Succi S 2001 *The Lattice Boltzmann equation for fluid dynamics and beyond* (Oxford:Clarend on Press)p. 308
[7] Mylonas D and Sayer P 2012 The hydrodynamic flow around a yacht keel based on LES and DES *Ocean Engineering* 46 pp. 18–32
[8] Constantinescu G S and Squires K D 2003 LES and DES investigations of turbulent flow over a sphere at Re¼10,000 *Flow, Turbulence Combustion* 70 pp. 267–98
[9] Georgiadis N J , Rizzetta D P and Fureby C 2010 Large-eddy simulation: current capabilities, recommended practices, and future research *AIAA J.* 48 (8) pp. 1772–84
[10] Hafez K A and El-Kot A A 2012 Comparative investigation of the stagger variation influence on the hydrodynamic interference of high speed trimaran *Alexandria Engineering Journal* 51 pp. 153-69

Evaluation of influence on shape design of floating offshore wind turbine substructures

Siwon Jang, Soonsup Lee, Donghoon Kang and Jonghyun Lee *

Department of Ocean System Engineering, Gyeongsang National University, Tongyeong, 650-160, Korea

*Email: gnujhlee@gnu.ac.kr

Abstract. The increase of demand and the depletion of energy resources have led to a rising demand of renewable energy. Among them, wind turbine has been researched because of wide generality with high resource potential. Despite these advantages, onshore wind turbine has a spatial constraint. To solve these problems, offshore wind turbine has been researched and the installation area of onshore structures has been expanded to offshore field. The design of the substructure is one of the most important issues in case of motion characteristics in the ocean rather than on the coast. Tri-pod type is attracting attention because it is easy to manufacture and efficient to reduce environmental load. However, the motion characteristics of these types of structures have not been clarified yet and are currently studied. In order to improve the float performance, it is necessary to apply the optimal float design such as size, weight, water plane area, and mooring to each design. In this study, we have described tri-pod type model and evaluated the model test to find the effects of the ratio of the diameter to draft in dynamic motions of the structures on the basis of the floating column design. Since each design variables are determined, experimental RAO results have analysed through motion analysis of 6 degrees of freedom.

1. Introduction

The increase of demand and the depletion of energy resources have led to a rising demand of renewable energy. Among them, wind turbine has been researched because of wide generality with high resource potential [1]. Despite these advantages, onshore wind turbine has a spatial constraint. To solve these problems, offshore wind turbine has been researched and the installation area of onshore structures has been expanded to offshore field [2]. In the case of wind turbine, the size of generators is becoming larger due to the formation of large-scale industrial complexes, and in order to increase the power generation efficiency, the need for floating type offshore wind turbine is gradually increasing.

The design of the substructure is one of the most important issues in case of motion characteristics in the ocean rather than on the coast. Tri-pod type is attracting attention because it is easy to manufacture and efficient to reduce environmental load. However, the motion characteristics of these types of structures have not been clarified yet and are currently studied. In the case of heave plates, it affects the vertical motion of the column by forcing the viscous damping by attaching to the structure [3]. In order to improve the float performance, it is necessary to apply the optimal float design such as size, weight, water plane area, and mooring to each design. To research the effect of tri-pod type structure on the column characteristics of the float for optimum design, the influence of the column diameter and draft on the design of the substructure shape is evaluated. In this study, the semi-

Content from this work may be used under the terms of the Creative Commons Attribution 3.0 licence. Any further distribution of this work must maintain attribution to the author(s) and the title of the work, journal citation and DOI.
Published under licence by IOP Publishing Ltd

submersible model of the NREL 5MW wind turbine generator was adopted as the standard model. The displacement was fixed and the diameter changed to about ±20%.

2. Data

2.1. Standard model

The model adopted in this study is NREL 5MW wind turbine generator. The main specifications are shown in table 1.

Table 1. Mooring system properties.

Rating	5MW
Rotor orientation, configuration	Upwind, 3 blades
Control	Variable speed, collective pitch
Drivetrain	High speed, multiple-stage gearbox
Rotor, hub diameter	126m, 3m
Hub height	90m
Cut-in, rated, cut-out wind speed	3m/s, 11.4m/s, 25m/s
Cut-in, rated rotor speed	6.9rpm, 12.1rpm
Rated tip speed	80m/s
Overhang, shaft tilt, precone	5m, 5 °, 2.5 °
Rotor mass	110,000kg
Nacelle mass	240,000kg
Tower mass	347,500kg
Coordinate location of overall CM	(-0.2m, 0.0m, 64.0m)

2.2. Model Experiment

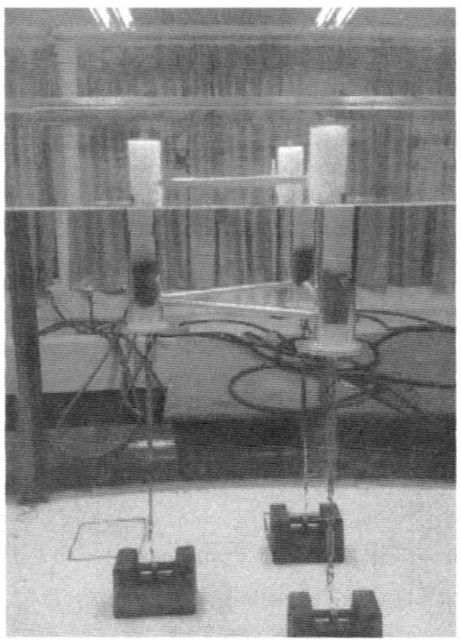

Figure 1. Model 1.

Figure 1 is a photograph of Model 1 used in this study. It was made in the form of Tri-pod semi-submersible.

Table 2 shows the main dimensions of the model made in figure 1 as dimensions, and the weight of the model was accommodated within the error range of 1%.

The purpose of this study is to evaluate the effect of the structure on the change of diameter and draft. Therefore, the displacement, thickness and size of heave plate are fixed.

Table 2. Main dimensions of model.

		M_1	M_2	M_3	M_4	M_5
Diameter	m	0.08	0.09	0.10	0.11	0.12
Length	m	0.396	0.314	0.255	0.211	0.178
Column interval	m	0.36	0.35	0.34	0.33	0.32
Weight	kg	3.180	3.120	3.170	3.150	3.150
Displacement	kg			6.024		
Heave Plate Diameter	m			0.16		
Heave plate thickness	m			0.03		

Table 3 shows the main materials for making the model. The structure was made of acrylic, the bracing was made of wood, and sand was used to match the draft. In order to prevent the internal structure from moving, the margins were fixed to the PE form.

Table 3. Material of the model.

	Material
Structure	Acrylic
Bracing	Wood
Draft setting	Sand
Blank space	PE Form

2.3. Experimental data

In the model experiment, 5 models were repeated 3 times. We tried to carry out the experiment with 3 to 20 period wave which is commonly used [4]. However, it was difficult to generate the short wave in the experimental environment. In the end, the experiment was conducted about 7 to 20 period wave.

Table 4 shows the values obtained by reducing to 1: 100 in consideration of the law of superiority [5].

For the mooring, the TLP method was used, and the spring constant K used for mooring was 809 N/m.

Table 4. Experimental data.

Period (s)	Frequency (rad/s)	Wave Height (m)	Water Depth (m)
0.7	8.97598	0.06	0.95
0.8	7.85398		
0.9	6.98132		
1.0	6.28319		
1.1	5.71199		
1.2	5.23599		
1.3	4.83322		
1.4	4.48799		
1.5	4.18879		
1.6	3.92699		
1.7	3.69599		
1.8	3.49066		
1.9	3.30694		
2.0	3.14159		

3. Analysis of experimental results

3.1. Six degrees of freedom analysis
In this study, we researched the 6 Degree of freedom motion of a model by generating a wave. The data were analyzed using a motive program. In the case of roll, pitch and yaw, Data could not be obtained due to error values. In case of surge, sway, and heave, position data were obtained within a small error range without error value.

3.2. RAO Analysis
Figures 2 to 4 show. The RAO of Heave and Pitch, which are the most influential motions in the ocean structure about the six degrees of freedom motion. And since it is an experiment for a wave in one direction, it also shows surge RAO. Due to the error in the rotation data, the pitch RAO is represented by two models.

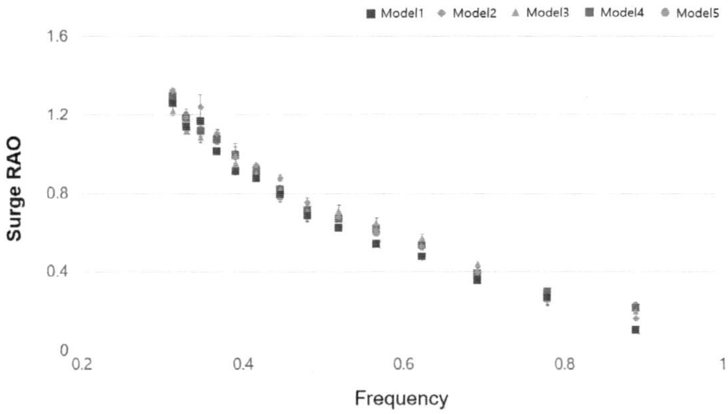

Figure 2. Surge RAO.

Figure 2 shows the Surge RAO of model 1 to model 5. Three repetition experiments were performed for each model, three points for each frequency are displayed in the error bar. In the case of surge RAO, there is no significant difference between the models.

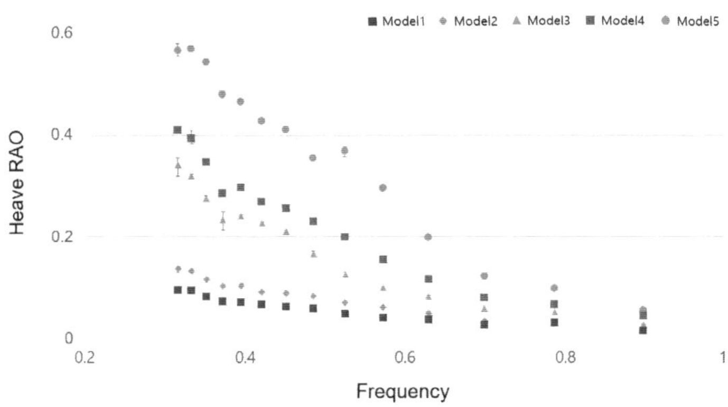

Figure 3. Heave RAO.

Figure 3 shows the Heave RAO of model 1 to 5. In the Heave RAO, three repetition experiments were performed for each model, three points for each frequency are displayed in the error bar. As the number of models increases, the data size also tends to increase.

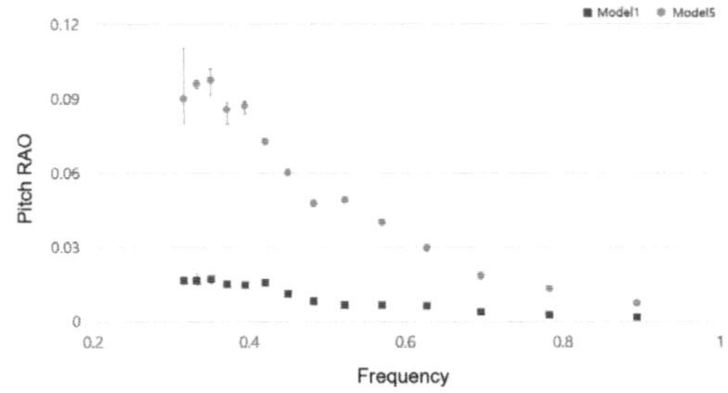

Figure 4. Pitch RAO.

Figure 4 shows the Pitch RAO of model 1 and model 5. In the Pitch RAO, three repetition experiments were performed for each model, three points for each frequency are displayed in the error bar. As the number of models increases, the data size also tends to increase. Rotation data was analyzed only by two models due to errors. However, the data shows that it is the same trend as the Heave RAO.

4. Conclusions and Future Research

In this study, the NREL 5MW wind turbine was adopted as a standard model and reduced to 1: 100 size, and the RAO of a tri-pod type semi-submersible substructure was researched. Model tests conducted on five models showed that the RAO of each model was researched by varying the diameter and draft of the column and fixing the displacement of the column. As a result, it was confirmed that the smaller the diameter of the column the more stable the motion.

Based on RAO that were confirmed as five model tests, we will simulate the floating offshore wind turbine using SW the optimum design is performed by comparing and analyzing the RAO and SW programming results of each motion obtained from the model experiment.

Review

This research was supported by the Korea Research Foundation's BK21 plus Program and the research was supported by 'Software Convergence Technology Development Program', through the Ministry of Science and ICT(S0503-17-1001).

References

[1] European Wind Energy Association 2013 *The next step for offshore wind energy*, (Brussel: EWEA)

[2] Han, K. M. et al., 2011. Optimal Design of Floating Offshore Wind Turbin Substructure, *Journal of the Society of Naval Architects of Korea Proceedings of the Annual Autumn Meeting*, SNAK, (mokpo), 3-4 November, 2011 pp. 541-546

[3] A.Subbulakshmi et al., 2015 Effect of Viscous Damping on Hydrodynamic Response of Spar with Heave Plate, *Aquatic Procedia* 4 (2015) 508-515

[4] R Bhattacharyya 1978 *Dynamics of Marine Vehicles, Director of Naval Architecture*: (Maryland)

[5] S. Nallayarasu & Senthil Kumar 2017 Experimental and numerical investigation on hydrodynamic response of buoy form Spar under random waves, (SHIPS AND OFFSHORE STRUCTURES) VOL 12 NO 5 pp734-7

IOP Conf. Series: Materials Science and Engineering **383** (2018) 012054 doi:10.1088/1757-899X/383/1/012054

Robust Wi-Fi fingerprinting-based positioning in the presence of lying identities

Wei-Chung Lu, Shih-Chun Yeh, Chung-Chih Chuang and Shih-Hau Fang

Yuan Ze University, Taiwan

*E-mail: a53051232000@yahoo.com.tw

Abstract. The lying identity of an access point (AP) is one of the most serious threat in Wi-Fi positioning because an adversary can easily acquire a valid address by monitoring the transmission and masquerade as another AP in the networks. This study proposes a robust Wi-Fi localization algorithm that can tolerate the liars instead of explicitly detecting them. The proposed algorithm considers all possible combinations of APs in an unionbased approach such that the adversaries cannot easily affect the positioning results by masquerading APs. On-site experimental results demonstrate that this approach apparently achieves more robust location estimation than the Bayesian approach and the cluster-based method in the presence of lying identities.

1. Introduction

Secure localization has gained considerable attention in the last several years because position estimations are often required for critical applications such as location-based authentication [1], [2], [3]. Today, many location systems utilize the base station's broadcasting power in wireless or cellular networks to infer the user's location due to the wide coverage and the widespread use of receivers [4], [5]. Among various approaches, fingerprinting is a popular location architecture, in which the user's location is estimated by comparing received signal strength (RSS) with the values pre-stored in a database [6], [7]. This architecture allows the location system to reuse existing wireless infrastructure [8], [9]. Recently, the massive deployment of Wi-Fi access points (APs), and the popular RSS sensing function on mobile devices make Wi-Fi a suitable technology for developing such location systems [10], [11]. However, due to the license-free spectrum (2.4GHz) and the shared nature of the wireless medium, the propagated radio waves in Wi-Fi networks are susceptible to attacks. Among various attacks, the lying identity of AP is definitely one of the most serious threat [12], [13].

In a typical 802.11 Wi-Fi environment, the MAC address is a unique identifier for an AP in Wi-Fi networks. This address is important for a location system because it indicates which AP transmits the signal. However, an adversary can easily acquire a valid MAC address and masquerade as another AP in the networks. Figure 1 depicts a typical scenario where the Wi-Fi network contains a lying identity. In this case, users in the Wi-Fi network cannot distinguish signals from AP1 and the lying AP. Such lying identities have a serious impact on a location system because the RSS value from AP1 mixes the lying data. That is, the adversary can easily fool the Wi-Fi location system by monitoring the transmission, masquerading APs, and changing the lying AP's broadcasting power.

The main contribution of this work is to use an enhanced probabilistic localization algorithm in [17] for building a robust Wi-Fi location system in the presence of lying identities. The proposed method of dealing with lying identities is to design dependable localization scheme that tolerates the liars instead

Content from this work may be used under the terms of the Creative Commons Attribution 3.0 licence. Any further distribution of this work must maintain attribution to the author(s) and the title of the work, journal citation and DOI.
Published under licence by IOP Publishing Ltd

of explicitly detecting them. Compared to the traditional probabilistic approaches which use the intersection of RSS from all APs, the proposed algorithm considers all possible AP combinations in an union-based approach such 2 that the adversaries cannot easily affect the positioning results by pretending APs. The main advantage of this approach is that a lying identity contributes little to the localization process, and thus, achieving more robust location estimations under security threats. `

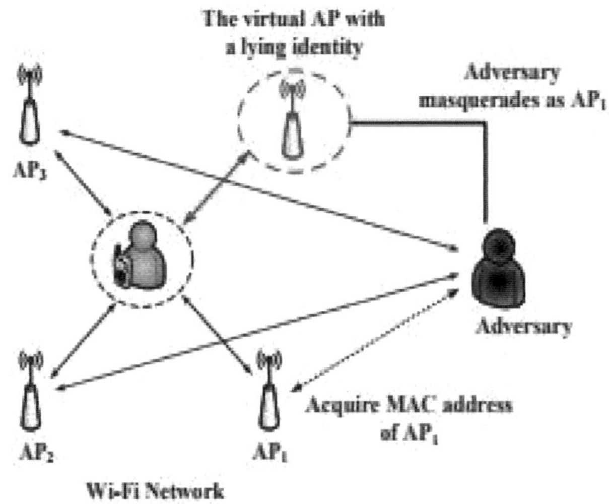

Figure 1. A Wi-Fi network contains a lying AP, where the RSS values from AP1 mix the lying data.

2. Background and related work

In this paper, we focus on the fingerprinting-based indoor location system that can effectively defend against lying identities in Wi-Fi networks. In a fingerprinting-based location system, the online stage is more vulnerable to attacks because the validation and detection schemes can be applied during the off-line stage. Moreover, the training data can be collected over various days to avoid corruption. Thus, this study assumes an attack-free condition in the off-line stage. Researchers have proposed many mechanisms to defend various attacks, including sybil [19], spoofing [15], [20], and transmit power attacks [13], [21].

Common approaches to securing wireless localization fall into two categories: attack detection and elimination, and robust localization schemes. The first method tries to detect the attacks and eliminate them from consideration during the positioning procedure. For instance, Liu et al. filtered out malicious beacon signals using the consistency among multiple beacon signals [22]. The second method of dealing with attacks is to design a robust localization scheme. For example, Li et al. developed two robust statistical methods, including a least median square position estimator for triangulation-based systems, and a median-based distance metric for fingerprinting location systems [23]. Liu et al. proposed a voting scheme to achieve attack-resistant location estimation [22].

3. Proposed algorithm

Assume a device senses N APs, then the observed RSS can be represented by a vector X with N elements as X=(x_1 , x_2 , · · · , x_N), where x_N means the observed RSS from the n-th AP. A fingerprinting location system regards X as an input and outputs the estimated location by a previously constructed RSS-location relationship. However, the presence of cheating APs may cause some xn totally unreliable and thus, result in a fake position of the user. The problem in robust localization is how to correctly locate the user given X which contains some lying x_N. In traditional probabilistic location systems, the joint probability of all sensed observations is estimated for all candidate locations to find the user's position. The main idea of such approaches is that each x_N is combined based on the "and" operator, denoted by $\wedge$. This is because traditional methods assume all APs behave honestly. Assuming each x_N is independent, the joint probability of measuring X at a certain location L equals the product of the individual probability of x_N as:

$$\begin{aligned} \mathrm{P}(X|L) &= P(x_1 \wedge x_2 \wedge \cdots \wedge x_N|L) \\ &= P(x_1|L)P(x_2|L)\cdots P(x_N|L) \end{aligned} \tag{1}$$

Because of the multiplication, x_N's small probabilities dominate the overall probability in Eq.1. This property is desirable in a normal condition because the multiplication allows the location system to effectively discriminate the neighboring locations. However, such characteristic makes the positioning algorithm sensitive to lying identities. An adversary can easily affect the location estimation by controlling the lying AP to produce a fake signal strength x_N The fake xn produces a small P(xn|L) due to a mismatch between the lying data and the lying-free database. Unfortunately, this small value destroys the location system's ability of generating a large probability $\mathrm{P}(x_N|L)$ at a correct location. This explains why the method in Eq.1 is susceptible to lying APs. Here we assume that the database is lying-free because many detection schemes can be applied to the data during the offline stage whereas the online lying detection is difficult.

This study proposes a general probabilistic approach to consider possible lying APs. Unlike the model in Eq.1, the individual probability of x_N is combined via an "or" model. This model assumes that the RSSs from reliable APs may be any x_N or any combination among x_N belongs to the complete set. This can be formulated by the "or" operator $\vee$ as: $X_\vee = x_1 \vee x_2 \cdots \vee x_N$, where X$\vee$ is a combined observation based on the "or", representing the possible true APs in X. In mathematics, the "or" is a logical operator that results in true whenever one or more of its arguments are true. Considering a three-AP case (N=3), X$\vee$ $= x_1 \vee x_2 \vee x_3$. These combinations characterize that the observed RSS vector encounters, respectively, two-lying, one-lying and lying-free conditions. The probability of this or model can be obtained by the probabilistic rules of the union of two random events: P(A$\vee$B)=P(A) +P(B)−P(A$\wedge$B). More specifically, the proposed algorithm computes P(X$\vee$) based on a recursion as:

$$\begin{aligned} \mathrm{P}\left(\vee_{n=1}^{m} x_n|L\right) &= P\left(\vee_{n=1}^{m-1} x_n|L\right) + P(x_m|L) \\ &\quad - P\left(\langle\vee_{n=1}^{m-1} x_n|L\rangle \wedge \langle x_m|L\rangle\right) \\ &\quad \mathrm{m} = 2,\cdots,\mathrm{N} \end{aligned} \tag{2}$$

Assuming that the observations are independent with each other, Eq.2 is simplified as:

$$\begin{aligned} \mathrm{P}\left(\vee_{n=1}^{m} x_n|L\right) &= P\left(\vee_{n=1}^{m-1} x_n|L\right) + P(x_m|L) \\ &\quad - \mathrm{P}\left(\vee_{n=1}^{m-1} x_n|L\right)\mathrm{P}(x_m|L) \\ &\quad \mathrm{m} = 2,\cdots,\mathrm{N} \end{aligned} \tag{3}$$

Finally, the estimated location coordinate can be calculated as the average of all reference locations by adopting their new probability as weights:

$$\hat{L} = \sum_{r=1}^{R} P(L_r|X_\vee) \cdot L_r \tag{4}$$

where L^ is the estimated location, R is the number of reference locations, L_r is the coordinate of the r-th reference location, and $\mathrm{P}(L_r|X_\vee)$ represents the union-based posterior probability of measuring X at the r-th reference location. The component of $\mathrm{P}(L_r|X_\vee)$, $\mathrm{P}(L_r|x_n)$ can be computed as

$$\mathrm{P}(L_r|x_n) = \frac{1}{\sqrt{2\pi}\sigma_{n,r}} exp\left(-\frac{(x_n-u_{n,r})}{2\sigma_{n,r}^2}\right) \tag{5}$$

where $u_{n,r}$ and $\sigma_{2n,r}$ represent the mean and variance of the n-th AP and at the r-th reference location. These parameters can be obtained by maximum likelihood estimation as

IOP Conf. Series: Materials Science and Engineering **383** (2018) 012054 doi:10.1088/1757-899X/383/1/012054

$$\widehat{\theta}_{n,r} = arg \max_{\widehat{\theta}_{n,r}} \prod_{k=1}^{K} P(x_n[k]|\theta_{n,r}) \quad (6)$$

where K is the number of training samples and $\theta_{n,r} = \{u_{n,r}, \sigma_{n,r}\}$ are the unknown parameters of the n-th AP and at the r-th reference location. A necessary condition that $\theta_{n,r}$ must satisfy is the gradient of Eq.6 with respect to $\theta_{n,r}$ to be zero. These parameters are estimated by using K training samples during the offline stage.

4. *Experimental setup and results*

4.1. Experimental Setup

We placed three WiFi APs in the target area and collected realistic RSS data at 27 different reference locations to build a fingerprinting location system. We used an Asus laptop with Windows XP as the mobile node, and NetStumbler network software to gather RSS. The size of the test-bed was 70 m2 and RSS was measured 200 times for each reference location. We then selected 15 locations, separated by a distance of 1.5 to 2 meters, as the training data. To simulate the lying identities, we choose specified MAC addresses as lying APs. Furthermore, RSS from the lying AP is changed to an alternative AP's data. That is, the RSS samples from normal APs used in positioning mixed with different ratios of the lying RSS. The experiments compared with the Bayesian and the securing cluster-based scheme [15], [16].

4.2. Performance Evaluation

The first experiment evaluates the positioning performance of our algorithm when two lying identities exist. Table I shows that our approach apparently improves the robustness in the presence of lying identities. All statistical errors show a significant reduction based on our approach. This means that a malicious adversary cannot easily fool our location system by masquerading APs. Numerical results show that the proposed algorithm outperforms the Bayesian and the cluster-based method in reducing the 50-th and the 67-th percentile localization errors by 5.42-26.76%, and 18.89- 34.24%, respectively. Table I also indicates that the Bayesian approach is the most sensitive to lying identities, as its 90-th percentile localization errors increased to almost six meters. This is because the probabilistic model that the Bayesian approach adopts is susceptible to lying identities.

The next experiment investigates the impact of different mixture ratios of lying data. Figure 2(a) shows the standard deviation and mean positioning error versus the mixture ratios for different algorithms. The lying ratio represents the power of the lying identities, where 0% represents a normal condition and 100% indicates the total replacement of actual data. Figure 2(a) shows that the proposed model significantly improved robustness compared with other methods. As the lying ratios increased, both the mean and SD error of the proposed algorithm remained almost fixed, whereas that of traditional approaches clearly increased. This again indicates that our location system is more secure and robust to tolerate the lying identities.

The next experiment considers an alternative scenario in which only one lying AP exists in the environment. Figure 2(b) compares the positioning performance in this case. This figures is generally consistent with figure 2. The main difference is that the proposed algorithm shows a mean-error improvement when the lying ratio exceeds 50%. In addition, the performance is better while reducing one identity, as compared figure 2(a) with figure 2(b).

Table 1. Five statistical errors comparison when two lying identities exist with a mixture rate 100%.

	Median error	Mean error	SD error	90-th percentile error
Bayesian	3.2149	3.3103	1.7328	5.8423
Cluster-based	2.4897	2.7693	1.5865	5.4099
Proposed algorithm	2.3547	2.4585	1.0036	3.5978

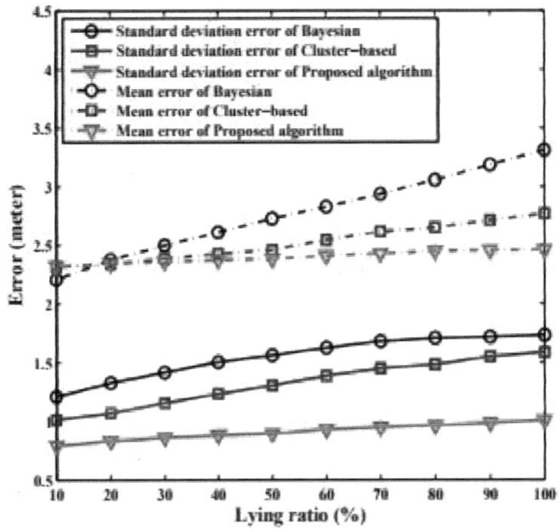

(a)

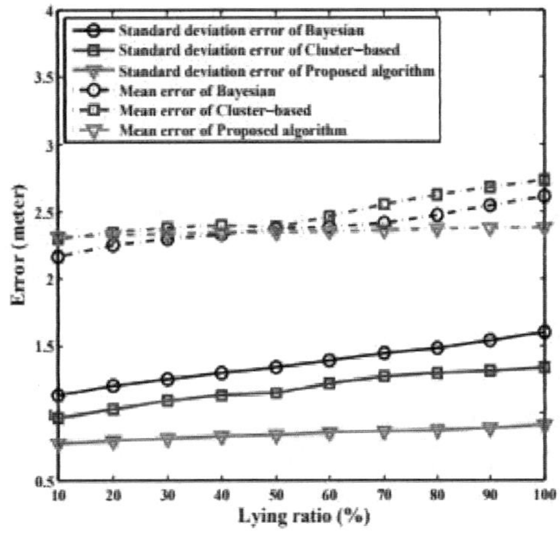

(b)

Figure 2. Standard deviation and mean positioning error versus the mixture ratio for different algorithms under two lying identities**(a)** and one lying identity**(b)**.

5. Conclusion

Secure wireless positioning is important because the location estimations are often input to some critical location-based services. However, due to the nature of the wireless medium, the broadcasting radio is vulnerable to a variety of malicious attacks such as the lying identity in Wi-Fi networks. This study proposes a robust fingerprinting localization algorithm that consider the existence of lying APs. By considering all possible AP combinations in an union-based approach, our location system allows a lying AP to contribute little to the localization process, and thus, achieving more robust location estimations under security threats. On-site experimental results demonstrate that our approach apparently improves the robustness in the presence of lying identities.

References

[1] C. Ardagna, M. Cremonini, S. De Capitani di Vimercati, and P. Samarati,"An obfuscation-based approach for protecting location privacy," IEEE Transactions on Dependable and Secure Computing, vol. **8**, no. 1, pp. 13–27, 2011.

[2] A. Boukerche, H. Oliveira, E. Nakamura, and A. Loureiro, "Secure localization algorithms for wireless sensor networks," IEEE Communications Magazine, vol. **46**, no. 4, pp. 96 –101, 2008.

[3] Y. Zhang, W. Liu, Y. Fang, and D. Wu, "Secure localization and authentication in ultra-wideband sensor networks," IEEE Journal on Selected Areas in Communications, vol. **24**, no. 4, pp. 829 – 835, 2006.

[4] M. Jadliwala, S. Zhong, S. Upadhyaya, C. Qiao, and J.-P. Hubaux,"Secure distance-based localization in the presence of cheating beacon nodes," IEEE Transactions on Mobile Computing, vol. **9**, no. 6, pp. 810–23, 2010.

[5] R. Ouyang, A.-S. Wong, and C.-T. Lea, "Received signal strength-based wireless localization via semidefinite programming: Noncooperative and cooperative schemes," IEEE Transactions on Vehicular Technology, vol. **59**, no. 3, pp. 1307 –1318, 2010.

[6] M. Bshara, U. Orguner, F. Gustafsson, and L. Van Biesen, "Fingerprinting localization in wireless networks based on received-signal-strength measurements: A case study on WiMAX networks," IEEE Transactions on Vehicular Technology, vol. **59**, no. 1, pp. 283–

294, 2010.
[7] Y. Jin, W.-S. Soh, and W.-C. Wong, "Indoor localization with channel impulse response based fingerprint and nonparametric regression," IEEE Transactions on Wireless Communications, vol. **9**, no. 3, pp. 1120 –1127, 2010.
[8] S.-H. Fang and T.-N. Lin, "A dynamic system approach for radio location fingerprinting in wireless local area networks," IEEE Transactions on Communications, vol. **58**, no. 4, pp. 1020–25, 2010.
[9] E. Chan, G. Baciu, and S. Mak, "Using the newton trust-region method to localize in WLAN environment," Wireless and Mobile Computing, Networking and Communications, pp. 363 –69, 2009.
[10] A. Ghosh, V. Kaul, and D. Famolari, "An approach to secure localization in WLANs," Wireless Communications and Networking Conference, pp. 3145 –50, 2008.
[11] S.-H. Fang and T.-N. Lin, "Projection-based location system via multiple discriminant analysis in wireless local area networks," IEEE Transactions on Vehicular Technology, vol. **58**, no. 9, pp. 5009 –19, 2009.
[12] Y. Chen, J. Yang, W. Trappe, and R. Martin, "Detecting and localizing identity-based attacks in wireless and sensor networks," IEEE Transactions on Vehicular Technology, vol. **59**, no. 5, pp. 1–1, 2010.
[13] K. Bauer, D. McCoy, E. Anderson, M. Breitenbach, G. Grudic, D. Grunwald, and D. Sicker, "The directional attack on wireless localization: how to spoof your location with a tin can," IEEE conference on Global telecommunications, pp. 4125–30, 2009.
[14] J. Joseph, B.-S. Lee, A. Das, and B.-C. Seet, "Cross-layer detection of sinking behavior in wireless ad hoc networks using SVM and FDA," IEEE Transactions on Dependable and Secure Computing, vol. **8**, no. 2, pp. 233 –45, 2011.
[15] J. Yang, Y. Chen, W. Trappe, and J. Cheng, "Determining the number of attackers and localizing multiple adversaries in wireless spoofing attacks," IEEE INFOCOM 2009, pp. 666 –74, 2009.
[16] Y. Chen, W. Trappe, and R. Martin, "Detecting and localizing wireless spoofing attacks," Sensor, Mesh and Ad Hoc Communications and Networks, pp. 193 –202, 2007.
[17] S.-H. Fang, C.-C. Chuang, and C. Wang, "Attack-resistant wireless localization using an inclusive disjunction model," IEEE Transactions on Communications, vol. **60**, no. 5, 2012.
[18] X. Li, Y. Chen, J. Yang, and X. Zheng, "Designing localization algorithms robust to signal strength attacks," in Proceedings IEEE INFOCOM, pp. 341 –45, 2011.
[19] J. Yang, Y. Chen, and W. Trappe, "Detecting sybil attacks in wireless and sensor networks using cluster analysis," Mobile Ad Hoc and Sensor Systems, pp. 834 –39, 2008.
[20] H. Chen, W. Lou, J. Ma, and Z. Wang, "TSCD: A novel secure localization approach for wireless sensor networks," Sensor Technologies and Applications, pp. 661 –666, 2008.
[21] A. Kushki, K. Plataniotis, and A. Venetsanopoulos, "Sensor selection for mitigation of RSS-based attacks in wireless local area network positioning," IEEE International Conference on Acoustics, Speech and Signal Processing, pp. 2065–68, 2008.
[22] D. Liu, P. Ning, A. Liu, C. Wang, and W. K. Du, "Attack-resistant location estimation in wireless sensor networks," ACM Transactions on Information System Security, vol. **11**, no. 4, pp. 1–39, 2008.
[23] Z. Li, W. Trappe, Y. Zhang, and B. Nath, "Robust statistical methods for securing wireless localization in sensor networks," Information processing in sensor networks, pp. 91–98, 2005.
[24] Y. Chen, K. Kleisouris, X. Li, W. Trappe, and R. P. Martin, "A security and robustness performance analysis of localization algorithms to signal strength attacks," ACM Transactions on Sensor Networks, vol. **5**, no. 1, pp. 1–37, 2009.

IOP Conf. Series: Materials Science and Engineering **383** (2018) 012055 doi:10.1088/1757-899X/383/1/012055

A Novel ICI Triple Cancellation Scheme in Fast Time-varying Environment for OFDM Systems

Jiongyao YE*, Hengxiao WANG, Weina YUAN, Nan WANG and Rui XU

School of Information Science and Engineering, East China University of Science and Technology, Shanghai, 200237, China

*E-mail: yejy@ecust.edu.cn

Abstract. In fast time-varying environment, orthogonal frequency division multiplexing (OFDM) systems lose the orthogonality between subcarriers due to the Doppler effect, resulting in inter-carrier interference (ICI), which greatly affects the performance of the communication system. Based on the analysis of the principle of ICI generation in time-varying environment, an ICI triple cancellation scheme is proposed to reduce the ICI of the OFDM systems in fast time-varying environment. In the scheme, the ICI self-cancellation modulation is performed at the transmitter of the OFDM system. Then, the time-varying channel is modeled as a linearly time-varying (LTV) model for ICI dual cancellation. Finally, the ICI self-cancellation demodulation is executed at the receiver of the system. Simulation experiment results show that the proposed ICI triple cancellation scheme can achieve a better effect of ICI cancellation than ICI self-cancellation scheme and ICI dual cancellation scheme, improving the performance of the communication system.

1. Introduction

Orthogonal frequency division multiplexing (OFDM) is widely used in wireless communication systems due to its advantages of high-speed data transmission capability, high spectral efficiency, robustness to the multi-path interference, flexible modulation mode and low price, etc., and it has become the main transmission technology of high-speed wireless communication systems and the core technology of many standards. Because the subcarriers of the OFDM are orthogonal to each other, the frequency spectrum of the subcarriers is allowed to overlap each other, so that OFDM has higher spectral efficiency.

However, in practical applications, the OFDM system is often in a mobile environment, and the resulting Doppler effect leads to the loss of orthogonality among the subcarriers of OFDM, which inevitably brings about inter-carrier interference (ICI). ICI seriously affects the performance of the communication system, especially in the high-speed mobile environment. Therefore, it is of great significance to study the ICI cancellation scheme in the OFDM system.

At present, the more commonly used ICI cancellation schemes mainly include frequency-domain equalization scheme [1], time-domain windowing scheme [2], and the ICI self-cancellation scheme [3]-[4], and so on. Frequency-domain equalization scheme is too complex to implement relatively. Time-domain windowing scheme has the defect of spectral leakage. The self-cancellation method is an easy and effective way to cancel ICI compared with other methods.

To further improve the performance of the OFDM systems in fast time-varying environment, an ICI triple cancellation scheme is proposed in this paper. ICI dual cancellation is achieved by modeling

Content from this work may be used under the terms of the Creative Commons Attribution 3.0 licence. Any further distribution of this work must maintain attribution to the author(s) and the title of the work, journal citation and DOI.
Published under licence by IOP Publishing Ltd

the time-varying channel as a LTV model, and ICI self-cancellation modulation and demodulation are performed at the transmitter and the receiver of the OFDM system, respectively, to enhance the ICI cancellation effect of the system. Simulation experiment results show that compared with the ICI self-cancellation scheme and dual cancellation scheme, the system after ICI triple cancellation can better suppress the impact of ICI and improve the performance of the OFDM systems in fast time-varying environment.

2. Related work

The ICI self-cancellation scheme is a simple way for ICI reduction. Zhao et al. [4] propose for the first time the ICI self-cancellation scheme, and its core idea is that the transmitted data and its opposite number are modulated onto the adjacent subcarrier at the transmitter to cancel part of the ICI, and then the data on the odd subcarriers are inverted and plused on the even subcarriers at the receiver to further cancel the ICI.

Since then, many more improved ICI self-cancellation schemes have been put forward in succession [5]-[7]. ICI self-cancellation scheme has the advantages of simple implementation and better effects of ICI cancellation, but its spectral efficiency is low. In order to improve the spectral efficiency of ICI self-cancellation algorithm, Liu et al. [8] put forward an ICI self-cancellation scheme based on differential coding, by differential encoding the data on all subcarriers at the transmitter and making the corresponding differential decoding at the receiver. The scheme can improve the spectral efficiency, but the differential encoding and decoding exacerbate the influence of channel Gaussian noise, so the performance of the scheme is poor when the signal to noise ratio (SNR) is low. Mostofi et al. [9] firstly come up with an ICI cancellation scheme based on LTV model, which is named as single cancellation scheme, but the effect of cancellation is poor. Kwak et al. [10] improve the single cancellation scheme, introducing the ICI pre-cancellation process by using the first sending OFDM header or the last OFDM symbol to obtain the channel information and pre-cancel the received signal by ICI, which improves the accuracy of subsequent cancellation.

In recent years, some ICI joint cancellation schemes have been proposed one after another. Liu et al. [11] combine the ICI self-cancellation scheme with the time-domain windowing scheme, by using the cosine window multiplied by the transmitted data after IFFT. ICI is effectively cancelled and the band-shaped frequency domain channel matrix is estimated by complex exponential basis expansion model (CE-BEM), which improves the performance of bit error rate (BER) of the system. Li et al. [12] put forward an ICI self-cancellation scheme based on symmetric subcarriers, which is combined with the time-domain cosine windowing scheme to improve the BER of the system. However, it has the disadvantage of poor performance in low SNR environment and low spectral efficiency. Singh et al. [13] come up with a joint cancellation scheme based on ICI self-cancellation and maximum likelihood estimation method. The performance of the system in this scheme has been improved compared with the ICI self-cancellation scheme and the maximum likelihood estimation scheme alone.

3. System Model

Assuming that the number of subcarriers in an OFDM system is N and the inter symbol interference (ISI) is canceled by inserting a cyclic prefix, the time domain form of the signal received at the receiver can be expressed as:

$$y = hx + w \tag{1}$$

where $x = [x_0, x_1, \dots, x_{N-1}]^T$, $y = [y_0, y_1, \dots, y_{N-1}]^T$, respectively, represent the collection of all transmitted data and received data in the OFDM symbol, w denotes the additive Gaussian white noise in the time domain, and h represents a time-domain channel matrix with a size of $N \times N$, which specifically can be expressed as:

$$h = \begin{bmatrix} h(0,0) & 0 & \cdots & h(0.2) & h(0,1) \\ h(1,1) & h(1,0) & \cdots & h(1,3) & h(1,2) \\ \vdots & \vdots & \vdots & \vdots & \vdots \\ h(L-1,L-1) & h(L-1,L-2) & \cdots & 0 & 0 \\ 0 & h(L,L-1) & \vdots & 0 & 0 \\ \vdots & \vdots & & \vdots & \vdots \\ 0 & 0 & \cdots & h(N-1,1) & h(N-1,0) \end{bmatrix} \tag{2}$$

where L represents the number of multi-path, and $h(n, l)$ denotes the channel response of the lth tap at the nth sampling point, $0 \le n \le N-1$, $0 \le l \le L-1$. With FFT on the Equ. (1), the input-output relationship in the frequency domain of the OFDM system can be represented as:

$$Y = Fy = FhF^{H}X + Fw = HX + W \tag{3}$$

where F, F^{H} represent the FFT matrix and IFFT matrix of the Nth point, respectively, and $(\blacksquare)^{H}$ represents the Hemet conjugate transpose of the matrix. The elements in F are given by:

$$F(n,m) = \frac{1}{\sqrt{N}} e^{-\frac{j2\pi nm}{N}} \tag{4}$$

If the channel is not time-varying, the condition $h(0, l) = h(1, l) = \cdots = h(N-1, l)$ is established. The channel frequency domain matrix H is a diagonal matrix with a size of $N \times N$, and the diagonal elements represent the frequency domain response of the channel, which means that at this time each subcarrier in the OFDM system is orthogonal without ICI. However, in practice, the assumption $h(0, l) = h(1, l) = \cdots = h(N-1, l)$ generally can not be established due to following reasons such as high-speed movement, so the channel comes to be a time-varying one. In this case, the channel frequency domain matrix becomes an approximately banded matrix, resulting in ICI.

For an approximately banded channel frequency domain matrix H, it can be decomposed into a channel matrix without ICI composed of diagonal elements of H and a channel matrix with ICI composed of off-diagonal elements of H. Without loss of generality, the frequency domain input-output relationship of the nth symbol in the OFDM system can be expressed as:

$$Y_n = H_n^{ave}H_n + H_n^{ICI}X_n + W_n \tag{5}$$

where H_n^{ave} denotes the no ICI channel matrix with a size of $N \times N$, its diagonal elements are equal to the diagonal elements of channel frequency domain matrix H, and its remaining elements are equal to zero. H_n^{ICI} denotes the ICI channel matrix with a size of $N \times N$, its diagonal elements are equal to zero, and its remaining elements are equal to the off-diagonal elements of channel frequency domain matrix H.

4. ICI Triple Cancellation Scheme

The basic principle of the ICI self-cancellation scheme is to perform ICI self-cancellation modulation on the transmitted signal at the transmitter end and perform corresponding ICI self-cancellation demodulation on the received signal at the receiver to suppress the impact of ICI. The classic ICI self-cancellation scheme is to perform the data-conversion on adjacent subcarriers, which can be expressed as:

$$\begin{cases} X_n^{'} = X_n, & n = 0, 2, 4, \cdots, N-1 \\ X_{n+1}^{'} = -X_n, & n = 0, 2, 4, \cdots, N-1 \end{cases} \tag{6}$$

Then many other self-cancellation schemes, such as the weighted conjugate of the complex number of adjacent data, and the conjugate of the symmetric data-conversion scheme, has been put forward in succession. Considering the existence of ICI, the received signal of the OFDM system can be represented as:

$$Y_m = \sum_{n=0}^{N-1} X_n S(n-m) + W_m \tag{7}$$

where N denotes the number of subcarriers, Y_n represents the received signal, X_n denotes the transmitted signal, W_m represents the noise interference, and $S(n - m)$ denotes the ICI coefficient, which represents the interference caused by the data of the nth subcarrier on the data of the mth subcarrier, $0 \leq m \leq N-1, 0 \leq n \leq N-1$. If data-conversion is performed on the adjacent subcarriers, only one symbol is transmitted for each pair of subcarriers, and the received signals on the mth subcarrier and the $(m+1)^{th}$ subcarrier can be derived as:

$$Y'_m = \sum_{n=0,2,4,\dots}^{N-2} X_n[S(n-m) - S(n+1-m)] + W_m \tag{8}$$

$$Y'_{m+1} = \sum_{n=0,2,4,\dots}^{N-2} X_n[S(n-m-1) - S(n-m)] + W_{m+1} \tag{9}$$

Therefore, the ICI coefficient after ICI self-cancellation modulation is given by:

$$S'(n-m) = S(n-m) - S(n+1-m) \tag{10}$$

Since the ICI coefficient *S(n - m)* has little change in adjacent subcarriers, the data on the $(m+1)^{th}$ subcarrier can better offset the ICI generated by the mth subcarrier.

Although ICI self-cancellation scheme is easy to implement and has a good effect of ICI cancellation, its spectral efficiency is very low due to the reason that one symbol is transmitted by two subcarriers. In this regard, ICI self-cancellation scheme based on differential coding can be utilized, it perform the transmitted data as differential encoding instead to perform data-conversion on its adjacent subcarriers at the transmitter of the system, and the transmitted signal can be expressed as:

$$X'_n = \begin{cases} X_0, & n = 0 \\ X_n - X_{n-1}, & 1 \leq n \leq N-1 \end{cases} \tag{11}$$

Then the signal received by the system can be represented as:

$$Y'_m = \sum_{n=0}^{N-1} X'_n S(n-m) = \sum_{n=0}^{N-1} S(n-m)(X_n - X_{n-1}) \tag{12}$$

Because the interference between adjacent subcarriers is the largest, the ICI can be suppressed to a certain extent by differential coding.

For the nth OFDM symbol, the relationship between the channel response $h_n(m, l)$ of the lth path at the mth sampling point and the channel response $h_n(m_0, l)$ of the lth path at the $m_0{}^{th}$ sampling point in the LTV model is given by:

$$h_n(m, l) = h_n(m_0, l) + a_n(l)(m - m_0) \tag{13}$$

where $a_n(l)$ represents the envelope of the channel, characterizing the time-varying characteristic of the channel.

In the LTV model, a better effect of ICI cancellation is obtained by modeling three consecutive OFDM symbols [10], then the time-domain input-output relationship of the OFDM system can be expressed as:

$$y_n = h_n^{mid} F^H X_n + (b_p a_{n,p} + b_f a_{n,f} + e_n) F^H X_n + w_n \tag{14}$$

where h_n^{mid} denotes the Toeplitz matrix composed of the channel response of all paths at the $(\frac{N}{2}-1)^{th}$ sampling point of the nth OFDM symbol, and the first column of the Toeplitz matrix is $[h_n\left(\frac{N}{2}-1, 0\right), h_n\left(\frac{N}{2}-1, 1\right), \cdots, h_n\left(\frac{N}{2}-1, L-1\right), 0, \cdots, 0]$, $a_{n,p}$ represents the time-varying characteristic of the channel from the midpoint of the previous symbol to the midpoint of the current symbol, similarly, $a_{n,f}$ denotes the time-varying characteristic of the channel from the midpoint of the current symbol to the midpoint of the next symbol, e_n represents the modeling error, b_p and b_f both are diagonal matrices, represent the interval between each sample point in the current OFDM symbol and the midpoint, the elements of which are given by:

$$b_p(m, n) = \begin{cases} m - \frac{N}{2} + 1, & 0 \leq m \leq \frac{N}{2} - 1 \\ 0, & else \end{cases} \tag{15}$$

$$b_f(m, n) = \begin{cases} m - \frac{N}{2} + 1, & \frac{N}{2} - 1 \leq m \leq N-1 \\ 0, & else \end{cases} \tag{16}$$

If the signal at the transmitter has been ICI self-cancellation modulated, then perform the FFT on Equ. (14). The frequency domain form of the received signal can be derived as:

$$Y_n^{'} = H_n^{mid} X_n^{'} + \left(B_p A_{n,p} + B_f A_{n,f} + E_n\right) X_n^{'} + W_n \tag{17}$$

where $X_n^{'}$ represents the data after ICI self-cancellation modulation, and $Y_n^{'}$ denotes the received data at the receiver.

In this paper, the ICI triple cancellation scheme uses the ICI channel matrix estimated from the previous OFDM symbol to perform ICI pre-cancellation at first. The first OFDM symbol can obtain the ICI channel matrix from the first transmitted header. The header in this paper is derived from [14], Schmidl and Cox propose a sequence for estimating time and frequency shift, and the data "0" is transmitted on the odd subcarriers and the PN sequences are transmitted on the even subcarriers. Next, the ICI channel matrix at this time is re-estimated by using the received signal after ICI pre-cancellation to cancel the ICI again. The structure of the OFDM frame in the ICI triple cancellation scheme is shown in figure 1.

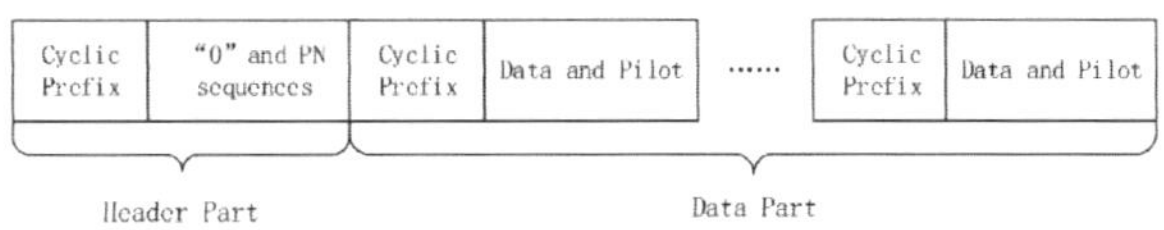

Figure 1. ICI triple cancellation scheme used in the structure of OFDM frame.

The ICI matrix estimated by the last OFDM symbol or header can be expressed as:

$$\widehat{H}_n^{ICI,pre} = C\widehat{A}_{n-1,p} \tag{18}$$

where C is Fb_pF^H whose (m, n) element is given by:

$$C(m,n) = \begin{cases} -\dfrac{1}{1-e^{-j\frac{2\pi(m-n)}{N}}}, & m \neq n \\ \dfrac{1}{2}, & m = n \end{cases} \tag{19}$$

The received signal after the ICI pre-cancellation is given as follows,

$$\begin{aligned} Y_n^{'IC,\,pre} &= Y_n^{'} - \widehat{H}_n^{ICI,pre}\widehat{X}_n^{'} \\ &= H_n^{ave} X_n^{'} - CA_{n-1,p}X_n^{'} \\ &\quad - C\widehat{A}_{n-1,p}\widehat{X}_n^{'} + W_n + E_n X_n^{'} \end{aligned} \tag{20}$$

where channel matrix without ICI $H_n^{ave} = \Phi\widehat{H}_n^{LS}$, Φ denotes an interpolation matrix, and $\widehat{H}_n^{LS}$ denotes a frequency domain channel response matrix at the pilot position, which can be obtained by using a transmitted pilot through least squares (LS) estimation. $\widehat{H}_n^{LS}$ can be expressed as:

$$\widehat{H}_n^{LS}(k) = \frac{Y_n^{'IC,pre}(\rho_k)}{X_n(\rho_k)} \tag{21}$$

Using the estimated channel matrix without ICI of three consecutive OFDM symbols, the envelope of the channel of the adjacent symbols can be derived as:

$$\widehat{A}_{n,p} = \frac{\widehat{H}_n^{ave} - \widehat{H}_{n-1}^{ave}}{N + N_g} \tag{22}$$

$$\widehat{A}_{n,f} = \frac{\widehat{H}_{n+1}^{ave} - \widehat{H}_n^{ave}}{N + N_g} \tag{23}$$

where N_g represents the length of the cyclic prefix. After the ICI pre-cancellation, ICI re-cancellation matrix can be expressed as:

$$\widehat{H}_n^{ICI,pos} = C_p\widehat{A}_{n,p} + C_f\widehat{A}_{n,f} \tag{24}$$

To achieve an accurate pre-demodulation at the receiver, the received signal after the ICI dual cancellation can be derived as:

$$Y_n^{'IC,\ pos} = Y_n^{'} - \widehat{H}_n^{ICI,pos}\widehat{X}_n^{'}$$
$$= H_n^{ave}X_n^{'} - (H_n^{ICI} - \widehat{H}_n^{ICI,pos})X_n^{'} \tag{25}$$
$$+ W_n + E_nX_n^{'}$$

The received signal $\widehat{X}_n^{IC}$ can be obtained by direct equalization with $Y_n^{'IC,pos}$.

The receiver then performs ICI self-cancellation demodulation on the received signal, and subtracts the data on the adjacent subcarriers to further cancel the ICI, which can be represented as:

$$\widehat{X}_m^{''} = \frac{1}{2}\left[\widehat{X}_m^{IC} - \widehat{X}_{m+1}^{IC}\right]$$
$$= \sum_{n=0,\ 2,4,\ldots}^{N-2} X_n[-S(n-m-1) + 2S(n-m) \ - S(n-m+1)] + W_m - W_{m+1} \tag{26}$$

The ICI coefficient after the ICI self-cancellation demodulation can be expressed as:

$$S^{''}(n-m) = \ -S(n-m-1) + 2S(n-m) - S(n-m+1) \tag{27}$$

For the three ICI coefficients, it has proved that $|S^{''}(\mathrm{n}-\mathrm{m})|$ is the smallest and $S(n-m)$ is the largest of the three [1]. Therefore, the signals are performed as the ICI self-cancellation modulation and the ICI self-cancellation demodulation in the OFDM system to better suppress ICI.

If the ICI self-cancellation modulation using a differential coding at the transmitter, then the signals should be performed as receiving decision and forward feedback to realize the differential decoding at the receiver, which is given as follows,

$$\widehat{X}_m^{''} = \begin{cases} Dec(\widehat{X}_m^{IC}), & m = 0 \\ Dec(\widehat{X}_m^{IC} + \widehat{X}_{m-1}^{''}), & 1 \le m \le N-1 \end{cases} \tag{28}$$

where $Dec(\blacksquare)$ represents the receiving decision of the data at the receiver, and $\widehat{X}^{''}$ represents the output data after the ICI triple cancellation. According to the Equ. (28), it can be seen that the noise on each subcarrier is superimposed by differential coding and decoding, which increases the interference of channel noise on the signal and reduces the performance of the scheme, especially at low SNR. However, the ICI self-cancellation scheme based on differential coding can improve the spectral efficiency of self-cancellation scheme and has better performance in high SNR environment.

Because the ICI self-cancellation scheme can choose the way of differential coding or adjacent data-conversion, so the ICI triple cancellation scheme includes two methods, one is the ICI self-cancellation modulation and demodulation module using differential coding and differential decoding (hereinafter referred to as method 1 proposed), and the other is that the ICI self-cancellation modulation and demodulation module adopting adjacent data-conversion and data-subtraction (hereinafter referred to as method 2 proposed). The remaining modules of the two methods are the same.

5. Simulation Experiment and Analysis

Table I shows the setting of simulation parameters.

Figure 2 depicts the symbol error rate (SER) comparison of each ICI cancellation scheme at different SNR. It can be seen that the performance of the differential coding self-cancellation scheme is the worst. According to Equ. (28), differential coding and decoding exacerbate the interference on the data by the noise and suffer more serious in the low SNR environment.

The performance of the single cancellation scheme is better than that of the differential coding self-cancellation scheme. However, owing to the lack of ICI pre-cancellation, the estimation accuracy of the channel matrix without ICI is low, which in turn influences the effect of subsequent ICI cancellation.

Table 1. Simulation parameters.

Parameter	Specification
Number of subcarriers	256
Length of cyclic prefix	30
Length of pilot	32
Modulation type	16QAM
Sampling interval	$5\mu s$
Number of multi-paths	5
Carrier frequency	2GHz
f_{nd} when the SNR is changed	0.1
SNR when the f_{nd} is changed	40dB

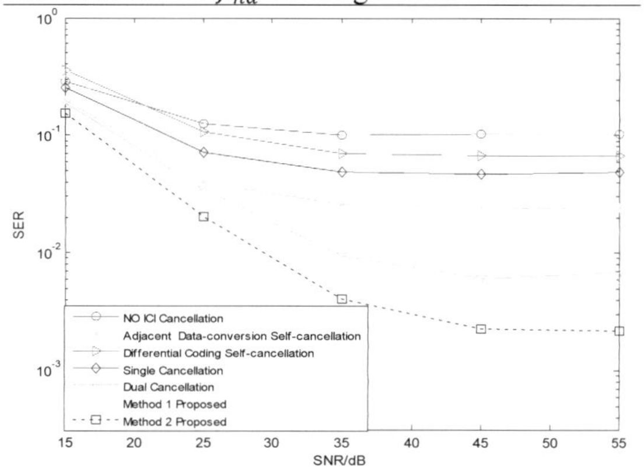

Figure 2. SER comparison at different SNR.

The performance of the adjacent data-conversion self-cancellation scheme is much better than the single cancellation scheme, but its spectral efficiency is low.

As a result of the addition of ICI pre-cancellation, the performance of dual cancellation scheme has been improved to some extent, especially in the high SNR environment, compared with the adjacent data-conversion self-cancellation scheme.

Due to the use of differential coding and decoding in self-cancellation modulation and demodulation, the performance of method 1 proposed in this paper is poor when the SNR is low, but its performance is obviously improved compared with the dual cancellation scheme when the SNR is more than *35dB*.

The spectral efficiency in method 2 proposed in this paper is low at the expense of spectral resources, but its performance is optimal among the ICI cancellation schemes.

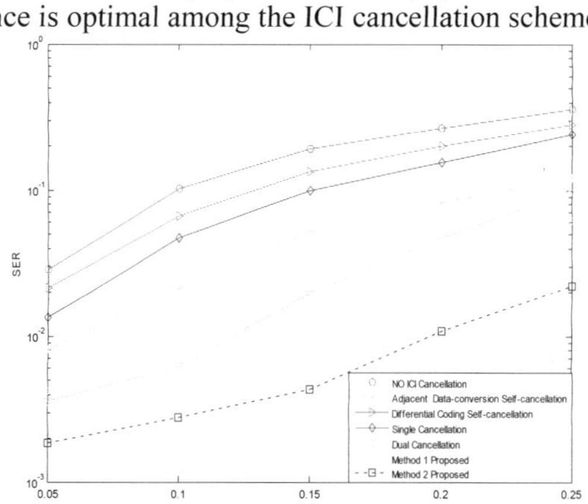

Figure 3. SNR comparison at different f_{nd}.

Based on the results shown in figure 3, it can be seen the SER comparison of each ICI cancellation scheme at different normalized Doppler frequency shift f_{nd}. Because the SNR is *40dB*, the method 1 proposed in this paper has less interference by noise, and its performance is better.

6. Conclusion

The OFDM system inevitably generates ICI under high-speed mobile environment due to Doppler effect, which greatly affects the performance of the communication system. In this paper, an ICI triple cancellation scheme is proposed to reduce the ICI of the OFDM systems in fast time-varying environment.

In view of the defect of low spectral efficiency of traditional ICI self-cancellation scheme, this paper introduces differential coding and decoding method into ICI self-cancellation modulation and demodulation, which improve the spectral efficiency of the system. Through the simulation experiment and analysis, it can be seen that when the spectral resources of the system are tight and the SNR is relatively high, method 1 can be adopted, and the ICI self-cancellation modulation and demodulation module using differential coding and differential decoding. When the spectral resources of the system are sufficient, method 2 proposed in this paper has a better effect of ICI cancellation, and the ICI self-cancellation modulation and demodulation module adopting adjacent data-conversion and data-subtraction.

Acknowledgement

This work was supported by the National Natural Science Foundation of China (NSFC) under Grant 61604054.

References

[1] W. G. Jeon, et al., ``An equalization technique for orthogonal frequency-division multiplexing systems in timing-variant multipath channels," IEEE Transactions on Communications, vol. **47**, no. 1, pp. 27-32, 2001.
[2] J. Armstrong, ``Analysis of new and existing methods of reducing intercarrier interference due to carrier frequency offset in OFDM," IEEE Transactions on Communications, vol. **47**, no. 3, pp. 365-369, 1999.
[3] Y. Peng, et al., ``Performance analysis of a new ICI-self-cancellation-scheme in OFDM systems," IEEE Transactions on Consumer Electronics, vol. **53**, no. 4, pp. 1333-1338, 2007.
[4] Y. Zhao and S. Häggman, ``Intercarrier interference self-cancellation scheme for OFDM mobile communication systems," IEEE Transactions on Communications, vol. **49**, no. 7, pp. 1185-1191, 2001.
[5] Q. Shi, Y. Fang, and M. Wang, ``A novel ICI self-cancellation scheme for OFDM systems," International Conference on Wireless Communications, Networking and Mobile Computing, pp. 1-4, 2009.
[6] A. Bishnu, A. Jain, and A. Shrivastava, ``A new scheme of ICI self-cancellation in OFDM systems," International Conference on Communication Systems and Network Technologies, pp. 120-123, 2013.
[7] S. Pareyani and P. Patel, ``An improved ICI cancellation method to reduce the impact of frequency offset in OFDM systems," International Conference on Computational Intelligence and Communication Networks, pp. 496-501, 2015.
[8] J. Liu, D. Yang, and Q. Chang, ``Research on inter-carrier interference cancellation based on differential coding in OFDM system," Journal of Electronics & Information Technology, vol. **29**, no. 7, pp. 1529-1532, 2007.
[9] Y. Mostofi and D. Cox, ``ICI mitigation for pilot-aided OFDM mobile systems," IEEE Transactions on Wireless Conmmunications, vol. **4**, no. 2, pp. 765-774, 2005.
[10] K. Kwak, et al., ``New OFDM channel estimation with dual-ICI cancellation in highly mobile channel," IEEE Transactions on Wireless Communications, vol. **9**, no. 10, pp. 3155-3165, 2010.

IOP Conf. Series: Materials Science and Engineering **383** (2018) 012055 doi:10.1088/1757-899X/383/1/012055

[11] T. Liu, et al., ``ICI self-cancellation with cosine windowing in OFDM transmitters over fast time-varying channels," IEEE Transactions on Wireless Communications, vol. **14**, no. 7, pp. 3559-3570, 2015.
[12] M. Li, et al., ``Combined algorithm for suppressing ICI in OFDM systems," Computer Engineering and Application, vol. **52**, no. 10, pp. 85-90, 2016.
[13] H. Singh, S. Sarowa, and S. Agrawal, ``A novel hybird ICI cancellation technique for OFDM optimization," International Conference on Recent Advances in Engineering & Computational Sciences, pp. 1-5, 2015.
[14] T. Schmidl and D. Cox, ``Robust frequency and timing synchronization for OFDM," IEEE Transactions on Communications, vol. **45**, no. 12, pp. 1613-1621, 1997

Sucker's reliability analysis of photovoltaic panel cleaning robot

Wang Jingshu[1,*], Xiao Xu[1], Gong Hengxiang[2] and Zou Zheng[1]

[1] College of Mechanical Engineering, Chongqing University of Technology; Chongqing, 400054, China
[2] School of Science, Chongqing University of Technology; Chongqing, 400054, China

*E-mail: donot@cqut.edu.cn

Abstract. The dust deposited on the surface makes a significant impact on the performance of photovoltaic modules. This paper mainly studies the sucker's reliability analysis of the photovoltaic panel cleaning robot. The research is carried from two aspects, the mechanical analysis and the finite element analysis. The reliability including anti-slipping, anti-overturning, and anti-twisting is researched based on the mechanical model, while the influence of the mechanical load is studied by the finite element model of the PV module. According to the constrain conditions, the sucker of the photovoltaic panel cleaning robot has been selected. Based on the prototype of PV cleaning robot, the reliability experiment is conducted to verify the sucker's reliability analysis.

1. Introduction

The dust deposition of microscopic particles in the outdoor environment affects the photoelectric conversion efficiency seriously [1, 2]. Many scholars had done out-door exposure tests in different areas. The results showed that the conversion efficiency reduced by 17% after 38 days exposure in Kuwait, while the reduction rate of efficiency is 40% after six months exposure in Sandi Arabia [3]. In Athens, the dust of the photovoltaic modules is 0.1-1g/m^2 after eight weeks exposure [4]. Therefore, it is necessary to clean the dust on the photovoltaic modules, especially for large solar power installation.

There are many types of mechanical devices aiming at cleaning the dust of PV modules, one of the fastest-growing devices is cleaning robot. In order to move on the angled surface, even vertical surface, the walking mechanisms applied by researchers include the tracked robot [5], the hybrid wheel-leg [6], and the magnetic absorption foot [7]. Nonetheless, the most widely used one is the sucker, which has advantages of simple structure, convenient control, and little influence on the surface [8, 9]. This paper focuses on the reliability analysis of the sucker of photovoltaic cleaning robot.

2. Suckers distribution of cleaning robot

The photovoltaic cleaning robot is capable of moving on the inclined surface of photovoltaic module. The dust will be cleaned when it reaches the destination. In our research, the walking mechanism of the cleaning robot is cruciform structure, while the suckers are installed on four terminals of the cruciform. The motion diagram of sucker based cleaning robot is shown in figure 1, in which the black sucker means it is attached to the surface, and the white sucker is not attached. Changing the statement of the suckers in figure 1 alternately, the cleaning robot can move along both the two directions of the cruciform.

Content from this work may be used under the terms of the Creative Commons Attribution 3.0 licence. Any further distribution of this work must maintain attribution to the author(s) and the title of the work, journal citation and DOI.
Published under licence by IOP Publishing Ltd

The features of photovoltaic cleaning robot result in different absorbability requirements of the suckers on four terminals. When the cleaning mechanism works, the attached suckers need strong adsorption property. When the robot is moving, the cleaning mechanism is not running, the absorbability requirement of the attached suckers is relatively low. Therefore, both the two terminals along one line of cruciform have four suckers, marked as sucker group 1 and 2, while the number of suckers is two for the terminals along another line.

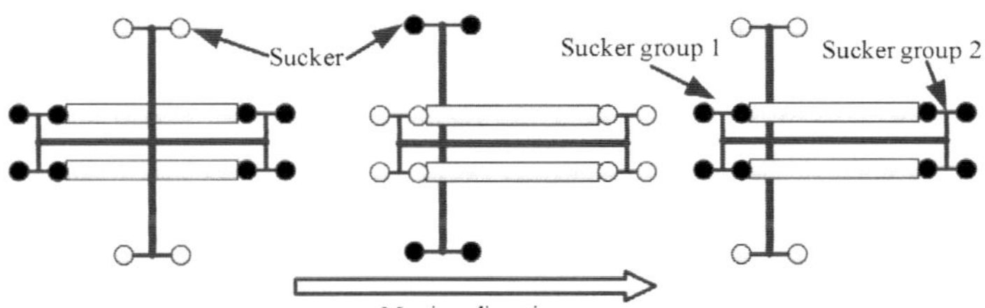

Figure 1. The motion diagram of sucker based cleaning robot.

Based on the suckers' distribution, the reliability analysis has be done from both theory and simulation aspects. Then, the prototype has also been made to verify the reliability.

3. Mechanical analysis of sucker reliability

For the cleaning robot should work on the slanting surface of PV module, the sucker reliability analysis is executed in three directions: anti-slipping, anti-overturning, and anti-twisting. The cleaning robot is described as a barycentre-sucker system shown in figure 2. In the simplified system, the centroid coordinate is (x_m, y_m, z_m) while the mass of the cleaning robot is denoted by m. The suckers are simplified as points in the system. The α indicates the tilt angle of the PV module.

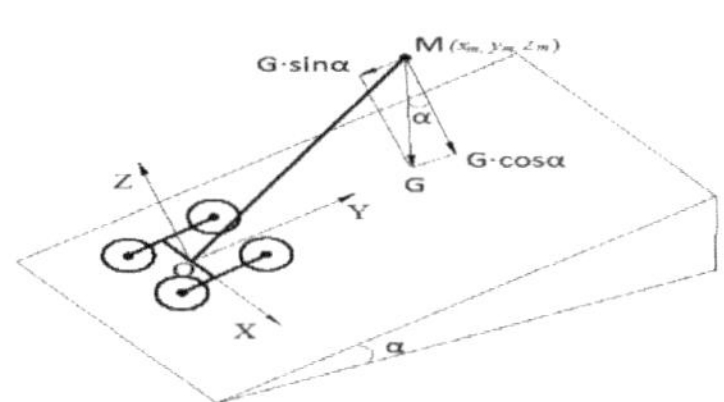

Figure 2. Analysis of mechanics force of robot.

3.1. Analysis of anti-slipping

The pressure difference and the adsorption area of the ith sucker are expressed by $\triangle P_i$ and S_i, respectively. When the cleaning robot works on the tilt surface without slippage, the maximum frictional force supported by the suckers should be greater than the gravity component. With the assumption that the friction coefficient of PV module is indicated by η, the anti-slipping constrain can be expressed as equation 1.

$$\eta\sum_{i=1}^{n} \Delta P_i \cdot S_i > G \cdot \sin\alpha, \quad i \geq 1 \tag{1}$$

3.2. Analysis of anti-overturning

In general, the suckers are distributed as line-style or rectangle-style. It is assumed that the coordinates of the diagonal points are j (x_j,y_j) and $k(x_k,y_k)$ for the rectangle-style. For the line-style distributed suckers, we can get a rectangle area based on the barycentre of sucker. Therefore, the overturn of the cleaning robot is the rotation around an edge or vertex of the rectangle.

The tendency of overturn is determined by the stress status and the centroid position of cleaning robot. Considering the equilibrium equations of forces and force moments, the anti-overturning constrains, including equations from (2) to (6), are listed in table 1.

Table 1. The anti-overturning constrains.

Equation number	Position of centroid	Stress status	Anti-overturning constrain
(2)	$x_m > x_k$	No constrain	$\sum_{i=1}^{n} \Delta P_i \cdot S_i \cdot \lvert x_i - x_k \rvert > G \cdot \cos\alpha \cdot z_m$
(3)	$x_m < x_j$	No constrain	$\sum_{i=1}^{n} \Delta P_i \cdot S_i \cdot \lvert x_i - x_j \rvert > G \cdot \cos\alpha \cdot z_m$
(4)	$y_m < y_k$	No constrain	$\sum_{i=1}^{n} \Delta P_i \cdot S_i \cdot \lvert y_i - y_k \rvert > G \cdot \sin\alpha \cdot z_m + G \cdot \cos\alpha \cdot \lvert y_m - y_k \rvert$
(5)	$y_m > y_j$	$G \cdot \cos\alpha \cdot \lvert y_m - y_k \rvert < G \cdot sin\alpha \cdot z_m$	$\sum_{i=1}^{n} \Delta P_i \cdot S_i \cdot \lvert y_i - y_k \rvert > G \cdot \sin\alpha \cdot z_m - G \cdot \cos\alpha \cdot \lvert y_m - y_k \rvert$
(6)	$y_m > y_j$	$G \cdot \cos\alpha \cdot \lvert y_m - y_j \rvert > G \cdot sin\alpha \cdot z_m$	$\sum_{i=1}^{n} \Delta P_i \cdot S_i \cdot \lvert y_i - y_j \rvert > G \cdot \cos\alpha \cdot \lvert y_m - y_j \rvert - G \cdot \sin\alpha \cdot z_m$

3.3. Analysis of anti-twisting

Based on the simplified barycentre-sucker system in figure 2, the force $G{\cdot}sin\alpha$ will drive robot to turn around Z-axis while the friction between suckers and surface will generate torque in opposite. Therefore, the anti-twisting constrain can be described as equation (7).

$$\eta \sum_{i=1}^{n} \Delta P_i \cdot S_i \cdot \lvert x_i \rvert > G \cdot \sin\alpha \cdot \lvert x_m \rvert \qquad i \geq 1 \tag{7}$$

In conclusion, the results of theory analysis for sucker reliability are comprised of equations from (1) to (7) and cover constrains of anti-slipping, anti-overturning, and anti-twisting.

4. Simulation analysis of sucker reliability

The walking motion of the cleaning robot is realized by the adsorption of the suckers, which will influences the deformation of the PV modules. However, as the key component of PV module, the main failure mode of silicon cell is crack which is hard to find out in experiment with the sealed PV module. Therefore, the simulation analysis of sucker reliability considering the deformation has been executed.

Based on standards in GB/T9535, the maximum stress and maximum deflection of PV module with the static load being 2400pa are applied as the reliable basises. If the stress and deflection of PV module with the cleaning robot working are less, the suckers of the cleaning robot are approved reliable.

The simulation parameters are listed in table 2. The dimension of PV module refers to the EP260C-60 produced by Ningbo yipai.

Table 2. The model materials of PV module.

Material	E(Gpa)	Poisson ratio	Thickness (mm)
stalinite	73	0.22	3.2
EVA film	0.655	0.3	0.4

silicon cell	144	0.3	0.3
TPT film	1.5	0.24	0.225
aluminum alloy	71	0.3	1

ANSYS is employed to conduct finite element analysis. The deformation and stress nephogram of silicon cell with load being 2400pa are illustrated in figure 3.

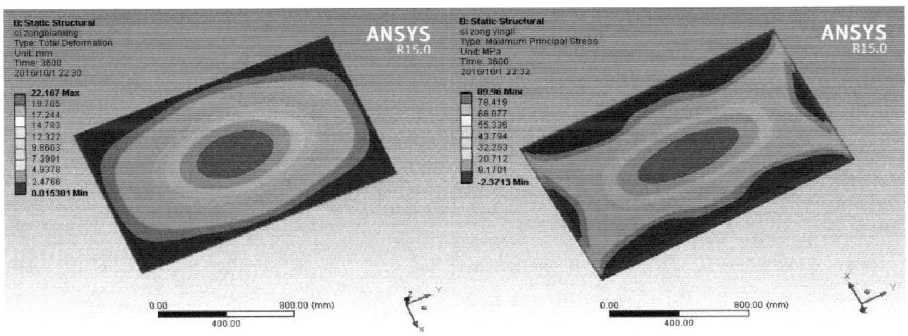

(a) Deformation of silicon cell. (b) Maximum stress nephogram of silicon cell.

Figure 3. Deformations of PV module subjected to load of 2400pa.

The simulation results present that the maximum stress of PV module is 109.01Mpa and the maximum deflection is 22.167mm. The values of silicon cell are 89.96Mpa and 22.167mm, respectively. According to figure 3, the maximum deformation occurs on the central area of PV module. The reason is that all the installation holes of PV module are distributed on the edges. Therefore, when validating the reliability of suckers, the values of maximum stress and deflection with the cleaning robot being working should less than the values of 89.96Mpa and 22.167mm, respectively.

5. Validation of sucker reliability

Before manufacturing the prototype of cleaning robot, the reliability validation of chosen suckers has been conducted. For practical purposes and economy reasons, the type of chosen sucker is FV50. The radius of sucker is 50mm while the radius of negative pressure chamber is 3mm. The type of matching vacuum pump is VBY7506, whose friction coefficient with PV surface is 0.6. Based on all the parameters of the sucker, the validation from both theory and simulation aspects are conducted as follow.

When only the sucker group 1 or sucker group 2 are absorbed on the PV panel, sucker failure and stress concentration may happen because the adsorption points are relatively centralized. Therefore, the validation is conducted for the situation when only the sucker group 1 or sucker group 2 are absorbed.

5.1. Theory analysis validation

Due to the symmetry of sucker group 1 and group 2, we choose arbitrary group. According to the parameters of suckers and coordinate system in figure 2, the coordinates of the four suckers in sucker group 1 are (50, 105), (-50, 105), (-50, -15), and (50, -15). The coordinates of barycentre of cleaning robot is (42.8, 415.1, 88.3).

When the tilt angle of the PV module α is 60°, the tendency to slip is the biggest. The force to slip and frictional force are calculated in equation (8).

$$\max F_{slip} = G \sin 60^{\circ} = 63.6\ N, \qquad \max F_{friction} = \eta \sum_{i=1}^{4} \Delta P_i \cdot S_i = 348.34N \tag{8}$$

When the tilt angle of the PV module α is 30°, the tendency to overturn is the biggest. The coordinates of barycentre can match the condition of Equation (6), the reacting force of PV surface can be calculated by equation (9).

$$\sum_{i=1}^{n}(F_{Ni}-F_i)\cdot|y_i-y_j| = G\cdot\sin 30°\cdot z_m - G\cdot\cos 30°\cdot|y_m-y_j| \qquad (9)$$
$$\Rightarrow F_{N1}+F_{N2}=18.36N$$

Based on the equation (7), the tendency to twist is the biggest when the tilt angle of the PV module α is 60°. The maximum torques of twist and anti-twist are

$$\max M_{twist} = G\cdot\sin 60^{o}\cdot x_m = 2.72N\cdot m, \qquad \max M_{anti} = \eta\sum_{i=1}^{4}\Delta P_i\cdot S_i\cdot|x_i| = 17.42N\cdot m \qquad (10)$$

The force to slip is less than the maximum frictional force supported by the suckers and the torque of anti-twist is bigger than the one of twist. The reacting force of PV surface is greater than zero. Therefore, the suckers are reliable from three aspects including anti-slipping, anti-overturning, and anti-twisting.

5.2. Simulation analysis validation

The finite element analysis of cleaning robot has been conducted with the condition that only sucker group 1 is absorbed. The weight of robot is supported by the four suckers. The results of simulation are illustrated in figure 4.

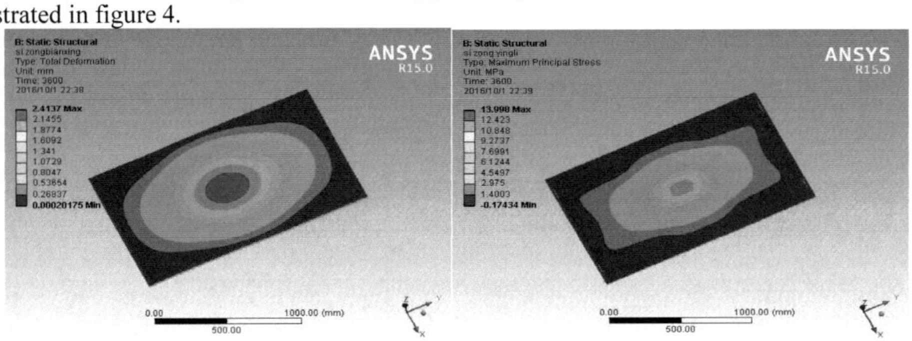

(a) Deformation of silicon cell. (b) Maximum stress nephogram of silicon cell.

Figure 4. Simulation results of PV module with cleaning robot being working.

It is obvious that the maximum stress of silicon cell is 14Mpa, less than 89.96Mpa, while the maximum deflection is 2.41mm, less than 22.167mm. Therefore, the simulation results can prove that the selected suckers are reliable.

6. Prototype experiment of cleaning robot

Based on the suckers validated in above section, the prototype of cleaning robot has been developed. The robot adsorption experiments with tilt angle being 30° and 60° have been conducted as figure 5. The working state is maintained as long as one minute to observe whether the cleaning robot slips, overturns, or twists. In order to improve experiment effect, the whole cleaning device of cleaning robot is hung outside of the PV module.

The same adsorption experiments have been repeated for several times. The suckers are stable reliable. Therefore, the results of theory analysis of sucker reliability are verified to be practical and feasible.

The deformations of cleaning robot in the working state are measured by dial indicator. Deformations of three points in the central region have been measured five times. The maximum deformation measured is 2.39 mm. It is close to the value of 2.41mm, which is the simulation result of figure 4. The experiment result is in good agreement with simulation result.

Figure 5. Adsorption experiment of prototype with tilt angle being 30°and 60°.

7. Conclusion

This paper focuses on the reliability analysis of the sucker of photovoltaic cleaning robot. After the theory analysis and simulation analysis of sucker reliability, a series of constrains have been obtained. Based on the suckers selected according to the reliable constraint conditions, a prototype of PV cleaning robot has been developed. The experimental results of the prototype show that the proposed reliability analysis of sucker is practical. The analysis methods of reliability provide a helpful reference for other sucker based walking mechanisms.

Acknowledgement

This project is supported by Foundation and Frontier Research Program of Chongqing Science and Technology Commission (Grant No. cstc2016jcyjA0538), and Society's Technological Innovation Project of Chongqing Science and Technology Commission (Grant No. cstc2016shmszx0606).

References

[1] CHEN W, WU Y Z, YUE Y F, et al., Science, 350, 944 (2015)
[2] W. Javed, Y. Wubulikasimu, B. Figgis,et al., Solar Energy, 142,123 (2017)
[3] Mohamed A O,Hasan A., Journal of Basic and Applied Scientific Research, 2, 11030 (2012)
[4] Kaldellis, J.K., Kokala, A., Energy, 35, 4862 (2010)
[5] XIONG Diao, LIU Yu-liang, Journal of electromechanical engineering, 32, 929 (2015)
[6] DONG Wei-guang, WANG Hong-guang, Journal of robot, 37, 264 (2015)
[7] XUE Sheng-xiong, REN Qi-le, CHEN Zheng-wen, et al., Journal of mechanical engineering, 47(21), 37 (2011)
[8] LIU Ming-qin, DAI Yong-xiong, Journal of Mechanical design and manufacturing, 05, 146 (2011)
[9] PENG Jin-min, LI Ji-ze, SHAO Yong,et al, Journal of china mechanical engineering, 23, 22 (2012)

Vehicle security system with monitoring of homeostatic processes and user validation via bluetooth

A. Valdez, C. Canales, D. Huamán, C. Pascual and M. Chauca y J. Grados*

Electric and Electronic Engineering Faculty, Universidad Nacional Del Callao-Perú

*Corresponding author's e-mail: mario_chauca@hotmail.com

Abstract. This project consist in evaluating a series of specific and general situations and thus to be able to collect data using some device that facilitate the reception, processing and sending of information that, when combined, can ignite the vehicle whose purpose is to reduce the rate of citizen insecurity in a significant in Peru and Latin American countries where the index of traffic accidents and thefts is critical. The project will avoid that people in an ethyl state of with cardiac diseases and thieves can drive since these situations have already been foressen and thus we have discovered a great security system making only the owner can access the vehicle.

1. Introduction

The study of the following project seeks to generate solutions to the various problems that today are presented in the transport sector, among the most prominent are: thefts and accidents.

The traffic accident rate worldwide and the percentage of vehicular accidents is obtained from figure 1.

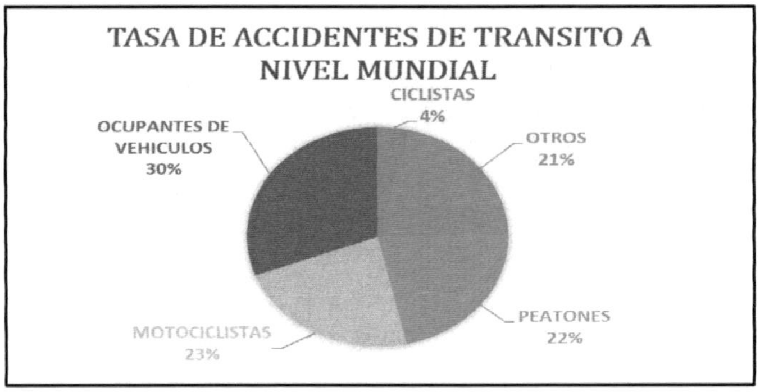

Figure 1. Deaths by traffic accidents according to the type of user (2013), by WHO región.

First, the thefts are the product of the low level of security that vehicles possess, even the most recent models, which is why access and use by multiple users is very practical. For example, the ignition of a car executed by a thief is simple since it is not subject to recognition to distinguish who monitors the vehicle. Secondly, accidents are mainly the result of excessive consumption of alcohol, narcotics and the handling of pilots with abnormal levels of cardiac pressure.

Because anyone can be a victim of the theft of their vehicle, there is a need to invest in a good security system that protects the owner and the vehicle. Of course you can always count on the

Content from this work may be used under the terms of the Creative Commons Attribution 3.0 licence. Any further distribution of this work must maintain attribution to the author(s) and the title of the work, journal citation and DOI.
Published under licence by IOP Publishing Ltd

IOP Conf. Series: Materials Science and Engineering **383** (2018) 012057 doi:10.1088/1757-899X/383/1/012057

traditional alarms, locks or bars for the steering wheel, pedals and naturally with the GPS; however, given that thieves vary their methods of robbery more and more, it has been thought with the present proposal to develop a security system that increases the robustness, using current technology. [1]

The data source of the problem to be solved was obtained from reliable pages (example: Ministry of transport and communications and the World Health Organization (OM). [2] The figure below shows that each year the number of accidents vehicular, which increases progressively and is reflected in the slope of the line after a linear regression.

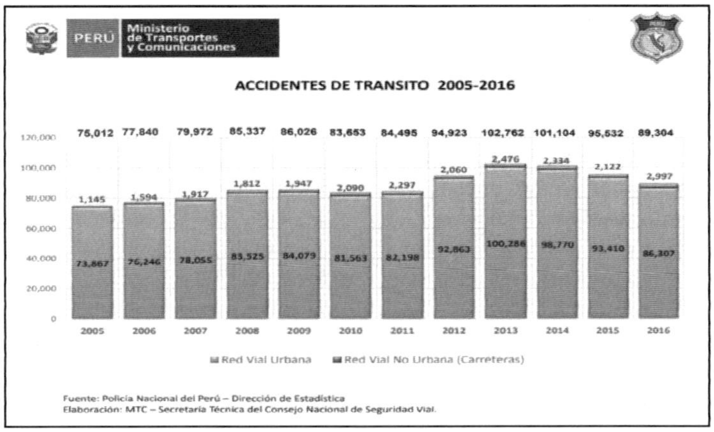

Figure 2. Statistical study of the increase in the number of accidents by vehicle according to the MTC.

The accelerated development of the road system of our country and the increasing use of motor transport has produced a significant increase in travel by road and urban roads, to the extent that drivers, motorcyclists, cyclists, passengers and pedestrians, who coexist in the road infrastructure , depend more and more on road signs and safety devices, for their protection and information. To satisfy this need, it is essential to standardize road signs at national and international levels, and install safety devices that minimize the severity of the damage caused by accidents. [3]. The data shown in figure 2 are passed to excel, where you can obtain a graph whose trend you can see in the following equation that is a linear regression of figure 3.

$$T(x) = 1912.98X - 376120.60 \quad (1)$$

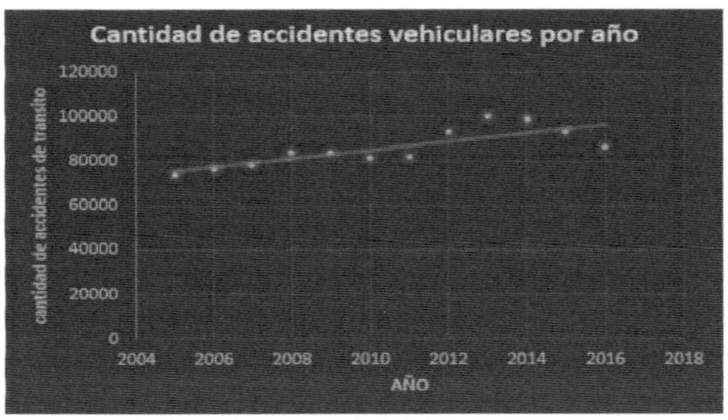

Figure 3. Graph of the trend of vehicle accidents.

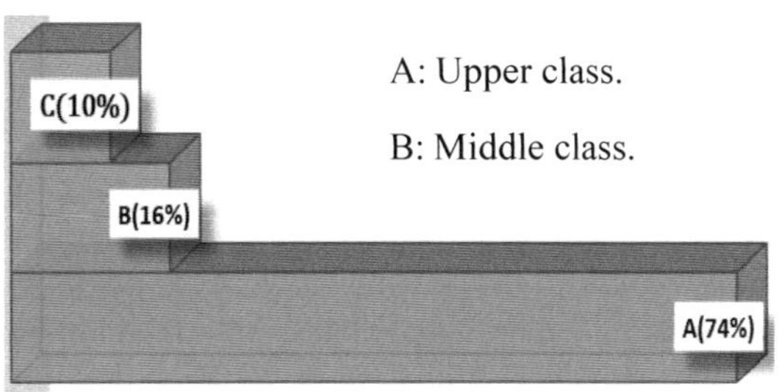

A: Upper class.

B: Middle class.

Figure 4. Percentage of the number of robberies according to social class.

The figure 4 shows the percent of robberies according to social class, having in the upper class with 74% as main objective, since they have a greater purchasing power of having modern vehicles. However, today most do not take into consideration the vehicular safety and the indifference of the different automotive companies towards that sector, since currently the vehicles have many sensors that only facilitate the visualization of certain parameters of control and comfort; but it should be realized that investments must be made in efficient and economical vehicle safety systems to reduce the percentage of thefts and traffic accidents.

Automobiles have a large number of sensors, the number of sensors varies between 60 and 70, which are necessary in the electronic management of the car. They can capture travel, position, rotations, speed, acceleration, vibrations, pressure, flow, temperature, among others and are used by the control units that manage safety, engine operation and vehicle comfort. The control and data processing units allow rapid evaluation of the aforementioned parameters. [4] However, it should be mentioned that these sensors have little average hope of avoiding accidents and robberies, before this a solution is sought. Currently, automotive companies, together with their research departments, focus on introducing new devices to the market that reduce the consequences of the impact in an accident and also reduce the likelihood of their occurrence. [5]

The total automation of driving and control over real vehicles has been considered a utopian idea that only some research groups around the world have dared to address. [6] The goal of machine learning is to design and develop algorithms that allow systems to use empirical data, experience and training methods to evolve and adapt to the changes that occur in their environment. [7]

Under this directive, this project seeks to contribute to the search and application of futuristic systems, which, although they are on the market, cannot all be used because of their high cost and the lack of knowledge of systems that provide safe driving for the driver. [8]

For the start of the project, it is sought to carry out a process of identification of the user to be driven and thus validate the process of action of the sensors. Being the user validation an important and decisive factor for any vehicle security system considering the access control.

The physical access control is focused on three questions: Who? When? and how?; that is, who is authorized to enter or leave? When will you enter or leave the area? and How will it be done? [9]

One of the fields that has registered the greatest evolution in recent years has been telecommunication systems, especially mobile telephony, which day after day offers more services to users. For example, IP telephony, data packets, Wi-Fi networks, text messages, multimedia messages, Bluetooth, music and video players, analog television, virtual banking, information systems, among others.

Within the communications landscape, wireless data transmission systems appear, these systems are particularly suitable for telemetry or portable computer applications, which allows mobility, with the advantages of being connected to a network. [10]

Nowadays the use of the mobile phone is such an integrated accessory and is so widespread and widespread, within the daily life of any user, since everyone has one, so that no one will have any inconvenience of use so no transitional behavior will be necessary nor will it be a radical or gradual change. [11] Therefore, the system is composed of a receiver activated from a transmitter by fingerprint of a mobile device through the Bluetooth connection to give access to the user's test. With the correct reception we can calculate the distance of the device for this we take the RSSI signal and perform the following test: [12]

$$\text{Distance} = 1 - 10\left(RSSI\right)^{5.4916} \qquad (2)$$

By analyzing the fingerprint as an identification method, it is possible to establish the importance that has been given within the security and business sector and in the proper administration of justice, avoiding impersonations and infiltrations that result in the leakage of information and loss of tangible resources intangible. [13]

After carrying out the identification process automatically, the sensors are activated. The operation consists of controlling the ignition of the vehicle as long as the sensors capture information that it considers optimal for driving. Among the components and sensors to be used is the ultrasonic sensor whose equation for determining the distance is as follows:

$$\text{Distance} = (duration / 2) / 29.1 \qquad (3)$$

$$f - 2f_e \frac{V}{V_s} \cos\theta = f_e - f_r \quad (4)$$

If what is desired is to know the time that the wave takes from leaving the emitter until it reaches the receiver, the following equation should be used: [14]

$$t = \frac{(d / sen\theta)}{V_s + V\cos\theta} \quad (5)$$

It is one of the most economic and smallest ultrasonic sensors that exists in the market and is still one of the best for its performance, it is stable, the error is so low reaching 3mm, it has a high precision and works with 5V (DC). [15]

A system used in almost all vehicles, is composed of a belt that goes over the hip and another that is arranged diagonally from the anchor point to the shoulder. [15] The abdominal and diagonal three-point belt is the safest and most commonly used in automobiles, trucks, minibuses, trucks and in the seats of bus drivers, while the two-point belt is the more usual for bus passengers. In recent years, seat belts have been integrated into the vehicle's general safety systems, which include devices such as pretensions, load limiters and airbags. [16]

In view of this, the pressure sensor is located in the band of the driver's chest, because the information provided by the pressure sensor when placed in the hip band would cause difficulty in receiving sensor data. To calculate the theoretical maximum heart rate (FCMT) of the person to drive, we resort to the following approximate expression [17]:

$$189.50 - 0.55 * age \quad (6)$$

With an average error of: $+/-14.957\, beats.* \min$.

And for the minimum determined by:

$$\text{FCmín} = 208 - (0.7 * age) \quad (7)$$

For which the sensor measures approximate higher values of the person's pulse in 700, due to this the same amount is subtracted to be the actual number of pulses, then:

$$pulse - 700 > 0 \quad (8)$$

Also, it will send an activation signal to the determined Arduino of the following equation, which will deny driving to the user.

$$I(\%) = 20 * V_{OUT}(v) \quad (9)$$

The purpose of this device is to find the successive moments of each heartbeat and calculate the interval of time between them called IBI (Inter-Beat Interval), making use of the predictable form and pulse wave pattern. [18]

Figure 5. Driver's blood alcohol level.

As we can see in the figure 5, from 0.6 of the level of alcohol in the blood the motor is affected, the movements are slowed down. There is a feeling of euphoria and confidence, aggressive and reckless driving by impulses without reasoning. And the following equation obtained from figure 5 is available [19].

$$N(x) = 1.018 \times (8.2396^{\cdot X}) \quad (10)$$

The MQ-3 sensor is a sensor that measures the concentration of alcohol in the air. It acts as a resistance that varies according to the amount of alcohol detected; the higher the alcohol concentration, the lower the resistance. [20] In this way for the sensor we programmed that, if the value of alcohol in the blood of the person subtracted to the limit value of alcohol that can possess is greater than zero, determined by equation (11), it will send a pulse that will disable the vehicle to be driven.

$$value\_alcohol - value\_\lim it > 0 \quad (11)$$

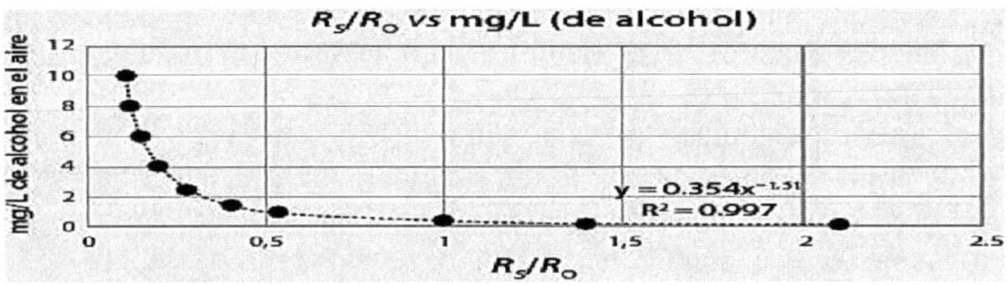

Figure 6. Graph and equation of the MQ-3 sensor.

By plotting some points of the previous alcohol curve of the MQ-3 sensor in the figure 6 and applying a linear regression, its characteristic equation is obtained: [21]

$$\frac{mg}{L} = 0.354 \times (\frac{R_s}{R_O})^{-1.518} \quad (12)$$

$$V_{OUT} = \frac{R_0}{R_0 + R_S} x V_{CC} \quad (13)$$

$$LevelOH = \frac{alcohol(g)}{weight(kg)x7} \quad (14)$$

This project developed to provide greater safety when driving, avoiding vehicle theft because it has an Android application for user validation by fingerprint and password that in turn communicates via Bluetooth with the microprocessor for the activation of the different sensors that they monitor those homeostatic processes to determine the driver's status and approve the ignition of the vehicle.

The range of the Bluetooth signal r is determined by:

$$X^2 + Y^2 - 2aX - 2bY + a^2 + b^2 + r^2 = 0 \quad (15)$$

2. Development

The project process operates under two stages: identification phase and data collection phase. The authentication phase consists in the activation of the circuit that will give access to the ignition of the vehicle under an interface connected via Bluetooth by means of a mobile device and the phase of data collection consists of the activation of the circuit with microcontroller to begin the measurements of proximity, alcohol and cardiac pressure.

The fingerprint, a reinforcement password and the measurements of the 3 sensors are used as process control variables to obtain an output function that in this case is the motor of the vehicle. Table 1 shows the aforementioned. Whose logic is denoted in table 2.

Table 1. Control and output variables.

input variables	output variables
fingerprint	Motor
password	
sensors	

Table 2. Logic of our output function.

Huella digital	Contraseña	Sensores	Motor
0	0	0	0
0	0	1	0
0	1	0	0
0	1	1	0
1	0	0	0
1	0	1	0
1	1	0	0
1	1	1	1

IOP Conf. Series: Materials Science and Engineering **383** (2018) 012057 doi:10.1088/1757-899X/383/1/012057

It is schematized by means of the Karnaugh map method in table 3, it is considered as a logical "1" in response to the correct conditions for the user to carry out the operation. In addition, the variable sensors are a simplified representation of the three sensors (proximity sensor, breathalyzer and heart pressure).

Table 3. Map of Karnaugh.

A		$\overline{A}$		
0	1	0	0	B
0	0	0	0	$\overline{B}$
$\overline{C}$	C		$\overline{C}$	

Since the circuit is very large, it is best represented by a block diagram that simplifies the entire process.

A detailed explanation is noted, with which it is easy to understand the operation of each material used in this project that will collect information from time to time to determine the safety of driving and will validate the user by means of a fingerprint and password, which without them will not light the vehicle.

To have a better understanding of this project it is necessary to have a flow diagram shown in figure 7.

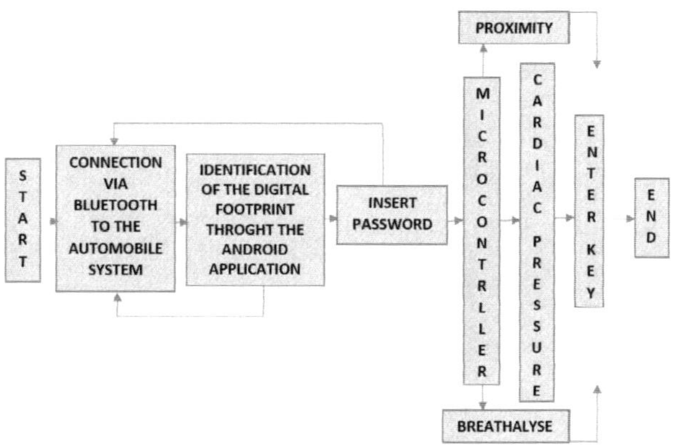

Figure 7. Block diagram.

The Arduino UNO has a series of facilities to communicate with a computer, another Arduino, another microcontroller or, as in this case, with a mobile device. The ATmega328 offers UART TTL (5V), a serial communication, which is available on digital pins 0 (RX) and 1 (TX). In addition, thanks to the Atmega16U2 firmware that uses the standard USB COM drivers, the board does not need any external driver and will appear on the computer as a virtual COM port.

3. Result

The result of this project is a large decrease in the rate of accidents and vehicle thefts whose expected range is 15% to 40%.

Table 4. Comparison of the cost of materials.

Sensor	Price	Others	Price
QM3	\$3.00	AL6000	\$30.00
HCSR04	\$4.00	PR3015DN	\$30.00
SEN0386	\$4.00	027i	\$15.00
	\$11.00		\$75.00

There is greater control and precision in the collection of data necessary for the monitoring of the homeostatic processes of the driver

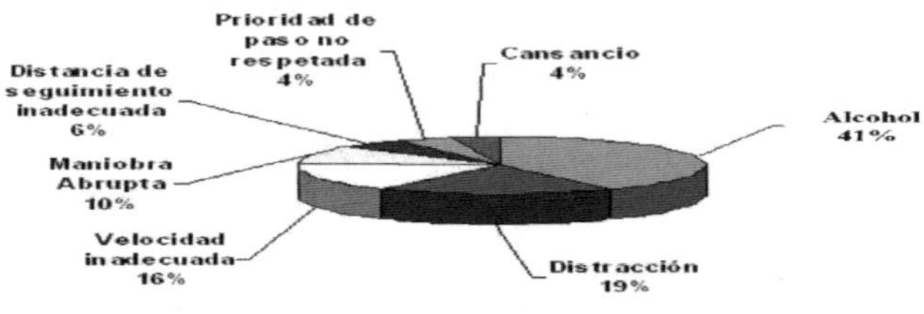

Figure 8. percentage of accidents causes.

4. Conclusions

It is possible to optimize any vehicle safety system by 41%, this is due to the considerable difference between the total number of accidents and the one that does not consider accidents due to alcohol in the blood, whose value is specified in figure 8.

It reduces costs by 76.4% and materials to obtain a system accessible to the general public obtained from table 4.

5. Observations

In the case of users who perform physical exercise and must use their car, the system will make it impossible to access the vehicle because it has a cardiac pressure that exceeds the limit considered normal.

If any incident related to cardiac pressure occurs, the system does not react in time.

6. Recommendations

Wait 5 minutes before driving the vehicle after exercising according to a study from the University of California.

Avoid calls and messages while driving the vehicle as this may cause concern or anger that alters the cardiac pressure.

References

[1] Cajas, M. (2017). *Diseño e implementación de un sistema de seguridad vehicular mediante reconocimiento facial atreves de visión artificial.* Universidad politécnica Salesiana. Cuenca, Ecuador.

[2] Organización Mundial de la Salud (2015). *Informe sobre la situación mundial de la seguridad vehicular.* Suiza.

[3] Secretaria de Comunicaciones y Transporte. (2014). *Manual de señalización vial.*

[4] La Rosa, K. (2012). *Diseño de sistema integral de seguridad vehicular.* Pontificia Universidad Católica del Perú. Lima, Perú.

[5] Casanovas, I. (2012). *Sistema de detección de situaciones riesgosas en la conducción de un automóvil mediante variables inherentes al conductor y de entorno.* Universidad tecnológica nacional facultad regional Buenos Aires. Autónoma, Buenos Aires.

[6] Pérez, J. (2012). *Agentes de control de vehículos autónomos en entornos urbanos y autovías.* Universidad Complutense de Madrid. Madrid, España.

[7] Qi, F. (2013). *Redes de sensores inteligentes.* New York: group Taylor.

[8] Bonilla, J. (2012). *Diseño y adaptación de un sistema de seguridad activo para estacionamiento vehicular y monitoreo continuo.* Escuela politécnica del ejército. Latacunga, Ecuador002E

[9] Morales, D. (2012). *Prototipo de control de acceso peatonal al campus de la Corporación Universitaria Lasallista.* Corporación Universitaria Lasallista. Caldas, Antioquia.

[10] Giraldo, Y. (2013). *Implementación, control y monitoreo de un sistema de seguridad vehicular*

por redes GSM/GPRS. Universidad tecnológica de Pereira. Pereira, Colombia.
[11] F. Javier García Ruiz, Eduardo J. Moya de la Torre, Alfonso Poncela Méndez (2015). *Detección e identificación de dispositivos bluetooth para control de acceso.* Instituto de las Tecnologías Avanzadas de la Producción, Valladolid – España
[12] Juan Carlos Torres Piedra (2015).Diseño de un Sistema de Localización de Sensores basado en Bluetooth y GPS.
[13] Maya, V. (2013). *Sistema biomédico de reconocimiento de huella dactilar. Bogotá.*
[14] Lasluisa, R. (2015). *Sistema electrónico de ayuda para el estacionamiento de vehículos livianos.* Universidad técnica de Ambato. Ambato, Ecuador.
[15] Arizaga, E. (2015). *Análisis de los sistemas faltantes de seguridad vehicular.* Universidad politécnica Salesiana. Cuenca, Ecuador.
[16] Foundation for the Automobile and Society (2014). *Cinturones de seguridad y sistemas de retención infantil.* Londres: FIA.
[17] Gandola F., Reeb P., Nucci S., Aguiar Muñoz S. (2005). *La ecuación "220-Edad" predice una frecuencia cariaca máxima teórica DESACERTADA.* Centro de Entrenamientos Adaptados, Universidad Nacional del Comahue, Neuquen, Argentina.
[18] Biendicho F. (2015). *Comunicación bluetooth entre Arduino UNO y Android aplicados a un detector de mentiras.* Universidad Politécnica de Valencia, Gandia.
[19] Escobar, B. (2012). *Sistema para la medición arterial.* Universidad CES, Medellín, Colombia.
[20] PCasanova, M. (2014). *Diseño, construcción e instalación de un alcoholímetro electrónico con dispositivo de bloqueo de vehículo.* Escuela superior politécnica de Chimborazo. Riobamba, Ecuador.
[21] Corona, L. (2014). *Sensores y actuadores.* México: Patria.

IOP Conf. Series: Materials Science and Engineering **383** (2018) 012058 doi:10.1088/1757-899X/383/1/012058

Reactive power characteristics analysis on emergency stop of doubly fed induction generator

Yuting Zhao [1, *], Yang Cui [1], Yaohua Tang [2] and Weimin Guo[2]

[1]School of Electrical Engineering, Northeast Electric Power University, 132012, Jilin, China
[2]State Grid Henan Electric Power Research Institute Zhengzhou 450052, Henan, China

*E-mail: bobtreeglasgow@163.com

Abstract. This paper analyzes the reactive power characteristics of the doubly-fed induction generator (DFIG) during the emergency off-grid process caused by the crowbar protection action. Since the action of the crowbar is much faster than the stator switching, the DFIG would be running with a short-term squirrel cage asynchronous operation. The mathematical model of the DFIG under above situation was proposed in this paper, based on which detailed analysis was given on reactive power demands during squirrel-cage asynchronous state of the DFIG. It is achieved that the amount of reactive power absorption was associated with the initial operating conditions such as the terminal voltage and speed before disconnection, which can be prescribed as the s-Q characteristic of DFIG. Finally, analysis on the measured data of the wind farm disconnected event verified the validity of the model.

1. Introduction

As the most principal means of generating power as non-hydro renewable energy generation presently, wind power is an important way to solve the world wide problems such as the crisis of fossil fuel, environmental deterioration, and so on.

Doubly Fed Induction Generator (DFIG) has been extensively employed in most wind farms for its high wind energy conversion efficiency and flexible active and reactive power control[1-3]. Nowadays, the DFIG in domestic wind farms were operated under unit power factor, which could not supply additional reactive power to the grid. In order to satisfy the generator operating voltage requirement, wind farm usually disposed reactive compensation according to the reactive requirement of the main transformer when the wind farm under rating output. However, which did not take the reactive requirement of the wind farm collect system into consideration, such as booster transformers of wind turbines and cluster lines within the farm.

It is generally considered that a lack of reactive power occurs when the wind farm output is in a high situation, and the reactive compensation disposition mentioned above can not satisfy the reactive power requirement of the main transformer and the collecting system simultaneously. It will cause the PCC voltage obviously declined and further the wind turbine low-voltage protection operating, which result in the generators disconnecting from grid. Therefore, the study aimed at improving the reactive level of wind farm presently, mostly considers excavating the reactive compensation ability of the DFIG, as shown in [4-9].

Content from this work may be used under the terms of the Creative Commons Attribution 3.0 licence. Any further distribution of this work must maintain attribution to the author(s) and the title of the work, journal citation and DOI.
Published under licence by IOP Publishing Ltd

However, by analyzing the measured data of one certain wind farm disconnecting from the power grid, it was recorded that the PCC voltage was under high level before the disconnecting, which would not trigger the low voltage protection of wind turbine (usually 0.8-0.9pu). It indicates that the above explanation can not explain the generators disconnecting from grid for the lack of reactive compensation.

This paper is to present another possible reason for the disconnecting event by analyzing the difference between the stator and the rotor circuit switching characteristics after the crowbar protection has been triggered.

2. Wind farm disconnecting event overview

The main electrical wiring of the wind farm occurring the disconnecting event is showed in fiigure 1. The wind farm includes two subsets (1#, 2#), the area is about 80 km2, and the total installed capacity is 400MW. Among them, 1# wind farm owns 251 Gamesa G58-850kW wind turbines, 2# wind farm owns 134 SL1500kW-77 wind turbines, all rated voltage 690V; wind farm collect system, including booster transformer and collecting lines, which rating voltage is 35kV.

The 1# wind farm is connected to 1# 220kV bus at the booster substation by four 6.3MVA booster transformers, and the 2# connected to 2# 220kV bus by two 10MVA booster transformers. The parameter of the line between the wind farm and the 1# 220kV substation is LGJ-240*2-100km, while the line between 1# substation and 2# substation is LGJQ-400-34.5km.

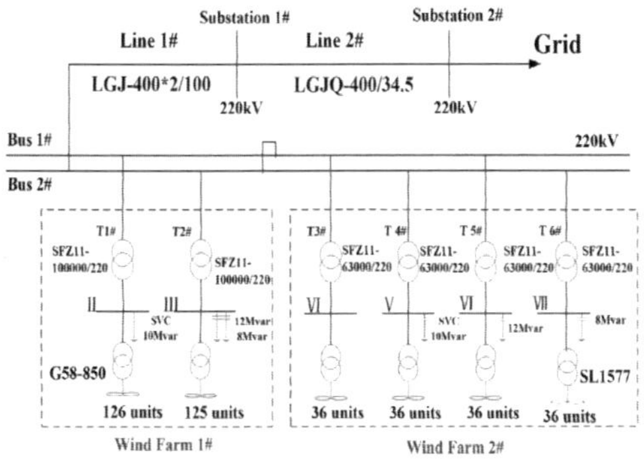

Figure 1. Main Wiring Diagram of the concerned wind farm.

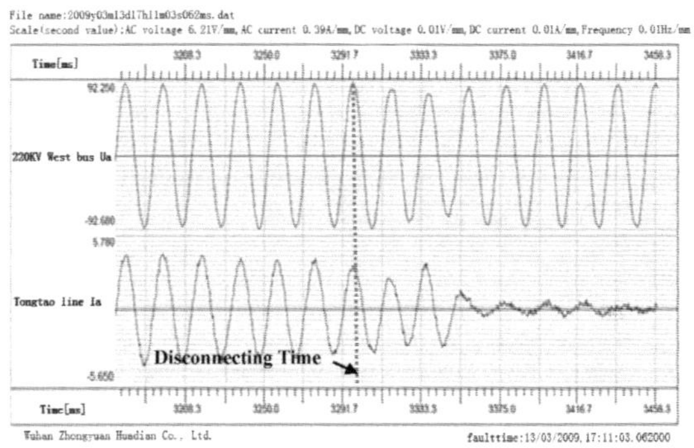

Figure 2. Record diagram of 220kV substation on current and voltage of phase A.

Under normal operation, every wind turbine is running with unit power factor. To ensure outlet bus voltage level of wind farm, reactive compensation equipment had been disposed at low voltage side (35kV) of each main transformer.

Figure 2 records the wind farm booster substation 220kV bus voltage and the changes of 1# line current, during the wind farm disconnecting from grid. The vertical line in figure 2 shows the disconnecting time. On the basis of the operators taking notes, the disconnecting event caused 1# wind farm, 2# main transformer and 125 units into outages. Short time later, all the turbines were off grid.

Outlet bus voltage of wind farm before the event was high to 1.1pu. The reactive compensation disposed could satisfy farm collect system, and the wind farm did not absorb reactive power from grid, which can be calculated by the data of figure 2. After the disconnecting time, the wind farm booster substation bus voltage appeared obvious fluctuation and three cycles later, all the wind farm turbines disconnected from grid.

The above analysis indicates that the reason reactive compensation lacking resulting in wind turbines disconnecting from grid can not explain the fact generators still disconnected from grid with high voltage (1.1pu). So a further analysis needs to be done to give reasonable explanation.

3. Squirrel-cage asynchronous operating characteristics of doubly fed induction generator

DFIG-based wind turbines mainly include wind turbine, doubly fed induction generator and four-quadrant converter and so on. Figure 3 is the schematic structure.

When the crowbar protection of rotor circuit has been triggered, the DFIG would get into emergency stop as following operating logic.

The three-phase rotor wiring of DFIG are short circuited after the crowbar protection started, meanwhile, the four quadrant converters quit. The execution element of the switch is the power electronic switch (Thyristor or IGBT) with microsecond-class action time.

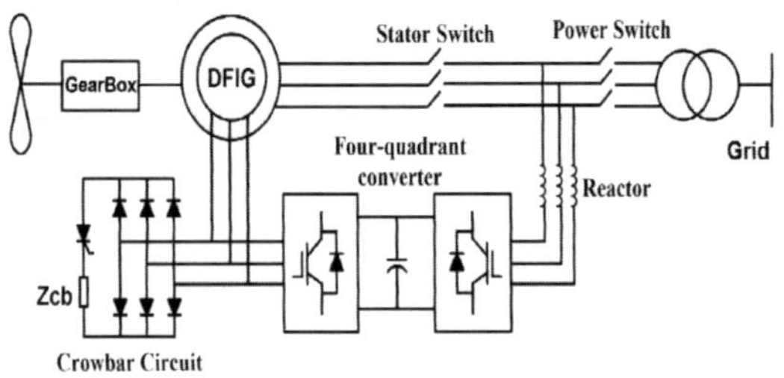

Figure 3. Configuration of DFIG.

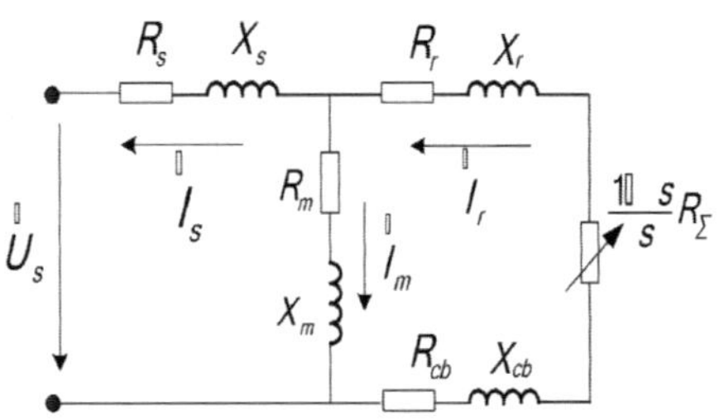

Figure 4. Equivalent circuits of a DFIG after crowbar action.

The stator circuit directly starts the stator grid-connected contactor and the execution element is mechanical switch, the action time is several milliseconds.

Therefore, as the crowbar protecting operates in very short time (microsecond-class) while the time for stator contactor to break off is longer (millisecond-class), the stators remain connecting with the grid for 50~70ms, after the crowbar protection started and the rotor converters been short circuited. At the moment, the DFIG was operating on squirrel-cage asynchronous state.

Nowadays, study concerning the mathematical model of DFIG-based wind turbines is various[10-12]. In order to analyze the operating characteristics after the physical construction of DFIG has changed, this paper builds the mathematical model of DFIG on squirrel-cage asynchronous state during disconnecting process, based on previous studies.

The equivalent circuit diagram of DFIG on asynchronous condition after the crowbar protection performed is showed in figure 4, and the corresponding model in dq coordinates, is showed as formulas (1) (2) (3).

$$\begin{cases} U_{sd} = R_s(-i_{sd}) - \omega_1 \psi_{sq} \\ U_{sq} = R_s(-i_{sq}) + \omega_1 \psi_{sd} \\ U_{rd} = 0 = R_r^{'} i_{rd} + d\psi_{rd} / dt - \omega_s \psi_{rq} \\ U_{rq} = 0 = R_r^{'} i_{rq} + d\psi_{rq} / dt + \omega_s \psi_{rd} \end{cases} \tag{1}$$

$$\frac{T_j}{N_p} \frac{d\omega_r}{dt} = T_m - T_e \tag{2}$$

$$\begin{cases} \psi_{sd} = L_s(-i_{sd}) + L_m i_{rd} \\ \psi_{sq} = L_s(-i_{sq}) + L_m i_{rq} \\ \psi_{rd} = L_m(-i_{sd}) + L_r i_{rd} \\ \psi_{rq} = L_m(-i_{sq}) + L_r i_{rq} \end{cases} \tag{3}$$

The formula $R'_r = R_r + R_{cb} + (R_r + R_{cb})(1-s)/s$; s is the slip ratio; ψ_{sdq}, ψ_{rdq} are respectively the magnetic flux linkage of the stator and the rotor of DFIG.

Owing to the converters that provide the generators with excitation have quitted, the generators must absorb plenty of reactive power for maintaining the present operating condition. According to asynchronous motor operating characteristics, the reactive power the grid needs to supply is as follows.

$$Q = \sqrt{3}\frac{U_s^2}{|Z_\Sigma|}\sin\varphi_S \tag{4}$$

where,

$$Z_\Sigma = Z_s + \frac{(Z_r + Z_{cb})Z_m}{(Z_r + Z_{cb}) + Z_m} \tag{5}$$

Formula (4) is the s-Q characteristics of DFIG on squirrel-cage asynchronous state during disconnecting process, which indicates that the Q is directly related to the grid voltage and rotational speed.

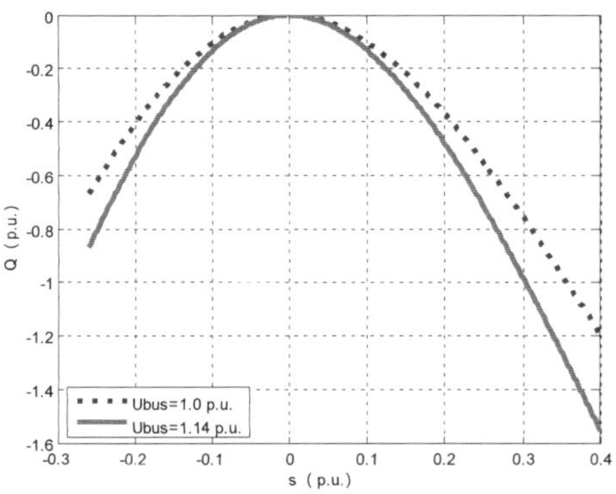

Figure 5. Slip-Reactive power characteristic of DFIG in disconnecting process with different voltage.

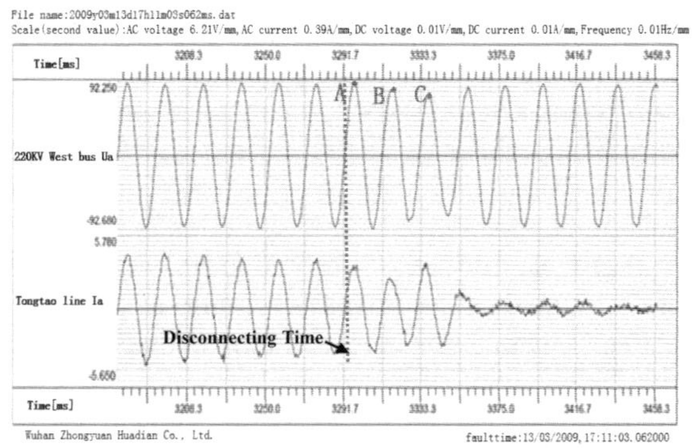

Figure 6. Postmortem analysis diagram of 220kV substation on current and voltage of phase A.

Taking Gamesa G58-850kW wind turbine as an example, since the machine may get into off-grid at different rotational speed or voltage level in practical operating, the reactive power absorbed could be calculated by DFIG at different initial conditions by using formula(4). The two curves in figure 5 respectively shows the s-Q characteristics of G58 generators during disconnecting process with different outlet voltage (1.0pu, 1.14pu) and rotational speed.

Figure 5 indicates the DFIG do not absorb reactive power from the grid only when the rotational speed is synchronous speed before disconnecting time, other rotational speed are opposite. The bigger the slippage is, the more reactive power is absorbed. Under the rated voltage, the slip ratio of Gamesa G58-850kW wind turbines is 0.4(corresponding starting speed 900 rpm) before disconnecting and the maximum reactive power absorbed from the grid reaches 1.2 times as much as generator rated rating.

4. Postmortem of wind farm disconnecting event

This section will make explanation for the process of the wind farm disconnecting event, based on the analysis result of the two sections above.

The obviously fluctuating process of the wind farm booster substation bus bar voltage was showed from A to B in figure 6.

Table 1 gives the recorded data during the adjacent three cycles (from A to C in figure 6) of the disconnecting process, which are the wind farm output active and reactive power, booster substation bus voltage and its phase-angle.

By the analysis of the data recorded by the data acquisition system OPEN3000, the output level of the wind farm was 351MW before the disconnecting event, which has already approached 88 percent of installed capacity (400MW). Taking the wind energy uneven distribution of large wind farm area into account, most of the wind turbine has been running close to or reach the speed limit at the moment.

Table 1. Wind farm output power and export bus voltage (the datum value: S_B=100MVA; U_B=220kV).

Time	P(pu)	Q(pu)	U(pu)	φ(dec)
A	2.1581	0	1.1411	0
B	1.0592	-0.7616	0.9889	-35
C	1.3967	-1.0148	0.9129	-36

The active power of Gamesa-G58-850kW generator is rated value 850kW at so high wind speed, and the generator rotational speed reaches 1900 rpm. The absorbed reactive power is 595Kvar during the disconnecting process by calculating based on Formula (4). The reason causing the reactive power shortfall during wind farm disconnecting process is analyzed as follows.

1) Form A to B the margin of the wind farm output active power to the power network is ΔP=1.0989pu. Besides PN=0.0085pu, it can be calculated that total 129 generators had disconnected from grid. Because the margin of the reactive power is ΔQ=0.7616pu since QN=0.0059pu, there are 129 units disconnecting from grid, and this moment, the wind farm absorbed reactive power from the grid.

2) Instantaneous large absorption of reactive power by the disconnecting wind generators caused local voltage decreased （ΔU=0.15pu）, and triggered the low-voltage protecting adjacent generators ceased. The development of voltage decline should be an avalanche-like process with an end of all the wind turbines disconnected.

The above analysis shows that the number of the wind turbines disconnecting from grid calculated by the measured data (129 units) is nearly identical to the result recorded by the operators (125 units). Therefore the correctness of the established model and the explanation for the wind turbines disconnecting reason could be verified. The calculation error may come from not taking the loss of the wind farm collect system into consideration.

5. Conclusion

This paper analyses the measured data about one certain wind farm disconnecting from grid under normal voltage level, then reveals one possible reason causing the lack of reactive power during the off-grid process. The conclusions are as follows:

1) After the crowbar protection triggered, there is nearly 50-70ms transient asynchronous time interval due to the different action characteristics between the stator and rotor circuit switches of the DFIG.

2) The asynchronous state of DFIG needs to absorb large reactive power during disconnecting process, and the maximum reactive power absorbed from the grid could reach 1.2 times as much as generator rated capacity.

3) The lack of reactive power caused by wind turbines disconnecting from grid will result in the voltage obviously declined instantly. If the wind farm voltage level is somewhat lower before disconnecting time, part of generators getting into outages will probably causes the adverse consequence with the generators off grid.

4) The analysis of the measured data verifies that the proposed explanation for the disconnecting event is reasonable and correct.

References

[1] Hao X, Jiang C, Wu L 2015 Based on the Power Factors of DFIG Wind Farm for Power Flow Optimization *Proceedings of the Second International Conference on Mechatronics and Automatic Control* Springer International Publishing 2015 pp 157-65

[2] Yao-Jia Li, Juan-Juan Wang, Zi-Lin Li and Chuang Fu 2016 VDCOL parameters setting influenced by reactive power characteristics of HVDC system *2016 International Conference on Smart Grid and Clean Energy Technologies* (ICSGCE) Proceedings of. IEEE 2016 pp 364-70

[3] Li Z, Yang H, Li W 2012 Reactive Power Optimization in Regional Power Grid Considering Voltage and Reactive Power Characteristics of Boundary Buses *Power and Energy Engineering Conference* IEEE 2012 pp 1-4

[4] Raboni P, Chaudhary S, Chen Z 2016 Design of reactive power regulator of synchronous generators by considering grid impedance angle for characteristic index objectives *Iet Generation Transmission & Distribution* 2016, 10(14) pp 3508-16

[5] Talavera I, Stepanescu S, Bennewitz F 2015 Vertical Reactive Power Flexibility based on Different Reactive Power Characteristics for Distributed Energy Resources *International ETG Congress 2015;Die Energiewende - Blueprints for the new energy age* Proceedings of. VDE 2015 pp 1-7

[6] Kumar R, Kumar A 2014 Reactive power cost characteristics for synchronous generator in deregulated electricity markets *Power India International Conference* IEEE 2014 pp 1-6.

[7] Kraiczy M, Lammert G, Stetz T 2016 Parameterization of Reactive Power Characteristics for Distributed Generators: Field Experience and Recommendations *Int. ETG Congress 2016;Die Energiewende - Blueprints for the new energy age* Proceedings of. VDE 2016 pp1-8

[8] Yan G, Wang M, Mu G 2008 Modeling of grid-connected doubly-fed induction generator for reactive power static regulation capacity study *Transactions of China Electrotechnical Society* 2008 23(7) pp 98-104

[9] Fang Y, Liu Q, Hou G 2012 Study of the Reactive Power Characteristics in the Double Fed Variable Speed Constant Frequency Wind Turbine *2012 Asia-Pacific Power and Energy Engineering Conference* IEEE 2012 pp 1-4

[10] Rezaei-Zare A 2015 Reactive Power Loss Versus GIC Characteristic of Single-Phase Transformers *IEEE Transactions on Power Delivery* 2015 30(3) pp 1639-40

[11] Fang X, Li F, Wei Y 2015 Reactive power planning under high penetration of wind energy using Benders decomposition *Iet Generation Transmission & Distribution* 2015 9(14) pp 1835-44

[12] Dai H, Wang Y, Li X 2012 Characteristic Analysis of Reactive Power Compensation Device at HVDC Converter Station *Asia-Pacific Power and Energy Engineering Conference* IEEE 2012 pp 1-5

IOP Conf. Series: Materials Science and Engineering **383** (2018) 012059 doi:10.1088/1757-899X/383/1/012059

An experimental investigation of performance characteristics of an energy-efficient air conditioning in vehicle

Wei YIN[1], Wen ZHANG[1], Ji ZHANG[1], Jian ZHAO[1], Xuexiang LIU[2] and Ning MEI[1,*]

[1]Marine Engineering, Ocean University of China, 238 Songling Road, Qingdao 266100, China
[2]Xiamen Donghai Institute, Xiamen, China

*E-mail: nmei@ouc.edu.cn

Abstract. In order to reduce the fuel consumption of air-conditioning in vehicle, an air conditioning system was designed based on the principle of direct evaporative cooling and liquid dehumidification. In this paper, main devices were conducted in the way of numerical simulation and experiment, the overall layout of the device was designed under the guidance of Solidworks. The results showed that the high temperature and humidity air could be changed into the livable air. In particular, the regenerator could be operated with waste heat from the engine for higher energy efficiency. It can achieve the effect of the temperature dropped by 5-7°C which need less energy than the traditional car air-conditioning while satisfy the comfort of the human body.

1. Introduction

Vehicle are usually equipped with air conditioning systems to provide comfortable cab environment for driver. The air conditioning system is expected to reduce temperature and dehumidify of air in the cab condition, both functions require the mechanical power delivered by the internal combustion engine [1]. Air conditioning need higher quality because of the wicked environment including sunlight exposure, densely populated, velocity variation [2]. At this stage, air conditioning in truck is composed of compressor, condenser, throttle element, evaporator, blower and the necessary control components [3]. The thermal efficiency of internal combustion engine is about 30% to 40%, of which less than half of the energy for the power output, 25% of the energy is lost in the form of cooling water, 10% of the fuel is used in the car air conditioning system [4].

In open loop air conditioning system, the refrigerant used for refrigeration must be non-toxic and harmless [5]. As a natural working substance, water is cheap, accessible and environmentally friendly. In addition, there is no compressor in this new equipment, that is, there is no rotating parts, which reduces noise energy consumption.

In recent years, people pay more attention to the research of membrane dehumidifier and dehumidifying technology which based on the penetrative and selective performances of membrane, in the whole process, there is always a partial pressure difference of water vapor between both sides of the membrane forming a continuous process [6].

The air-conditioning system of vehicle is facing energy consumption problems. It is necessary to reduce the fuel consumption of air conditioning while making driver's working environment well. The

Content from this work may be used under the terms of the Creative Commons Attribution 3.0 licence. Any further distribution of this work must maintain attribution to the author(s) and the title of the work, journal citation and DOI.
Published under licence by IOP Publishing Ltd

IOP Conf. Series: Materials Science and Engineering **383** (2018) 012059 doi:10.1088/1757-899X/383/1/012059

research is of great significance to develop an energy-saving, environmentally friendly and economical air-conditioning system.

2. Experiment section

2.1. Feedstock materials

Polylactic acid(PLA) was used for most part of the device, acrylic(PMMA) were used as the feedstock for parts to be subjected to high temperatures. Water was the refrigerant in this air conditioning, lithium bromide solution with a concentration of 50% was used as liquid dehumidifier. Polyvinylidene fluoride(PVDF) membrane was used in dehumidifier.

2.2. Machine layout and workflow

The air conditioning system was consisted of four main units: the ejecting humidification device, two dehumidifiers and the regeneration of lithium bromide solution device. The theory of this air conditioning system was shown in figure 1. The whole circulation could be divided into the gas circulation and the liquid circulation. The high temperature and humidity air A1 was blown into the primary dehumidifier 5 under the action of the blower 1, while the wet air passed through the air flow path in the primary dehumidifier 5, water was absorbed by the lithium bromide solution through the PVDF membrane. High temperature and low humidity air A2 was gained. The ultrasonic atomization device 3 constantly atomized the water in the reservoir 4 into a mist, Negative pressure is formed at the nozzle of the ejector 2 while the air A2 entered, which could inhale the mist into the ejector 2. The water mist evaporated in the ejector 2 also absorbed heat. The low temperature and humidity air A3 was obtained and discharged from the nozzle outlet of the ejector 2. Air A3 need to be dehumidified through the secondary dehumidifier 6 for an appropriate temperature and humidity.

The lithium bromide solution was pumped from the reservoir 8 through the pumps 7 and 12 into the dehumidifier, it would be circulated back to the reservoir 8 after dehumidification. A density detection which monitored the height difference between the float and the liquid level at different concentrations based on the relationship between the concentration and the buoyancy was placed in the reservoir 8. Liquid circulation began when the solution concentration was below 40%, low concentration solution was inhaled into regenerate device 9, pump 10 pumped the solution to the cooling tank 11 while finished regeneration. The solution would be pumped into the reservoir 8 after cooling down.

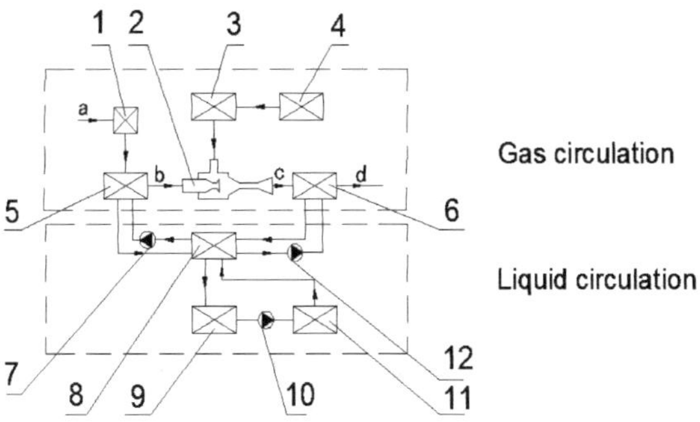

Figure 1. The theory and entity graph of the air conditioning system. (1) blower; (2) ejector; (3) ultrasonic atomization device; (4) water storage device; (5) primary dryer ;(6) secondary dryer; (7) Pump; (8) Solution storage device; (9) regeneration device; (10) pump; (11) Cooling pool; (12).

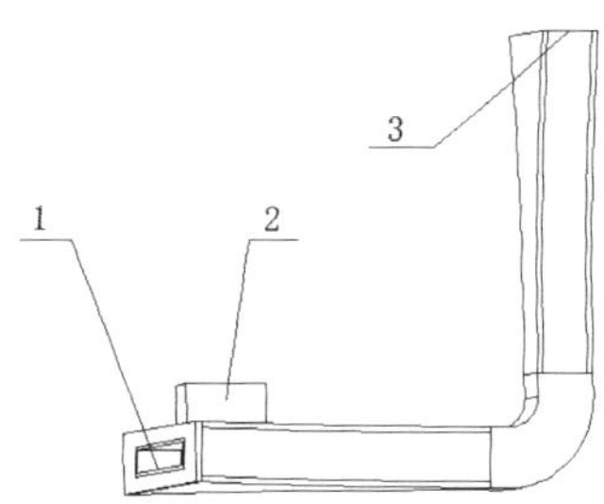

Figure 2. Model of ejector. (1) Eject inlet; (2) Ejected inlet; (3) Air outlet.

IOP Conf. Series: Materials Science and Engineering **383** (2018) 012059 doi:10.1088/1757-899X/383/1/012059

2.3. Ejecting humidification process

The ejecting humidification is shown in figure 2. The direct evaporative refrigeration process in ejector was the core of the circulation. The injection process of ejector was simulated by Fluent. Figure 3 shows the internal pressure and flow velocity while the air inlet velocity of the ejector was 3m/s. It can be seen that the water mist produced by the ultrasonic atomizer was inhaled into the ejector under the action of negative pressure.

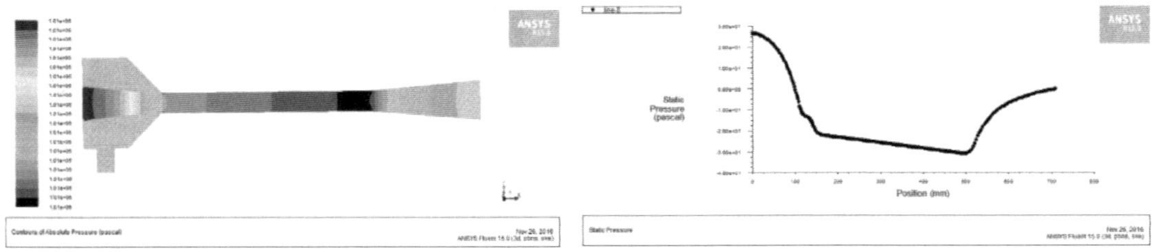

(a) Pressure distribution in ejector (b) Pressure distribution on the central axis of ejector

Figure 3. Internal pressure and flow velocity of ejector.

2.4. The dehumidfifier

The primary dehumidifier is shown in figure 4. PVDF membrane was supported by several columns. lithium bromide solution was pumped into three spray tube, which had certain quantity of holes symmetrically, the lithium bromide solution flowed out from the holes and sprayed onto the PVDF membrane. Water molecules were absorbed by lithium bromide solution through the membrane, then the solution flowed back to the reservoir.

The structure of the secondary dehumidifier is similar to that of the primary dehumidifier. The difference was that the secondary dehumidifier did not need obvious dehumidification effect.

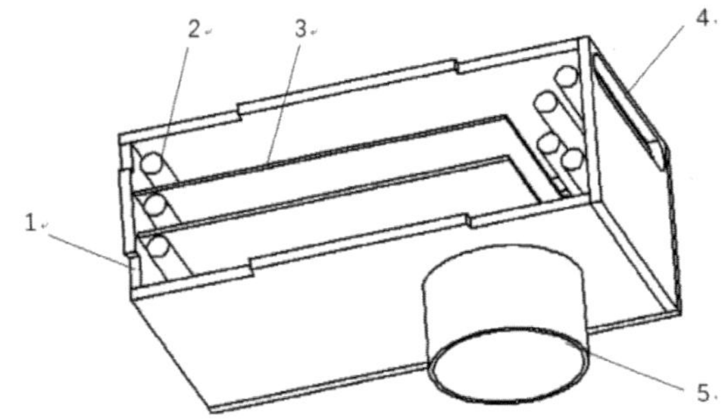

Figure 4. The model of primary dehumidifier. (1) dehumidifier housing; (2) Membrane support column; (3) Baffle; (4) Air inlet; (5) Air outlet.

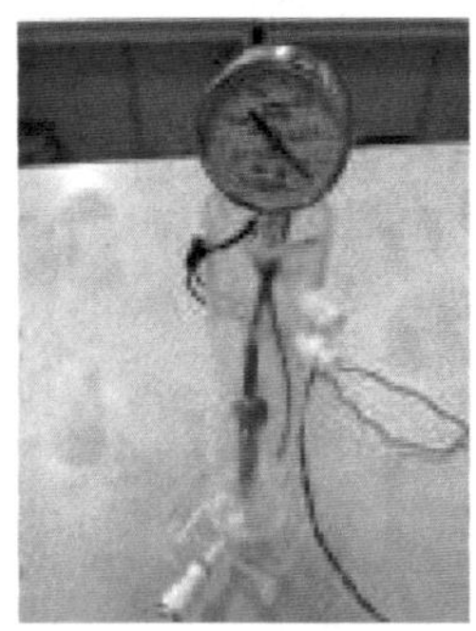

Figure 5. Regeneration device of LiBr.

2.5. Regeneration of lithium bromide solution

The concentration of lithium bromide solution would be at a low level while working, water in the solution must be evaporated in order to ensure dehumidification effect. The escaping water was discharged by vacuum pump at the same time. The regeneration device of lithium bromide solution was shown in figure 5.

3. Result and discussion

3.1. Machine test

The size of the device was 600x550x400mm. The experimental device is provided as figure 6.

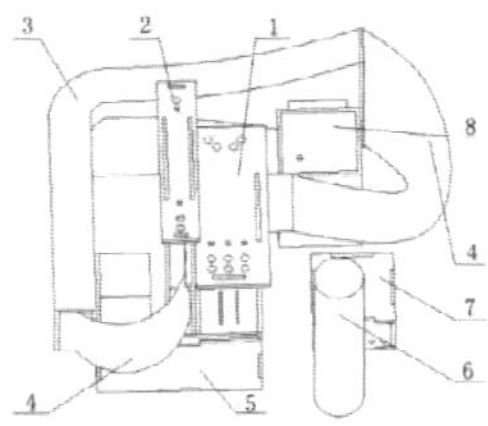

Figure 6. Test device of experiment. (1) primary dryer; (2) secondary dryer; (3) Ejector; (4) Ventilation ducts; (5) Reservoir; (6) Regeneration device; (7) Cooling pool.

The experiment was carried out under the temperature 32°C and relative humidity 80%. When the system was stable, air conditions at the inlet and outlet were measured. An anemometer and a hand-held hygrothermograph shall be provided at the air inlet (primary dehumidifier air inlet) and the air outlet (secondary dehumidifier air outlet) to measure the outlet air flow rate, temperature and humidity.

3.2. Power consumption

The following table shows the power consumption of this device.

Table 1. Summary of power consumption equipment.

Equipment	Located	Power (W)
Blower(x2)	air inlet of Primary dryer	2
Ultrasonic atomizer	Ultrasonic atomizer device	12
Pump(x4)	Circulation of LiBr	120
Small vacuum pump	regeneration device of LiBr	1.2
Electric heating bar	regeneration device of LiBr	50

While the entire device was fully operational, the total power was 185.2W.

3.3. Evaluation criteria

Both environmental factors and the body's own factors involved in the body heat balance equation. In the multi-factor interaction, human body would have different thermal comfort, it could be evaluated through the indicators or comprehensive indicators [7]. Direct indicator referred to the environmental parameters that could be measured directly by the meters. Comprehensive indicator referred to the theoretical derivation and experimental verification, the various environmental parameters and the integration of the relevant variables obtained by the comprehensive indicators, it could more truly reflect the human body's thermal sensation.

At present, PMV (Predicted Mean Vote) indicator proposed by ISO (International Organization for Standardization) is used to describe and evaluate the thermal environment, it means the majority of people's feeling in a certain environment, which involves six parameters, the following table 2 describes the scope of these six parameters.

Table 2. Applicable Scope (ISO 7730).

Temperature (°C)	Average radiation temperature (°C)	Water vapor partial pressure (Pa)	wind velocity (m/s)	Human activity (W)	Clothing thermal resistance ($m^2 \cdot K/W$)
10~30	10~40	0~2700	0~1	46.4~232	0~0.31

The value of PMV can be calculated by the following formula.

$$PMV = \left[0.303e^{(-0.036M)} + 0.028 \right] S \tag{1}$$

Which M means body energy metabolism rate, the values of M can be found in the table 3 below; S means body heat storage rate.

Table 3. Adult Men's Energy Metabolic Rate.

Activity intensity	Energy metabolism rate(W/m^2)
Lying down	46
Sitting and rest	58
Standing and rest	70
Sitting and active	70
Slightly labor	93
Heavily labor	116
Hard labor	165

The value of PMV is shown in the following table 4.

Table 4. PMV thermal sensation standard.

Thermal Sensation	Hot	Warm	Slightly warm	Moderate	Slightly cool	Cool	Cold
PMV	3	2	1	0	-1	-2	-3

PMV thermal sense scale is determined in the steady state environment, when the PMV is greater than 2 or less than -2, the result is distortion, it can be more accurately indicate thermal sensation when its value is between -2 to 2 [8].

3.4. Experimental results

The following table 5 shows the experimental results.

Table 5. The experimental data of entire device.

Serial number	Air inlet condition			Air outlet condition			COP	PMV
	Temperature (°C)	Relative humidity (%)	Wind velocity (m/s)	Temperature (°C)	Relative humidity (%)	Wind velocity (m/s)		
1	32	80	3	26.3	64.3	0.9	1.04	0.39
2	32	65	3	25.8	63.7	0.8	0.63	0.35
3	30	80	3	25.4	65.2	0.9	0.82	0.16
4	32	80	5	26.5	74.2	2.9	2.72	/

We obtained the export air state, cooling capacity, PMV value, COP value under different conditions of the initial state of the air.

$$COP = \frac{Q_t}{P} \tag{2}$$

Which

$$Q_t = m_a \Delta h \times 1000 = \rho_a v_{a,out} s_{out} \left(h_{in} - h_{out}\right) \times 1000 \tag{3}$$

Q_t —Total power of the device;

$v_{a,out}$ —Air speed of the outlet;

s_{out} —Air flow area of the outlet;

h_{in} —Enthalpy of inlet;

h_{out} —Enthalpy of outlet;

In the first three conditions, the air outlet temperature and humidity were comfortable, it could basically meet the requirements of human comfort considered from the point of view of PMV. Although the fourth working condition had a large cooling capacity and lower temperature, the relative humidity was at a high level and the wind speed reached 3m/s, which was beyond the limit which shown in the table 3. From the energy consumption point of view, the power of the entire device is 185.2W.

3.5. Uncertainty analysis of the experimental data

The uncertainty analyses of the experimental measurements are shown as follows:

(1) Temperature and humidity

The temperature and humidity of the air outlet was measured by the Rotronic HP21.

According to its measuring range and accuracy mentioned above in table 1, its uncertainty was 0.28 % and 1 %.

(2) Velocity

The velocity of air was measured by the SiWei FF-45A, its uncertainty was 5%.

4. Conclusion

During the working process, the air flow rate played a more critical role in the final air condition, and the wind speed of blower should be adjusted reasonably. The COP value of the air conditioning is about 1, which can satisfy the heat exchange of the human body surface. However, this device is better in the energy saving effect, it has a certain practical significance from point of view of fuel cost. At the air conditions including temperature 32°C, relative humidity 80%, the final air conditions of outlet can reach temperature 26.2 °C, relative humidity 64%, it means that under the premise of reducing fuel consumption, the new type of air conditioning device can meet the comfort requirements of the human body and it has a certain application prospect in the cars.

Acknowledgments

Financial support came from the National Natural Science Foundation of China (51679225 and 51706214), Shandong Provincial Natural Science Foundation, China (ZR2017BEE011), Key Basic Research of Shandong Provincial Natural Science Foundation (ZR2017ZA0203), Central University of basic scientific research business fees special of China (201713034) and Post-Doctoral Application Research Projects in Qingdao, Special Fund for Agro-scientific Research in the Public Interest (201503135-04).

References

[1] Kessels, J. T. B. A., and P. P. J. Van Den Bosch. Springer International Publishing. *Model Development for Air Conditioning System in Heavy Duty Trucks*. (2016)

[2] Wang, Wu, and Sun Wentao. Automobile Applied Technology.*Heavy-duty truck air conditioning system design*. **8**, 25-31. (2012)

[3] Fu Baixue, Ma Biao, and Pan Xufeng. *Modern Automotive Electronics Technology*, Beijing Institute of Technology Press. (2008)

[4] Ji Wenjing, Yuan Han, and Chen Guoxing. Autobile&Parts. *Design of Air Conditioning System with Waste Heat Truck*. **35**, 54-56.(2012)

[5] You ShiJun. Journal of HV&AC. Open loop dehumidification cooling air conditioning system. **2**, 3-5. (1992)

[6] Ol Crabtree. Electronic Theses.A Numerical Study of Nonlinear Static and Dynamic Behavior of a Square Thin Plate Membrane Structure for Application to a Mems Micro-generator. (2004)

[7] Shi Xiaokun. Dong Hua University. Numerical Simulation of the Influence of Clothing on Human Thermal Comfort. (2004)

[8] Ji Xiuling. Chinese Center For Disease Control And Prevention. Evaluating and Prognosticating Human Thermal Sensation inside the Human Habitation. (2003)

IOP Conf. Series: Materials Science and Engineering **383** (2018) 012060 doi:10.1088/1757-899X/383/1/012060

Design and implementation of the electrochlorination filter model based on solar photovoltaic for daily watering systems at Eco-Garden-SD IT Al-Uswah Surabaya

O Penangsang, A Soeprijanto, N K Aryani, D F U Putra, Rony Seto W, Suyanto, Arseto Y B, Alfan P, M Ali Fikri*, A Riangga, S B Panuntun, A Saad D, Rinthon B A and Elpha Aulia A

Department of Electrical Engineering, Faculty of Electrical Technology, Institut Teknologi Sepuluh Nopember ITS, Surabaya Indonesia, 60111

*E-mail: fikrielits14@gmail.com

Abstract. Al-Uswah Elementary School holds an Eco-Garden program, but in the process, Al-Uswah has an issue with the lack of water to fulfil the daily watering needs. So we assumed the solution to this problem is filtering the wastewater in Al-Uswah into clean water, which is mainly available in Al-Uswah. Filtering of wastewater to supply clean water for the eco-garden and hydroponic system using electro-chlorination water filter integrated with Solar PV is one of the best ways to provide daily hydroponic water needs. This paper presents a method for modelling and sizing electro-chlorination water filter integrated with Solar PV for an eco-garden and hydroponic system. The model of electro-chlorination described how to convert wastewater into clean water, predict flow rate to meet the daily need, and provide an experimental result to calculate the accuracy of the model.

1. Introduction

The world's population is expected to grow from 7 billion to 9.3 billion in 2050, and the urban population from 3.6 billion to 6.3 billion, a 72% increase [1]. Its effect to availability of clean water requirements for daily watering such as agricultural need. Agriculture in semi-arid and arid regions is highly dependent on water for irrigation; however, water availability is already problematic in many regions with imbalances between renewable resources and total demands, which jeopardise irrigated agriculture and its resilience. The fulfilment of clean water at Surabaya city very depends on PDAM, because the water source of Surabaya city is very polluted, and beside it, the population density is also caused clean water scarcity. Individual water use in Surabaya is estimated at 150 litres per day, while the availability of clean water by PDAM is 10,000 litres per second, or the production capacity of PDAM is 21 million to 22 million cubic meters per month, which means that PDAM production can only serve 70% of Surabaya people [2]. Meanwhile, since 2010 the Surabaya city government promotes eco-garden and urban farming program to overcome environmental problem and food shortages, but in the implementation of this program, the water scarcity is the main problem of this program implementation. In order to overcome the problem of water scarcity and environmental problem in SD IT Al-Uswah Surabaya, Indonesia, we design electro-chlorination filter integrated with solar PV system to supply clean water for eco-garden and hydroponic system for daily watering. In this case wastewater can converted to clean water for daily watering eco-garden and hydroponic

Content from this work may be used under the terms of the Creative Commons Attribution 3.0 licence. Any further distribution of this work must maintain attribution to the author(s) and the title of the work, journal citation and DOI.
Published under licence by IOP Publishing Ltd

system using electro-chlorination water filter integrated with Solar PV System as renewable energy source.

2. Literature review

Chemical disinfection is a common unit process used in water supply and wastewater treatment. Traditionally, chlorination is the most dominant method of disinfection. However, there are serious safety concerns and great ecological risks involved in the use of chlorine. Other methods, such as ozonation, UV radiation and ClO_2 application, are still more expensive or less convenient than chlorination. It has been reported that electro-chlorination can destroy a wide variety of microorganisms from viruses through bacteria and algae to larger species, such as *Euglena* [3].

An electro-chlorination system consists of electrodes, an electrolyte, and a circuit, and anodes are a significant factor for chlorine generation. According to the property of anodic materials, anodes are divided into active and nonactive electrodes; IrO_2 and RuO_2 are the former and boron-doped diamond (BDD) is the latter. Among various electrodes such as IrO_2, RuO_2, and BDD, the IrO_2 electrode has better efficiency than the others to produce chlorine in the presence of NaCl. Furthermore, the chlorine solution generated in an electrochemical system is slightly more efficient for inactivating microorganisms than free chlorine under certain pH conditions [4].

Wastewaters that contain heavy metals, organic complexing agents, and suspended matters from metal finishing industries are harmful to public health. Electro-chlorination filter was experimented by Khelifa, Aoudj Moulay, and De Petris-Wery to involve the indirect electrooxidation process of the active chlorine generation and promotes the oxidation of organic pollutants efficiently. Dimensionally stable anodes (DSA) are commonly used in wastewater treatment as the result of its numerous advantages. In addition, it has shown that the efficiency of high chlorine production may be achieved in an electro-chlorination cell by DSA ruthenium oxide coated titanium anode consider graphite anodes and platinized titanium. The previous research has resulted that NI, Zn, Pb, Cu/CN can be efficiently removed from wastewater by using a two-compartment electrolytic cell before the filtration of brine. But the process is hindered by the use of platinum as its anode and the requirement of ion permeable. Khelifa has shown that electrochlorination is a promising way to replace the classical technologies and quite interesting as it is very suitable for wastewaters of the metal finishing industry [5].

3. Research methods

3.1. Sampling and analysis

At the beginning, the wastewater is collected in the 1st-floor reservoir of SD IT Al-Uswah which include the wastewater from the 1st to the 3rd-floor and will be tested in the laboratories that are the physical and the chemical test. From physical test some results are obtained that are colour, smell, turbidity, solid particle, temperature, and electrical conductivity. While the chemical test includes pH, total hardness, the chemical components that are chloride, sulphate, notate, ammonia, nitrite, iron, manganese, lead, zinc, chromium, fluoride, arsenic, mercury, cadmium, selenium, cyanide, kmnO4, and detergent.

3.2. *Planning filter frototype*

After the water test results are known, an electro-chlorination filter is used to filter the water because electro-chlorination is the most effective way to eliminate the organic content contained in wastewater. This electro-chlorination method uses 2 electrodes i.e. carbon plate on the cathode and stainless steel on the anode. The design of electro-chlorination filter prototype shown in figure 1.

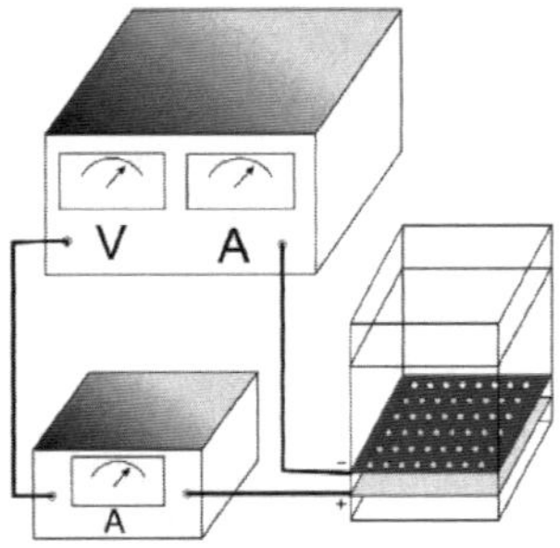

Figure 1. Electro-chlorination water filter prototype.

3.3. Lab-scale electro-chlorination test

The filter is injected with current and DC voltage until the value of the current and the voltage are obtained so that it can break the organic contents of wastewater ablution. In addition, also observing changes in water samples at some time variations to obtain the best results.

3.4. Making electro-chlorination filters

In this phase, the design and implementation of electro-chlorination filter are adapted to the laboratory test results considering the input-output of the wastewater and photovoltaic power supply, including that the DC converter dimension of the filter is adjusted to the planning of the filter placement and combined with a conventional filter as zeolite, sand, and gravel.

3.5. Running filter and integration with PV

At this stage experiments and implementation of electro-chlorination filter using the supply of photovoltaic in SD IT Al-Uswah and integrated with hydroponic system as shown in figure 2, in accordance with experimental results from the filter in the laboratories.

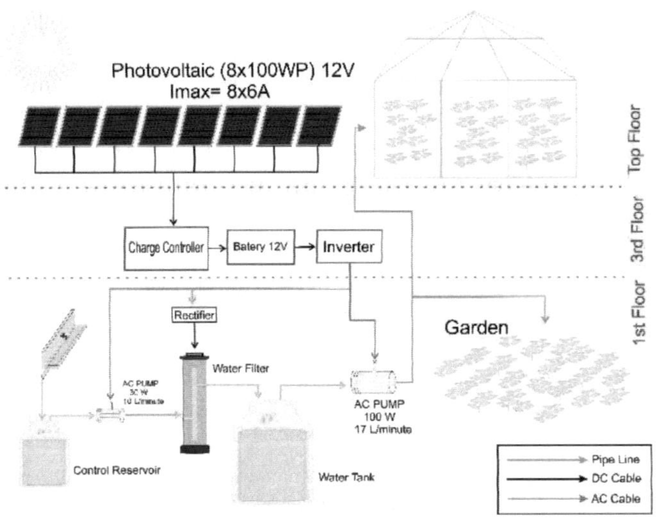

Figure 2. SD IT Al-Uswah Eco-Garden system integrated with Solar PV and Filter.

4. Analysis and result

The wastewater test is divided into 2 sections, the first section was tested without electro-chlorination water filter, and the second section was tested with the filter. On the first section the wastewater tested to find out the wastewater content. The result of the first section shown in table 1.

IOP Conf. Series: Materials Science and Engineering **383** (2018) 012060 doi:10.1088/1757-899X/383/1/012060

Table1. wastewater result analysis.

Parameter	unit	Clean Water Requirement (PERMENKES RI No.416 MENKES/PER/IX/90)	Analysis Result	Analysis Method
1. PHYSICS TEST				
Colour	PtCO	50	80	Spectrophotometry
Taste	-	No taste	No taste	-
Aroma	-	No aroma	No aroma	-
2. CHEMICAL TEST				
a. inorganic Chemistry				
pH	-	6.5 – 9.0	7.70	pH meter
Hardness Total	mg/L $CaCO_3$	500	185.71	Complexometry
Chloride	mg/L Cl-	600	56	Argentometry
Sulphate	mg/L SO_4	400	640.36	Spectrophotometry
Nitrate	mg/L NO_3-N	10	2.05	Spectrophotometry
Nitrite	mg/L NO_2-N	1	0.03	Spectrophotometry
Ammonia	mg/L NH3-N	-	1.41	Spectrophotometry
Fluoride	mg/L F	1.5	3.56	Spectrophotometry
b. organic Chemistry				
$KMnO_4$	mg/L$KMnO_4$	10	17.95	Oxidase
Detergent	mg/L	0.5	0.28	Spectrophotometry

From the test result, in chemical test from wastewater exceeds of organic and inorganic content that doesn't needed by crop such as sulphate, fluoride and $KMnO_4$. Furthermore, in physical test the wastewater colour exceeds of determined limit at 50 PtCO.

The second section is filtering wastewater by using electro-chlorination test. The purpose of this test is to remove organic, and inorganic content of the wastewater, and also to purify the wastewater. After this process the output of the water is sufficient to use by hydroponic.

From the result given in figure 3, the optimal injection of DC voltage and DC current are 30VDC and 0.24A, which takes 25 minutes to obtain 100% efficiency. By optimizing the voltage and current value to supply electro-chlorination filter, it will decrease the filtering time. Power consumption for filtering process will decrease along with the filtering time process reduction.

So, two step of filtering process the wastewater become sufficient for eco-garden used. After finishing the prototype, the actual design filter depends on the field requirements. Next the actual design of the electro-chlorination filter is integrated with hydroponic system and solar PV.

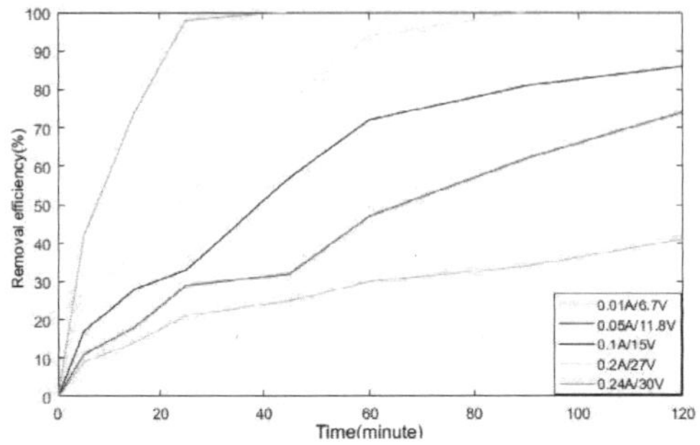

Figure 3. removal efficiency (%) as time function.

5. Discussion and conclusion

Installation of the electro-chlorination filter is integrated with solar PV as power supply and hydroponic system. The electro-chlorination filter used to supply clean water requirements for the eco-garden at SD IT Al-Uswah Surabaya. It obtains optimal removal efficiency of organic and inorganic contents by injecting DC voltage 30V and 0.24A of DC current, it takes 25 minutes time delay to filter wastewater into clean water.

Acknowledgement

This part of this research is founded by Ministry of Research, Technology, and Higher Education and Institut Teknologi Sepuluh Nopember Surabaya (ITS).

References

[1] Jakob Skøt and et all 2016 The State of Food and Agriculture *Food Agric. Organ. U. N.* xi
[2] Munfarida I 2017 Evaluasi Kinerja Teknis PDAM Tirta Kepri Provinsi Kepulauan Riau *Al-Ard J. Tek. Lingkung.* **2** 53–61
[3] Ghernaout D and Ghernaout B 2010 From chemical disinfection to electrodisinfection: The obligatory itinerary? *Desalination Water Treat.* **16** 156–75
[4] Choi J, Shim S and Yoon J 2013 Design and operating parameters affecting an electrochlorination system *J. Ind. Eng. Chem.* **19** 215–9
[5] Khelifa A, Aoudj S, Moulay S and De Petris-Wery M 2013 A one-step electrochlorination/electroflotation process for the treatment of heavy metals wastewater in presence of EDTA *Chem. Eng. Process. Process Intensif.* **70** 110–6

IOP Conf. Series: Materials Science and Engineering **383** (2018) 012061 doi:10.1088/1757-899X/383/1/012061

Design and implementation of seawater eco distillatory using parabolic solar concentrator and photovoltaic pump for isolated island

A Soeprijanto, D F U Putra, Suyanto, A R Hidayatullah, A Hadi*, A Q Yusrina, F S Wicaksana, E W Adinda, M A Choiri and M Baihaqi

Department of Electrical Engineering, Faculty of Electrical Technology, Institut Teknologi Sepuluh Nopember ITS, Surabaya Indonesia, 60111

*E-mail: amirulhadiw@gmail.com

Abstract. Water is a primary human need. Of the total fresh water in the world, less than 1% are ready to drink and 11% of the total human population cannot achieve it. WHO/UNICEF by 2015 reports there are nearly 750 million people in the world who still have inadequate drinking water. The current condition is also occurs in the Ay Island which located in Banda Naira Islands, Central Maluku, Indonesia. Ay Island residents have to travel for 4 hours to the main island of Banda Naira to buy clean water. To overcome the current situation, the government have tried to provide clean water by building a $4 billion dam project in 2016. However, the dam project is not able to meet the needs of the resident because of the polluted water. In this research is proposed an Eco Water Treatment by using Parabolic Solar Concentrator and Photovoltaic Pump. This method consists of two main parts, PV Sea Water Pump and Solar PV-Distillator. The PV Pump is used to deliver water from the sea to the boiler, while the solar-powered distillator converts sea water into fresh water. Fresh water generated from the distillator have the same content as the Indonesian Minister of Health's standard. The system can provide clean water for 50 liter continuously based on capacity of the boiler each day depend on the weather and the temperature. From the experiment, proven that solar concentrator could increase the boiler temperature up to 300°C to boil the seawater and produce clean water.

1. Introduction

Water is the primary needs of living beings. In [1] said that 97.5 % of all water on earth is salt water, leaving only 2.5% as freshwater. As the result, less than 1% of the water on earth is available for consumption. However, [2] shared a fact that only 89% of the global population (6.5 billion people) used at least a basic drinking-water service, which is an improved drinking water source within a round trip of 30 minutes to get the water.

Indonesia is an archipelagic country that has an area of 1,910,931 km^2 [3], two-thirds of its area is the ocean. It is ironic because, with the affluence of the seawater, there are still millions of Indonesian inhabitants that having lack of clean water. For example, a case that occurred in Banda Naira islands, Maluku Tengah. To buy clean water, Ay Island residents have to cross the sea that takes approximately 4 hours to reach the Banda Island. In order to overcome the condition of lack clean

Content from this work may be used under the terms of the Creative Commons Attribution 3.0 licence. Any further distribution of this work must maintain attribution to the author(s) and the title of the work, journal citation and DOI.
Published under licence by IOP Publishing Ltd

IOP Conf. Series: Materials Science and Engineering **383** (2018) 012061 doi:10.1088/1757-899X/383/1/012061

water, one of the sources that could potentially be used as clean water is the seawater which can be transformed to clean water by desalination process.

The process of desalination can be made with different techniques such as vapour pressure, chemical processes, membranes processes (reverse osmosis and electrodialysis), solvents extraction, and evaporation [4]. The desalination process could be generated by many kind of methods, one of the method is solar thermal energy or solar desalination. Considering the condition of Indonesia that crossed by the equator line made Indonesia has superfluous solar energy potential, so desalination using solar thermal is the right method to be applied in Banda Naira Island. Based on the report of Indonesian ministry of energy and mineral resources, Indonesia has solar energy potential of 112 GW. There are two modes of conversion in solar desalination: flat plate and solar concentrator. In this paper, the PV panel is using solar concentrator to multiply the reflected sunlight into one focus point better than flat plate.

2. Literature Review

The global community has begin to replace fossil fuel source into solar energy through photovoltaic (PV) system. As in this paper, the solar energy is used to generate electricity to pump clear water in some remote areas. The system planning for water pumping system depends on the estimation of solar irradiation and the water requirements as shown in [5], while in literature [6] provides the calculation of inverter to obtain high-efficiency, high-lifetime, and low-cost converter for an autonomous photovoltaic water pumping system.

The seawater desalination is an alternative way to provide clean water in the coastal area [7]. Seawater intakes can be obtained in two ways, by directly channel the seawater from the sea and the other method is indirectly from the sea through subsurface systems. In this paper, the water will be obtained directly from the sea to the boiler using PV pump. Based on the literature [8], the concentrator shown in figure 1, the system can be modeled into simple mathematical modeling and set up to get the stable point and efficient parabolic dish. The boiler material has been discussed in [9] with parabolic geometry shown in figure 1.

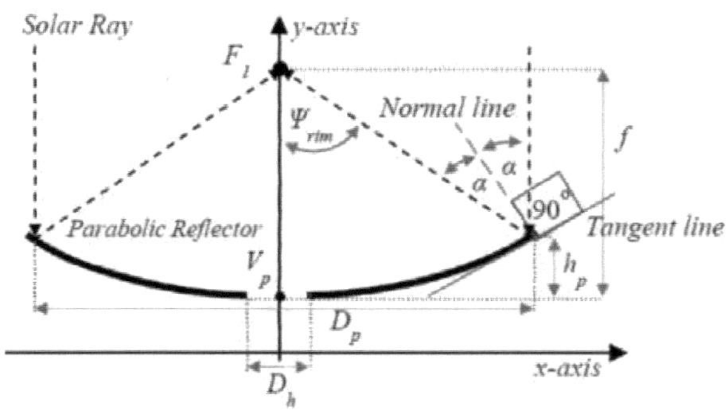

Figure 1. Parabolic Geometry.

This proposed system uses alumunium as a boiler material due to its excellent corrosion protection and good heat conductivity that suits the coastlines environmental conditions. This aims to accelerate the heat-conduction process of the boiler that heated using reflected solar radiation through the solar concentrator.

3. Research Method

The data of current water condition in Banda Island obtained by interviewing local coastal communities. The methodology of this reseaech is presented in figure 2.

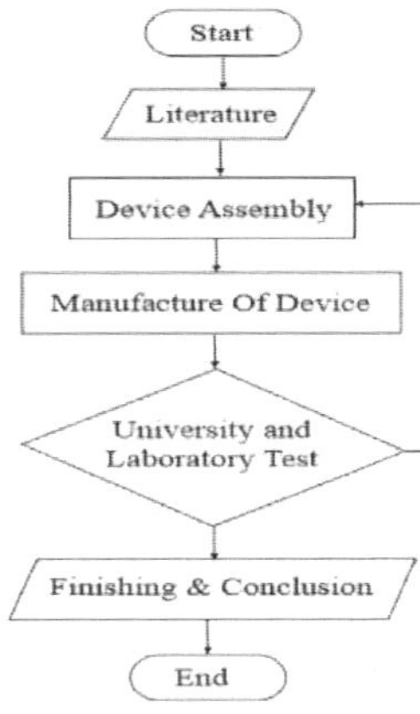

Figure 2. Research Methods.

The literature review used for reinforcing the compiled hypothesis [10], [11]. It is related to the water problem condition in Banda Island. Devices were implemented by references retrieved from the literature review and study field. Basically, the system consists of three main components that illustrated in figure 3; which are photovoltaic, concentrator, and the desalinator. Photovoltaic is used to convert photon radiation, concentrator to convert thermal radiation, and desalinator to remove salt ion and other wasteful materials.

The energy converted by photovoltaic will be stored in a battery with a charge controller and an inverter. Then followed by concentrator suspension construction. The suspension is placed above the parabolic solar concentrator coated with chrome. The boiler is installed on the focal point of the solar concentrator. The process continued with the creation of desalination component consisting of a heater tank, hose pipe and condenser tank (for the cooling process). After all of the component is finished, it is coupled into single entity directly in the field.

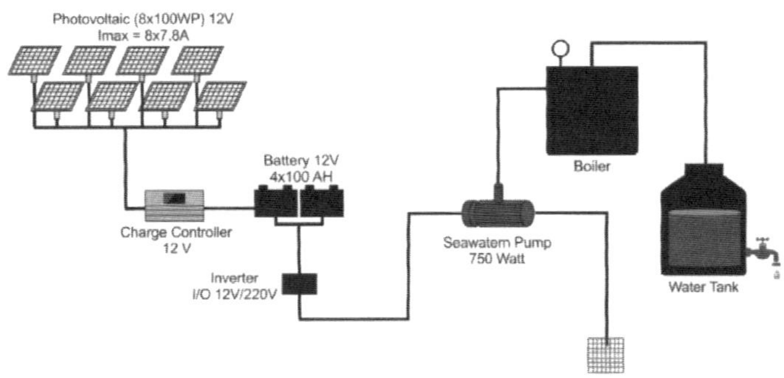

Figure 3. Design to be implemented.

In the laboratory scale trials, the electric components of photovoltaic, the parabolic solar desalination process and concentrator were done in the open area of ITS. The result of the desalination process will be tested again in ITS Environmental Engineering Laboratory to check the content of water desalination. The next trials are on university level through the exchange funding scheme from the Ministry of Research and Technology of higher education.

Evaluation and analysis of the test results are used to improve the performance of the equipment. During the final stage of research method, the conclusion is carried out after the research is finished. The withdrawal of the conclusion is including the work performance of the photovoltaic, parabolic solar concentrator and the condenser

4. Analysis & Results

4.1. Photovoltaic and Pump Performance

In this experiment, we test the characteristic of solar cell installed on Power System Simulation Laboratory and the performance of the seawater pump. The inverter is controlled by DVC method with 50 Hz frequency and close loop MPPT model. In this experiment, the results of the experiment obtained in no load condition.

The current input of the inverter is 62.33 A, while the batteries capacity is 400Ah. Therefore, the needed current is 60A (by the assumption of maximum charging time) and the solar cell needed is 8 PV 100 WP.

To protect system electricity, the emergency switch is designed by taking data from measured voltage and current. Table 1 explains the amount of DC voltage from solar cell output. DC voltage and DC current are the output from solar cell transferred to batteries and inverter, while AC power is the output from the inverter to the electric load.

Table 1. Solar cell voltage and current data experiment.

Time	Charge Controller 12V PV to Battery				Power (W)
	Input (V)	Input (A)	Output (V)	Output (A)	
08:00	20.9	14.03	13.58	43.18	293.22
09:00	20.9	17.50	13.58	53.86	365.75
10:00	20.9	20.12	13.58	61.92	420.50
11:00	20.9	20.15	13.58	62.02	421.13
12:00	20.9	19.15	13.58	58.62	398.14
13:00	20.9	16.69	13.58	51.36	348.82
14:00	20.9	13.12	13.58	40.38	274.20
15:00	20.9	9.9	13.58	30.46	206.91
16:00	20.9	6.1	13.58	18.72	127.49
17:00	20.9	5.8	13.58	17.84	121.22
Total	**209**	**142.56**	**135.80**	**440**	**2977.41**

Based on data experiment of 4 PV 100 WP, the system could produce 2,977.41Wh/day. Meanwhile, based on our system we use 8 PV 100WP so it can produce 5,954.82 Wh/day.

4.2. Parabolic Solar Concentrator

The Parabolic solar concentrator is used to focusing the solar thermal radiation to one-point focus as shown in figure 4. The solar thermal energy is used to heat the boiler. So we need to determine the focal point of the solar concentrator in order to get the maximum temperature.

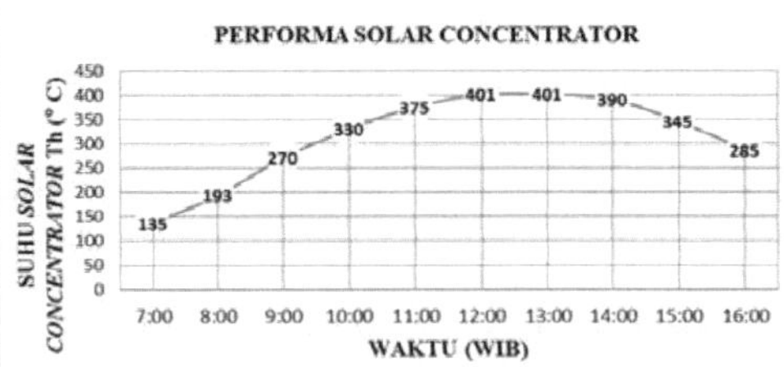

Figure 4. Parabolic Solar Concentrator Experiments.

By using the mathematical formula, the calculated focal point of the parabolic solar concentrator is 67.5 cm by assuming the reflectivity of material and copper conductivity is 90%. So, the surface area of the concentrator is 1.788 m^2. Therefore, the q reflected is 810W/m2 with the power input is 1528.74W.

4.3. Distillator

The specification of the distillatory consists of Stainless SS401 material and 40 cm of boiler surface diameter. By using a mathematical formula, the surface of the boiler is $0.125m^2$ and 57.216 of concentration ratio. Thus, the total energy produced is 1528.74 W. The boiling temperature of seawater is 110 while its rated temperature is 32^0C, therefore, 78^0C more is needed to boil the seawater. The volume of the seawater is produced by calculating the seawater mass and the volume, it is 51,3 kg. While the Q needed is 15.605.460 Joule. In this experiment, the water needs 45 minutes and 20 seconds to be boiled.

4.4. The Power Efficiency and Result

The power efficiency to boil the water is the ratio between the output power and input power, the output power is 5,737.3 watts by dividing the energy and time. So, the efficiency is 93.86% as the result of dividing the output power and input power times the parabolic components.

The chemical reaction process is separated material of the seawater by boiling the seawater until meeting its' boiling point. Seawater boiling point is 110^oC or 379^oK. The Result will be steam and the salt material. The salt material will dispose of through the disposal valve.

Table 2. Composition of Distilled Water.

pH		**Conductivity**	
Before	**After**	**Before**	**After**
8.01	**6.015**	**36.006**	**441.9**

Based on the experiment, the average pH of seawater is 8.01 while the average PH of seawater after distilled is 6.015. From the table 2 is known that the composition of salt is decreasing. The electrical conductivity test is to see how much the level of seawater salinity, if the conductivity is low, so the salinity is decreasing. And the salinity composition average before distillation is 36,006 ms and after distillation process is 441.9 ms. In this research, the pH level of water after distillation process is decreasing by 25%, while the conductivity of water is decreasing to 81.46%.

5. Conclusion

This paper shows application of PV pump and solar concentrator combination to provide clean water in a coastal area with water crisis by taking advantages from solar energy and seawater in Indonesia as the largest archipelagic country in the world. Within the experimental data with power efficiency greater than 80%, can be concluded that this system is decent to be applied. The system can provide clean water for 50 liter continuously based on capacity of the boiler each day depend on the weather and the temperature. The solar concentrator increases the boiler concentrator up to 300^oC which enough to boil the seawater and produce clean water.

References

[1] U. S. Geological Survey, "Earth Where Water. [Online]. Available: https://water.usgs.gov/edu/earthwherewater.html," November 10, 2017.

[2] W. H. Organization, "Drinking-water. [Online]. Available: http://www.who.int/mediacentre/factsheets/fs391/en/," November 10, 2017.

[3] E. Britannica, "Earth Where Water. [Online]. Available: http:// https://www.britannica.com/place/Indonesia," November 10, 2017.

[4] Zhiwen Lin, Jinghuan Ma, Jianpeng Shi and Ying Liu, 2011. A pilot study of ultrafiltration pretreatment for seawater reverse osmosis desalination in Bohai Bay.In 2011 Second International Conference on Mechanic Automation and Control Engineering, (pp. 33-36). IEEE.

[5] Nabila, L., Khaldi, F. and Aksas, M., 2014, March. *The design of photo voltaic pumping system using water tank storage for a remote area in Algeria.* In Renewable Energy Congress (IREC), 2014 5th International (pp. 1-5). IEEE.

[6] Caracas, Victor, J., Teixeira, L. F. M., Farias, G. D. C. and Ribeiro, L. A. D. S., 2012. Implementation of a high efficiency and low-cost converter for a photovoltaic water pumping system. In Applied Power Electronics Conference and Exposition (APEC), 2012 Twenty-Seventh Annual IEEE (pp. 2080-86). IEEE.
[7] A. T. Jones and R. L. Campbell, 2005. Seawater desalination: a generalized model for feedwater intakes. Proceedings of OCEANS 2005 MTS/IEEE, (pp. 2647-51). IEEE.
[8] Asepta Surya Wardhana, Heri Suryoatmojo and Mochamad Ashari, 2016. Design of parabolic solar concentrator toimprove the optical efficiency for thermal engine generator using dual reflector Gregorian method. In 2016 International Seminar Intelligent Technology and Its Applications (ISITIA), (pp.457-64). IEEE
[9] Lazić, V., et al. (2016). "Selection and Analysis of Material for Boiler Pipes in a Steam Plant." Procedia Engineering 149(Supplement C): 216-23.
[10] Acakpovi, A., Xavier, F.F., and Awuah-Baffour, R., 2012, October. The analytical method of sizing photovoltaic water pumping system. In 2012 IEEE 4th International Conference on Adaptive Science & Technology (ICAST) (pp. 65-69). IEEE.
[11] Kavitha, B., Karthikeyan, S. and Iswarya, B., 2014. The design of solar PV water pumping system using BLDC drives using the sensorless method. The International Journal Of Engineering And Science (IJES) Volume, **3**, pp.41-46.

IOP Conf. Series: Materials Science and Engineering **383** (2018) 012062 doi:10.1088/1757-899X/383/1/012062

The smart meter design for total waste of electricity usage costs on household sector calculating using artificial neural network method

D F U Putra, A Soeprijanto, M A Fikri, A R Hidayatullah* , R L Praseto, C H Sihotang, S Zahra and E Budiardi

Department of Electrical Engineering, Faculty of Electrical Technology, Institut Teknologi Sepuluh Nopember ITS, Surabaya Indonesia, 60111

* E-mail: rizkihidayatullah@gmail.com

Abstract. This research proposed a Smart Meter design that have a feature to calculate the total wasted energy from electricity usage. This feature obtained by monitoring and scheduling any installed electricity equipment. With this feature, every Smart Meter user expected to be able to manage the use of electronic equipment that installed in each home optimally. In this research artificial neural network with a radial basis network learning algorithm (RBFNN) used to identify any connected device. This method is one of the controlled learning methods and have fast learning speed. To identify the electronic equipment, it used the RMS current rating and the peak value of the wavelet transformation of the RMS current. From the transformation will become input for the RBFNN. Experiment and simulation results show that the Smart Meter with radial basis network method able to identify each load well, where the average training accuracy is 96.76% and the average testing accuracy is 85.08%.

1. Introduction

Today, energy conservation is a challenging issue as it exponentially increases energy demand. Fossil energy resources are limited and estimated global energy demand will doubled by the end of 2030 [1]. The economy, climate change, and energy crisis in a country are directly affected by the growth of energy consumption. Steps to reduce electricity wastage can be achieved by monitoring the electricity consumption and relaying back the information to the consumer [2].

Several studies have shown that maximum electrical energy savings can be achieved using a direct feedback mechanism providing electrical energy usage real-time information [3]. In the future, the load monitoring system will focus on strategy development techniques to minimize the amount of instrumentation equipment using the Non-Intrusive Load Monitoring (NILM) system [4]. Several methods have been used to identify loads based on transient conditions using wavelet transforms and steady state conditions that use Fast Fourier Transforms, with artificial intelligence methods that have multi-layer perceptron classification capability, basis radial function, and support vector machine, in real time [5].

This research proposes a smart meter simple concept with one monitoring sensor and identification of electric energy consumption in household sector by utilizing current waveform from every electronic equipment. The load identification process will be done using radial basis network method in real time. The concept of the smart meter offers a low-cost technology but has more features than

Content from this work may be used under the terms of the Creative Commons Attribution 3.0 licence. Any further distribution of this work must maintain attribution to the author(s) and the title of the work, journal citation and DOI.
Published under licence by IOP Publishing Ltd

analog and digital kWh meters installed in consumer house, such as: Details of total loads used by electric energy users, total cost of electricity usage, and total wasted energy from electricity usage.

2. Literature Review

The NILM system requires only sensors voltage and current, and this sensor not installed at each load. These sensors transmit information on the use of electrical energy for home energy management systems for monitoring and load control. Through this technique, the energy usage of each load can be monitored and controlled [6].

The wavelet transform is a linear transformation that is almost like the Fourier transform. However, there is an important difference between wavelet and Fourier transform that the wavelet transforms allow the time placement of the different frequency components from the signal that given. Using the wavelet transform, the frequency information and timing information of the occurrence of that frequency on the signal can be determined by precision [7]. In this research, the signal of the simulation will transform from time domain to frequency domain using wavelet transformation, on simulation we use wavelet transformation tool box in LabVIEW, and we use the signal for the input of transformation. The peak of signal wave of the result from wavelet transformation is use as the parameter in the input of radian based network

The neural networks with a radial basis is the learning algorithm that supervised and requires more neurons compared to feedforward networks. This network works well if the input data quite a lot. In this basis radial network, the inputs of the vector distance between the weight vectors and the input vectors multiplied by the bias weights will be processed using radial-based activation functions. While in the output layer, Purelin activation function will be used [8]. Training algorithm of the radial basis network is a feedforward stage. There are two input for neurons, first is root mean square current and the second is peak value of wavelet transformation. While the neuron output affirms the electronics equipment status. The identification status of electronics equipment is a neuron output labelled by binary numbers (0 and 1). Neurons in a single hidden layer used for all different electronics devices that connected in the house. The number of these neurons depends on the amount of training data. The more training data used, the output data will be better due to the complexity of mapping factors on inputs and outputs on every electronic equipment.

In this study, the radial basis network used as a method to identify load names and the combinations of electronic equipment on and off status conditions based on real time load data measurement. The training progress of radial basis network is using MATLAB software with input and output parameter that we get from wavelet transformation.

3. Research Method

3.1. General Structure of Load Identification

The general structure of the load identification system shown in figure 1. The main parameters used in load identification systems are voltage and current wave samples. Measurement of the current in system identification can be performed by measure the currents on each single load separately. This method requires a relatively high cost because the current sensor installed on each load.

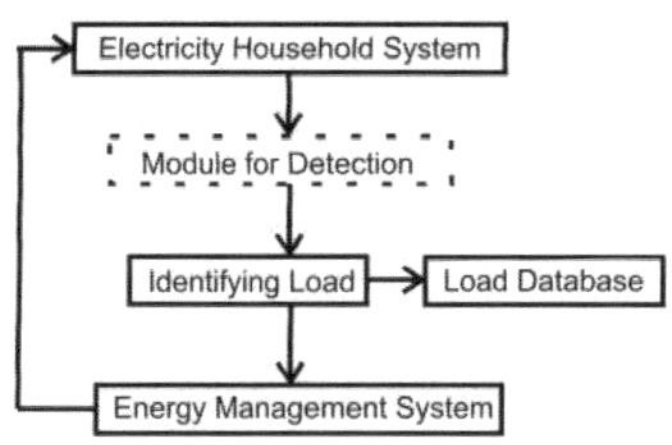

Figure 1. General Structure of load identification.

The data acquisition module serves to obtain the measurement signal under steady state and transient conditions. Sampling frequency to obtain signals in steady state and transient condition regulated through data acquisition module. The event detection module will detect if any load is active or inactive. In addition, this module is also applied to decide whether the load is connected to the system or released from the system. If the load is connected, the sample shapes of the voltage and current waves will be analysed by the load identification module. In the subsequent process, the identified load information sent to energy management for the evaluation of the energy wastage rate.

3.2. Load Identification Method

Methods for monitoring and identifying loads specified into two categories that are method under the steady-state and the transient conditions. The method under steady state conditions uses a constant signal parameter when the electrical load operates in a stable state. While the method in transient conditions relies on transition switching signal parameter from the load. A method that applied to identify the load is the method based on the waveform characteristics [9]. In addition, the peak current characteristics, base current, and RMS current (root mean square) can be used as the parameters of the load identification

3.3. Schedule of Each Home Appliance

In this study, 4 equipment is used, there are blender, hairdryer, LED lamp 11 Watt and ETL Lamp 35 Watt. Each equipment will pass through the several same processes with previous process: Measuring the RMS wavelet current in each equipment, wavelet Transform of current wave in each equipment, training using radial basis network, testing several LABVIEW feature that already designed. The data sample and schedule can be seen in table 1 and 2.

Table 1. Sorting of radial basis network sample data appliance.

No.	Electronic Device Combination	Amount of Training Data Sample	Amount of Data Sample Testing
1	Blender	3	2
2	Hairdryer	3	2
3	LED Lamp 11-Watt	3	2
4	ETL Lamp 35-Watt	3	2
5	Blender + Hairdryer	3	
6	Blender + LED lamp 11-Watt	3	2
7	Blender + ETL Lamp 35-Watt	3	2
8	Hairdryer + LED Lamp 11-Watt	3	2
9	Hairdryer + ETL Lamp 35-Watt	3	2
10	LED Lamp 11-Watt + ETL Lamp 35-Watt	3	2
11	Blender + Hairdryer + LED Lamp 11-Watt	3	2
12	Blender + Hairdryer + ETL Lamp 35-Watt	3	2
13	Blender + LED lamp 11-Watt +	3	2

Table 2. Schedule of each home.

Equipment	Working time
Blender	10.00 am – 01.00 pm
Hairdryer	07.00 am – 09.00 am and 03.00 pm – 05.00 pm
LED Lamp 11W	05.00 pm- 06.00 am
ETL Lamp 35W	05.00 pm – 10.00 pm

IOP Conf. Series: Materials Science and Engineering **383** (2018) 012062 doi:10.1088/1757-899X/383/1/012062

	ETL Lamp 35-Watt		
14	Hairdryer+ LED lamp 11-Watt + ETL Lamp 35-Watt	3	2
15	Blender + Hairdryer + LED lamp 11-Watt + ETL Lamp 35-Watt	3	2

4. Analysis & Results

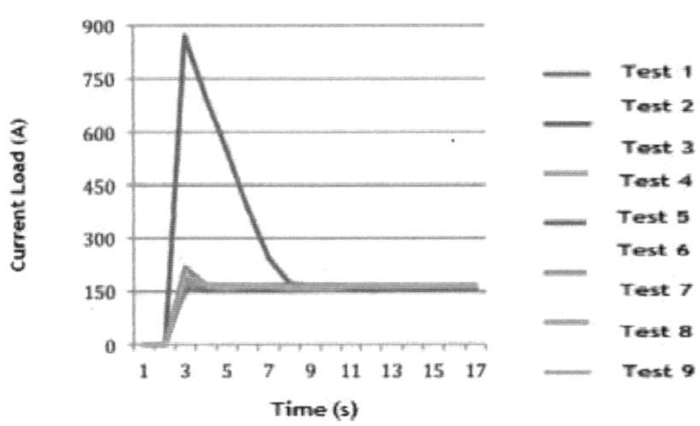

Figure 2. Current Root Mean Square Plot Result Lamp Appliance ETL 35 watt.

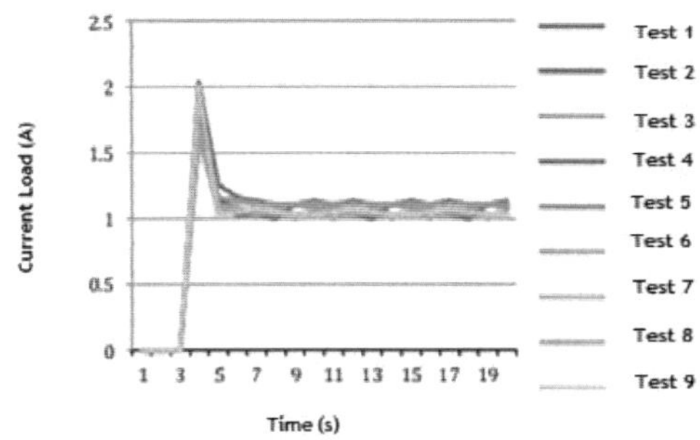

Figure 3. Current Root Mean Square Plot Result Blender Appliance.

Figure 2 show the simulation result of RMS current from Energy Thrift Lamp (ETL) 35 Watt, it shows that the RMS current of ETL 35 watt is around 150 mA. Figure 3 show the simulation result of RMS current from blender, it shows that the RMS current of blender reach up to the same point about 1 A.

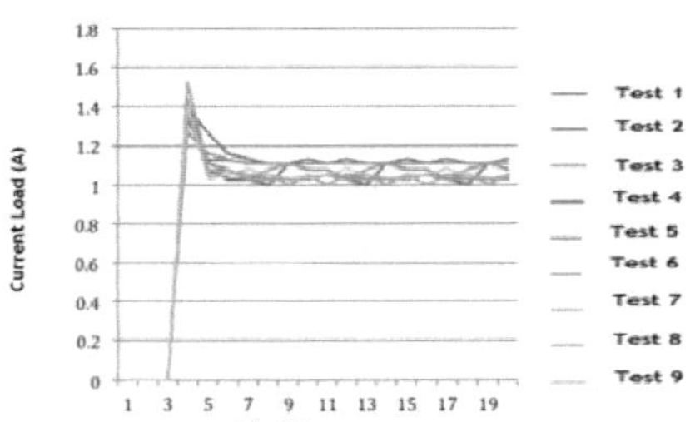

Figure 4. Current Root Mean Square Plot Result Hair Dryer Appliance.

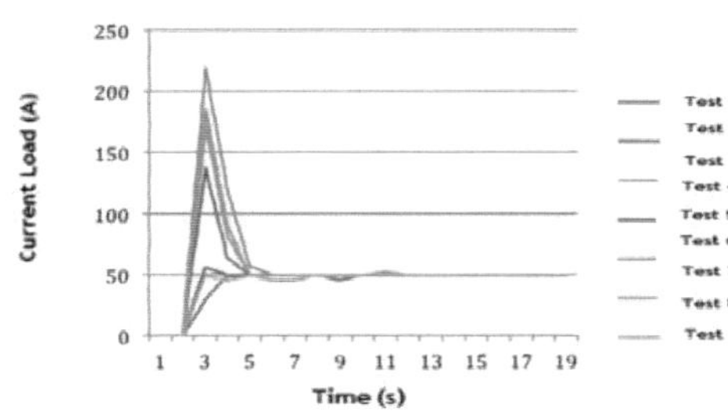

Figure 5. Current Root Mean Square Plot Result Lamp Appliance LED 11.

While figure 4 show the simulation result of RMS current from hair dryer, it shows the RMS current of hair dryer is about 1 A to 1.2 A. It shows that the result of RMS current from ETL 35 watt is under from blender and hairdryer. Figure 5 show the simulation result of LED 11 watt, it shows that the RMS current from LED 11 is at the same point at 50 mA, it is under the result of ETL 35-Watt simulation.

The result of weight and bias that we get from radial basis network training with MATLAB will send to LabVIEW. In this progress, training is not required, smart meter that we designed can identify blender, hair dryer, ETL 35 watt, and LED 11 watt.

Table 3. Load Identification Result for Mixed Electronics Appliance.

No	Electronic appliance combination	Training Accuracy (%)	Testing Accuracy (%)	No	Electronic appliance combination	Training Accuracy (%)	Testing Accuracy (%)
1	Blender	100	78.86604131	9	Hairdryer + ETL 35 Watt	100	68.0596407
2	Hairdryer	100	76.81270273	10	LED 11 Watt + ETL 35 Watt	100	91.06881559
3	LED 11 Watt	100	97.79629975	11	Blender + hairdryer + LED 11 Watt	100	84.22691221
4	ETL 35 Watt	100	99.0198632	12	Blender + hairdryer + ETL 35 Watt	100	80.61592912
5	Blender + Hairdryer	100	91.56898881	13	Blender + ETL 35 Watt + LED 11 Watt	100	73.40062406
6	Blender + LED 11 Watt	100	78.23230885	14	Hairdryer + ETL 35 Watt + LED 11 Watt	100	87.93839088
7	Blender + ETL 35 Watt	100	63.65545735	15	Blender + hairdryer + ETL 35 Watt + LED 11 Watt	100	95.11031818
8	Hairdryer + LED 11 Watt	100	81.71946239				

The result of the radial basis network training can be seen in the table 3. The radial basis network is able to identify the load well. From the accuracy of trial and training also given in table 3. Whole electronic device that we try, the features of trial from smart meter and the detail energy waste application can be seen at figure 6, figure 7 and figure 8 below.

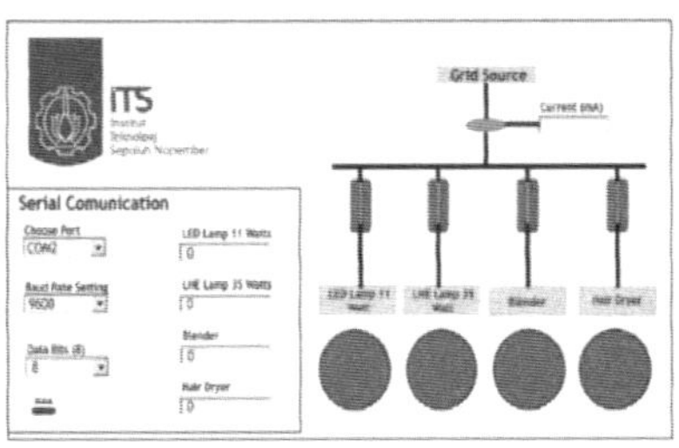

Figure 6. Smart Meter Indicator.

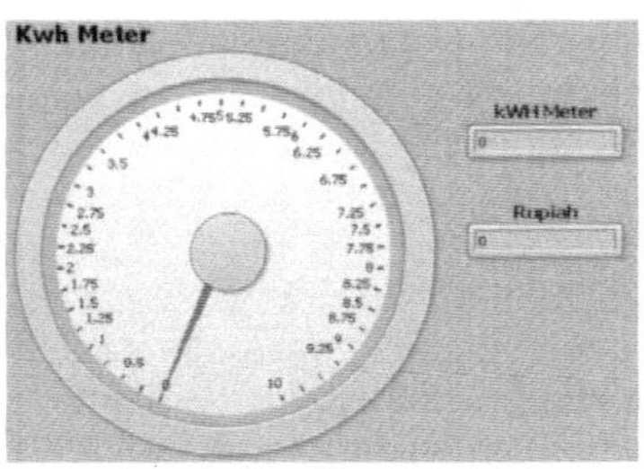

Figure 7. Electrical Energy Usage and Billing Actual Measure.

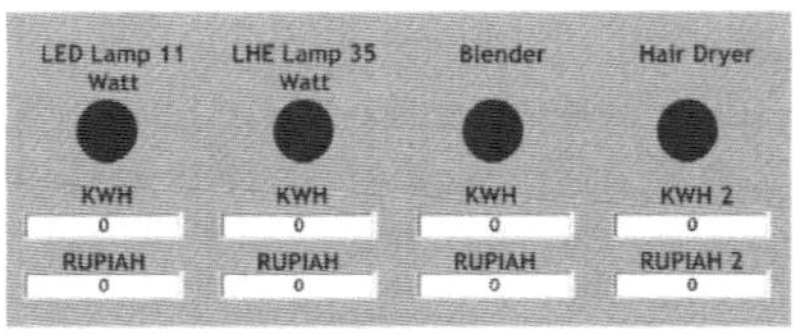

Figure 8. Detailed Waste Energy Usage Interface.

5. Discussion & conclusion

Radial basis network is used to identify the load at real time for smart meter application. From the experiment analysis, it can be conclude that the purposed method is faster than backpropagation neural network for the computation training and testing work speed. The accuracy of training and testing using radial basis network show that the method is identify the load at real time with 100% accuracy average for training and 83.2 % accuracy average for testing. By utilizing the RMS data and the peak value of wavelet transformation from each electronic device as input in radial basis network show the good reliability. It can be seen from testing analysis from the variety of voltage at different level, the changes of power consumption and different load operational. Next research step is wavelet transformation RMS current as parameter to identify electronic device or load.

References

[1] J. Uteley, and L. Shorrock, “Domestic Energy Fact File 2008”, Building Research Establishment: Garston, UK, Tech. Rep. 2008.

[2] Hosseini, S. S., et al. "Non-intrusive load monitoring through home energy management systems: A comprehensive review." *Renewable and Sustainable Energy Reviews* vol. **79,** pp. 1266-74, 2017.

[3] K. E Martinez, K.A Donelly, and J.A Laitner, “Advanced Metering Initiatives and Residential Feedback Programs: A Meta-Review for Households Electricity-Saving Opportunities”, American Council for an Energy-Efficient Economy (ACEE), USA, Tech. Rep. E105, 2010.

[4] G. W. Hart, “Nonintrusive Appliance Load Monitoring”, in *Proc. IEEE*, Vol. **80**, No. 12, December 1992.

[5] H. Y Lin, and M.S Tsai, “Applications of Hierarchical Support Vector Machines for Identifying Load Operation in Nonintrusive Load Monitoring Systems”, *Proc. World Congress on Intelligent Control and Automation*, 21-25 June 2011.

[6] J. G. Roos, I. E. Lane, E. C. Lane, and G. P. Hanche, “Using neural networks for non-intrusive monitoring of industrial electrical loads,” in *Proc. IEEE Instrumentation and Measurement Technology Conference*, 10-12 May 1994.

[7] Jian Liang, Simon K. K. Ng, Gail Kendall, and John W. M. Cheng,” Load Signature Study—Part I: Basic Concept, Structure, and Methodology,” *IEEE Transactions on Power Delivery*, vol **25**, pp. 551-60, April. 2010.

[8] Kusumadewi, S, *Membangun Jaringan Syaraf Tiruan Menggunakan MATLAB & EXCEL LINK*, Graha Ilmu, 2004

[9] Yi Du, Liang Du, Bin Lu, Ronald Harley, and Thomas Habetler,” A Review of Identification and Monitoring Methods for Electric Loads in Commercial and Residential Buildings,” in proc. IEEE Energy Conversion Congress and Exposition, 12-16 Sept. 2010.

IOP Conf. Series: Materials Science and Engineering **383** (2018) 012063 doi:10.1088/1757-899X/383/1/012063

A numerical investigation on the voltage harmonics from multi-level converters in smart grid system

Masamu Kamaga

Bina Nusantara University, Kemanggisan Ilir III No. 045, Palmerah, West Jakarta, 11480, Indonesia

*E-mail: masamu.kamaga@binus.edu

Abstract. The economic growth in Asia is now remarkable. Because of this, the energy consumption has also increased drastically. Under such condition, it is necessary for efficient use of energy to introduce smart grid. At that time, electromagnetic interference (EMI) generated by the voltage harmonics from power converters which are connected to the power system becomes one of the biggest issues. Therefore, to suppress voltage harmonics in the power converters is required. Although it can be realized by connecting passive filters, it leads to circuit enlargement. On the other hand, recently there are some suggestions that the multilevel converters will be miniaturized and integrated in the future. If such a converter is introduced, it is considered that the smart grid system with renewable energy sources can provide high-quality electricity without EMI. Therefore, in this paper, the voltage harmonics to the grid system by assigning multi-level converters is investigated.

1. The necessities of power converters in smart grid

Demand of energy has increased all over the world [1]. Especially in Asian countries, due to its highly economic development, the energy consumption has also increased drastically. Under such condition, the efficient use of energy is essential.

As one of the ways to satisfy the demand of energy, it is very important to introduce renewable energy sources, such as photovoltaic energy and wind energy. However, it is very difficult to manage the energy from these energy sources. Because of that, Smart Grid is expected as one of the key solutions.

However, the form of the output power of the renewable energy sources and that of electricity in the power system are different. Therefore, power converters are assigned to the renewable energy sources.

Currently, as for the power converters, converters with using switching are widely used. For this reason, the output voltage waveform is pulsed and contains many harmonics as shown in figure 1 (a). Simultaneously, its switching speed (dv/dt, di/dt) has become higher. Therefore, the electromagnetic interference (EMI) has been increased. Output LC filters and EMI filters (passive filters) are inserted to reduce these harmonics and EMI, however, the passive filters are composed of inductors and capacitors, therefore the filters are bulky [2,3].

On the other hand, there are some circuit topologies called multi-level converters. As shown in figure 2, there are 3 typical circuit topologies of multi-level converters; series-connected H-bridge converter, diode-clamped converter, and flying capacitor converter. Although there are many comparative investigations for the circuit topologies, in both cases, there is a feature that the output voltage can be

Content from this work may be used under the terms of the Creative Commons Attribution 3.0 licence. Any further distribution of this work must maintain attribution to the author(s) and the title of the work, journal citation and DOI.
Published under licence by IOP Publishing Ltd

IOP Conf. Series: Materials Science and Engineering **383** (2018) 012063 doi:10.1088/1757-899X/383/1/012063

multi-leveled as shown in figure 1 (b). Although there are many investigations of 3-level converters, there are few investigations of the converters with the large number of levels [4,5].

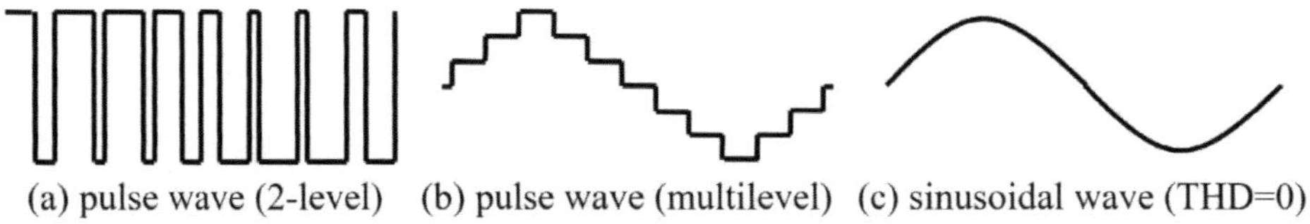

Figure 1. Waveforms.

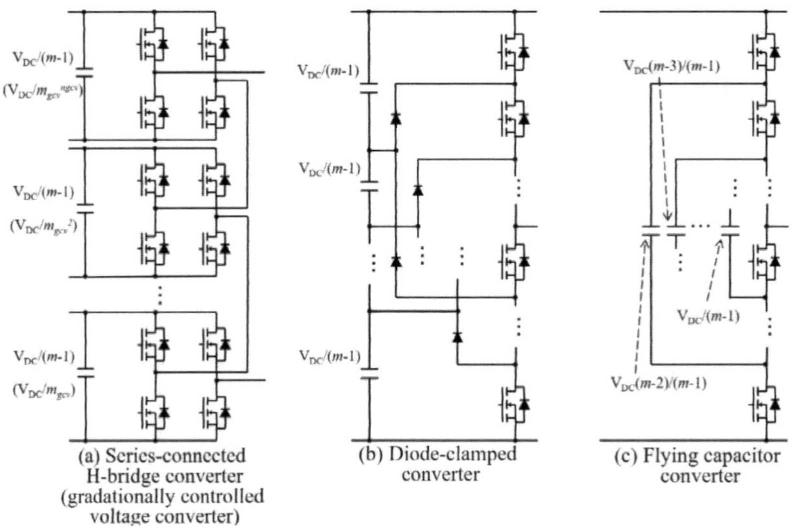

Figure 2. Multi-level circuit topologies.

Regarding the grid-connected power converters, inverters (DC/AC converters) are usually assigned. Therefore, in this paper, 2-level and multi-level inverters are investigated.

2. Voltage harmonics

2.1. Total harmonic distortion (THD) of the output voltage waveform

Because of the switching, there is harmonics in its output voltage waveform. This distortion factor is called total harmonic distortion (THD). If the form of waves is complete sinusoidal wave like figure 1 (c), THD is zero. THD of the output voltage waveform can be calculated from the harmonic distortion in a voltage and is defined as the ratio of the sum of the voltages of all harmonic components to the voltage of the fundamental frequency. Therefore, THD can be calculated by equation (1).

$$THD = \frac{\sqrt{\sum_{n=2}^{\infty} V_n^{\,2}(\theta)}}{V_1(\theta)} \tag{1}$$

2.2. 2-level and multi-level inverters

As shown in figure 1, the multilevel inverters can reduce the voltage fluctuation in each switching by increasing the number of levels of the output voltage. It can be also predicted that the output waveform

IOP Conf. Series: Materials Science and Engineering **383** (2018) 012063 doi:10.1088/1757-899X/383/1/012063

becomes closer to the sinusoidal wave and the harmonic component can be reduced by assigning the multi-level inverters.

3. Analysis of the voltage harmonics of 2-level and multi-level converters

To control the inverters, pulse width modulation (PWM) is usually applied. As for multi-level inverters, it depends on the circuit topologies or applications [4,5]. Therefore, both the conditions are investigated.

3.1. The voltage harmonics of 2-level and multi-level converters (without PWM switching)

THD without PWM switching is calculated. The calculation of THD without PWM switching is based on the Fourier series expansion as shown in figure 3.

3.2. The voltage harmonics of 2-level and multi-level converters (with PWM switching)

THD with PWM switching is calculated. The calculation is based on the simulated result using circuit simulation PSIM. The simulated waveforms and FFT analysed results are shown in figure 4.

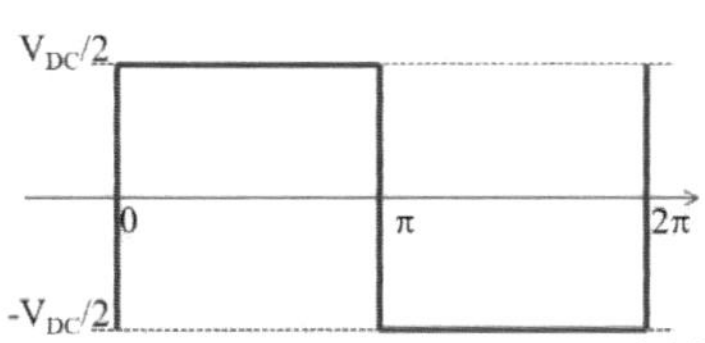

$$V_n=\frac{2}{T}\int_0^T v(t)\sin n\omega t\,dt$$
$$=\frac{2}{2\pi}\int_0^{\pi}\frac{V_{DC}}{2}\sin n\omega t\,dt+\int_{\pi}^{2\pi}\left(-\frac{V_{DC}}{2}\right)\sin n\omega t\,dt$$
$$=\frac{4}{n\pi}\frac{V_{DC}}{2}\qquad (n=2k+1,k=1,2,3,\cdots)$$

(a) THD (2-level, without PWM switching)

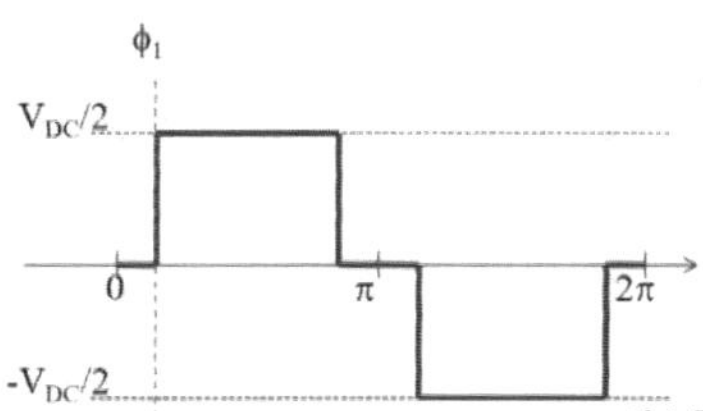

$$\phi_1=\arcsin\left(\frac{1}{2}\frac{V_{DC}}{2}\right)$$
$$V_n=\frac{2}{T}\int_0^T v(t)\sin n\omega t\,dt$$
$$=\frac{4}{n\pi}\frac{V_{DC}}{2}\sin n\theta\cos n\phi_1$$

(b) THD (3-level, without PWM switching)

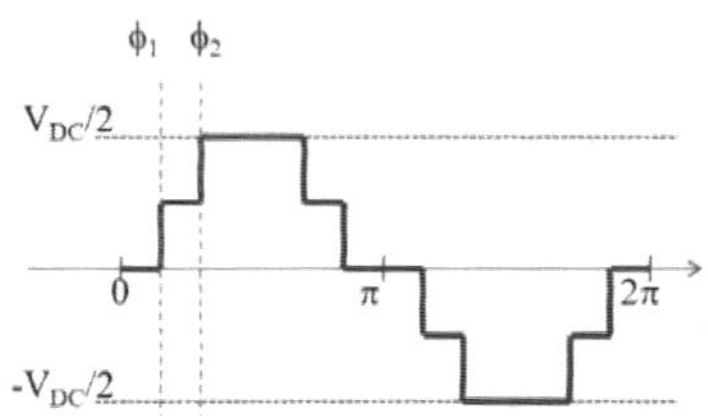

$$\phi_1=\arcsin\left(\frac{1}{4}\frac{V_{DC}}{2}\right),\phi_2=\arcsin\left(\frac{3}{4}\frac{V_{DC}}{2}\right)$$
$$V_n=\frac{2}{T}\int_0^T v(t)\sin n\omega t\,dt$$
$$=\frac{2}{n\pi}\frac{V_{DC}}{2}\sin n\theta\cos(n\phi_1+n\phi_2)$$

(c) THD (5-level, without PWM switching)

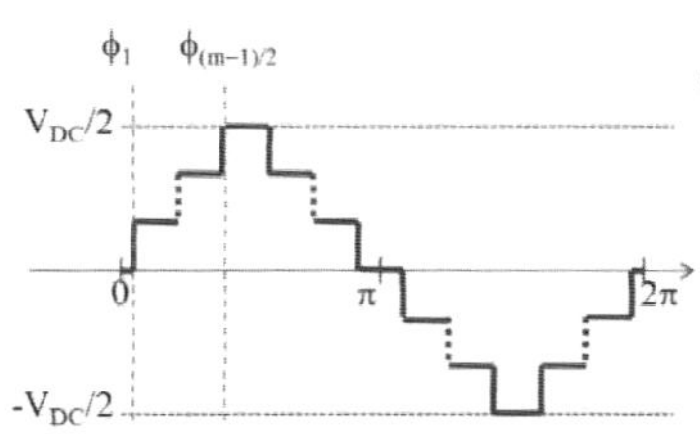

$$\phi_1=\arcsin\left(\frac{1}{m-1}\frac{V_{DC}}{2}\right),\phi_2=\arcsin\left(\frac{3}{m-1}\frac{V_{DC}}{2}\right),\cdots,\phi_{\frac{m-1}{2}}=\arcsin\left(\frac{m-2}{m-1}\frac{V_{DC}}{2}\right)$$
$$V_n=\frac{2}{T}\int_0^T v(t)\sin n\omega t\,dt$$
$$=\frac{8}{(m-1)n\pi}\frac{V_{DC}}{2}\sin n\theta\cos\left(n\phi_1+n\phi_2+\cdots+n\phi_{\frac{m-1}{2}}\right)$$

(d) THD (m-level, without PWM switching)

Figure 3. THD of 2-level and multi-level converters (without PWM switching).

IOP Conf. Series: Materials Science and Engineering **383** (2018) 012063 doi:10.1088/1757-899X/383/1/012063

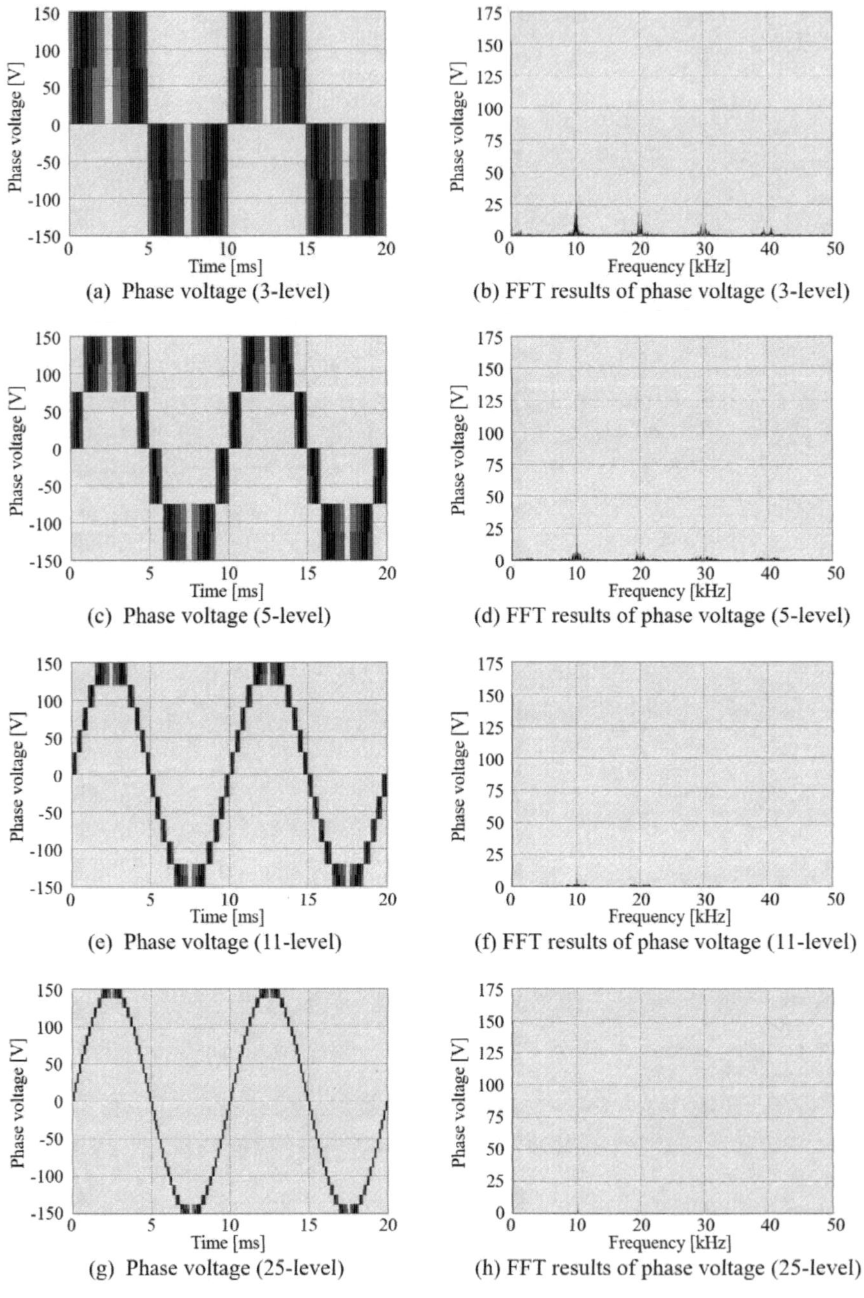

Figure 4. THD of 2-level and multi-level converters (with PWM switching).

4. Discussions

Figure 5 shows the relationship between the number of levels and THD of the output voltage. This result can be explained as following.

- THD can be reduced by increasing the number of the levels. The reduction of THD is approximately inversely proportional to m-1 (m means the number of levels). This is because the voltage value of the harmonic is inversely proportional to m-1.

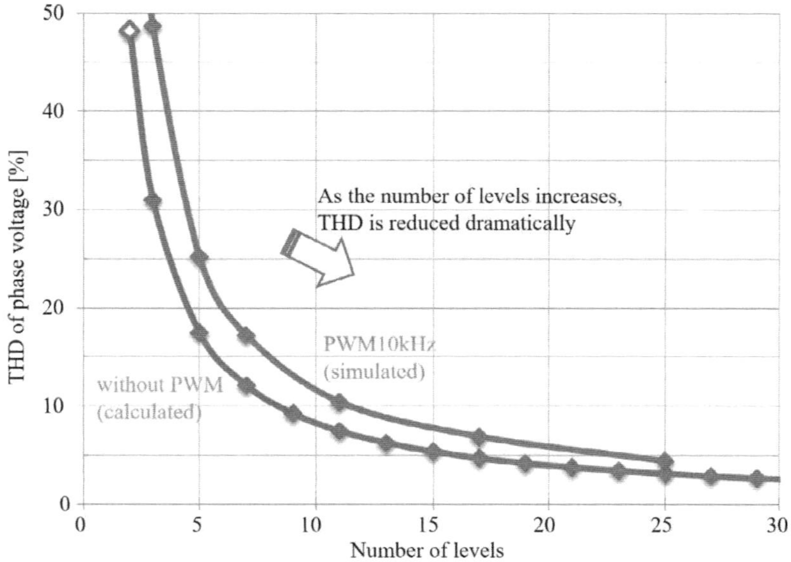

Figure 5. The relationship between the number of levels and THD.

- Comparing the case of using PWM switching and the case of not using PWM switching, the former case is better. However, when the PWM switching is used, since the harmonic order increases in proportion to the PWM carrier frequency, the output LC filters can become small even though the output LC filters are still assigned.
- The output voltage with less than 5 % THD can be realized when the number of levels is larger than or equals to 17-levels without using PWM switching or 25-levels with using PWM switching. For example, according to the regulations of the power system in Japan, this can be directly connected to the power system without the output LC filters [2].

5. Conclusion

In this paper, the voltage harmonics to the power system by assigning multi-level converters is numerically investigated.

As a result, the relationship between the number of levels and THD is shown. It can be realized that THD which is less than 5% with assigning 17-levels or more when not using PWM switching or 25-levels or more when using PWM switching.

References

[1] "Annual report of energy 2015", White Paper, Agency for Natural Resources and Energy in Japan 2015 (2015-7) (in Japanese)

[2] "General purpose inverter: current status of energy saving effect, performance improvement, and environmental harmonization", IEEJ Technical Report, No. 807 (2000-11) (in Japanese)

[3] "Power Conversion Circuits for Utility Power Line Interface", IEEJ Technical Report, No. 1096 (2007-7) (in Japanese)

[4] J.S. Lai, and F.Z. Peng: "Multilevel converters-a new breed of power converters", IEEE Transactions on Industry Applications, VOL. **32**, NO. 2, pp. 509-517 (1996-5)

[5] J. Rodriguez, J. Lai, and F.Z. Peng: "Multilevel Inverters: A Survey of Topologies, Controls, and Applications", IEEE Transactions on Industrial Electronics, VOL. **49**, NO. 4, pp.724-738 (2002-8)

IOP Conf. Series: Materials Science and Engineering **383** (2018) 012064 doi:10.1088/1757-899X/383/1/012064

Experimental study on the heat gain of water in hybrid solar collector integrated with fin-and-tube heat exchanger with respect to air and water flow rate

H.U. Choi[1], R. Fatkhur[1], Y.B. Kim[2], C.H. Son[3], J.I. Yoon[3] and K.H. Choi[3, *]

[1] Graduate school of Refrigeration and Air-conditioning Engineering, Pukyong National University, Busan 608-739, Republic of Korea
[2] Dept. of Mechanical System Engineering, Pukyong National University, Busan 608-739, Republic of Korea
[3]Dept. of Refrigeration and Air-conditioning Engineering, Pukyong National University, Busan 608-739, Republic of Korea

*E-mail: choikh@pknu.ac.kr

Abstract. Solar assisted heat pump system is the system using solar thermal energy that can lead to the improvement of performance of heat pump by transferring solar thermal energy obtained from solar collector to evaporator of the heat pump for evaporation of refrigerant. In traditional system, solar collector can get a thermal energy only from solar radiation. So, the collector is hard to be used when the solar radiation is not enough such as cloudy day or night even though these collector take up much space. Thus the hybrid solar collector that has fin-and-tube heat exchanger have been developed for getting a thermal energy from not only solar radiation but also ambient air. Due to the fin-and-tube heat exchanger, this collector can get a thermal energy from ambient air for heating chilled water by evaporation of refrigerant in heat pump system. At this time, heat gain of water from ambient air by the collector needs to be confirmed before combining with real heat pump system. Thus, in this study, heat gain of water in the hybrid solar collector was investigated experimentally with respect to air and water flow rate on the various temperature difference between inlet air and water. As a results, heat gain of water was shown maximum value of 900W and it was increased with increment of air and water flow rate. Also, heat gain of water was changed linearly with increment or decrement of temperature difference between inlet air and water on the specific air and water flow rate. Furthermore, relationship between heat gain of water and operating conditions was also confirmed that can be used for decision of collector area and heat pump capacity for designing solar assisted heat pump integrated with this collector.

1. Introduction

Solar assisted heat pump system is the system using solar thermal energy by combining traditional heat pump system and solar collector. Thermal energy obtained from solar collector is used for evaporation of refrigerant in the heat pump system and the evaporation temperature can be increased than traditional heat pump system that can lead to the increment of C.O.P.(coefficient of performance). Also, solar collector can be operated with the higher thermal efficiency than traditional solar collector due to the lower operating temperature. Thus the many researches related to solar assisted heat pump system have been conducted for improving the performance of these systems [1-3]. But solar collector

Content from this work may be used under the terms of the Creative Commons Attribution 3.0 licence. Any further distribution of this work must maintain attribution to the author(s) and the title of the work, journal citation and DOI.
Published under licence by IOP Publishing Ltd

in the traditional solar assisted heat pump system is hard to operating when the solar radiation is not enough even though these collector take up much space.

Thus, the hybrid solar collector that can retrieve thermal energy from not only solar irradiance but also ambient air has been developed [4-6]. This collector is flat plate solar collector that has fin-and-tube heat exchanger installed under the absorbing plate. Thus the circulated water chilled by evaporating of refrigerant in the heat pump system can get a thermal energy from ambient air similar with the outdoor unit of air source heat pump system by the collector when the solar radiation is not enough. Consequently, heated water by ambient air in the collector transfer thermal energy to evaporator of heat pump system as a heat source for evaporation of refrigerant. At this time, heat gain of water from ambient air in the collector needs to be confirmed before apply the hybrid solar collector to the heat pump system.

Thus, in this study, heat gain of water from ambient air in the collector was investigated experimentally with respect to air and water flow rate on the various temperature difference between inlet air and water and the relationship between heat gain of water and inlet fluid temperature difference in the collector was also confirmed on the each air and water flow rate.

2. Experimental apparatus and method

2.1. Experimental apparatus

Hybrid solar collector that has fin-and-tube heat exchanger is composed of glass cover, absorbing plate, fin-and-tube heat exchanger contact with absorbing plate, aluminum case and two open air side for inlet and outlet air as shown in figure 1 and fin-and-tube heat exchanger in the collector was shown in figure 2. Because of this heat exchanger, circulated water chilled by evaporation of refrigerant in the heat pump system can get a thermal energy from the ambient air even though the solar radiation is not enough to heating water.

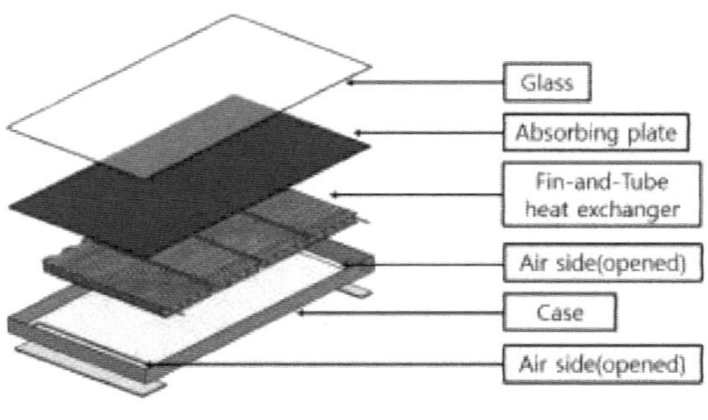

Figure 1. Exploded view of hybrid solar collector.

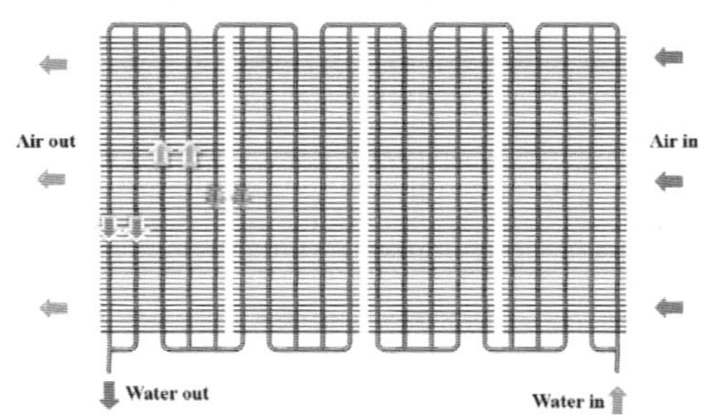

Figure 2. Fin-and-tube heat exchanger in hybrid solar collector.

Actual view of this collector and schematic diagram of experimental setup were shown in figure 3 and 4 respectively. Main component of experimental apparatus are hybrid solar collector, water tank, chiller and the collector was installed with 30 degree of horizontal angle considering solar heating mode at the Campus of University of Pukyong National University in Busan(at 36°6.98'N and 129°5.39E).

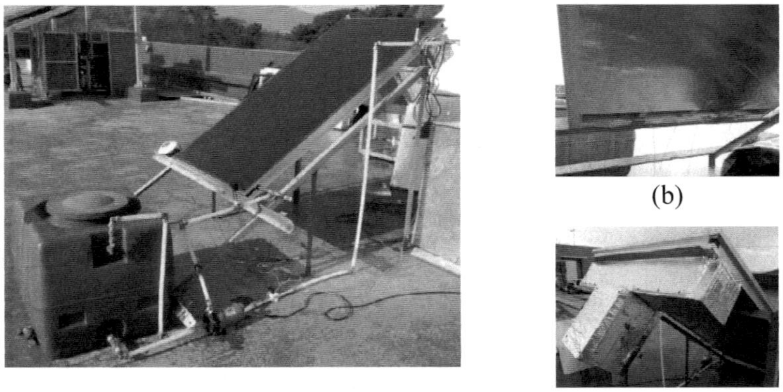

(a) (b) (c)

Figure 3. Actual view of experimental setup; (a) hybrid solar collector, (b) Inlet air side, (c) outlet air side.

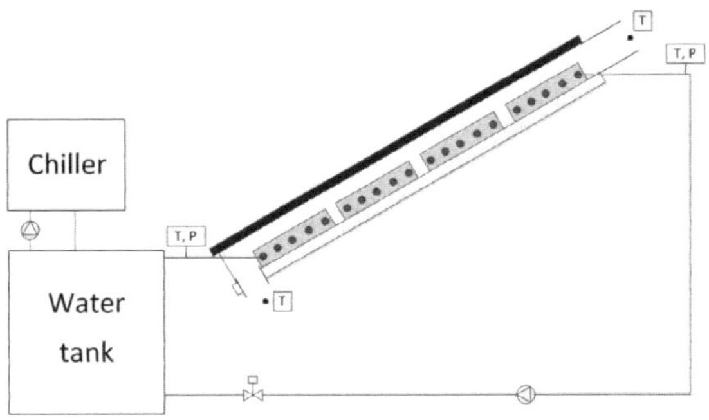

Figure 4. Schematic view of experimental setup.

2.2. Experimental method

Experiment was conducted at night for preventing the effect of solar radiation and data was acquired until the temperature difference between inlet air and water reach to almost 0°C from 16°C.

Table 1. Experimental conditions.

Parameter	Value
Collector size[(Length) x (width) x (height)]	1810 x 800 x 800 mm
Installation angle	30°
Height of air channel in fin-and-tube heat exchanger	50mm
$T_{air,in}$-$T_{w,in}$	4°C - 16°C
Water flow rate	2L/min, 4L/min, 6L/min
Air flow rate	0.07kg/s, 0.09kg/s, 0.11kg/s

In case of air, ambient air was used directly and water was chilled by water chiller. Air and water flow rate were 0.07 kg/s, 0.09 kg/s, 0.11kg/s and 2L/min, 4L/min, 6L/min, respectively. Temperature of inlet air, outlet air, inlet water and outlet water were measured by T-type thermocouple and all of data were recorded by a data acquisition at an internal of 20s. More detail experimental conditions are tabled in table 1.

3. Results and discussion

Heat gain of water from ambient air by the collector with respect to water flow rate and inlet fluid temperature difference was shown in figure 5 with the air flow rate of 0.07 kg/s. Heat gain of water obtained from the collector was shown from almost 0 to maximum value of 600W and it was increased with increment of water flow rate and inlet fluid temperature. Also, linear change of heat gain of water with the change of inlet fluid could be confirmed.

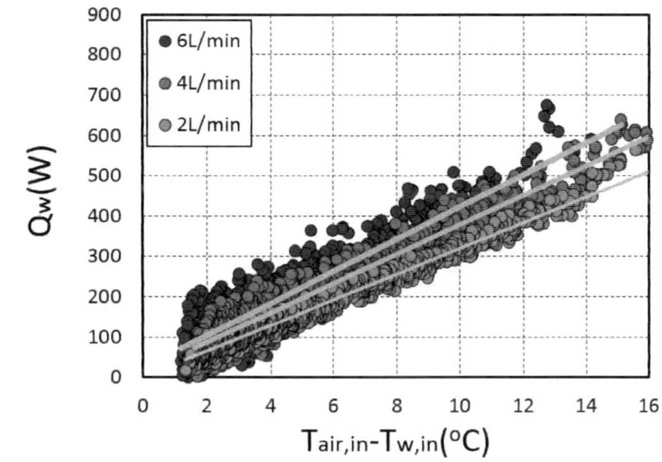

Figure 5. Heat gain of water with respect to water flow rate and inlet fluid temperature difference($\dot{m}_{air}$=0.07kg/s).

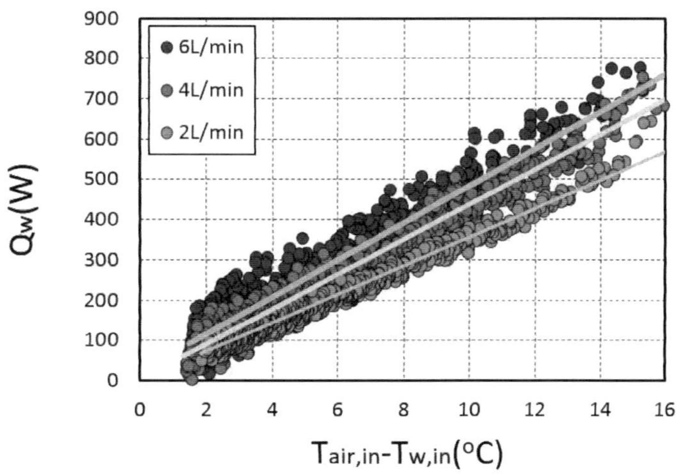

Figure 6. Heat gain of water with respect to water flow rate and inlet fluid temperature difference($\dot{m}_{air}$=0.09kg/s).

In case of the air flow rate of 0.9 kg/s, all of heat gain of water by the collector was increased compare with the case of air flow rate of 0.7 kg/s on the same conditions as shown in figure 6 and maximum value of 800W was obtained. In addition, the improvement of the heat gain of water by the collector with increment of inlet fluid temperature difference was also confirmed.

Heat gain of water with the air flow rate of 0.11 kg/s was shown in figure 7. Maximum value of the heat gain of water was shown 900W and all of heat gain of water by the collector was higher than other lower air flow rate conditions on the same water flow rate and inlet fluid temperature difference similar with previous results. But the lower effect of water flow rate on the heat gain of water by the collector was also confirmed from the decrement of increase ratio of heat gain of water to increment of water flow rate.

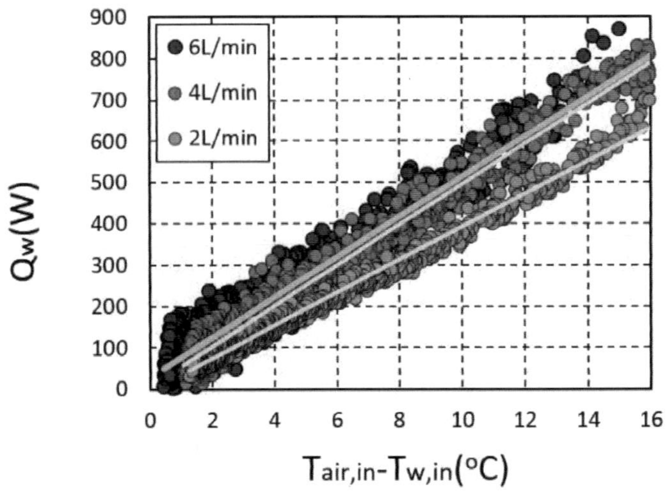

Figure 7. Heat gain of water with respect to water flow rate and inlet fluid temperature difference($\dot{m}_{air}$=0.11kg/s).

At this time, relationship between heat gain of water by the collector and operation conditions such as air and water flow rate, inlet fluid temperature difference needs to be confirmed for designing heat pump system integrated with this collector.

Table 2. Correlation of heat gain of water with inlet fluid temperature difference.

$\dot{m}_{air}$ \ $\dot{m}_w$	2 L/min	4 L/min	6 L/min
0.07 kg/s	$Q_w=31.84(T_{air,in}-T_{w,in})+2.11$ (R^2=0.984)	$Q_w=36.43(T_{air,in}-T_{w,in})+14.44$ (R^2=0.967)	$Q_w=39.82(T_{air,in}-T_{w,in})+23.51$ (R^2=0.844)
0.09 kg/s	$Q_w=35.42(T_{air,in}-T_{w,in})+1.48$ (R^2=0.987)	$Q_w=43.40(T_{air,in}-T_{w,in})+3.62$ (R^2=0.981)	$Q_w=46.10(T_{air,in}-T_{w,in})+22.85$ (R^2=0.923)
0.11 kg/s	$Q_w=39.78(T_{air,in}-T_{w,in})-4.54$ (R^2=0.990)	$Q_w=49.20(T_{air,in}-T_{w,in})+4.43$ (R^2=0.974)	$Q_w=48.67(T_{air,in}-T_{w,in})+28.13$ (R^2=0.976)

Thus the relationship between heat gain of water and inlet temperature difference was confirmed on the specific air and water flow rate as shown in table 2 and it was expected that these equations obtained from experiment can be used for decision of the collector area, heat pump capacity of solar assisted heat pump system integrated with this hybrid solar collector that has fin-and-tube heat exchanger.

4. Conclusion

In this study, heat gain of water from ambient air by the hybrid solar collector that has fin-and-tube heat exchanger was investigated experimentally with respect to air and water flow rate with various inlet fluid temperature difference. As a result, heat gain of water was shown maximum value of 900W and it was increased with increment of air and water flow rate and inlet fluid temperature difference.

IOP Conf. Series: Materials Science and Engineering **383** (2018) 012064 doi:10.1088/1757-899X/383/1/012064

Also linear change of heat gain of water with increment or decrement of inlet fluid temperature difference was confirmed on the specific air and water flow rate.

Furthermore, from these results, the relationship between heat gain of water and inlet fluid temperature difference was confirmed on the specific air and water flow rate and it was expected that the relationship obtained from this study contribute to better design for solar assisted heat pump system integrated with this hybrid solar collector including decision of collector area, heat pump capacity and so on.

Acknowledgments

This work was supported by the Korea Institute of Energy Technology Evaluation and Planning(KETEP) and the Ministry of Trade, Industry & Energy(MOTIE) of the Republic of Korea(No.20153030081190).

References

[1] Raghad S. Kamel, Alan S. Fung and Peter R.H. Dash 2015 *Solar systems and their integration with heat pumps: A review*, Energy and Building **87** 395-12

[2] Mahmut Sami Buker and Saffa B. Riffat 2016 *Solar assisted heat pump systems for low temperature water heating applications: A systematic review,* Renewable and Sustainable Energy Reviews **55** 399-13

[3] Zhanyuan Wang, Peng Guo, Haijing Zhang, Wabsheng Yang and Sheng Mei 2017 *Comprehensive review on the development of SAHP for domestic hot water*, Renewable and Sustainable Energy Reviews **72** 871-81

[4] H.W. Choi, R. Fatkhur, N.J. Lyu, C.H. Son, J.I. Yoon and K.H. Choi 2016 *Numerical Analysis on Heat Gain of Liquid from Ambient Air with Various Fin Heights and Pitches of Fin-and-Tube Heat Exchanger in Hybrid Solar Collector*, Journal of the Korean Solar Energy Society **36** 53-61

[5] H.W. Choi, R. Fatkhur, Y.B. Kim, C.H. Son, J.I. Yoon and K.H. Choi 2016 *A Numerical Analysis on Heat Transfer Between the Air and the Liquid in a Hybrid Solar Collector*, Lecture Notes in Electrical Engineering **415** 175-83

[6] H.W. Choi, R. Fatkhur, Y.B. Kim, C.H. Son, J.I. Yoon and K.H. Choi 2017 *A Study on the Heat Exchange Performance of Hybrid Solar Collector with Air to Water Heat Exchanger Type*, Lecture Notes in Electrical Engineering **465** 967-75

IOP Conf. Series: Materials Science and Engineering **383** (2018) 012065 doi:10.1088/1757-899X/383/1/012065

Design of a solar dryer with fins and baffles for rice-cracker drying

Panuwat Pawakote* and Atit Koonsrisuk

School of Mechanical Engineering, Institute of Engineering, Suranaree University of Technology, Nakhon Ratchasima, 30000, Thailand

*E-mail: jay.panu@hotmail.com

Abstract. The rice-cracker is one of the popular snacks in Thailand. To produce it, drying is a necessary preparative step. This research aims to investigate the performance of a solar dryer with fins and baffles installed in the air flow channels of the dryer for rice crackers. The cover made of a 4 mm thickness poly methyl methacrylate (PMMA). A mathematical model was developed and used to design a solar dryer. The simulations revealed that a minimum of 2.25 m^2 solar collector area is required for solar radiation of 628.93 W/m^2 and the mass flow rate of 34 kg/hr and outlet temperature of 64.9°C are obtained. In addition, the collector efficiency is 52.01%.

1. Introduction

The rice-cracker is one of the popular snacks in Thailand. It is made of sticky rice soaked in water for 6 hours, steamed, and mixed with watermelon juice or muskmelon juice. After that it is shaped in a mould and then dried under open sun as shown in figure 1. The open sun drying provides drying with low-cost, however, it relies on the weather. Also, the products may contaminate by dust. The production is suffered by slow drying rate.

Figure 1. Rice cracker drying under open sun.

The solar drying, which the products are kept in a cabinet and dried by heated air created by greenhouse effect in a solar collector, is one of the promising alternative for rice-cracker drying. Yeh et al. [1] conducted a solar drying using a solar collector flat plate. They studied 3 cases: 1) solar collector with fins, 2) solar collector with fins and baffles, 3) solar collector without fins and baffles. The tested solar intensity were at 830 W/m^2 and 1100 W/m^2. It was found that the efficiency of Case 1 was 59%, Case 2 was 72%, and Case 3 was 47%. Abene et al. [2] also did an experiment with grape

Content from this work may be used under the terms of the Creative Commons Attribution 3.0 licence. Any further distribution of this work must maintain attribution to the author(s) and the title of the work, journal citation and DOI.
Published under licence by IOP Publishing Ltd

drying. They compared 3 cases: 1) without obstacle (WO), 2) with obstacle of type WDL1, 3) with obstacle of type TL (transverse-longitudinal). It should be noted that, although calling with different names, Case 3 is the solar collector with fins and baffles. The experiment reveals that Case 1 needed 13 hours and 20 minutes, Case 2 needed 6 hours and 53 minutes, and Case 3 needed 5 hours and 50 minutes for drying. It is obvious that again the case with fins and baffles provides the fastest rate of drying.

AHMED-ZAID et al. [3] studied the drying of onions. They compared 3 cases: 1) without obstacle (WO), 2) with obstacle of type WDL1, 3) with obstacle of type TL (transverse-longitudinal). The experiment reveals that Case 1 needed 13 hours and 30 minutes, Case 2 needed 7 hours, and Case 3 needed 5 hours and 48 minutes for drying. Meanwhile, the study of [4] shows that drying of apple sliced using a solar collector without fins and baffles needs longer time. Again, the study of [5] reveals that the efficiencies of the solar collector with fins and without fins were 40.02% and 34.92%, respectively. The study of [6] indicates that the different cover materials provide different performance. These cited articles provide the impression that drying using the solar collector with fins and baffles is a very promising option.

As a result, the objective of this study is to design a solar collector with fins and baffles for rice-cracker drying. First, a mathematical model was developed and used to calculate the required dimensions for the collector that provide the heated air temperature of 60°C. Then a prototype of solar collector was built and tested.

2. Design concept of the prototype

In the literature, it is evident that solar collectors equipped with fins and baffles provide a high temperature comparatively. Accordingly, this study aims to design a rice-cracker dryer equipped with fins and baffles. The proposed design consists of a transparent cover, black-color absorber, insulation, and fan. To help the heated air flows easily in zig-zag pattern as shown in figure 2, a fan is a part of the prototype proposed as shown in figure 3.

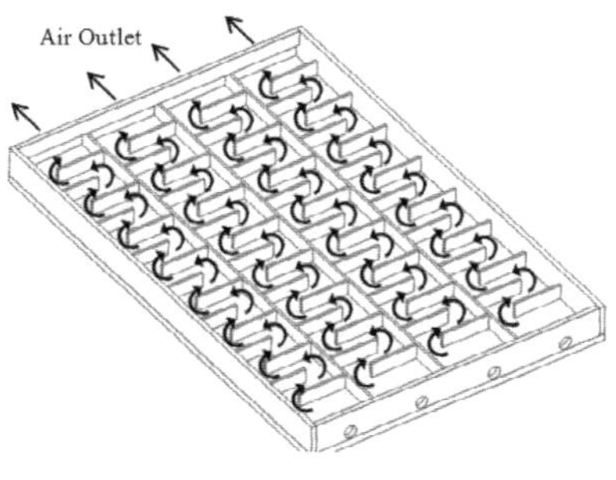

Figure 2. Airflow paths in the solar collector.

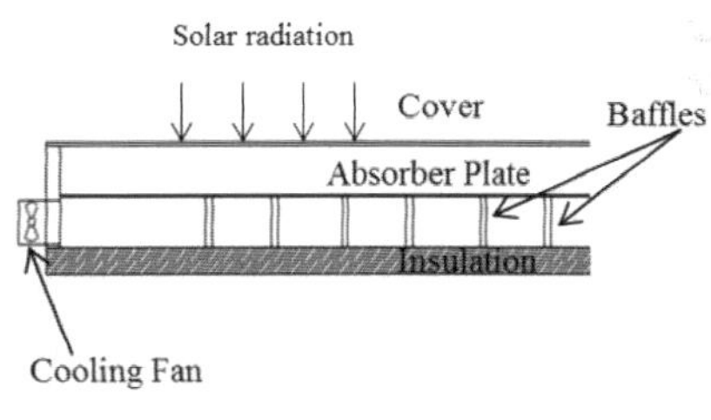

Figure 3. Schematic diagram of proposed solar collector with fins and baffles.

3. Mathematical Modeling

A code was developed using Matlab to predict the performances of the proposed prototype. A set of equations used are as follows [7-10]:

$$A_C = BL \tag{1}$$
$$A_f = 2H(L - 0.4) - (15W_2t)n \tag{2}$$
$$A_b = 15(2W_2W)n \tag{3}$$
$$Re = 2m/\mu(B + H) \tag{4}$$
$$Nu = 0.0158(Re^{0.8}) \tag{5}$$
$$D_{e,f} = 2H/(1 + (H/w_1)) \tag{6}$$
$$h = (kNu)/D_{e,f} = h_1 = h_2 \tag{7}$$
$$h_w = 5.7 + 3.8V \tag{8}$$

where $D_{e,f}$ refers to the hydraulic diameter and h is the heat transfer coefficients between the air and wall. The fin efficiency (η_f) and baffle efficiency (η_b) can be calculated from equations (9) and (10) .

$$\eta_f = (\tanh(Mw_2))/Mw_2 \quad (9)$$

where $M = \sqrt{2h_2/kt}$

$$\eta_b = 26.361(w/D_{e,f})^{-0.454}(L/l)^{-0.634} \quad (10)$$

$$\phi = 1 + (A_f/A_c)\eta_f + (A_b/A_c)\eta_b \quad (11)$$

$$h_r \approx 4\sigma T_{f,m}^3/[(1/\varepsilon_P) + (1/\varepsilon_g) - 1] \quad (12)$$

where $\alpha_p=\varepsilon_p=0.95$, $\varepsilon_R=\varepsilon_g=0.94$ in equation(12).

It should be noted that, in equation(12), the average temperature of the surface absorbing solar radiation ($T_{p,m}$) is assumed to be equal to the flow average temperature($T_{f,m}$).

The heat loss coefficient from the top surface of the solar collector (U_t) can be calculated using equation (13) [3, 4].

$$U_t = \left\{ \frac{2(T_{P,m}-520)}{\left[\frac{\left(\frac{T_{P,m}}{T_a}\right)}{2+(1+0.089h_w-0.1166h_w\varepsilon_P)(1+0.07866x2)}\right]^{0.43\left(1-\frac{100}{T_{P,m}}\right)}} + \frac{1}{h_w} \right\}^{-1} \quad (13)$$

$$+ \sigma(T_{P,m}+T_a)(T_{P,m}^2+T_a^2)/\{(\varepsilon_P + 2x0.00591h_w)^{-1}$$
$$+ [2x2 + (1+0.089h_w - 0.1166h_w\varepsilon_P)(1+0.07866x2) - 1 + 0.133\varepsilon_P]/\varepsilon_g - 2\}$$

where

$$U_b = k_s/l_s \approx 0$$

$$U_L = \frac{(U_b+U_t)(h_1h_2\phi+h_1h_r+h_2\phi h_r)+U_bU_t(h_1+h_2\phi)}{h_1h_2+h_2\phi U_t+h_2\phi h_r+h_1h_2\phi} \quad (14)$$

$$F' = \frac{h_rh_1+h_2\phi U_t+h_2\phi h_r+h_1h_2\phi}{(U_t+h_r+h_1)(U_b+h_2\phi+h_r)-h_r^2} \quad (15)$$

$$F_R = (mC_P/A_CU_L)\{1-\exp[-(U_LF'A_c/mC_P]\} \quad (16)$$

The efficiency and outlet temperature of the solar panel could be estimated using equations (17) and (18) respectively [8],

$$\eta = \frac{Q_u}{A_cI_0} = F_R\left[\alpha_P\tau_g^2 - U_L(T_{f,i}-T_a)/I_0\right] \quad (17)$$

$$T_{f,o} = F_RA_C(I_o\tau_g\alpha_P - U_L(T_{f,i}-T_a))/mC_P + T_{f,i} \quad (18)$$

where $T_a=T_{f,i}$ and $\tau_g=0.875$.

4. Dimensions of the proposed prototype

The expected outlet temperature of the proposed prototype is 60°C. Using the mathematical model proposed, the dimensions of the prototype are as shown in table 1, where the layout of the prototype is displayed in figure 4.

Table 1. Dimensions of the prototype.

Parameter	Value	Unit
Ta (ambient temperature)	40	°C
I (incident solar radiation)	628.93	W/m^2
L (collector length)	1,840	mm
B (collector width)	1,220	mm
W (width of baffles)	190	mm
W1(distance between fins)	287.5	mm
W2 (height of fins)	60	mm
t (thickness of baffles)	10	mm
l (distance between baffle to baffle)	20	mm

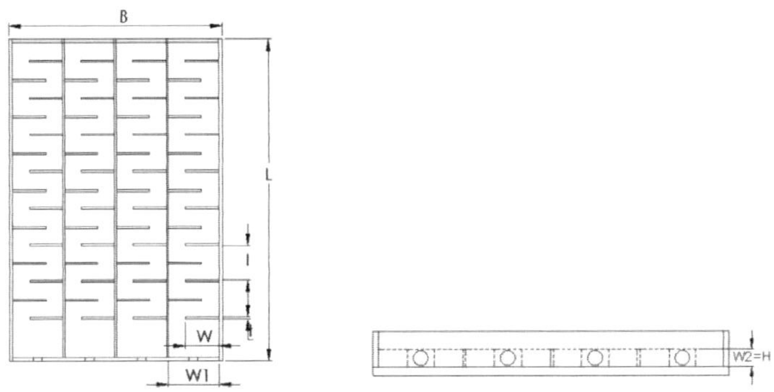

(a) (b)

Figure 4. (a) Top view and (b) front view of the prototype.

5. The prototype

According to the dimensions shown in table 1, the prototype was built. Its fan is powered by a PV panel next to the collector as depicted in figure 5. The measuring instruments were calibrated and equipped. The specifications of the instruments is listed in table 2.

(a) (b)

Figure 5. Prototype built.

The details of the prototype are as follows:

(1) A cover was made from a poly methyl methacrylate (PMMA) sheet with thickness of 4 mm.

(2) An absorber plate was made from a 1-mm-thick iron sheet painted black on its top side. Its absorptivity and emissivity are 0.95 and 0.95, respectively.

(3) A fin, baffle, and insulation were made from a plastwood sheet with emissivity of 0.91-0.93 and thermal conductivity of 0.19 W/mK.

(4) A cooling fan with mass flow rate of 18.3 cfm (NMB Model 06015KA-24N).

Table 2. Specifications of the measuring instruments.

Equipment	Description	Accuracy
Thermometer (OEM TER040)	Temperature Gage Meter	±1°C
Humidity/Temperature (Honeywell HIH6131-021-001)	Measurement Humidity/Temperature	±4%RH,±1°C
Thermometer /Hygrometer	Measurement Humidity/Temperature	±4%RH,±1°C

6. Results

The ambient, experimental outlet temperature, and predicted outlet temperature are displayed in figure 6 and figure 7. It can be seen that the trends and values of the experimental and the predicted temperatures are agree well. The errors are 12.90% and 12.23 % on 25 and 26 November 2017, respectively. Their discrepancy could be due to heat losses from the walls of the air passages, where these losses are not included in the modelling. It should also be noted that the trends of the outlet temperatures are in accordant to the ambient temperature.

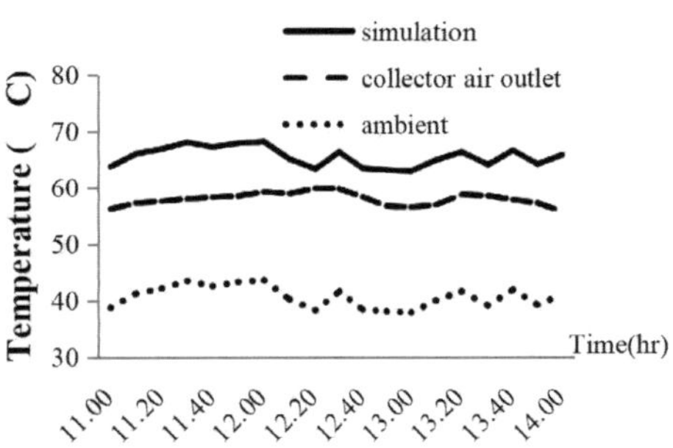

Figure 6. Comparisons on 25 Nov 2017.

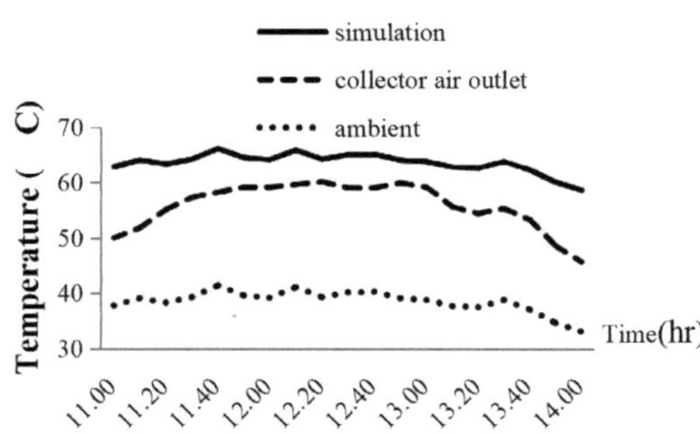

Figure 7. Comparisons on 26 Nov 2017.

The system efficiencies are shown in figure 8. It is obvious that the efficiencies on both days are approximately the same and the values are relatively constant. This is in accord with intuition as the outlet temperature changes with the ambient temperature.

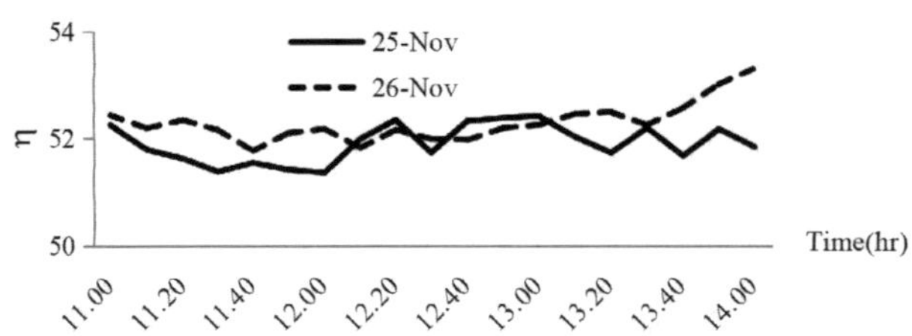

Figure 8. System efficiency versus time.

The Nusselt number (Nu) and Reynolds number (Re) of the system are shown in figure 9 and figure 10. It can be seen that both numbers have the same trend. Their trends are opposite to the trend of the ambient temperatures shown in figure 6 and figure 7. This could be due to the change of the flow properties with the ambient temperature.

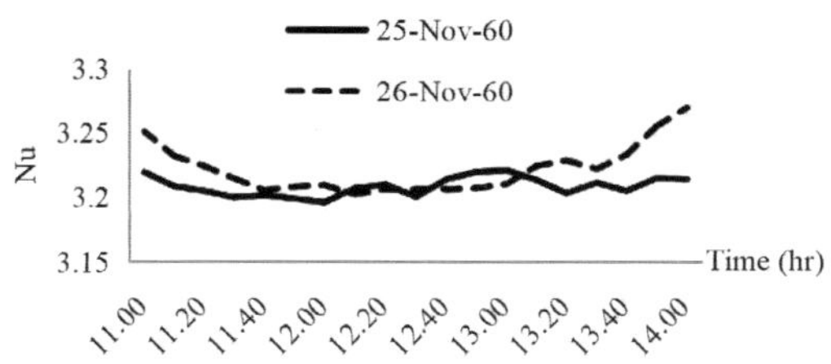

Figure 9. Nusselt number versus time.

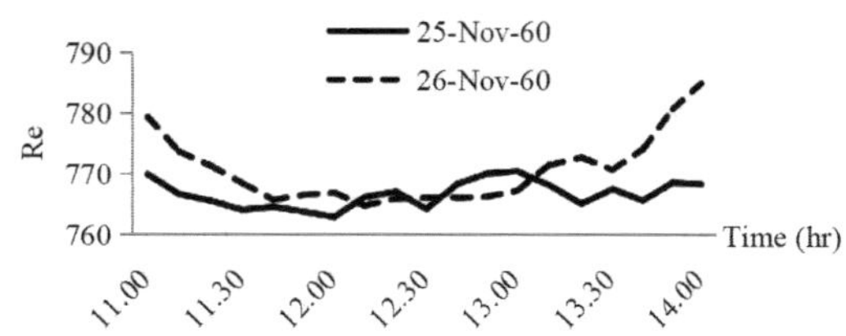

Figure 10. Reynolds number versus time.

7. Conclusion

This study investigates the performances of a solar collector with fins and baffles numerically and experimentally. It was found that the experimental efficiency and outlet temperature are about 47.27% and 60.1°C, respectively. While the predicted efficiency and outlet temperature are about 52.01% and 64.9°C, respectively. The discrepancy could be due to heat losses from the walls of the air passages.

References

[1] Yeh,H.M. Chii-Dog ho and Chi-yen lin (1998) the influence of collector of collector aspect ratio on the collector efficiency of baffled solar air heaters, Energy Vol. **23**, No.1.pp. 11-16, 1998

[2] A. Abene, V.Dubois, M.Le Ray and A. Ouagued J. (2004) Study of a solar air flat plate collector: use of obstacles and application for the drying of grape, *Journal of Food Engineering* 65 (2004) 15-22.

[3] A.Ahmed-zaid, H messaoudi, A.Abene, M.Le ray, J. Y. desmons and B. Abed J. (1999) experimental study of thermal performance improvement of a solar air flat plate collector througr the use of obstacles application for drying of 'yellow onion' Int. *J. Energy Res.,* **23**, 1083-1099 (1999)

[4] Maundu Nicholas Musembi, Kosgei Sam Kiptoo, Nakajo Yuichi (2016) Design and Analysis of Solar Dryer for Mid-Latitude region, Energy Procedia 100(2016) 98-110

[5] Foued Chabane, Noureddine Moummi and Said Benramache J. (2013) Experimental analysis on thermal performance of a solar air collector with longitudinal fins in a region of Biskra,Algeria, *Journal of Power Technologies* 93 (1) (2013) 52-58

[6] Ramadhani Bakari, Rwaichi J.A. Minja and Karoli N. Njau J. (2014) Effect of Glass Thickness on Performance of Flat plate Solar Collectors for Fruits Drying, *Journal of Energy* Volume 2014, Article ID 247287, 8 pages

[7] McAdams WH. Heat transmission, 3rd ed. New York: McGraw-Hill, 1954.

[8] Duffie JA, Beckman WA. Solar engineering of thermal processes. 3rd ed. New York: Wiley, 1977, p.223, 238,245,705.

[9] Yeh HM.Energy-The Internation *Journal 1992*; 17:697.

[10] Chou WH. A study on collector efficiency of baffled solar air heaters. M.S. thesis, Department of Chemical Engineering., Cheng Kung University, Tainan, Taiwan, 1987.

Nomenclature

A_b	Total surface area of baffles (m^2)
A_C	Surface area of collector (m^2)
A_f	Total surface area of fins (m^2)
C_P	Specific heat of air at constant pressure J /kg.K)at 313 K
$D_{e,f}$	Equivalent diameter of conduit with fins, (m)
F'	Collector efficiency factor of solar air heater
F_R	Heat-removal factor for solar air heater

H Height of air tunnel in solar collector (m)
h, h_1, h_2 Convective heat transfer coefficient for fluid flowing over flat plate (W/m^2 k)
h_r Radiant heat transfer coefficient between two parallel plates (W/m^2 k)
h_w Convective heat transfer coefficient for air flowing over outside surface cover (W/m^2 k)
k, k_s Thermal conductivity of air, insulator (W/m^2 k)
ℓ_s Thickness of insulator (m)
M Quantity defined (m)
m Mass flow rate of air (kg/s)
N_U Nusselt number
n Number of fins on absorbing plate
Q_u Useful gain of energy carried away by air per unit time (W)
R_e Reynolds number
T_f Fluid temperature (K)
$T_{f,i}, T_{f,o}$ Inlet and outlet of solar air heater (K)
$T_{f,m}$ Average value of Tf (K)
T_P, T_R Temperature of absorbing plate, bottom plate (K)
$T_{P,m}$ Average value of Tp (K)
U_b Loss coefficient from surface of edges and bottom of solar collector to ambient (W/m^2 k)
U_L Overall loss coefficient (W/m^2 k)
U_t Loss coefficient from top of solar collector to ambient (W/m^2 k)
v Air velocity in the flow channel (m/s)
V Wind velocity (m/s)

Greek Letters
α_p Absorptivity of absorbing plate
η Collector efficiency
η_b Baffle efficiency
η_f Fin efficiency
ϕ Dimensionless quantity defined
τ_g Transmittance of glass cover
$\varepsilon_g, \varepsilon_p, \varepsilon_R$ Emissivity of glass cover, absorbing plate, bottom plate
a_p Absorptivity
ρ Air density (1.127)(kg/m^3) at 313 K
μ Air viscosity (19.15×10^{-6}) (kg/m.s) at 313 K
σ Stefan-Boltzmann constant (5.68×10^{-8}) ($W/m^2 K^4$)

IOP Conf. Series: Materials Science and Engineering **383** (2018) 012066 doi:10.1088/1757-899X/383/1/012066

Design and implementation of real-time monitoring based on Android and HMI of PV tower system for laboratory scale

O Penangsang, A Soeprijanto, N K Aryani, D F U Putra, Suyanto, A Q Yusrina, A Hadi, A Riangga*, S B Panuntun, A P Winata, G Manuella and R M Azmi

Department of Electrical Engineering, Faculty of Electrical Technology, Institut Teknologi Sepuluh Nopember ITS, Surabaya Indonesia, 60111

*E-mail: rianggaaden@gmail.com

Abstract. The performance of photovoltaic (PV) cannot be described by the user, so it is difficult to anticipate when it's undesirable or fault condition occurs. Therefore, the monitoring of PV output parameters is needed. In this case, the parameters that can be displayed are current and voltage value. The parameter data are obtained from PV tracking of the sun position by MPPT Perturb and Observe (P&O) method used in a mechanical system of the dual axis solar tracker. In this experiment, there are two monitoring methods used, which are Real-time Online Monitoring Based on Android system and Human Machine Interface (HMI) display monitoring which is controlled by Arduino Mega 2560. Both methods are expected to show the performance of PV in real-time, remotely or not. The results of this experiment are Android apps and HMI that can monitor the PV output condition. To maintain the PV system in real-time, they keep the device under the supervision and extend its lifetime for more prolonged use.

1. Introduction
Solar energy is one of a renewable energy source that can be converted into electrical energy. Nowadays this energy source is getting a lot of world attention because it can replace the sources of energy that cannot be renewed resulting in pollution, one of them fossil fuel. To convert solar energy into electrical energy is to use photovoltaic cells (PV). Photovoltaic generates direct current (DC) and can be converted to alternating current (AC) using an inverter. The PV system demonstrates green technology that is environmentally friendly, easy installation, low maintenance cost, and also durable [1]. Therefore, this technology is growing very rapidly and become a large-scale system [2]. This large system of course will be found many problems in the future, generally they are under voltage, overvoltage, overcurrent or output power of PV. To solve this problem thus needed Human Machine Interface (HMI) [3]. HMI connect the user and the PV as manually or by visualization on devices. So the user can interact with the PV in real-time, they keep the device under the supervision and extend its lifetime for more prolonged use.

2. Literature review
Human Machine Interface commonly known as HMI referred to the user interface, terminal or operator panel. HMI functions are monitoring or visualizing, controlling, and managing device processes [4]. HMI implementation is based on Internet of Things (IOT) that provides access to users

Content from this work may be used under the terms of the Creative Commons Attribution 3.0 licence. Any further distribution of this work must maintain attribution to the author(s) and the title of the work, journal citation and DOI.
Published under licence by IOP Publishing Ltd

to be able to adjustly manage the device parameter in realtime.[5]. In the application, it can manage a system by setting the value of the motor, valves, temperature values, and display the data realtime data in a portable GUI. By using the GUI, the user can set the valve status open /closed.

A solar cell can be briefly described as pn (positive-negative) junction diode. When the light exposed the positive-negative junction there will be electricity generation from converted light energy. Where photon energy is enough to free the outer bonding of material's electron. PV produced open circuit voltage (Voc) before it connected to the load when PV terminal is being shorted short circuit current (Isc) will generate. When the PV connected with real load there will be real power generated depends on value and characteristic of voltage and current. Peak point from the end of V-I is the result of multiplying maximum voltage with maximum current and this called maximum power value. By the literature [6] the PV resulting optimum output value using the solar tracker. The solar tracker could follow the direction of sun motion and optimize the position of PV.

Arduino Mega 2560 is a microcontroller based on ATmega2560. It has 54 digital input-output pins there are 14 pins can be used for PWM outputs, a 16 MHz crystal oscillator, 16 analog inputs with larger space so this gives plenty room and opportunity, 4 UARTs, a power jack, a USB connection, an ICSP header and a reset button. The Mega 2560 has 256 KB flash memory for storing the code of which 8 KB used by the bootloader, 4 KB EEPROM and 8 KB SRAM. To get started is by connecting it to a computer with USB cable or power it with battery or AC-DC adapter [7].

Blynk is a Platform for Android and iOS apps to control microcontroller (hardware) Arduino, Raspberry Pi, Sparkfun, Particle and similar hardware on the internet. Blynk can build a graphic interface for the project by dragging and dropping widgets. The graphic interface can display sensor data, store data, and visualize it. It is supporting any hardware whether Arduino, Raspberry Pi or more as long as it is linked to the internet over Wi-Fi, Ethernet or ESP8266 chip, Blynk will get online and ready for the Internet of Thing (IoT)[8]. Blynk implantation as an Android application platform has been adapted as field schedule and plant growth monitoring in agriculture domain. Blynk in this study used to display commands from Arduino using Internet-based cloud system, by detecting the temperature, humidity, soil moisture, liquid float level, and magnetic float sensors [9].

3. Research method

3.1. Prototype design

The power plant experiment consists of two identical PV panels mounted in parallel with output power specification for 200+5%. The Human Machine Interface (HMI) is designed to monitor PV output condition using android app and Liquid-Crystal Display (LCD). Where both LCD and android app allows the operator to interact with the machine in the graphical user interface. With controls and graphical user interface displayed on the screen the user can control the machinery. The HMI used in machine and process control to connect the sensors, actuators and machines on the PV tower to input/output control system. This HMI contain a solar panel with dual axis solar tracker system, current and voltage sensors, Arduino Mega 2560, LCD and also android app platform, Blynk. PV will generate electricity (voltage and current) and then the sensors will detect the voltage and current before sent them to microcontroller Arduino Mega 2560. Arduino Mega 2560 process the data and give it to LCD and Blynk to be displayed. The HMI of PV Tower System for Laboratory Scale showed in figure 1.

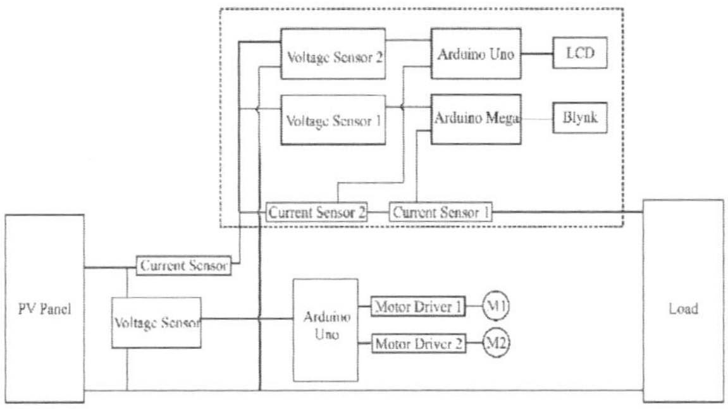

Figure 1. The HMI of PV Tower System for Laboratory Scale.

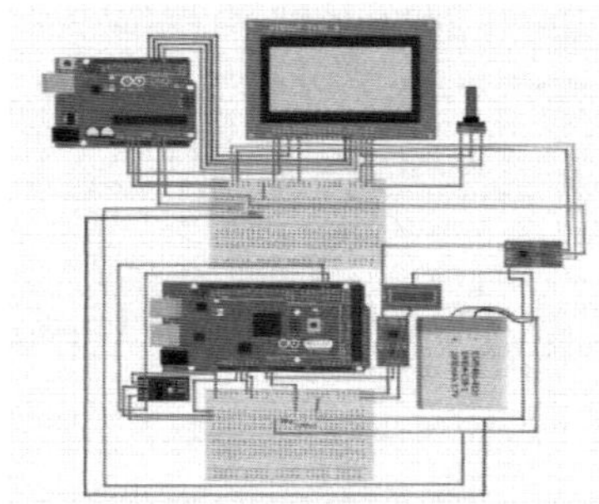

Figure 2. Schematic Design of the Monitoring System.

Monitoring system that is showed in figure 2 divided into two blocks, remote monitoring system and non-remote monitoring system. The remote monitoring system is using Blynk application, where the users could manage the data from any sensors to be displayed (voltage and a current sensor). To acquire the expected data, both the Blynk app and the Arduino device must connect to the Blynk server by internet connection. The Android application connects to the internet by Wi-Fi or cellular data, while the Arduino device is using Wi-Fi module ESP8266 connected to the local wireless network. The non-remote monitoring system is using LCD as the interface. The current value is retrieved using an ACS712 sensor with a max rating of 20A in series. While the voltage sensor is using a voltage divider circuit with a certain resistance value, so the Arduino can measure the voltage 0-50V. To connect to a local wireless network using ESP8266 which is integrated with Arduino Mega 2560.

3.2. Flowchart of the Human Machine Interface system

Start with the acquisition of current data then the PV output voltage flow to the charge controller and to the battery. Current and voltage sensors begin to detect data values to be sent to the microcontroller. Arduino Uno is the LCD display controller to process and give the command to display the data on the LCD. On the other hand, the display on smartphone application controlled by Arduino Mega 2560 to command Blynk server to display the data on smartphone application using internet connectivity. Flowchart of the HMI system is showed in figure 3.

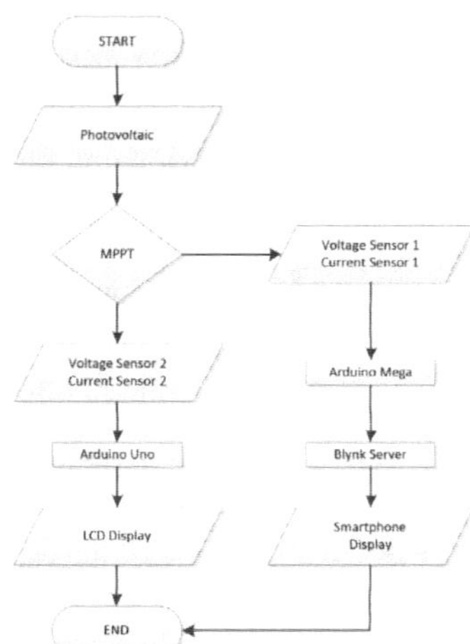

Figure 3. Flowchart of the Human Machine Interface system.

4. Analysis & results

The experiments were performed for three days in cloudy weather. The result showed that the PV produces fluctuate power value shown in table 1 from November 5 until November 7, 2017 with time interval 1-hour start from 6:00 until 19:00. Then, the result of LCD interface and Blynk platform could be seen in figure 4 and figure 5.

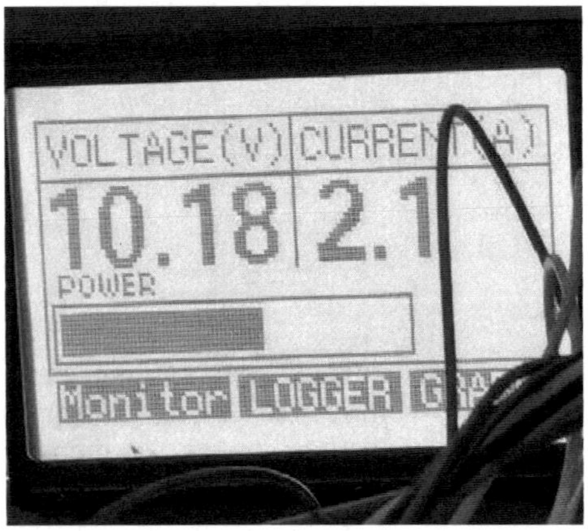

Figure 4. LCD display.

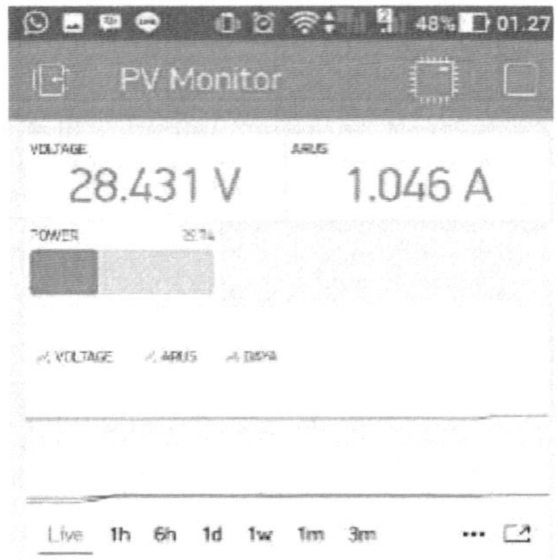

Figure 5. Blynk display.

Table 1. The result of tracking PV displayed by Blynk app with time sampling 1-minute. To minimize the table, the time interval is changed into 1-hour.

	November 5, 2017			November 6, 2017			November 7, 2017		
Hour	Voltage (Volt)	Current (Ampere)	Power (Watt)	Voltage (Volt)	Current (Ampere)	Power (Watt)	Voltage (Volt)	Current (Ampere)	Power (Watt)
6:00	32.67	0.30	9.80100	32.65	0.98	31.9970	38.31	1.23	47.1213
7:00	32.86	1.19	39.1034	32.78	1.50	49.9950	40.80	1.67	68.1360
8:00	32.98	3.20	105.536	33.02	3.12	103.022	40.13	2.31	92.7003
9:00	33.07	5.40	178.578	33.33	4.46	148.652	40.14	5.78	232.009
10:00	32.87	6.68	219.572	32.26	7.13	230.014	40.29	7.37	296.937
11:00	35.28	4.15	146.408	35.20	3.19	112.288	40.16	7.21	289.554
12:00	32.98	1.93	63.6514	34.29	1.57	53.8353	39.97	5.29	211.441
13:00	32.80	1.20	39.3600	33.83	1.38	46.6854	40.52	4.54	183.961
14:00	32.37	0.36	11.6532	33.69	0.95	32.0055	40.42	4.37	176.635
15:00	32.49	0.43	13.9707	32.17	0.64	20.5888	31.44	2.01	63.1944
16:00	31.27	0.22	6.87940	30.63	0.35	10.7205	6.900	0.72	4.96800
17:00	13.78	0.17	2.34260	14.69	0.13	1.90970	2.380	0.22	0.52360
18:00	1.520	0.19	0.28576	2.200	0.14	0.30800	1.900	0.18	0.34200
19:00	1.400	0.19	0.26600	2.200	0.14	0.30800	1.860	0.10	0.18600

The first experiment performed on the first day, start when PV in the initial condition on 06:00 generated 9.8 W of power value, 32 V of voltage current, and 0.3 A of current value. In subsequent hours, the resulting current is higher with voltage value tends to be stable which causes the resulting power is also greater. From the graph, we can obtain that the PV will generate the greatest power value in the time range of 09:00 until 12:00 in the peak hour at 10:00 with a power value of 219.572 Watt.

After reach the peak power value, the power generated by PV will slowly decrease. The second experiment performed on the next day with results that resemble the first experiment. In the initial condition at 06:00, PV produces a voltage of 32.65 V, a current of 0.98 A, and generate power of 31.997 Watt. Power will increase over time and reach the peak of 230.014 Watt at 10:00. The third day also shows similar results based on the overall data, this can be seen in figure 6.

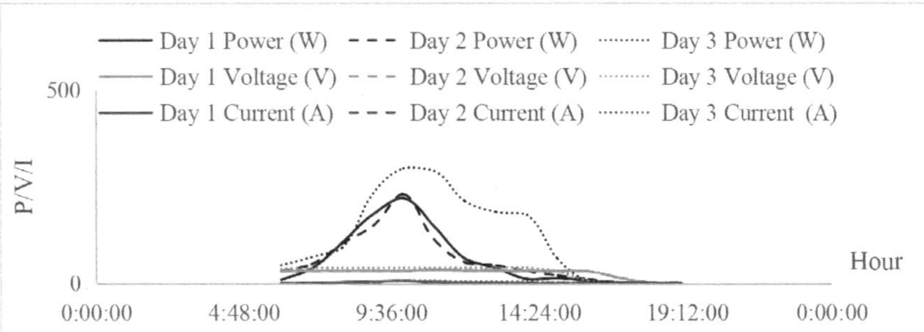

Figure 6. Comparison graph of output current, voltage and power value generated by PV during November 5 – November 7, 2017.

5. Discussion & conclusion

In this paper, design and implementation of real-time monitoring based on android and HMI of PV tower system for laboratory scale has been shown. HMI design is implemented in the PV tower monitoring system where the process starts by detecting voltage and current value using the sensor in the PV output. Then the data is processed and displayed in LCD using Arduino Uno controller device. While the display on the smartphone application controlled by Arduino Mega 2560 to commands Blynk server via the internet before it is displayed on smartphone applications.

The result retrieved by three-days experiment shown that the real-time monitoring based on android and HMI of PV tower system can be applied and displayed the voltage and current value of PV tower output. Power value displayed on the experiments shown overall similar result at 10:00. Within this result, it can be concluded that real-time monitoring based on android and HMI could be implemented in PV tower for laboratory scale to optimized the device using.

References

[1] Pranesh K A and Saranya K 2015 Solar Tracking System using DC Gear Motor *J. Appications Inf. Commun. Eng. Febr.* **1** 20–3

[2] Tyagi V V, Rahim N A, Rahim N A, Jeyraj A and Selvaraj L 2013 Progress in solar PV technology: Research and achievement *Renew. Sustain. Energy Rev.* **20** 443–61

[3] F. Brito, I. Fonseca, J. T. Farinha and F. M. Barbosa 2016 Integrating low cost platforms on electrical power systems for control and condition monitoring *2016 51st International Universities Power Engineering Conference (UPEC)* 2016 51st International Universities Power Engineering Conference (UPEC) pp 1–5

[4] Anon 2014 *Human Machine Interface (HMI) Guide* (Texas Instruments Incorporated)

[5] Saini H S and Daruwala R D 2016 Human Machine Interface in Internet Of Things system *Computing Communication Control and automation (ICCUBEA), 2016 International Conference on* (IEEE) pp 1–4

[6] Penangsang O, Suyanto, Aryani N K, Putra D F U and Soeprijanto A 2016 Optimization of Panel PV Position Using Dual Axis Solar Tracker With Perturb and Observe Method in PV Tower - Laboratory Scale Microgrid System *International Conference on Sustainability, Green Buildings, Environmental Engineering & Renewable Energy (SGER 2016)* NZAAR

(Science Publication)
[7] Kusriyanto M and Putra B D 2016 Smart home using local area network (LAN) based arduino mega 2560 *Wireless and Telematics (ICWT), 2016 2nd International Conference on* (IEEE) pp 127–31
[8] Blynk Inc. 2017 *Blynk* [Online] Available : http://docs.blynk.cc/
[9] Mohanraj I, Ashokumar K and Naren J 2016 Field Monitoring and Automation Using IOT in Agriculture Domain *Procedia Comput. Sci.* **93** 931–9

IOP Conf. Series: Materials Science and Engineering **383** (2018) 012067 doi:10.1088/1757-899X/383/1/012067

Simulation of economic effect on electricity democracy based on clean energy at UMKM'S IPP in Indonesia

Supriadi Legino*, Irma Wirantina Kustanrika, Ranti Hidayawanti and Iriansyah Sangadji

Sekolah Tinggi Teknik PLN, Jakarta, Indonesia

*E-mail: supriadi@sttpln.ac.id

Abstract. The facts show that most large-scale power projects in Indonesia cannot be completed on schedule. The reason is related to the dynamic constraints such as legal, political, economic, social and cultural issues. In general, there is a potential of community empowerment approach using a small size plant has not been done. Such approach may give a lot of advantages including shorter construction time and easier to use clean energy like solar, wind, and waste or biomass which is available around us. Constructing 1000 units of 1 MW power plant simultaneously will have similar capacity with the construction of one 1000 MW power plant unit that had so many obstacles. Another important advantage is that this kind of plant can be self-managed under the ownership of local communities that can empower local businesses to become small- scale IPP developers with simple technology and small capital costs. This paper will examine the economical impact of local people's electrical self-management known as IPP UMKM by simulating several scenarios.

1. Background

Until recently, we are still relying on conventional power system, aimed at optimizing eficiency and reliability. Conventional system is typically large centralized power plants, which are interconnected one into others and delivers the power to the customers using transmission and distribution lines. However, more and more problems emerged to hinder the construction of big power plant as well as its associated transmission lines, including land acquisition, right of way permit, environmental and social obligation. Many big power plants and transmission lines projects are delayed because facing the above mentioned challenges and its cost may offset the benefit of the economies of scale and interconnection system. Moreover, most conventional system plants are dominated by fossil fuel while there are huge resources of renewable energy available surrounding any communities. Other drawbacks of the large scale power plants under the conventional system include the need of large amount of money, highly dependent of imported equipment and spare parts, and mostly owned by a small group of giant investor. Those are conflicting with the energy security law (UU No 30/2007), stating that fulfilling energy need should be involving people ability and caring about environment.

"Listrik Kerakyatan" is the Indonesia term, means people owned electricity, could become an alternate solution to answer the conventional power system dilemma. It will be discussed further in this paper using the principle of math commutative law, 1000 x 1 = 1 x 1000, This expression is illustrated in figure 1[1].

Most of private power plants under the IPP (Independent Power Producer) scheme are usually large scale and owned by the giant investors. Furthermore, the large scale plants are still dominated by fossil

Content from this work may be used under the terms of the Creative Commons Attribution 3.0 licence. Any further distribution of this work must maintain attribution to the author(s) and the title of the work, journal citation and DOI.
Published under licence by IOP Publishing Ltd

energy, which harmed the environment. In order to weigh up the situation, the government has to consider local people involvement as a small Independent Power Producer (IPP), which plants should use renewable energy. Listrik Kerakyatan (LK) is adopting the concept of distributed generation, which is small scale distributed power plants using sun, biomass, and wind energy resources, which always available close to the community.

UMKM is the abbreviation of "Usaha Mikro Kecil dan Menengah," an Indonesia term of Micro Small and Medium Enterprise (MSME). According to Sarwono and Hartadi [2]. Micro, Small and Medium Enterprises (MSMEs) have an important and strategic role in national economic development. UMKM is a very potential market for the financial services industry, especially banks to distribute financing. UMKM therefore becomes essential to the economic growth and life quality improvement. He also describes that typically, UMKM is a self-managed business and self-funded business and its assets is relatively small. It has only small number of employees and usually market its product or service locally. UMKM business may vary from a start-up informal sector like livelihood activities and micro enterprise like home craft industry, small dynamic enterprise, and fast moving enterprise. Based on the type and ownership level, UMKM consist of three levels from the smallest Micro (Mikro) with net asset IDR50 million, Small (Kecil) with net asset IDR50 million to IDR500 million, and Medium (Menengah) level with net asset of IDR 500 million to IDR2500 million.

UMKM is getting special incentives from the government including low interest credit, simple and easy business permits, and various business development support from related government agencies. Based on its characteristics and various advantages given by the government, UMKM can be very potential to own and run electricity business, compliment to the large size conventional electricity companies, which are dominated by a limited group of large enterprise.

The small size of LK, provides opportunity for local business players to own small scale power plant or waste to energy plant because the investment cost per package of LK is relatively low, less than one billion rupiah for the unit size of 30 kW to 50 kW. Such small size plant also needs only small area (200-1000 m2) and flexible to be located any where around school, housing, market, mall, hotel, and office. For remote areas that have no electricity, LK can be managed locally as a hybrid system.

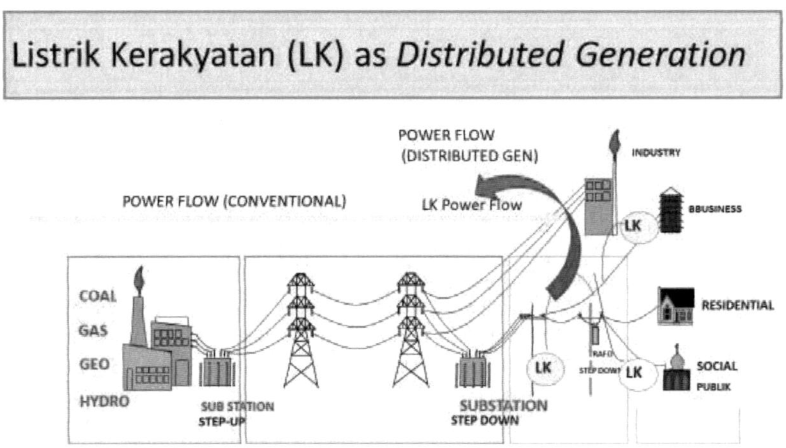

Figure 1. Diagram of LK over the conventional power system(Source: Legino & Arianto , 2017 [1]).

As shown in figure 1, LK plants location are scattered around the communities or electricity customers. Therefore, technically, LK will reduce the transmission losses and the problem of excessive transformer overload and drop voltage. From the energy balance aspect, LK offers the additional renewable energy capacity to the system without any transmission lines construction.

On top of the above advantages, LK will be the concrete way to use both local material and man power, especially for biomass power plant. If LK is expanded nationally, the government target to have 100 percent electrification ratio and to reach renewable portion of 23% by 2025 can be achieved in less

IOP Conf. Series: Materials Science and Engineering **383** (2018) 012067 doi:10.1088/1757-899X/383/1/012067

than three years. Unlike, the giant project that needs foreign funding, LK can be funded by local people equity and social support allocation for villages. In addition, LK also has opportunity to get a low interest credit from national bank under the UMKM scheme.

2. Objective

The objective of this paper is to examine whether UMKM can feasibly apply small-scale model electricity business economically under the corridor of existing regulation. This paper will undertake a financial simulation using cash-flow analysis to determine the profit by calculating NPV (Net Present Value), Payback period, and Internal Rate of Return (IRR). The main criteria is the electricity price regulation in which the tariff offers by UMKM IPP should be lower than the Electricity Production Cost of the region

3. Literature review

According to IEA[3] : Energy security is "the uninterrupted physical availability at a price which is affordable, while respecting environment concerns." In this case, any effort to increased energy efficiency, energy storage, CO_2 capture , maximizing renewables, and nuclear power plants will all be important. According to Ohja, Kuldeep[4]: The centralized power system ignores energy needs of the rural areas and poor and has also led to environmental degradation, whereas decentralized energy planning model is in the interest of efficient utilization of resources. They also depict decentralized renewable energy as non-polluting plants is essential. In the other opinion, Allan, Grant Eromenko, Igor Gilmartin, Michelle Kockar, Ivana McGregor, Peter[5] said that decentralized electricity system is recognized as one means of achieving efficient and renewable energy provision, as well as addressing concerns over ageing electricity infrastructure and capacity constraints.

In other study by I Wayan Dipta [6], small and medium enterprises have been recognized as highly strategic and important not only for economic growth but also for equitable revenue sharing. Due to its strategic and important role, Indonesia pays special attention to their developments, including fostering an environment with a conducive business climate, facilitating and providing access to productive resources and strengthening entrepreneurship and competitiveness. The role of government and other support agencies is strategic and important, for instance by means of providing facilities and support such as financial assistance and other needs to bridge the partnership between the two parties.

To determine financial viability [7], one of the major topics which is taught in the field of Finance is the rules of capital budgeting, including the Payback Period and the Net Present Value (NPV). Bhandari, Shyam B [8], introduces the discounted payback period criterion for small business capital budgeting decisions; properties and limitations of the traditional payback technique; discussion of the discounted payback approach; Comparison of the payback methods with other capital budgeting techniques

According to Kierulff, Herbert [9], the article explains significant problems with the measurement of reinvestment issues, multiple IRRs, timing problems, problems of choice among unequal investment opportunities, and practical difficulties with multiple discount rates. And according to Bhattacharyya, Nalinaksha[10], establishes the validity of the IRR interpretation as returns earned on funds still internally invested in the project. Using this interpretation, IRR can be viewed as a tool to evaluate the risk of capital budgeting proposals.

$$IRR = i_1 + \frac{NPV_1}{(NPV_1 - NPV_2)} (i_2 - i_1) \quad (1)$$

where: IRR = Internal Rate of Return

i_1 = The resulting Discount Rate NPV+

i_2 = The resulting Discount Rate NPV-

NPV_1 = Net Present Value positive

IOP Conf. Series: Materials Science and Engineering **383** (2018) 012067 doi:10.1088/1757-899X/383/1/012067

NPV2 = Net Present Value negative

Payback period is the period of return of investments that have been incurred through the profits derived from a project that has been made. According to [11] To support the specific systems' sustainable character one should be able to ensure minimum period of energy pay-back.

$$Payback\ Period = n + \frac{(a-b)}{(c-b)} \times 1\ year \tag{2}$$

n = The last year in which the amount of cash flow still cannot close the initial investment

a = The amount of initial investment

b = The cumulative sum of cash flows in year n

c = The cumulative sum of cash flows in year n + 1

4. Data

For the purpose of simulation, the data for capital expenditure (Capex), operational expenditure (Opex), and estimated revenues are taken from actual cost incurred during the pilot projects done by STT PLN at two locations, Pondok Kopi, and Duri Kosambi regencies as shown in table 1. Capex including civil work, briquetting equipment, and electric equipment; Opex including operators salary, material, and other overhead. Revenue condition 1 is based on Feed in Tariff IDR 850 for Java Island, tipping fee of IDR100,000, briquette price of IDR500,- and other valuable material of IDR400/kg. Revenue conditon 2 is based on average outside Jawa Feed in Tariff IDR 1,500, tipping fee only IDR20,000, briquette price of IDR500, and no income from valuable material.

Table 1. Capex, Opex, and estimated Revenues of UMKM LK plant (3 ton waste per day).

No	Description	IDR	USD	USD/today*
1.	Capital Expenditure	668,500,000	51,423	17,141
2.	Operational expenditure	355,248,000	27,327	9,109
3.	Annual Rev (Java)	511,920,000	39,378	13,126
4.	Annual Rev (Out. Java)	410,400,000	31,569	10,523

*1USD: IDR13000

5. Result

The results of the simulation are calculated using two scenarios by using cash flow analysis and the above criteria. As seen in table 2, the project in Java-Bali with the Feed in tariff of IDR 850 is financially viable (IRR>i) by using UMKM 7% credit scheme, if there is capex subsidy.

Table 2. Financial simulation result for Java and Bali areas.

Scenario	i (%)	Capex Subsidy	NPV (IDR)	IRR (%)	Viability
1	7	0	407,579,681	11	Moderate
2	7	10%	474,429,681	14	Good
3	14	40%	398,862,478	20%	Good

For outside Java areas, the Feed in Tariff are vary from IDR 1100 to around IDR 2200. Since people income is lower than that of Java, it is assumed that no income from valuable waste and the tipping fee is assumed only IDR20,000, much lower than that of Java. The result in table 3 indicates that the project UMKM outside Java-Bali is viable under the UMKM credit scheme of 7%.

Table 3. Financial simulation result for outside Java Bali areas.

Scenario	Feed in Tariff	i (%)	Capex Subsidy	NPV (IDR)	IRR (%)	Viability
1	IDR 1100	7	80%	107,978,194	15	Good
2	IDR 1100	14	85%	80.012.940	17	Moderate
3	IDR 2000	7	60%	247,355,044	16	Good
4	IDR 2000	14	70%	182.540.516	18	Good

6. Discussion

The government of Indonesia provides incentive for UMKM by giving special low interest credit (around 7%) compare to the commercial credit of 14%. In addition, the government allocates special budget for rural area around one billion IDR per village. On the other hand, the income from electricity sales is limited by the ceiling value of the production cost of the Electricity State Owned Enterprise (PLN) as stipulated in Ministry of Energy and Mineral Resources decree (Kepmen ESDM No. 1404.k/20/MEM/2017. Therefore the financial viability for UMKM IPP can still be attained if the government allocates some portion of rural funding to support UMKM capital expenditure.

As simulated in table 2 for Java-Bali island and table 3 for Outside Java Bali island, the UMKM can still get good profitability to sustain their business. If there are averagely 2 UMKM run 100 kW IPP business in each villages, the nation will get additional renewable plant capacity of 14,000 MW from 70,000 villages all over the country.

7. Conclusion

Based on the result of simulation, the idea to apply self-management of electricity democracy, particularly by using UMKM's IPP scheme can be used as alternatives to solve the on-going dilemma that challenges the construction of large scale conventional electricity system. By using the available instruments, such as special budget and soft credit incentive for rural people, the Indonesia government can apply the UMKM's IPP model nationally. By applying this model, the national energy policies including 100 percent electrification ratio, 100 percent rural electrification, and 23 percent of renewable energy portion can be achieved as fast as two years completion time.

References

[1] S. Legino and R. Arianto, "Solving large scale unit dilemma in electricity system by applying commutative law," in *ICOMPAC*, 2017.(Presented)

[2] H. A. Sarwono, "Profil Bisnis Usaha Mikro, Kecil dan Menengah (UMKM)," *Bank Indones. dan LPPI*, pp. 5–57, 2015.

[3] IEA, "Energy Technology Perspectives 2012 Pathways to a Clean Energy System," *Energy Technol. Perspect. 2012 Pathways to a Clean Energy Syst.*, pp. 1–12, 2012.

[4] F. Rubik and G. Scholl, "Integrated Product Policy (IPP) in Europe - A development model and some impressions," *J. Clean. Prod.*, vol. **10**, no. 5, pp. 507–15, 2002.

[5] G. Allan, I. Eromenko, M. Gilmartin, I. Kockar, and P. McGregor, "The economics of distributed energy generation: A literature review," *Renewable and Sustainable Energy Reviews*, vol. **42**. pp. 543–56, 2015.

[6] I. W. Dipta, "Strategi Penguatan Usaha Mikro, Kecil Dan Menengah (UMKM) Melalui Kerjasama Kemitraan Pola CSR," *INFOKOP*, vol. **16**, pp. 62–75, 2008.

[7] K. Ardalan, "Payback Period and NPV: Their Different Cash Flows," *J. Econ. Financ. Educ.*, vol. **11**, no. 2, pp. 10–16, 2012.

[8] S. B. Bhandari, "Discounted Payback: A Criterion For Capital Investment Decisions.," *J. Small Bus. Manag.*, vol. **24**, no. 2, pp. 16–22, 1986.

[9] H. Kierulff, "IRR: A blind guide," *Am. J. Bus. Educ.*, vol. **5**, no. 4, pp. 417–26, 2012.
[10] N. Bhattacharyya, "Why Do Managers Prefer IRR," *SSRN Electron. J.*, 2004.
[11] J. K. Kaldellis, M. Simotas, D. Zafirakis, and E. Kondili, "Optimum autonomous photovoltaic solution for the Greek islands on the basisof energy pay-back analysis," *J. Clean. Prod.*, vol. **17,** no. 15, pp. 1311–1323, 2009.

IOP Conf. Series: Materials Science and Engineering **383** (2018) 012068 doi:10.1088/1757-899X/383/1/012068

Weight assessment for a blended wing Body-Unmanned aerial vehicle implementing boundary layer ingestion

E A Valencia*, V H Alulema and V H Hidalgo

School of Mechanical Engineering, Escuela Politécnica Nacional, Ladrón de Guevara E11-253, Quito 170517, Ecuador.

* Email: esteban.valencia@epn.edu.ec

Abstract. Boundary Layer Ingestion (BLI) has been studied extensively for implementation in Blended Wing Body (BWB) transport aircrafts, and it has been reported that BLI offers potential benefits mainly in terms of energy, improving the propulsive efficiency of a propulsion system and reducing the fuel consumption. This paper examines potential benefits of BLI related with a reduction on the Take-Off Gross Weight (TOGW) for a BWB-Unmanned Aerial Vehicle (UAV). It is discussed the methodology for weight estimation of a BWB-UAV through parametric models. To evaluate the benefits of BLI in terms of weight reduction, it has been taken into account the removal of pylons and the nacelle weight reduction when embedding nacelles into the airframe. The validation of the parametric models, and the case of study have been carried out for the aircraft NASA X-48B. The results show a reduction in TOGW between 8% and 19% when removing pylons and embedding nacelles into the airframe.

1. Introduction

The benefits of boundary layer ingestion (BLI) have been evaluated extensively with a special focus on Blended Wing Body (BWB) transport aircrafts. Most of previous studies about BLI for transport aircrafts have focussed mainly on design of inlets, reduction on fuel consumption, improvement of propulsive efficiency, ram and viscous drag reduction, aeroacoustics effects, and flow control at duct inlet. In 1998, Liebeck [1] introduces BLI in the design of a subsonic BWB transport aircraft, and the results show a reduction of 13% on thrust to weight ratio, and a reduction of 27% on fuel burned, compared to a conventional aircraft operating at the same conditions.

Rodriguez [2] performed a multidisciplinary design optimization (MDO) of BLI inlets, and estimated a reduction of 3.7% in fuel burn rate, and 2.3% in total drag coefficient. In 2004, a design optimization of a BWB subsonic transport was developed by Liebeck [3]; in his work, BLI was evaluated experimentally, simulating the boundary layer over a flat plate and testing different duct geometries. A. Plas [4] assessed the performance of a BLI propulsion system; to evaluate the benefits of BLI in terms of energy, Plas considered power saving coefficient as figure of merit. For the "Silent aircraft", Plas estimated a power saving between 3% and 4% when embedding nacelles into the BWB airframe [4].

Campbell [5] and Carter [6] designed and tested 2 scale models of a BWB aircraft: a clean wing BWB, and a BWB with BLI nacelles. These studies determined aerodynamic coefficients for both configurations and evaluated the Propulsor-Airframe Integration (PAI) through computational fluid dynamics (CFD) and wind tunnel experiments at high Reynolds numbers about 75 million and a Mach number of 0.85 [6].

Content from this work may be used under the terms of the Creative Commons Attribution 3.0 licence. Any further distribution of this work must maintain attribution to the author(s) and the title of the work, journal citation and DOI.
Published under licence by IOP Publishing Ltd

Whilst, Hardin [7] estimated the fraction of viscous drag that could be ingested, and determined that BLI propulsors can reduce from 3% up to 5% the fuel burn rate, for BLI fractions between 10% and 12% (Drag ingested/ Total Drag). All the aforementioned works were conducted on baseline configurations of BWB transport aircrafts. The results from those works represents a quantifiable evidence of the benefits of BLI in terms of energy mainly. By now, the reduction of fuel burned, total drag coefficient, and the increment on range and propulsive efficiency are some of the performance benefits accrued from BLI.

For small UAVs, the incorporation of BWB and BLI have not been extensively documented, which is related to their military past and hence different design criteria for their development. In 2006, Berends and van Tooren [8] performed a MDO of a BWB-UAV; however, the focus of their work was the development of a multi-objective distributed optimization tool, and the aerodynamics of a BWB-UAV was used as case of study. Lee et al. [9] also carried out a MDO of the aerodynamics of a BWB Unmanned Combat Aerial Vehicle (UCAV). The figure of merit of their study was the aerodynamic efficiency. Chandrasekhara [10] assessed experimentally, in a wind tunnel, the aerodynamics of a UCAV (wingspan = 9 in). Jo et al. [11] evaluated, through numerical simulations, the aerodynamic characteristics of a BWB-UCAV (wingspan = 0.7 m). The same BWB-UCAV used by Jo [11], has been tested experimentally in a wind tunnel in Shim and Park's work [12]. These works have focussed on the aerodynamics characterization of BWB-UAVs and UCAVs; however, they neither address the propulsion system integration with the airframe, nor BLI integration features.

As mentioned in [4,5,7]; BLI requires a deeper and broader analysis in aspects such as PAI, engines installation, flow control, inlet design, and mainly, it is required to analyse the effect of BLI in the performance of the whole aircraft, not only in the propulsion system.

This work examines, at a preliminary design stage, another potential benefit of BLI. Taking Take-Off Gross Weight (TOGW) as a figure of merit, a reduction assessment on TOGW for a BWB-UAV, implementing BLI, is presented. This works explores the integration of BLI for BWB-UAVs taking into account integration aspects such as the removal of pylons, nacelle embedding percentage, and propulsors location. In order to evaluate the benefits of BLI in terms of weight reduction, a parametric model for TOGW estimation for BWB-UAVs is also presented.

2. Weight estimation for a BWB – UAV integrating BLI

To estimate the weight for a BWB-UAV aircraft, a Class II method was used. As described by Roskam [13] and Elham [14], Class II method is based on empirical equations and is appropriate to be used at a preliminary design stage, when a weight breakdown and enough geometrical information for the aircraft in analysis is available [15].

2.1. BWB-UAV: weight breakdown

Weight breakdown is required to estimate aircraft weight using Class II method [13]. In this work, the following weight groups for a general baseline BWB-UAV have been considered:

2.2. Weight group 1: wtructural group

Structural group is integrated by: BWB-Wing, a generic landing gear, and control surfaces along the wing trailing edge.

BWB Wing: As mentioned in [16], the following parametric empirical equation, developed by Beltramo [17] is accurate for wing weight estimation at preliminary design stage. This equation has been used by Liebeck for the design of the BWB subsonic transport [3].

$$W_{Wing} = 4.24 \times I_W + 0.57 \times S_W \tag{1}$$

where:

$$I_W = \frac{n_{load} \times (AR)^{1.5} \times (ZFW / TOGW)^{0.5} \times (1+2\lambda) \times (TOGW / S_W) \times S_W^{\ 1.5} \times (10^{-6})}{(t/c) \times (\cos \Lambda_{1/4}) \times (1+\lambda)} \tag{2}$$

In equation , n_{load} is the ultimate load factor, AR is the aspect ratio, λ is the taper ratio of the wing, S_W is the wing reference area, $\Lambda_{1/4}$ is the quarter chord sweep angle, t/c is the thickness to chord ratio, ZFW is the Zero Fuel Weight, and $TOGW$ is the Take-Off Gross Weight.

To estimate the wing weight, the BWB has been considered as a composition of small wing sections, as it is shown in figure 1 (a). Every wing section must be defined geometrically according to figure 1 (b). For every wing section, it is estimated the corresponding weight, and the total wing weight for a BWB (W_{BWB_Wing}) is defined as follows.

$$W_{BWB\_Wing} = \sum_{i=1}^{nws} W_i \tag{3}$$

In equation , W_i is the weight of the wing section i, and nws is the number of wing sections of a BWB-UAV.

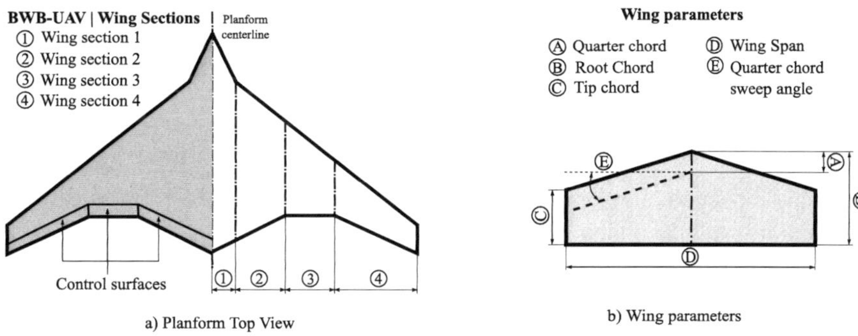

Figure 1. BWB geometrical description.

Landing gear: The weight of landing gear (W_{LG}) was estimated with equation , which was used by Ko [16] in a design optimization of a BWB aircraft.

$$W_{LG} = 0.0135 \times TOGW^{1.1} \tag{4}$$

Control surfaces: Equation is used to estimate the weight of control surfaces (W_{CS}). This equation is presented in [18].

$$W_{CS} = 0.44 \times TOGW^{2/3} \tag{5}$$

2.2.1. Weight group 2: Propulsion system, Fuel and fuel system

Engine, nacelle, pylon, fuel and fuel system are considered in this group. The following empirical equations are presented in Isikveren [19].

Engine: $$W_{engine} = 0.0177 \times T_0^{1.0572} \tag{6}$$

Nacelle: $$W_{nacelle} = 0.32775 \times W_{engine} \tag{7}$$

Pylon: $$W_{pylon} = 0.574 \times W_{engine}^{0.736} \tag{8}$$

In equation (6), T_0 is the maximum sea level static thrust.

IOP Conf. Series: Materials Science and Engineering **383** (2018) 012068 doi:10.1088/1757-899X/383/1/012068

Fuel and fuel system: According to Kundu [20] fuel can be accounted as 20% of the $TOGW$. Whilst, Gundlach [21] mentions that fuel system can be estimated as 10% of the fuel weight.

2.2.2. Weight Group 3: Sub-systems
In this weight group; flight control system, electrical system, instrumentation, avionics, and electronics are included. Based on empirical equations developed by Roskam [13], equation presents a reduced expression to estimate weight for this group.

$$W_{Subsystems} = 0.055 \times TOGW \quad (9)$$

2.2.3. BWB-UAV Take-Off Gross Weight (TOGW) estimation
To estimate the TOGW for a BWB-UAV, considering the weight groups described above, it has been used the "Unity equation", equation , presented in Torenbeek [22].

$$TOGW = \frac{\sum_{i=1}^{n} W_{fix} i}{1 - \sum_{j=1}^{m} ((W_{var} j) / TOGW)} \quad (10)$$

In equation , W_{fix} groups the components whose weight does not depend on $TOGW$, and W_{var} groups the components whose weight does depend on $TOGW$. To solve equation , the secant numerical method has been used as in [21].

2.3. Boundary layer ingestion assessment in terms of weight.
To assess the benefits of BLI, TOGW has been taken as a figure of merit; however, the TOGW reduction for BLI configurations is related directly with the reduction on propulsion system weight when embedding nacelles into the airframe and removing pylons. The weight of a pylon-mounted propulsion unit is defined by equation .

$$W_{Prop\_Unit}^{No\text{-}BLI} = W_{engine} + W_{pylon} + W_{nacelle}^{No-BLI} \quad (11)$$

And the weight of a BLI propulsion unit can be accounted with equation , which includes the effects of removal of pylons and embedding nacelles into the airframe.

$$W_{Prop\_Unit}^{BLI} = W_{engine} + W_{nacelle}^{BLI} \quad (12)$$

The reduction on propulsion system weight, for BLI configurations due to nacelle embedding and removal of pylons, is characterized with equation . In this equation, ζ is the embedding percentage and represents the fraction of nacelle diameter that is embedded into the airframe, assuming a symmetrical nacelle. This assumption allows to relate directly the embedding percentage with a reduction in nacelle wetted area and hence nacelle weight.

$$W_{nacelle}^{BLI} = (1 - \zeta) \times W_{nacelle}^{No\_BLI} \quad (13)$$

Figure 2 (a) and 2 (b), show schematics representations of a pylon mounted nacelle and a BLI nacelle, respectively. Whilst in figure 2 (c), it is appreciated a detail view of an embedded nacelle and the fraction of embedding (ζ), which allows to model the nacelle weight reduction for BLI configurations using equation .

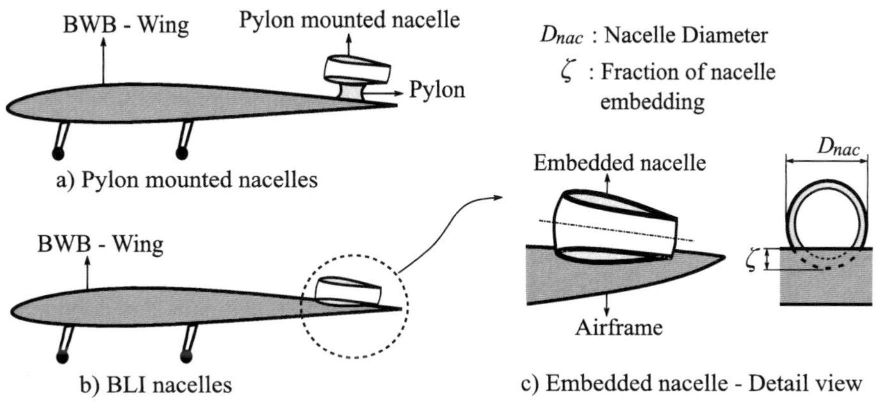

Figure 2. Nacelles configurations on a BWB-UAV [23].

3. Results and discussion

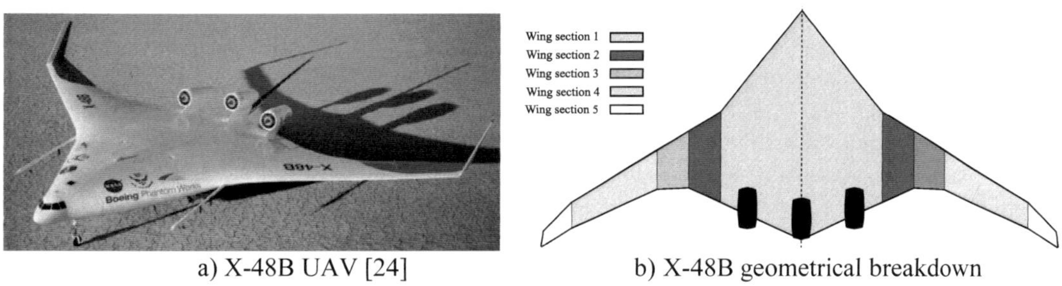

Figure 3. NASA X-48B (8.5% scale model) [25].

The validation of the methodology and the case of study were carried out on the unmanned aircraft NASA X-48B BWB, figure 3 (a). Figure 3 (b) shows the geometrical breakdown in wing sections, which was used to estimate the BWB wing weight using equation (3). Table 1 presents the specifications of the NASA X-48B BWB-UAV.

Table 1. X-48B BWB-UAV specifications [26].

Parameter	Value
Wing span	6.22 (m)
Wing reference area	9.34 (m^2)
TOGW	227 (kg)
Maximum speed	60 (m/s)
Ceiling	3000 (m)
Endurance	40 (min)
Propulsion	3 JetCat P200 Turbojet 0.23 kN thrusts each

To estimate the engine, nacelle and pylon weight, T_0 has been assumed to be equal to the maximum thrust provided by the three jet engines used by the X-48B aircraft. In this study, the engine weight remains constant for No BLI and BLI configurations. The propulsors are assumed to be distributed symmetrically rear of aircraft's centre of gravity on the BWB's centre body.

Table 2 presents the validation of the methodology used in this work, the actual value of TOGW, for the X-48B BWB, was taken from [26], and the actual weight of a JetCat P200 engine is given in [27]. To reduce the error incurred in TOGW estimation, it is required a more detailed weight breakdown to integrate some minor aircraft components that could have been passed over in this work.

Table 2. Validation results.

Parameter	Actual value (kg)	Estimated value (kg)	Error (%)
TOGW	227	218.5	3.74
Engine Weight	2.507	2.45	2.99

3.1. TOGW reduction due to BLI on NASA X-48B

To assess the TOGW reduction due to BLI on NASA X-48B, it was considered 3 BLI configurations with the following values of embedding fraction. $\zeta = 0$: This embedding fraction characterizes the removal of pylons. $\zeta = 0.5$ simulates a 50% of nacelle embedding, and $\zeta = 1.0$ represents the disappearance of nacelles and a fully embedded configuration, which would require to use special shaped ducts to ingest the airframe boundary layer as presented by Anabtawi [28]. Table 3 presents a comparison of the estimated TOGW for pylon-mounted nacelles and BLI configurations. The reduction on TOGW ($\Delta TOGW_{BLI}$) is in the range of 8% when removing pylons (0% BLI) and 19% for fully embedded configurations. When embedding nacelles, the propulsion system weight (W_{PS}) is reduced directly. However, the reduction on TOGW is not equal to the reduction on propulsion system weight (ΔW_{PS-BLI}), this is because the estimation of TOGW not only depends on propulsion system and the estimation of TOGW uses an iterative numerical method. TOGW reduction due to BLI is considerable and could be beneficial to improve cruise range, reduce fuel consumption, and other benefits in terms of aircraft performance.

Table 3. $TOGW$ and W_{PS} reduction due to BLI for NASA X-48B BWB-UAV.

Parameter	Pylon Mounted nacelles	$\zeta = 0$ 0% Embedded BLI	$\zeta = 0.5$ 50% Embedded BLI	$\zeta = 1.0$ 100% Embedded BLI
$TOGW$ (kg)	218.50[a]	200.47	187.35	175.09
$\Delta TOGW_{BLI}$ (%)	-	8.25	14.25	19.86
W_{PS} (kg)	24.61[b]	21.42	18.62	16.19
ΔW_{PS-BLI} (%)	-	12.96	24.33	34.17

[a] Reference value to evaluate the reduction on TOGW
[b] Reference value to evaluate the reduction on propulsion system weight.

3.2. Non-conventional propulsion system configurations assessment

3.2.1. Distributed propulsion system configurations

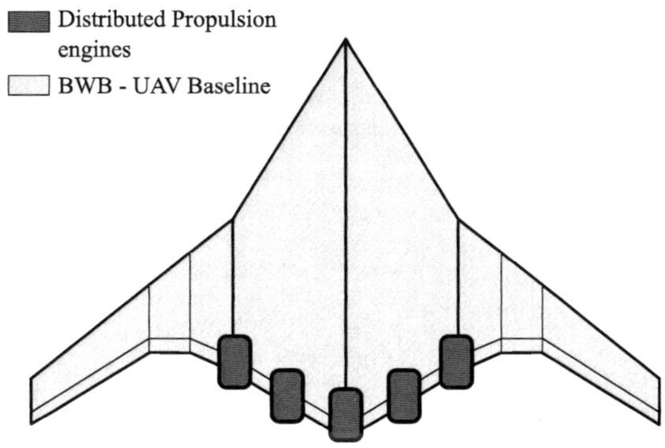

Figure 4. Distributed propulsion system configuration.

Additionally, the effect of increasing the number of engines on TOGW has been also evaluated. Figure 4 presents a non-conventional propulsion system with 5 propulsors distributed symmetrically on the BWB's centre body. To estimate engine, nacelle and pylon weight for distributed propulsion configurations, it was also assumed that T_0 is equal to the maximum thrust provided by the three engines used by the X-48B aircraft. This thrust has been divided for the number of propulsors, in analysis, to determine the required thrust from each engine. Figure 5 shows the benefits of BLI in terms of $TOGW$ and W_{PS} reduction, when increasing the number of propulsors.

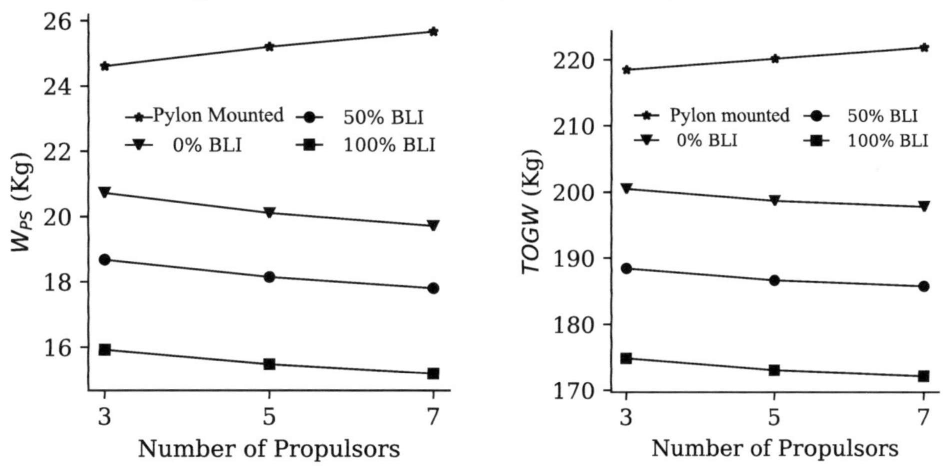

a) TOGW **b)** Propulsion system Weight

Figure 5. $TOGW$ and W_{PS} reduction due to BLI and higher number of propulsors.

In figure 5 (a), it is observed that for BLI configurations, TOGW decreases when the number of propulsors increases, this is because for a higher number of propulsors, both engine and nacelle weight are reduced. However, the reduction in TOGW for BLI configurations, increasing the number of propulsors, is not so considerable. Now, for pylon-mounted configurations, increasing the number of propulsors increases TOGW due to pylons. Figure 5 (b) presents the same trend for the propulsion system weight, which shows that the main driver on TOGW is the propulsion system weight. Figure 6 shows the reduction on the wing weight for the X-48B when applying BLI and increasing the number of propulsors. Fuel weight, for the corresponding TOGW, also decreases with a higher number of propulsors for BLI propulsion systems. The reduction on wing and fuel weight is because the empirical relations used to estimate their weight depend directly on the TOGW, so if the TOGW increases, wing and fuel weight are increased proportionally.

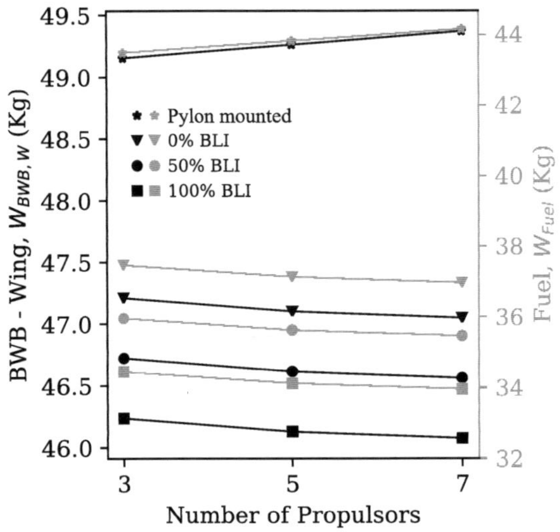

Figure 6. BWB wing and fuel weight reduction due to BLI and higher number of engines.

4. Conclusions

A weight assessment, at a preliminary design stage, for a BWB-UAV implementing BLI has been presented. TOGW has been taken as figure of merit and it has been determined that the implementation of BLI can reduce the TOGW between 8% and 19%. The reduction on TOGW due to BLI depends on how much a nacelle is embedded into the BWB airframe. TOGW decreases when the number of engines of a BLI propulsion system is increased; however the reduction is not so considerable. The reduction on TOGW due to BLI will improve range and endurance of a BWB-UAV, overcoming the current performance limitation, this makes more attractive and promotes the research of BWB-UAVs with BLI for civil applications, such as Site Specific Monitoring (SSM). Another benefit of BLI has been addressed, and the results obtained in this work encourages a more detailed research about the effect of TOGW reduction due to BLI, on aircraft performance, and propulsion-airframe integration.

Acknowledgements

The authors gratefully acknowledge the financial support provided by Escuela Politécnica Nacional for the development of this work, which is part of the research project PIMI 15-03.

References

[1] Liebeck R, Page M and Rawdon B 1998 Blended-wing-body subsonic commercial transport *36th AIAA Aerospace Sciences Meeting and Exhibit* (American Institute of Aeronautics and Astronautics)

[2] Rodriguez D L 2001 *A multidisciplinary optimization method for designing boundary layer ingesting inlets* PhD Thesis (Stanford University)

[3] Liebeck R H 2004 Design of the Blended Wing Body Subsonic Transport *J. Aircr.* **41** 10–25

[4] Plas A (Angélique P 2006 *Performance of a boundary layer ingesting propulsion system* PhD Thesis (Massachusetts Institute of Technology)

[5] Campbell R, Carter M, Pendergraft O, Friedman D and Serrano L 2005 Design and Testing of a Blended Wing Body with Boundary Layer Ingestion Nacelles at High Reynolds Numbers (Invited) *43rd AIAA Aerospace Sciences Meeting and Exhibit* (American Institute of Aeronautics and Astronautics)

[6] Carter M B, Campbell R L, Pendergraft O C, Friedman D M and Serrano L 2006 Designing and Testing a Blended Wing Body with Boundary-Layer Ingestion Nacelles *J. Aircr.* **43** 1479–89

[7] Hardin L, Tillman G, Sharma O, Berton J and Arend D 2012 Aircraft System Study of Boundary Layer Ingesting Propulsion *48th AIAA/ASME/SAE/ASEE Joint Propulsion Conference & Exhibit* (American Institute of Aeronautics and Astronautics)
[8] Berends J P T J, Tooren M J L and Belo D N V 2006 A Distributed Multi-Disciplinary Optimisation of a Blended Wing Body UAV Using a Multi-Agent Task Environment *47th AIAA/ASME/ASCE/AHS/ASC Structures, Structural Dynamics, and Materials Conference* (American Institute of Aeronautics and Astronautics)
[9] Lee D-S, Gonzalez L F, Srinivas K, Auld D and Wong K C 2007 Multi-Objective and Multidisciplinary Design and Optimisation of Blended Wing Body UAV via Advanced Evolutionary Algorithms *Faculty of Built Environment and Engineering* 2nd Australasian Unmanned Air Vehicles Conference (Melbourne)
[10] Chandrasekhara M S and McLain B K 2013 Aerodynamic studies over a manoeuvring UCAV 1303 configuration *Aeronaut. J.* **117** 445–65
[11] Jo Y-H, Chang K, Sheen D-J and Park S H 2015 Numerical simulation of aerodynamic characteristics of a BWB UCAV configuration with transition models *Int. J. Aeronaut. Space Sci.* **16** 8–18
[12] Shim H J and Park S O 2013 Low-speed Wind-tunnel Test Results of a BWB-UCAV Model *Procedia Eng.* **67** 50–8
[13] Roskam J 1985 *Airplane Design: Component weight estimation* (Roskam Aviation and Engineering Corp.)
[14] Elham A 2012 But how much does it weight? ; structural weight estimation of aircraft lifting surfaces *Am. J. Bot. - AMER J BOT*
[15] Elham A 2010 Structural Design: Weight Estimation *Encycl. Aerosp. Eng.*
[16] Ko Y-Y A 2003 *The multidisciplinary design optimization of a distributed propulsion blended-wing-body aircraft* PhD Thesis (Virginia Tech)
[17] Beltramo M N T 1977 *Parametric study of transport aircraft systems cost and weight*
[18] Torenbeek E 2013 *Synthesis of Subsonic Airplane Design: An introduction to the preliminary design of subsonic general aviation and transport aircraft, with emphasis on layout, aerodynamic design, propulsion and performance* (Springer Science & Business Media)
[19] Isikveren A T 2002 *Quasi-analytical Modelling and Optimisation Techniques for Transport Aircraft Design* PhD Thesis (Institutionen för flygteknik)
[20] Kundu A K, Price M A and Riordan D 2016 *Theory and Practice of Aircraft Performance* (John Wiley & Sons)
[21] Gundlach J 2012 *Designing unmanned aircraft systems: A comprehensive approach* (American Institute of Aeronautics and Astronautics)
[22] Torenbeek E 2013 *Advanced Aircraft Design: Conceptual Design, Technology and Optimization of Subsonic Civil Airplanes* (John Wiley & Sons)
[23] Valencia E A, Saa J, Alulema V and Hidalgo V 2018 Parametric study of aerodynamic integration issues in highly coupled Blended Wing Body configurations implemented in UAVs *2018 AIAA Information Systems-AIAA Infotech @ Aerospace* (Kissimmee, Florida: American Institute of Aeronautics and Astronautics)
[24] Kim J 2009 NASA X48-B planform view
[25] Vicroy D NASA X-48 B Testt flight report
[26] Merino-Martinez R 2014 *Design and Analysis of the Control and Stability of a Blended Wing Body Aircraft* MSc thesis (Stockholm, Sweden: Royal Institute of Technology)
[27] Anon 2017 P200-RX Turbine Engine, 52 lbs Thrust, by JetCat *ChiefAircraft.com*
[28] Anabtawi A, Blackwelder R, Lissaman P and Liebeck R 1999 An experimental investigation of boundary layer ingestion in a diffusing S-duct with and without passive flow control *37th Aerospace Sciences Meeting and Exhibit* (American Institute of Aeronautics and Astronautics)

Effect of nickel content and austempering temperature on microstructure and mechanical properties of austempered ductile iron (ADI)

P Sellamuthu[1], D G Harris Samuel[1,*], D Dinakaran[1], V P Premkumar[2], Zushu Li[3] and S Seetharaman[3]

[1]Department of Mechanical Engineering, Hindustan Institute of Technology and Science, Chennai – 603103, India
[2]Research and Development, Nelcast Ltd, Chennai – 600018, India
[3]Warwick Manufacturing Group, University of Warwick, Coventry CV4 7AL, United Kingdom

*E-mail: dgharris@hindustanuniv.ac.in

Abstract. In this article, the effect of two variables namely nickel content and austempering heat treatment temperature on mechanical and microstructural properties of ADI were investigated. Four Nickel content levels of 0.0, 0.6, 0.8 and 1% and three austempering temperatures of 320, 340 and 360°C were selected. Prior to austempering, all the samples were austenitized for 30 min at 840°C. The microstructure of austempered alloys mainly consisted of a mixture of needle like acicular ferrite and carbon stabilized austenitic matrix. The increase in Ni content improved the strength and hardness and it reached its peak at 0.8%. The microstructures also became finer with increase in Ni content as confirmed by Impact energy and elongation. Nodularity was found to be above 90% for all alloys at all austempering temperatures. However, the nodule counts increased with increase in Ni content up to 0.8% and then dropped. XRD analysis was done to estimate the retained austenite content present in the ausferrite matrix and it was concluded that for any given Nickel content level, the retained austenite content increases with austempering temperature. Nickel of 0.8% was found to be optimum for all austempering temperatures.

1. Introduction

Austempered ductile iron (ADI) is emerging materials which is presently being used extensively in automobile, agricultural and mining applications due to its high strength, toughness and wear resistance [1-4]. Through suitable heat treatments or alloying additions, wear and mechanical properties of ADI could be improved by obtaining appropriate final matrix structure.

Additions of alloying elements to ADI enhance the mechanical properties and make it more suitable for commercial use. Unlike other alloying elements, nickel does not affect the carbon solubility in austenite during austenitization and in fact, aids in beneficial mechanical properties for castings [5]. Nickel was found to increase the impact toughness in ADI [6]. Previous researchers have observed that phase transformation temperature is changed by nickel addition. Nickel has the potential to eliminate the secondary carbides precipitation in the upper ausferritic range [7]. The present authors have reported elsewhere the influence of austempering temperature of 300, 320,340 and 360°C on

Content from this work may be used under the terms of the Creative Commons Attribution 3.0 licence. Any further distribution of this work must maintain attribution to the author(s) and the title of the work, journal citation and DOI.
Published under licence by IOP Publishing Ltd

mechanical, wear properties and energy consumption [8].Since not many published research articles are available on nickel, the current work is carried out at different Ni levels and different austempering temperature.

2. Experimental Procedure

2.1. Casting of ductile iron

Charging materials were calculated and melted in an induction furnace with a capacity of 1.5 MT. Figure 1 shows the dimensions of Y-blocks in which nodular iron was cast. The Y- blocks were cast along with the regular commercial products to make sure that the same treatment is given to ensure validation. The resultant chemical composition of all alloys were estimated using an optical emission spectrometer and is given in Table 1. Suitable Nickel additions to get Ni levels of 0.6%, 0.8% and 1.0% were added and they were given codes as Alloy A, Alloy B and Alloy C respectively.

Table 1. Base Chemical Composition for all Alloys.

C	Si	Mn	S	P	Mg
3.64	2.56	0.19	0.002	0.017	0.050

2.2. Austempering heat treatment of ductile iron

The heat treatment schedule for austempering of as-cast ductile iron samples is shown in the Figure 2. A preheating furnace was used to eliminate the impurities on the sample surface by heat treating the samples at 450°C for 10 min. In a closed austempering furnace, samples were heated to 840°C for 30 min to ensure complete austenitization. Finally austempering were carried out by quenching in a molten salt bath consisting of mixture of potassium and sodium nitrates maintained at three different temperatures of 320, 340 and 360°C. Rinsing the castings in a machine with hot water was used to wipe out the residual salt on the casting surface after austempering. The authors in their earlier study had concluded that though austempering at 300°C gives good tensile, it results in low elongation % and low impact strength and do not meet requirements of grade 1 ADI as per ASTM A897/A897M-16 specification and hence higher temperatures of 320, 340 and 360°C are selected for this study.

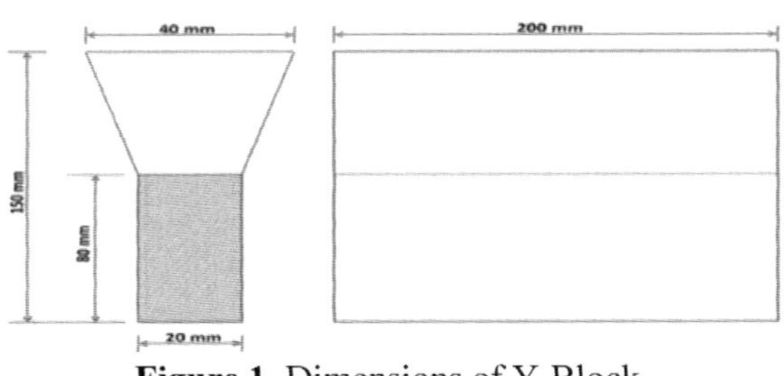

Figure 1. Dimensions of Y-Block.

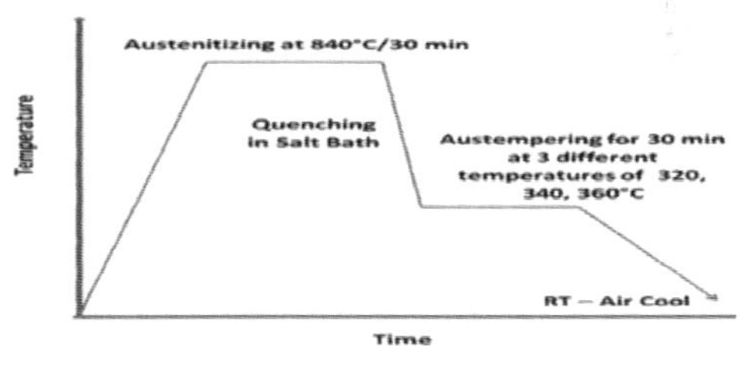

Figure 2. ADI Heat Treatment Cycle.

2.3. Mechanical testing and optical microscopy

Hardness Test: The tests were done directly on the surface of the Y blocks after rough grinding. The load applied was 3000 kg and a steel ball with 10 mm diameter was used. Tensile Test: Samples with 14 mm gauge diameter and 70 mm gauge length were machined from the lower portion of the Y blocks. Impact Test: A Charpy type impact tester was used to find out the impact toughness. The rectangular samples were machined without any notch (unnotched) with dimensions of 55 x 10 x 10 mm. Metallography and Optical Microscopy: Samples were prepared for metallographic investigation by cutting, mounting, grinding, polishing and etching (using 2% Nital) to capture the images at a magnification of 100X (before etching) and 400X (after etching) using an inverted metallurgical microscope with Envision software.

2.4. X-Ray diffraction analysis

The phase analysis of samples was done in X-Pert PRO PANalytical goniometer. Samples were analysed using Cu-Kα radiation at 30 kV and 10 mA. The retained austenite content available in the matrix was calculated using direct comparison method.

3. Results and discussions

3.1. Effect of nickel on mechanical properties

The effect of different nickel content on mechanical properties of alloys austempered at 320°C is shown in Figure 3. There is an increase in yield and tensile strength when Ni is added and at 0.8% it reaches a peak and then there is a drop in value when Ni reaches 1%. Impact energy and elongation also gradually increases with increase in Ni content and reaches a maximum at 0.8% and then decreases. The same trend is observed in all alloys austempered at 340°C and 360°C as shown in Figure 4 and Figure 5 respectively.

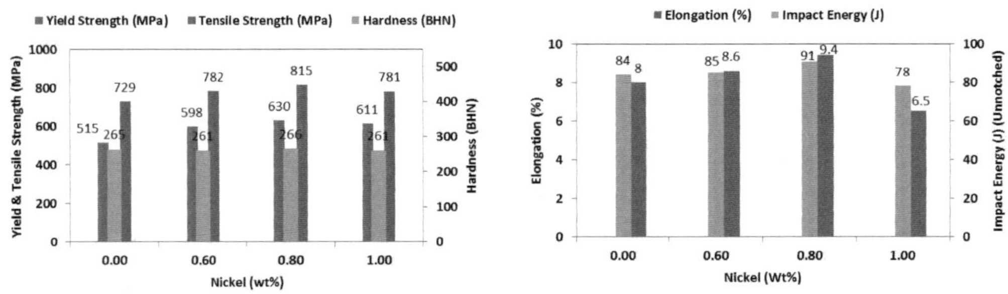

Figure 3. Effect of Nickel Content on Mechanical Properties of Alloys Austempered at 320°C.

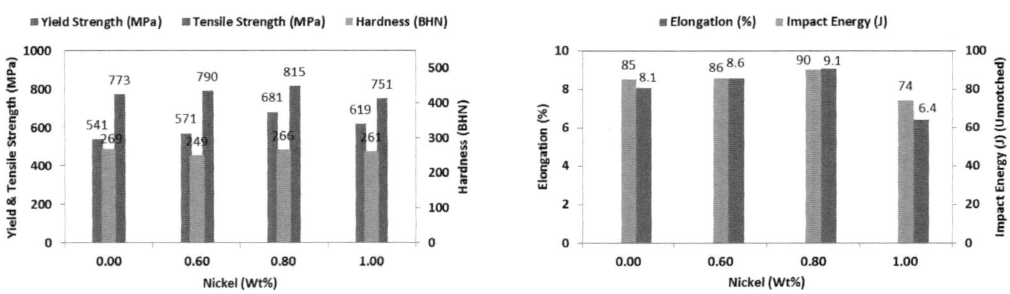

Figure 4. Effect of Nickel Content on Mechanical Properties of Alloys Austempered at 340°C.

3.2. Effect of nickel on microstructure

The effect of nickel content on microstructure of alloys austempered at 320°C, 340°C and 360°C is shown in Figures 6, 7 and 8 respectively. The microstructure of all alloys consists of a mixture of acicular ferrite in a carbon stabilized/enriched austenite matrix and this mixture is called as ausferrite. From the micrographs, it is known that the structure is changing from coarse to fine with increasing Ni content when austempering temperature is kept constant. However, with increase in austempering temperature, the microstructure becomes finer with more amount of ausferrite. The nodularity and nodule count of all alloys austempered at three different temperatures are shown in Table 2.

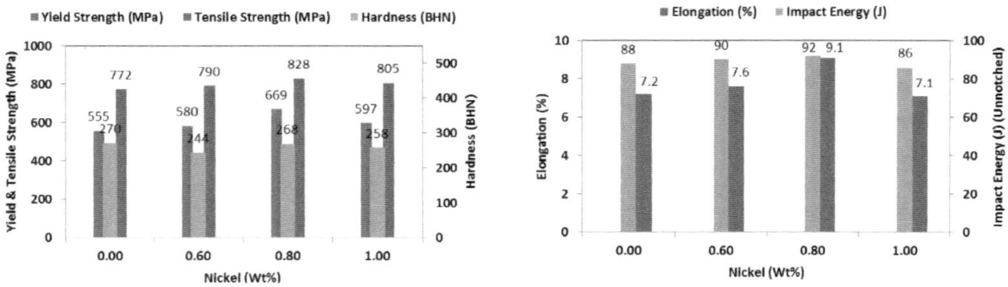

Figure 5. Effect of Nickel Content on Mechanical Properties of Alloys Austempered at 360°C.

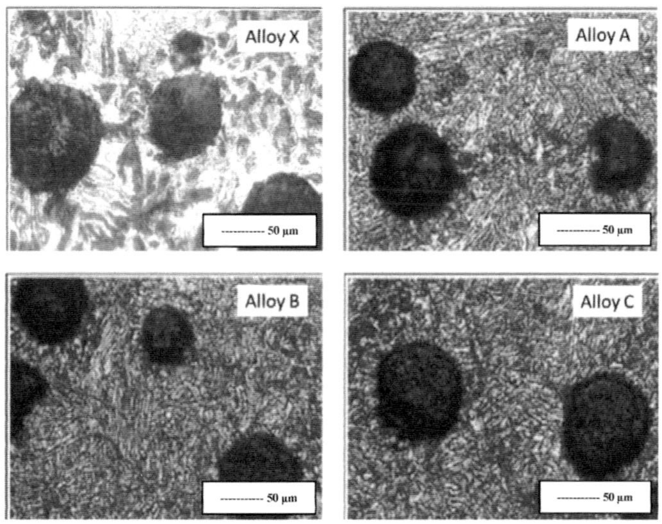

Figure 6. Effect of Nickel Content on Microstructures (400X) of Alloys Austempered at 320°C.

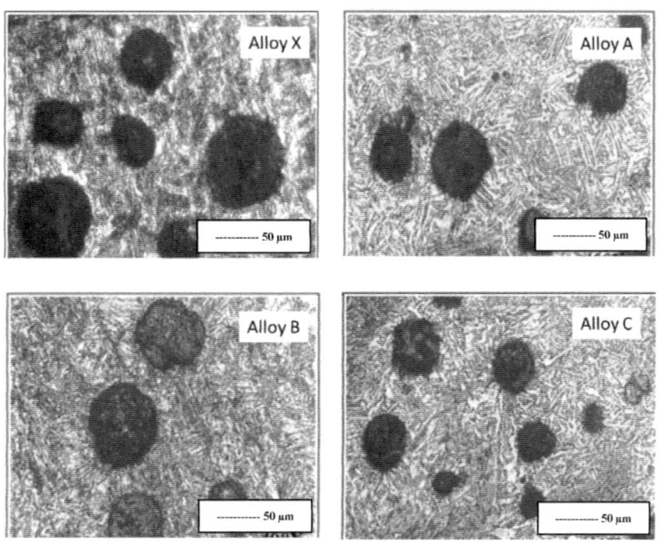

Figure 7. Effect of Nickel Content on Microstructures (400X) of Alloys Austempered at 340°C.

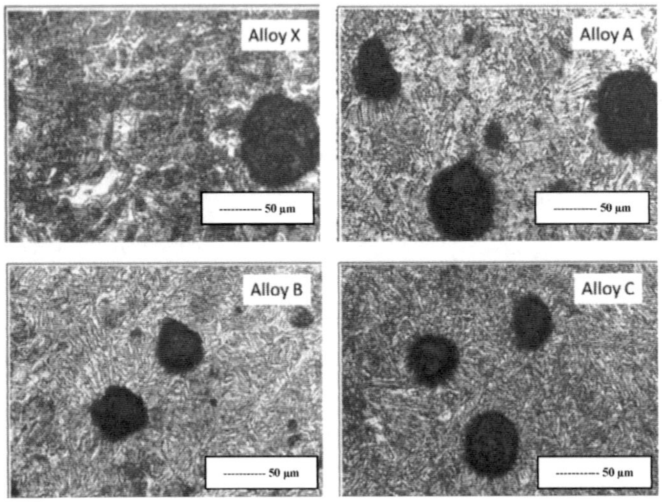

Figure 8. Effect of Nickel Content on Microstructures (400X) of Alloys Austempered at 360°C.

Table 2. Nodularity and Nodule Count.

Nickel (Wt %)	Nodularity (%)			Nodule Count (/mm^2)		
	Austempered at			Austempered at		
	320°C	340°C	360°C	320°C	340°C	360°C
0.0 [Alloy X]	92.7	96.2	94.6	141	139	130
0.6 [Alloy A]	95.5	93.3	90.6	159	156	145
0.8 [Alloy B]	95.4	95.1	95.6	237	208	200
1.0 [Alloy C]	97.3	93.1	97.9	178	174	133

In the case of nodularity%, no trend is observed except that it is constantly above 91% for all alloys austempered at various temperatures. However, the nodule counts decreases with increase in austempering temperature and the value is maximum for 0.8% Ni addition as shown in Figure 9. This is reflected in the mechanical properties of all alloys.

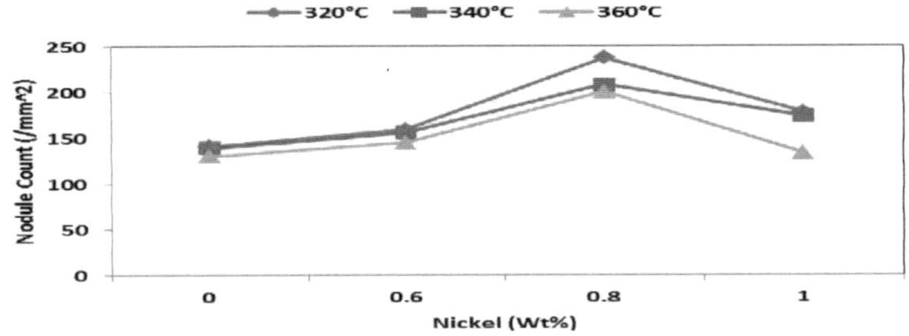

Figure 9. Effects of Nickel Content and Austempering Temperature on Nodule Count.

3.3. Effect of Nickel on Retained Austenite

The XRD patterns were studied to identify the intensities and peak positions of austenite {111} plane and ferrite {110} plane at different nickel content as shown in Figure 10, 11 and 12. The retained austenite content were estimated using the intensities of the above planes using the equation (1) [9],

$$\frac{I\gamma(hkl)}{I\alpha(hkl)} = \frac{R\gamma(hkl)}{R\alpha(hkl)}\frac{X\gamma}{X\alpha} \quad (1)$$

Where Iγ(hkl) is the integrated intensity of a plane in austenite (γ) phase; Iα(hkl) is the integrated intensity of a plane in ferrite (α) phase; Rγ(hkl) and Rα(hkl) are constants.

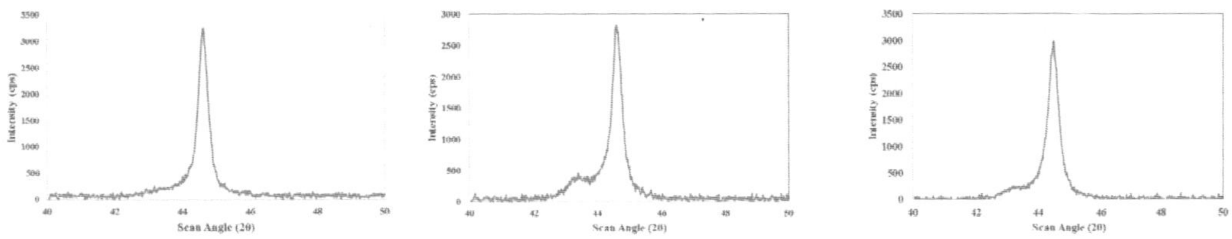

(a) 0.6% (b) 0.8% (c) 1.0%

Figure 10. XRD patterns of samples with different nickel content austempered at 320°C.

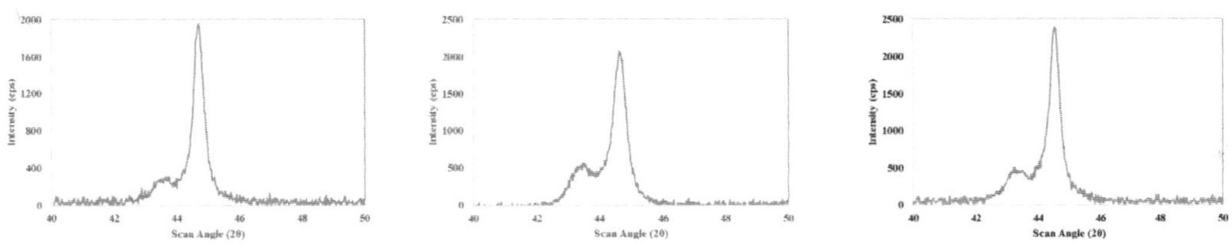

(a) 0.6% (b) 0.8% (c) 1.0%

Figure 11. XRD patterns of samples with different nickel content austempered at 340°C.

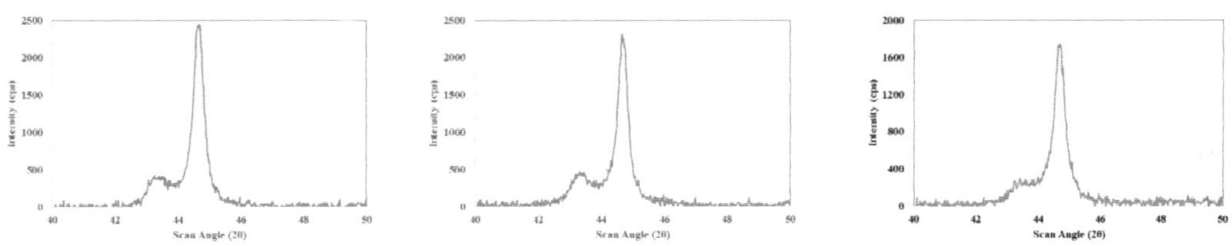

(a) 0.6% (b) 0.8% (c) 1.0%

Figure 12. XRD patterns of samples with different nickel content austempered at 360°C.

Figure 13 shows a plot of the retained austenite content against nickel content at different austempering temperature. The retained austenite content was found to reach a maximum value at 0.8% Ni irrespective of austempering temperature.

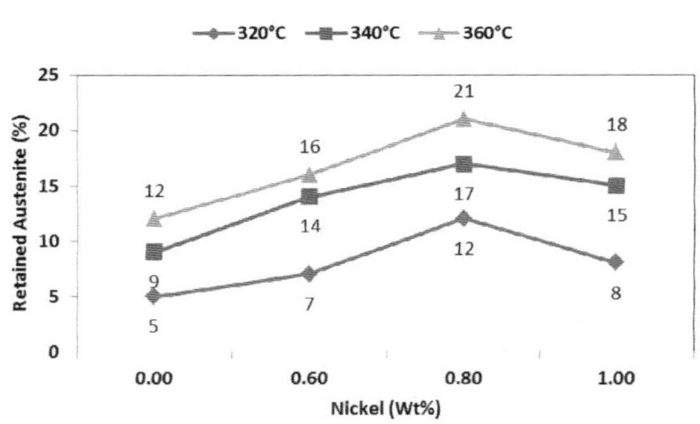

Figure 13. Effect of nickel content on retained austenite at different austempering temperature.

From the results, it is evident that all properties are increasing with nickel content up to 0.8% and then it decreases which is reflected in both nodule count and amount of retained austenite present in the matrix. This could be due to the fact that nickel, as a solute element, dissolved in the matrix interacts with dislocations and elements segregated near dislocations by forming Cottrell atmosphere. This Cottrell atmosphere increases the dislocation motion resistance and pinning constant which reflects in the brittleness of the ductile iron with nickel content above 0.8% [10].

4. Conclusions

- The strength and hardness of alloys increases with increase in Ni content and it is maximum at 0.8% and then decreases. Impact energy and elongation also exhibited the same behavior at all austempering temperatures.
- The microstructure of austempered alloys were a mixture of acicular ferrite and carbon stabilized austenite. The structure becomes finer with respect to increase in Ni content.
- The nodularity is above 90% for all alloys irrespective of austempering temperature. The nodule counts increase with increase in Ni content up to 0.8% and then it decreases with increase in austempering temperature.
- Without affecting impact, elongation and hardness, Ni addition of 0.8% is optimum and results in giving 20% increase in yield strength and 7% increase in tensile strength.
- The study of X-ray diffraction of austempered samples reveal that the retained austenite reaches maximum at 0.8%Ni in all 3 austempering temperatures studied and then it reduces on further increase of Nickel.
- Among the three austempering temperatures, 340°C gives optimum properties with respect to yield and tensile strength without affecting impact and elongation.

Acknowledgments
Authors acknowledge The Royal Academy of Engineering, London (UK) for providing financial support, under its Higher Education Partnership – India Newton Bhabha Scheme.

References

[1] Jack R. *Laub Proc., of the world Con., on austempered ductile iron* March 12-14 1991 Bloomingdale (Chicago IL, USA) pp 622-37
[2] Eric O, Sidjanin L, Miskovic Z, Zec S and Jovanovic M T 2004 *Materials Letters*. **58** pp 2707-11
[3] Sohi M H, Ahmadabadi M N and Vahdat A B 2004 *Journal of Materials Processing Technology*. **153-154** pp 203-08
[4] Alagarsamy A *Proc., of the world Con., on austempered ductile iron* March 12-14 1991 Bloomingdale (Chicago IL, USA) pp 408-23
[5] Gundlac R and Janowak J *Proc., of the 2nd Int. Con. on austempered ductile irons*, March 17-19 1986 University of Michigan (Ann Arbor MI, USA) pp 23-30
[6] Kovacs B V *Proc., of the world Con., on austempered ductile iron* March 12-14 1991 Bloomingdale (Chicago IL, USA) pp 241-70
[7] Uma T R, Simha J B and Narasimha Murthy K 2011 *Wear*. **271** pp 1374-78
[8] Sellamuthu P, Harris Samuel D G, Dinakaran D, Premkumar V P, Li Z and Seetharaman S 2018 *Metals*. **8** pp 53
[9] Cullity B D 1974 *Elements of X-ray Diffraction Reading*, MA: Addition-Wesley pp 411-12
[10] Sun Y, Hu S, Xiao Z, You Z, Zhao J and Lv Y 2012 *Materials and Design*. **41** pp 37-42

AUTHOR INDEX

AUTHOR INDEX

AUTHOR INDEX

Institute of Physics
Dirac House, Temple Back
Bristol BS1 6BE UK

ISSN: 1757-8981
ISBN 978-1-5108-6752-9

CURRAN ASSOCIATES INC.
proceedings.com